Technology of
Engineering Materials

Mathew Philip
Bill Bolton

BUTTERWORTH
HEINEMANN

OXFORD   AMSTERDAM   BOSTON   LONDON   NEW YORK   PARIS
SAN DIEGO   SAN FRANCISCO   SINGAPORE   SYDNEY   TOKYO

Butterworth-Heinemann
An imprint of Elsevier Science
Linacre House, Jordan Hill, Oxford, OX2 8DP
225 Wildwood Avenue, Woburn, MA 01801-2041

First published 2002

British Library Cataloguing in Publication Data
A catalogue record for this book is available from the British Library

ISBN 0 7506 5643 3

For information on all Butterworth-Heinemann publications
visit our website at www.bh.com

Printed and bound in Great Britain

Contents

Series Preface

'There is a time for all things: for shouting, for gentle speaking, for silence; for the washing of pots and the writing of books. Let now the pots go black, and set to work. It is hard to make a beginning, but it must be done' – Oliver Heaviside, *Electromagnetic Theory*, Vol 3 (1912), Ch. 9 'Waves from moving sources - Adagio, andante, allegro moderato.'

Oliver Heaviside was one of the greatest engineers of all time, ranking alongside Faraday and Maxwell in his field. As can be seen from the above excerpt from a seminal work, he appreciated the need to communicate to a wider audience. He also offered the advice 'So be rigorous: that will cover a multitude of sins. And do not frown.' The series of books that this prefaces takes up Heavisides's challenge but in a world which is quite different to that being experienced just a century ago.

With the vast range of books already available covering many of the topics developed in this series, what is this series offering which is unique? I hope the next few paragraphs help to answer that; certainly no one involved in this project would give up their time to bring these books to fruition if they had not thought that the series is both unique and valuable.

The motivation for this series of books was born out of the desire of the UK's Engineering Council to increase the number of incorporated engineers graduating from Higher Education establishments, and the Institution of Incorporated Engineers' (IIE) aim to provide enhanced services to those delivering Incorporated Engineering Courses. However, what has emerged from the project should prove of great value to a very wide range of courses within the UK and internationally – from Foundation Degrees or Higher Nationals through to first year modules for traditional 'Chartered' degree courses. The reason why these books will appeal to such a wide audience is that they present the core subject areas for engineering studies in a lively, student-centred way, with key theory delivered in real world contexts, and a pedagogical structure that supports independent learning and classroom use.

Despite the apparent waxing of 'new' technologies and the waning of 'old' technologies, engineering is still fundamental to wealth creation. Sitting alongside these are the new business focused, information and communication dominated, technology organisations. Both facets have an equal importance in the health of a nation and the prospects of individuals. In preparing this series of books, we have tried to strike a balance between traditional engineering and developing technology.

The philosophy is to provide a series of complementary texts which can be tailored to the actual courses being run – allowing the flexibility for course designers to take into account 'local' issues, such as areas of particular staff expertise and interest, while being able to demonstrate the depth and breadth of course material referenced to a common framework. The series is designed to cover material in the core texts which approximately corresponds to the first year of study with module texts focusing on individual topics to second and final year level. While the general structure of each of the texts is common, the styles are quite different, reflecting best practice in their areas.

Another set of factors which we have taken into account in designing this series is the reduction in contact hours between staff and students, the evolving responsibilities of both parties and the way in which advances in technology are changing the way study can be, and is, undertaken. As a result, the lecturers' support material which accompanies these texts, is paramount to delivering maximum benefit to the student.

It is with these thoughts of Voltaire that I leave the reader to embark on the rigours of study:

'Work banishes those three great evils: boredom, vice and poverty'

Alistair Duffy
Series Editor
De Montfort University, Leicester, UK

Further information on the IIE Textbook Series is available from
bhmarketing@repp.co.uk
www.bh.com/iie

Please send book proposals to:
rachel.hudson@repp.co.uk

Other titles currently available in the IIE Textbook Series

Mechanical Engineering Systems	0 7506 5213 6
Business Skills for Engineers and Technologies	0 7506 5211 1
Design Engineering	0 7506 5211 X
Mathematics for Engineers and Technologists	0 7506 55444
Systems for Planning and Control	
in Manufacturing	0 7506 49771

Summary

Engineers are involved with materials, having to select and use them and consider their behaviour in use. This part of the book is an introduction to the range of engineering materials, their properties, sources of data on such properties and details of the testing methods used to obtain such data.

Objectives

By the end of this chapter, the reader should be able to:

- appreciate the range of materials encountered by engineers and their characteristic properties;
- recognise and use the terms needed to describe the basic properties of engineering materials;
- use data sources to obtain the relevant data on the properties of materials;
- know the principles of the basic methods used for the testing of materials to obtain properties data.

1.1 Materials and properties

This section is an introduction to the range of materials that exists and their general properties; later chapters take up in more detail the properties, how they can be determined and how they are related to the internal structure of a material and can be changed.

Materials are usually classified into four main groups. Here we consider the grouping in terms of their properties; however, it should be recognised that the properties arise from the differences in internal structures of the materials and this point is considered in later chapters.

- *Metals*

  If you touch a piece of material and it feels cold then it is likely to be a metal; this is because it is a good conductor of heat. If a lump of material is heavy then it is likely to be a metal since

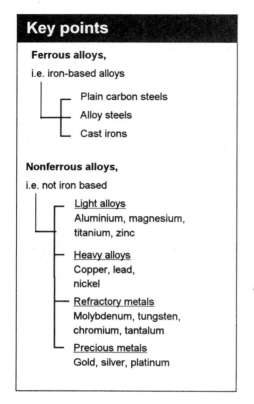

Key points

Ferrous alloys,

i.e. iron-based alloys

- Plain carbon steels
- Alloy steels
- Cast irons

Nonferrous alloys,

i.e. not iron based

- Light alloys
  Aluminium, magnesium, titanium, zinc
- Heavy alloys
  Copper, lead, nickel
- Refractory metals
  Molybdenum, tungsten, chromium, tantalum
- Precious metals
  Gold, silver, platinum

metals tend to have a high density. If you can bend a strip of a material into a new shape then it is likely to be a metal since metals are ductile. If you connect a piece of material into an electrical circuit and it conducts electricity then it is likely to be a metal. In general, metals have high thermal conductivity, a relatively high density, ductility, a relatively high stiffness and strength and high electrical conductivity.

Engineering metals are generally alloys, *alloys* being metallic materials formed by mixing two or more elements. For example, mild steel is an alloy of iron and carbon, stainless steel an alloy of iron, chromium, carbon, manganese and possibly other elements. Elements are added to metals to improve their properties. Thus carbon improves the strength of iron and the chromium in stainless steel improves the corrosion resistance. Metal alloys are classified as ferrous alloys for those which are iron-based and non-ferrous for those that are not. The Key points show how metals are classified.

- ### Polymers

A *plastic* consists of a *polymer* plus various additives such as dye, fillers, fire retardents, etc. Polymers have low electrical conductivity and low thermal conductivity, hence their use for electrical and thermal insulation. Compared with metals, they have lower densities, expand more when there is a change in temperature, are generally more corrosion resistant, have a lower stiffness, stretch more and are not as hard. When loaded they tend to creep, i.e. the extension gradually changes with time. Their properties depend very much on the temperature so that a polymer which may be tough and flexible at room temperature may be brittle at 0°C and show considerable creep at 100°C.

Polymers can be classified as either *thermoplastics* or *thermosets*. Thermoplastics soften when heated and become hard again when the heat is removed. The term implies that the material becomes 'plastic' when heat is applied. Thus thermoplastic materials can be heated and bent to form required shapes, thermosets cannot. Thermoplastic materials are generally flexible and relatively soft. Polythene is an example of a thermoplastic, being widely used as films or sheets for such items as bags, 'squeezy' bottles, and wire and cable insulation. Thermosets do not soften when heated, but char and decompose; they are rigid and hard. Phenol formaldehyde, known as Bakelite, is a thermoset. It is widely used for electrical plug casings, door knobs and handles. The term *elastomers* is used for polymers which by their structure allow considerable extensions that are reversible, e.g. rubber bands.

- ### Ceramics

When the word ceramics is used the tendency is to think of cups and saucers. However, to an engineer, the term ceramics covers far more materials than just those used for cups and saucers. Ceramics can be grouped as shown in the Key points.

Key points

Glasses	—	Soda lime glasses, borosilicate glasses, pyroceramics
Domestic ceramics	—	Porcelain, vitreous china, earthenware, stoneware, cement
Engineering ceramics	—	Alumina, carbides, nitrides
Natural ceramics	—	Rocks
Electronic materials	—	Ferrites, semi-conductors, ferroelectrics

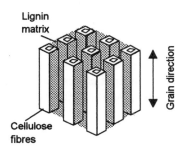

Figure 1.1 *Wood as a composite with cellulose fibres in a lignin matrix*

Ceramics tend to be brittle, relatively stiff, stronger in compression than tension, hard, chemically inert and bad conductors of electricity and heat. Because of their hardness and abrasion resistance, engineering ceramics are widely used as the cutting edges of tools.

- *Composites*
  Composites are materials composed of two different materials bonded together. For example, there are composites involving glass fibres or particles in polymers, ceramic particles in metals (cermets), and steel rods in concrete (reinforced concrete). Wood is a natural composite consisting of tubes of cellulose in a polymer called lignin (Figure 1.1). Composites made with fibres embedded, all aligned in the same direction in some matrix, will have properties in that direction markedly different from properties in other directions. Composites can be designed to combine the good properties of different types of materials while avoiding some of their drawbacks.

1.1.1 Properties

Consider the materials used for fizzy soft drink containers. The properties that might be considered in selecting the material for a container are:

- Physical properties: low density so that it is not too heavy.

- Mechanical properties: rigid, so that the container does not deform unduly under the weight of the drink, and strong enough to withstand the weight of the drink and cope with forces likely to be incurred in use without breaking.

- Chemical properties: resistant to chemical attack by the drink and able to keep the 'fizz' in the drink, i.e. not allow the gas to escape through the walls of the container.

In addition it needs to be capable of being processed to the required shape, cheap and we might consider there to be a requirement for it to be capable of being recycled and so reduce the demands on the earth's resources.

You can buy such drinks in aluminium cans, in glass bottles and in plastic (polyethylene terephthalate) bottles, these being materials that engineers have selected as being solutions to the above issues.

In selecting a material for a product engineers have to decide:

- What properties are required and whether the required properties will be maintained during the service life of the product.

- Whether the material can be processed/manufactured to the required shape and dimensional accuracy.

- Whether it will cause environmental problems.

- Whether the material can be economically used to produce the product.

The basic properties

The basic properties of materials can be grouped as:

- *Physical properties*
  These can be considered to include density and melting point.

- *Mechanical properties*
  These are the properties displayed when a force is applied to a material and include strength, stiffness, hardness, ductility, toughness and wear resistance.

- *Electrical properties*
  These are the properties displayed when the material is used in electrical circuits or components and include resistivity, conductivity and resistance to electrical breakdown.

- *Thermal properties*
  These are displayed when there is a heat input to a material and include expansivity, heat capacity, thermal conductivity and melting point.

- *Optical properties*
  These are the properties involved when light passes through the material. They include such properties as the refractive index and transmissivity.

- *Chemical properties*
  These are, for example, relevant in considerations of corrosion and solvent resistance.

- *Magnetic properties*
  These are relevant when a material is considered as a magnet or part of an electrical component, such as an inductor which relies on such properties.

In addition to the above we can also consider economic properties, i.e. raw materials and processing costs, availability, and aesthetic properties, i.e. appearance, texture, colour.

In discussing the properties of materials it is important to recognise that they are often markedly changed by the temperature at which a material is being used and any treatments the materials undergo. For example, a plastic may be relatively stiff at room temperature but far from stiff at the boiling point of water. A steel may be ductile at 20°C but become brittle at temperatures below −10°C. Steels can have their properties changed by heat treatment, such as annealing which involves heating to some temperature and slowly cooling. This renders the material soft and ductile. Heating a steel to some temperature and then quenching, i.e. immersing the hot material in cold water, can be used to make a steel harder, stronger and less ductile. Materials can also have their properties changed by working. For example, if you take a piece of carbon steel and permanently deform it, perhaps by bending it, then it will have different mechanical properties to those existing before that deformation. It is said to be work hardened.

1.1.2 Physical properties

An important physical property of a material is its density. The *density* of a material is the mass per unit volume, i.e.

$$density = \frac{mass}{volume} \qquad [1]$$

It has the unit of kg/m³. It is a property that often is required in conjunction with a mechanical property. Thus, for example, an aircraft undercarriage is required to be not only strong but of low mass and so what is required is a high value of strength/density, this quantity being referred to as the *specific strength*.

1.1.3 Mechanical properties

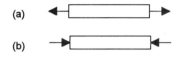

Figure 1.2 *(a) Tension: the forces extend the material, (b) compression: the forces cause the material to contract*

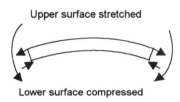

Figure 1.3 *Bending*

When a material is subjected to external forces which stretch it and make it extend, then it is said to be in *tension* (Figure 1.2(a)). When a material is subjected to forces which squeeze it and make it contract, then it is said to be in *compression* (Figure 1.2(b)). An object, in some situations, can be subject to both tension and compression, e.g. a beam (Figure 1.3) which is being bent, the bending causing the upper surface to contract and so be in compression and the lower surface to extend and be in tension.

In discussing the application of forces to materials, an important aspect is often not so much the size of the force itself as the size of the force applied per unit area. If we stretch a strip of material by a force F applied over its cross-sectional area A, then the force applied per unit area is F/A (Figure 1.4(a)), this being termed the *stress*:

$$stress = \frac{force}{area} \qquad [2]$$

The area used in calculations of stress is generally the original area that existed before the application of the forces, not the area after the force has been applied. This stress is thus sometimes referred to as the *engineering stress*, the term *true stress* being used for the force divided by the actual area existing in the stressed state.

When a material is subject to tensile or compressive forces, it changes in length. The term *strain* (Figure 1.4(b)) is defined as:

$$strain = \frac{change\ in\ length}{original\ length} \qquad [3]$$

Since strain is a ratio of two lengths it has no units. Strain is frequently expressed as a percentage.

$$Strain\ as\ a\ \% = \frac{change\ in\ length}{original\ length} \times 100 \qquad [4]$$

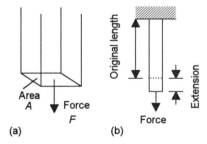

Figure 1.4 *(a) Stress is force/area, (b) strain = extension/original length*

Thus the strain of 0.01 as a percentage is 1%, i.e. the change in length is 1% of the original length.

Example

A bar of material with a cross-sectional area of 50 mm² is subject to tensile forces of 100 N. What is the tensile stress?

The tensile stress = force divided by the area

$$= 100/(50 \times 10^{-6}) = 2 \times 10^6 \text{ Pa} = 2 \text{ MPa.}$$

Example

A strip of material has a length of 50 mm. When it is subject to tensile forces it increases in length by 0.020 mm. What is the strain?

The strain is the change in length divided by the original length and is thus 0.020/50 = 0.004. Expressed as a percentage, the strain is 0.04%.

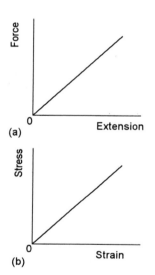

Figure 1.5 *Hooke's law: force proportional to extension or stress proportional to strain*

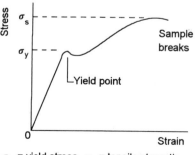

σ_y = yield stress σ_s = tensile strength

Figure 1.6 *Stress–strain graph for mild steel*

Stress–strain graphs

The behaviour of materials subject to tensile and compressive forces can be described in terms of their stress–strain behaviour. If gradually increasing tensile forces are applied to, say, a strip of mild steel, then initially when the forces are released the material springs back to its original shape. The material is said to be *elastic*. If measurements are made of the extension at different forces and a graph plotted, then the force needed to produce a given extension is found to be proportional to the extension and the material is said to obey *Hooke's law*. Figure 1.5(a) shows a graph when Hooke's law is obeyed. Such a graph applies to only one particular length and cross-sectional area of a particular material. We can make the graph more general so that it can be applied to other lengths and cross-sectional areas of the material by dividing the extension by the original length to give the strain and the force by the cross-sectional area to give the stress. Then we have, for a material that obeys Hooke's law:

stress ∝ strain   [5]

The stress–strain graph (Figure 1.5(b)) is just a scaled version of the force–extension graph in Figure 1.5(a).

Figure 1.6 shows the type of stress–strain graph which would be given by a sample of mild steel. Such graphs are described by the following terms:

Table 1.1 *Typical tensile strengths*

Strength (MPa)	Material
<10	Polymer foams
2 to 12	Woods perpendicular to grain
2 to 12	Elastomers
6 to 100	Woods parallel to grain
10 to 40	Bulk thermoplastics
60 to 100	Engineering polymers
20 to 60	Concrete
20 to 60	Lead alloys
80 to 300	Magnesium alloys
160 to 400	Zinc alloys
100 to 600	Aluminium alloys
80 to 1000	Copper alloys
250 to 1300	Carbon and low-alloy steels
250 to 1500	Nickel alloys
500 to 1800	High-alloy steels
100 to 1800	Engineering composites
1000 to >10 000	Engineering ceramics

Key points

The tensile strength is often quoted in text books as indicating the maximum useful loading of a material. However, the yield stress or proof stress is more valuable to the mechanical designer as it is the stress that should not be exceeded if permanent damage is to be avoided. The yield stress is not, however, so easy to measure as the tensile strength.

The modulus of elasticity is more commonly called the stiffness of a material.

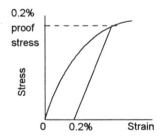

Figure 1.7 *0.2% proof stress*

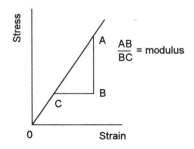

Figure 1.8 *Modulus of elasticity*

- *Limit of proportionaility*
Initially the graph is straight line and the material obeys Hooke's law. The point at which the straight line behaviour is not followed is called the *limit of proportionality*.

- *Elastic and plastic behaviour*
With low stresses the material springs back completely to its original shape when the stresses are removed, the material being said to be *elastic*. At higher forces this does not occur and the material is then said to show some *plastic* behaviour. The term plastic is used for that part of the behaviour which results in permanent deformation. This point often coincides with the point on a stress–strain graph at which the graph stops being a straight line, i.e. the *limit of proportionality*. The stress at which the material starts to behave in a non-elastic manner is called the *elastic limit*.

- *Strength*
The term *tensile strength* is used for the maximum value of the stress that the material can withstand without breaking (Figure 1.6), the *compressive strength* being the maximum compressive stress the material can withstand without becoming crushed. Table 1.1 shows the types of tensile strengths that might be expected of materials.

- *Yield stress*
With some materials, e.g. mild steel, there is a noticeable dip in the stress–strain graph at some stress beyond the elastic limit and the strain increases without any increase in load. The material is said to have yielded and the point at which this occurs is the *yield point*.

- *Proof stress*
Some materials, such as aluminium alloys (Figure 1.7), do not show a noticeable yield point and it is usual here to specify *proof stress*. The 0.2% proof stress is obtained by drawing a line parallel to the straight line part of the graph but starting at a strain of 0.2%. The point where this line cuts the stress–strain graph is termed the 0.2% yield stress. A similar line can be drawn for the 0.1% proof stress.

- *Modulus of elasticity*
The *stiffness* of a material is the ability of a material to resist bending. When a strip of material is bent, one surface is stretched and the opposite face is compressed, as was illustrated in Figure 1.2. The more a material bends, the greater is the amount by which the stretched surface extends and the compressed surface contracts. Thus a stiff material would be one that gave a small change in length when subject to tensile or compressive forces. This means a small strain when subject to tensile or compressive stress and so a large value of stress/strain and hence a steep initial gradient of the stress–strain graph (Figure 1.8). This gradient is called the *modulus of elasticity* (or *Young's modulus*), symbol E:

Table 1.2 *Tensile modulus values*

Tensile modulus (GPa)	Material
<0.2	Polymer foams
<0.2	Elastomers
0.2 to 10	Woods parallel to grain
0.2 to 3	Bulk thermoplastics
3 to 10	Engineering polymers
2 to 20	Woods perpendicular to grain
10 to 11	Lead alloys
20 to 50	Concrete
40 to 45	Magnesium alloys
50 to 80	Glasses
70 to 80	Aluminium alloys
43 to 96	Zinc alloys
110 to 125	Titanium alloys
100 to 160	Copper alloys
200 to 210	Steels
80 to 1000	Engineering ceramics

$$\text{modulus of elasticity } E = \frac{\text{stress}}{\text{strain}} \qquad [6]$$

The units of the modulus are the same as those of stress, since strain has no units. With 1 GPa = 10^9 Pa, typical values are about 200 GPa for steels and 70 GPa for aluminium alloys. For most engineering materials, the modulus of elasticity is the same in tension as in compression. Table 1.2 shows typical values of modulus of elasticity for materials.

When a material is subjected to load, its dimensions change. For example, in tension, the material will stretch along the direction of the load. However, the lateral dimensions will then decrease (Figure 1.9). This is measured by the *Poisson's ratio* for the material. This is the ratio of the strain in the lateral direction to the strain along the direction of the applied stress.

$$\text{Poisson's ratio} = \frac{\left(\dfrac{w_1 - w_0}{w_0}\right)}{\left(\dfrac{l_1 - l_0}{l_0}\right)} \qquad [7]$$

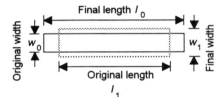

Figure 1.9 *Poisson's ratio*

Activity

Get hold of strips of different materials, e.g. 30 cm long rulers made from polystyrene, from wood and from steel. Using the terms introduced so far in this chapter, suggest reasons for the following:

- Why is the steel ruler the thinnest of the three?
- Why would you not use a ruler made from rubber?
- How much would you expect each of the rulers to bend before it failed and why?
- How would it fail and why?

Examples of stress–strain graphs

- **Cast iron**

  The stress–strain graph for cast iron (Figure 1.10) is virtually just a straight line with virtually all elastic behaviour and little plastic deformation. The slight curved part at the top of the graph indicates a small departure from straight-line behaviour and a small amount of plastic behaviour. The graph gives the limit of proportionality as about 280 MPa, the tensile strength about 300 MPa and the modulus of elasticity about 200 GPa.

- **Glass**

  The stress–strain graph for glass (Figure 1.11) has a similar shape to that for cast iron, with virtually all elastic behaviour and little plastic deformation. The graph indicates the limit of proportionality as about 250 MPa, the tensile strength about 260 MPa and the modulus of elasticity about 70 GPa.

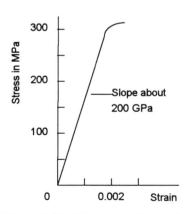

Figure 1.10 *Stress–strain graph for cast iron*

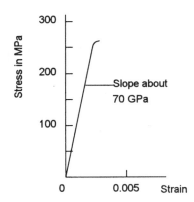

Figure 1.11 *Stress–strain graph for glass*

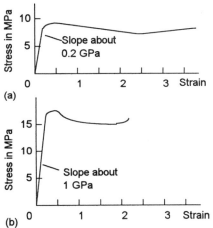

Figure 1.12 *Stress–strain graphs for polyethylene: (a) low density, (b) high density*

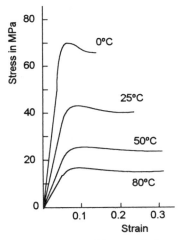

Figure 1.13 *Stress–strain graph for celullose acetate*

- ### *Mild steel*

  The stress–strain graph for mild steel (see Figure 1) shows a straight-line portion followed by a considerable amount of plastic deformation. Much higher strains are possible than with cast iron or glass, i.e. mild steel stretches much more. The limit of proportionality is at about 240 MPa, the tensile strength about 400 MPa and the modulus of elasticity about 200 GPa.

- ### *Plastics*

  Stress–strain graphs for polyethylene (Figure 1.12) shows only a small region where elastic behaviour occurs and a very large amount of plastic deformation is possible. Very large strains are possible, a length of such material being capable of being stretched to almost four times its initial length. Low density polyethylene has a strength of about 8 MPa and a modulus of elasticity about 0.2 GPa; high density polyethylene a strength of about 20 MPa and a modulus of elasticity of about 1 GPa.

  The stress–strain properties of plastics depend on the rate at which the strain is applied, unlike metals where the strain rate is not usually a significant factor, and the properties change significantly when there is a change in temperature (Figure 1.13) with both the modulus of elasticity and the tensile strength decreasing with an increase in temperature.

- ### *Rubber*

  The stress–strain graph for rubber (Figure 1.14) typically shows a tensile strength of about 25 MPa. Very large strains are possible and the material shows an elastic behaviour to very high strains. The modulus of elasticity is not so useful a quantity for elastomers as it can refer to only a very small portion of the stress–strain graph. A typical modulus would be about 30 MPa.

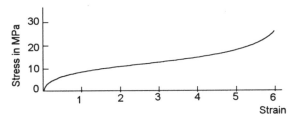

Figure 1.14 *Stress–strain graph for rubber*

Because for many polymeric materials there is no initial straight-line part of the stress strain graph and a value for the modulus of elasticity cannot be arrived at, the *secant modulus* is sometimes quoted, this being the stress/strain value at 0.2% strain (Figure 1.15).

Table 1.3 gives typical values of yield stress or 0.2 % proof stress, tensile strength and modulus of elasticity for a range of materials.

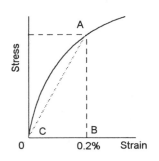

Figure 1.15   *Secant modulus*
is AB/BC

Example

A sample of an aluminium alloy has a tensile strength of 140 MPa. What will be the maximum force that can be withstood by a rod with a cross-sectional area of 1 cm²?

Since 1 cm² = 10^{-4} m², the cross-sectional area of the rod is 1×10^{-4} m². Since the tensile strength is maximum force/area:

maximum force = $140 \times 10^6 \times 1 \times 10^{-4}$

$$= 140 \times 10^2 = 14.0 \text{ kN}.$$

Table 1.3 *Typical tensile properties of materials at 20°C*

Material	Yield or 0.2% proof stress MPa	Tensile strength MPa	Modulus of elasticity GPa
Metals			
Mild steel	230	400	200
Chromium alloy steel	500–600	700–1000	200
Cast iron	150–600	150–600	120–170
Copper alloys	60–300	160–600	120–170
Aluminium alloys	50–300	100–400	70
Thermoplastics			
Polyvinyl chloride		35–60	2–4
Polyethylene		30	1
Cellulose acetate		13–62	0.5–2.8
Thermosets			
Phenol formaldehyde		50–55	5–6
Epoxy, cast		200–420	3.2
Elastomers			
Natural rubber		30	
Neoprene		28	

Example

A material has a yield stress of 200 MPa. What tensile forces will be needed to cause yielding with a bar of the material with a cross-sectional area of 100 mm²?

Since stress is force/area, then:

yield force = yield stress × area

$$= 200 \times 10^6 \times 100 \times 10^{-6} = 20\ 000 \text{ N} = 20 \text{ kN}.$$

Example

For a material with a tensile modulus of elasticity of 200 GPa, what strain will be produced by a stress of 4 MPa?

Provided the stress does not exceed the limit of proportionality, since the modulus of elasticity is stress/strain:

$$\text{strain} = \frac{\text{stress}}{\text{modulus}} = \frac{4 \times 10^6}{200 \times 10^9} = 0.000\,02$$

Expressed as a percentage, the strain is 0.002%.

Example

The following are tensile modulus values for some plastics: ABS 2.5 GPa, polycarbonate 2.8 GPa, polypropylene 1.3 GPa, PVC 3.1 GPa. Which is the stiffest?

The stiffest plastic is the one with the highest tensile modulus and so is the PVC.

Example

A 200 mm length of a material has a percentage elongation of 10%, by how much longer will a strip of the material be when it breaks?

$$\text{Change in length} = \frac{\% \text{ elongation} \times \text{original length}}{100}$$

$$= \frac{10 \times 200}{100} = 20 \text{ mm}$$

Brittleness/ductility

For some materials, the difference between the elastic limit stress and the stress at which failure occurs is very small, as with the cast iron in Figure 1.10, and very little plastic deformation occurs. Thus the length of a piece of cast iron after breaking is not much different from the initial length. Such materials are said to be *brittle*. A material which suffers a considerable amount of plastic strain before breaking, is said to be *ductile*, e.g. the mild steel in Figure 1.6. If you drop a glass, a brittle material, and it breaks, then it is possible to stick all the pieces together again and restore the glass to its original shape. If a car is involved in a collision, the bodywork of mild steel is more likely to dent and show permanent deformation, i.e. plastic deformation. Ductile materials permit manufacturing methods which involve bending them to the required shapes or

Key point

A mechanical designer would prefer to use a material with some ductility although the design limit is going to be the yield stress. This is because, if the yield stress is exceeded unexpectedly, a ductile material will still carry the applied stress whereas a brittle material would fail catastrophically.

using a press to squash the material into the required shape. Brittle materials cannot be shaped in this way. The *percentage elongation* of a test piece after breaking is used as a measure of ductility:

$$\text{percentage elongation} = \frac{\text{final length} - \text{initial length}}{\text{initial length}} \times 100\% \quad [8]$$

A reasonably ductile material, such as mild steel, will have a percentage elongation of about 20%, a brittle material such as a cast iron less than 1%. Thermoplastics tend to have percentage elongations of the order of 50 to 500%, thermosets 0.1 to 1%. Thermosets are brittle materials, thermoplastics generally not.

Example

The material 80–20 brass has a percentage elongation of 50%, 70–30 brass 70% and 60–40 brass 40%. Which is the most ductile?

The most ductile material is the one with the largest percentage elongation, i.e. the 70–30 brass.

Example

A sample of a carbon steel has a tensile strength of 400 MPa and a percentage elongation of 35%. A sample of an aluminium–manganese alloy has a tensile strength of 140 MPa and a percentage elongation of 10%. How does the mechanical behaviour of the materials compare?

The higher value of the tensile strength of the carbon steel indicates a stronger material; for the same cross-sectional area, a bar of carbon steel could withstand higher tensile forces than a corresponding bar of the aluminium alloy. The higher percentage elongation of the carbon steel indicates that the material has a greater ductility than the aluminium alloy. The steel is stronger and more ductile.

Toughness

The materials in many products may contain cracks or sharp corners or other changes in shape that can readily generate cracks. A tough material can be considered to be one that, though it may contain a crack, resists breaking as a result of the crack growing and running through the material. Think of trying to tear a sheet of paper or a sheet of some cloth. If there is an initial 'crack' then the material is much more easily torn. In the case of the paper, the initial 'cracks' may be perforations put there to enable the paper to be torn easily. In the case of a sheet of cloth, it may be the initial 'nick' cut in the edge by a dressmaker to enable it to be torn easily. In the case of, say, the skin of an aircraft where there may be holes, such as

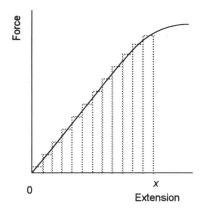

Figure 1.16 *Area under the force–extension graph is the work done in producing the extension*

windows or their fastenings, which are equivalent to cracks, there is a need for cracks not to propagate. A tough material is required.

Toughness can be defined in terms of the work that has to be done to propagate a crack through a material, a tough material requiring more energy than a less tough one.

Consider a length of material being stretched by tensile forces. When a length of material is stretched by an amount x_1 as a result of a constant force F_1 then the work done is the force × distance moved by point of application of force and thus:

$$\text{work} = F_1 x_1$$

Thus if a force–extension graph is considered (Figure 1.16), the work done, when we consider a very small extension, is the area of that strip under the graph. The total work done in stretching a material to an extension x, i.e. through an extension which we can consider to be made up of a number of small extensions with $x = x_1 + x_2 + x_3 + \ldots$ with F_1, F_2, F_3, $\ldots$ the average values of the forces corresponding to each small extension, is thus:

$$\text{work} = F_1 x_1 + F_2 x_2 + F_3 x_3 + \ldots$$

and so is the area under the graph up to x. If we divide both sides of this equation by the volume, i.e. the product of the cross-sectional area A of the strip and its length L, we have:

$$\frac{\text{work}}{\text{volume}} = \left(\frac{F_1}{A} \times \frac{x_1}{L} \right) + \left(\frac{F_2}{A} \times \frac{x_2}{L} \right) + \left(\frac{F_3}{A} \times \frac{x_3}{L} \right) + \ldots$$

But the term in each bracket is just the product of the stress and strain. Thus the work done per unit volume of material is the area under the stress–strain graph up to the strain corresponding to extension x. The area under the stress–strain graph up to some strain is the energy required per unit volume of material to produce that strain. For a crack to propagate, a material must fail. Thus the area under the stress–strain graph up to the breaking point is a measure of the energy required to break unit volume of the material and so for a crack to propagate. A large area is given by a material with a large failure stress and high ductility. Such materials can thus be considered to be tough.

An alternative way of considering toughness is the ability of a material to withstand shock loads. A measure of this ability is obtained by *impact tests*, such as the Charpy and Izod tests (see Section 1.3) in which a test piece is struck a sudden blow and the energy needed to break it is measured. A brittle material will generally require less energy than a ductile material. The results of such tests are often used as a measure of the brittleness of materials.

Hardness and wear

The *hardness of a material* is a measure of the resistance of a material to abrasion or indentation. A number of scales are used for hardness, depending on the method that has been used to measure it

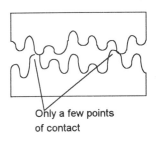

Only a few points
of contact

Figure 1.17 *Principle of two surfaces in contact with only a few points on the surfaces touching*

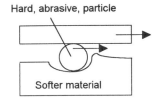

Hard, abrasive, particle

Softer material

Figure 1.18 *Principle of abrasive wear, material is removed by 'ploughing out' of the softer material by the harder material*

(see Section 1.3 for discussions of test methods). The tensile strength for a particular material is roughly proportional to the hardness (see Section 1.3); thus the higher the hardness of a material, the higher is likely to be the tensile strength.

Wear is the progressive loss of material from surfaces as a result of sliding or rolling contact between surfaces or from the movement of fluids containing particles over surfaces. Because wear is a surface effect, surface treatments and coatings play an important role in improving wear resistance. Lubrication can be considered to be a way of keeping surfaces apart and so reducing wear.

A number of different mechanisms for wear have been identified:

- ***Adhesive wear***

  On an atomic scale, even smooth surfaces appear rough and thus when two surfaces are brought together, contact is made at only a few points (Figure 1.17). As a consequence, the forces holding the surfaces together can result in very high stresses at the few very small areas of contact. Surface projections thus become plastically deformed by the pressure and can weld together. Sliding thus involves breaking these welded bonds, the breaks resulting in cavities being produced on one surface, projections on the other and frequently tiny abrasive particles. The term *adhesive wear*, or scoring, galling or seizing, is used for this type of wear when two solid surfaces slide over one another under pressure. If the hardnesses of the two surfaces are high, the wear rate can be reduced. Also high strength, high toughness and ductility all contribute to reducing such wear, preventing the tearing of material from the surfaces.

- ***Abrasive wear***

  The term *abrasive wear* is used when material is removed from a surface by contact with hard particles, sliding resulting in the 'ploughing out' of the softer material by the harder material (Figure 1.18). Such wear is common in machinery used to handle abrasive materials. Materials with a high hardness, high toughness and high strength are most resistant to such wear.

- ***Corrosive wear***

  When rubbing between surfaces takes place in a corrosive environment, surface reactions can take place and reaction products form on the surfaces. These generally poorly adhere to the surfaces and the rubbing removes them. The process thus involves the repeated forming of reaction products and their removal by the rubbing. Lubricants can be used to separate surfaces and protect the surfaces from the corrosive environment.

- ***Surface fatigue***

  Adhesive and abrasive wear depends on direct contact between surfaces and can be prevented by separating the surfaces with a lubricant film. However, with rolling with bearings wear can still occur though the surfaces are separated. This is because, although direct contact between the surfaces does not occur, the opposing surfaces experience large stresses transmitted through the lubricant film. As rolling proceeds, the stresses can become alternating and fatigue failure thus becomes possible for surface protrusions. Hence wear can occur.

1.1.4 Electrical properties

The electrical *resistivity* ρ is a measure of the electrical resistance of a material, being defined by:

$$\rho = \frac{RA}{L} \qquad [9]$$

where R is the resistance of a length L of a material of cross-sectional area A (Figure 1.19). The unit of resistivity is the ohm metre (Ω m). An electrical insulator such as a ceramic will have a very high resistivity, typically of the order of 10^{10} Ω m or higher. An electrical conductor such as copper will have a very low resistivity, typically of the order of 10^{-8} Ω m. The term *semi-conductor* is used for materials which have resistivities roughly half way between conductors and insulators, i.e. of the order of 10^2 Ω m.

The electrical *conductance* G of a length of material is the reciprocal of its resistance and has the unit of Ω^{-1}. This unit is given a special name, the siemen (S). The electrical *conductivity* σ is the reciprocal of the resistivity:

$$\sigma = \frac{1}{\rho} = \frac{L}{RA} = \frac{LG}{A} \qquad [10]$$

The unit of conductivity is thus Ω^{-1} m^{-1} or S/m. Since conductivity is the reciprocal of the resistivity, an electrical insulator will have a very low conductivity, of the order of 10^{-10} S/m, while an electrical conductor will have a very high conductivity, of the order of 10^8 S/m. Semiconductors have conductivities of the order of 10^{-2} S/m.

Table 1.4 shows typical values of resistivity and conductivity for insulators, semiconductors and conductors. Pure metals and many metal alloys have resistivities that increase when the temperature increases; some metal alloys do, however, show decreases in resistivities when the temperature increases. For semiconductors and insulators, the resistivity decreases with an increase in temperature.

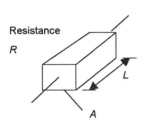

Figure 1.19   *Resistivity*

Key point

In general, the resistivities of conductors increases with temperature; for semiconductors and insulators it tends to decrease.

Example

Using the value of electrical conductivity given in Table 1.2, determine the electrical conductance of a 2 m length of nichrome wire at 20°C with cross-sectional area 1 mm².

Using the equation $\sigma = L/RA$, with the conductance $G = 1/R$, then we have $\sigma = LG/A$ and so:

$$G = \frac{\sigma A}{L} = \frac{0.9 \times 10^6 \times 1 \times 10^{-6}}{2} = 0.45 \text{ S}$$

Table 1.4 *Typical resistivity and conductivity values at about 20°C*

Material	Resistivity Ω m	Conductivity S/m
Insulators		
Acrylic (a polymer)	$> 10^{14}$	$< 10^{-14}$
Polyvinyl chloride (a polymer)	10^{12}–10^{13}	10^{-13}–10^{-12}
Mica	10^{11}–10^{12}	10^{-12}–10^{-11}
Glass	10^{10}–10^{14}	10^{-14}–10^{-10}
Porcelain (a ceramic)	10^{10}–10^{12}	10^{-12}–10^{-10}
Alumina (a ceramic)	10^{9}–10^{12}	10^{-12}–10^{-9}
Semiconductors		
Silicon (pure)	2.3×10^{3}	4.3×10^{-4}
Germanium (pure)	0.43	2.3
Gallium arsenide	0.05	20
Conductors		
Nichrome (alloy of nickel and chromium)	108×10^{-8}	0.9×10^{6}
Manganin (alloy of copper and manganese)	42×10^{-8}	2×10^{6}
Nickel (pure)	7×10^{-8}	14×10^{6}
Copper (pure)	2×10^{-8}	50×10^{6}
Silver	1.6×10^{-8}	63×10^{6}

Example

Suggest a material that could be used for the heating element of an electric fire?

The heating element must be a conductor of electricity. The power dissipated by the element is V^2/R, thus the lower the resistance R the greater the power produced by a given voltage V. The material must also be able to withstand high temperatures without melting or oxidising. Nichrome wire is commonly used. The wire is wound on a spiral around an insulating ceramic support.

Example

A manufacturer of fine electrical connectors for liquid crystal displays is considering printing fine conducting tracks on a polymer sheet. The choice of material is between copper and silver for the electrical conductors. If the maximum resistance allowed is 10 Ω, what would be the thickness of a track of square cross-section if it is 4 cm long?

The cross-sectional area $A = t^2$, where t is the thickness. Thus, $t^2 = \rho L/R$ and so:

$$t = \sqrt{\frac{\rho L}{R}}$$

For copper, $\rho = 2 \times 10^{-8}$ Ω m and so:

$$t = \sqrt{\frac{2 \times 10^{-8} \times 0.04}{10}} = 8.9 \times 10^{-6} \text{ m} = 8.9 \text{ } \mu\text{m}$$

For silver, $\rho = 1.6 \times 10^{-8}$ Ω m and so:

$$t = \sqrt{\frac{1.6 \times 10^{-8} \times 0.04}{10}} = 8.0 \times 10^{-6} \text{ m} = 8.0 \text{ } \mu\text{m}$$

The minimum thickness with copper is thus 8.9 μm and with silver 8.0 μm.

Dielectrics

When a pair of parallel conducting plates are connected to a d.c. supply (Figure 1.20), charge flows onto one of the plates and off the other plate. One of the plates becomes positively charged and the other negatively charged. The amount of charge Q on a plate, be it negative or positive, is proportional to the potential difference V between the plates. Hence:

$$Q = CV$$

where C is the constant of proportionality, called the *capacitance*. The unit of capacitance is the farad (F) when V is in volts and Q in coulombs. The factors determining the value of the capacitance are the plate area A, the separation d of the plates and the medium between them:

$$C = \frac{\varepsilon A}{d} \qquad [11]$$

where ε is the factor, called the *absolute permittivity*, which relates to the medium between the plates. A more usual way of writing the equation is, however, in terms of how the permittivity of a material compares with that of a vacuum. Thus:

$$C = \varepsilon_r \varepsilon_0 \frac{A}{d} \qquad [12]$$

where $\varepsilon = \varepsilon_r \varepsilon_0$. ε_0 is called the *permittivity of free space* and has a value of 8.85×10^{-12} F/m. ε_r is called the *relative permittivity*. It has no units, merely stating the factor that must be used to multiply the permittivity of free space in order to obtain the permittivity of some material. For a vacuum the relative permittivity is 1, for plastics it is between about 2 and 3, for glass between 5 and 10. The relative permittivity is often termed the *dielectric constant* and the material between the conducting plates

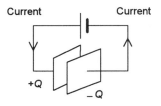

Figure 1.20 *Charging a capacitor involves charge moving onto one plate and off the other*

Current   Current

+Q   −Q

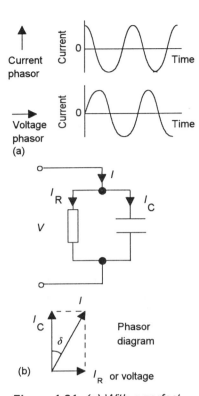

Figure 1.21 *(a) With a perfect dielectric, the current leads the voltage by 90°, (b) with a lossy dielectric the current leads the voltage by 90° − δ*

the *dielectric*. The relative permittivity, or dielectric constant, is the term used to describe the property of a material to store charge. The higher it is, the greater the amount of charge stored for a particular potential difference.

If the potential difference between two plates separated by a dielectric is too high or the thickness of the dielectric is too small, the dielectric breaks down and the electrical charge can move through it between the two plates. The *dielectric strength* is a measure of the highest voltage that an insulating material can withstand without electrical breakdown. It is defined as:

$$\text{dielectric strength} = \frac{\text{breakdown voltage}}{\text{insulator thickness}} \qquad [13]$$

The units of dielectric strength are volts per metre. Polyethylene has a dielectric strength of about 4×10^7 V/m. This means that a 1 mm thickness of polyethylene will require a voltage of about 40 000 V across it before it will break down.

When an alternating current is applied to two plates separated by a dielectric, a fraction of the energy is lost each time the current alternates. With a perfect dielectric, the current leads the voltage by 90° (Figure 1.21(a)). However, a useful model we can adopt is of the capacitor with the lossy dielectric as being represented as a capacitor with a perfect dielectric in parallel with a resistor giving the power dissipation (Figure 1.21(b)). The current now leads the voltage by 90° − δ, where δ is termed the *dielectric loss angle*. From the phasor diagram:

$$\tan \delta = \frac{I_R}{I_C}$$

The current through the resistor $I_R = V/R$ and the current through the capacitance $I_C = V/X_C$, where the capacitive reactance $X_C = 1/\omega C$ and the angular frequency $\omega = 2\pi f$ with f being frequency. Thus:

$$\tan \delta = \frac{I_R}{I_C} = \frac{X_C}{R} = \frac{1/\omega C}{R} = \frac{1}{\omega RC}$$

The power loss in the parallel resistor is V^2/R and thus the power loss per cycle of alternating current is $(V^2/R)T = (V^2/R)(2\pi/\omega)$, where T is the periodic time. The maximum energy stored by the capacitor is $\frac{1}{2}CV_{max}^2$, with $V_{max} = \sqrt{2}V$ for a sinusoidal waveform. The fraction of the maximum energy lost each cycle divided by 2π is termed the *loss factor* and is thus given by:

$$\text{loss factor} = \frac{(2\pi V^2/R\omega)}{2\pi(\frac{1}{2}C2V^2)} = \frac{1}{\omega RC} = \tan \delta \qquad [14]$$

Table 1.5 shows some typical values of dielectric constant, dielectric strength and loss factor tan δ.

Table 1.5 *Typical dielectric values*

Material	Relative permittivity at 50 Hz	at 10^6 Hz	Dielectric strength 10^6 V/m	Loss factor $\tan \delta$ at 10^6 Hz
Alumina	9	6.5	6	0.0002–0.01
Glass (Pyrex)	4.3	4	14	0.01–0.02
Mica	7	7	40	0.001
Polyethylene	2.3	2.3	20	0.0002–0.0005
Polystyrene	2.3	2.3	20	0.0001–0.001
Titanium dioxide		100	6	0.0002–0.005
Barium titanate		3000	12	0.0001–0.02

Example

An electrical capacitor is to be made with a sheet of polythene of thickness 0.1 mm between the capacitor plates. What is the greatest voltage that can be connected between the capacitor plates if there is not to be electrical breakdown? The dielectric strength is 4×10^7 V/m.

The dielectric strength is defined as the breakdown voltage divided by the insulator thickness, hence:

breakdown voltage = dielectric strength $\times$ thickness

$$= 4 \times 10^7 \times 0.1 \times 10^{-3} = 4000 \text{ V}$$

Example

A 0.1 µF capacitor has a dielectric with a loss factor of 0.003. What will be the power loss when an alternating voltage of 240 V, 50 Hz is connected across it?

The power loss in the parallel resistor, resulting from the dielectric being lossy, is given by, V^2/R and, since $\tan \delta = 1/\omega RC$:

power loss = $V^2 \omega C \tan \delta$

$$= 240^2 \times 2\pi \times 50 \times 0.1 \times 10^{-6} \times 0.003$$

$$= 5.4 \times 10^{-3} \text{ W}$$

1.1.5 Thermal properties

Thermal properties that are generally of interest in the selection of materials include how much a material will expand for a particular change in temperature; how much the temperature of a piece of material will change when there is a heat input into it; and how good a conductor of heat it is.

The *linear expansivity a* or *coefficient of linear expansion* is a measure of the amount by which a length of material expands when the temperature increases. It is defined as:

$$a = \frac{\text{change in length}}{\text{original length} \times \text{change in temperature}} \qquad [15]$$

and has the unit of K^{-1}.

The term *heat capacity* is used for the amount of heat needed to raise the temperature of an object by 1 K. Thus if 300 J is needed to raise the temperature of a block of material by 1 K, then its heat capacity is 300 J/K. The *specific heat capacity c* is the amount of heat needed per kilogram of material to raise the temperature by 1 K, hence:

$$c = \frac{\text{amount of heat}}{\text{mass} \times \text{change in temperature}} \qquad [16]$$

It has the unit of $J\ kg^{-1}\ K^{-1}$. Because metals have smaller specific heat capacities than plastics, weight-for-weight metals require less heat to reach a particular temperature than plastics, e.g. copper has a specific heat capacity of about 340 J kg^{-1} K^{-1} while polythene is about 1800 J kg^{-1} K^{-1}.

The *thermal conductivity λ* of a material is a measure of the ability of a material to conduct heat. There will only be a net flow of heat energy through a length of material when there is a difference in temperature between the ends of the material. Thus the thermal conductivity is defined in terms of the quantity of heat that will flow per second divided by the temperature gradient (Figure 1.22), i.e.:

$$\lambda = \frac{\text{quantity of heat/second}}{\text{temperature gradient}} \qquad [17]$$

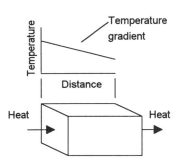

Figure 1.22 *Thermal conductivity is a measure of the ability of a material to conduct heat*

and has the unit of W m^{-1} K^{-1}. A high thermal conductivity means a good conductor of heat. It means a small temperature gradient for a particular rate of influx of heat. Metals tend to be good conductors, e.g. copper has a thermal conductivity of about 400 W m^{-1} K^{-1}. Materials that are bad conductors of heat have low thermal conductivities, e.g. plastics have thermal conductivities of the order 0.3 W m^{-1} K^{-1} or less. Very low thermal conductivities occur with foamed plastics, i.e. those containing bubbles, i.e. cavities, full of air. For example, foamed polymer polystyrene, known as expanded polystyrene and widely used for thermal insulation, has a thermal conductivity of about 0.02 to 0.03 W m^{-1} K^{-1}.

Table 1.6 *Thermal properties*

Material	Material	Linear expansivity 10^{-6} K^{-1}	Specific heat capacity J kg^{-1} K^{-1}	Thermal conductivity W m^{-1} K^{-1}
Metals	Aluminium	24	920	230
	Copper	18	385	380
	Mild steel	11	480	54
Polymers	Polyvinyl chloride	70–80	840–1200	0.1–0.2
	Polyethylene	100–200	1900–2300	0.3–0.5
	Epoxy cast resin	45–65	1000	0.1–0.2
Ceramics	Alumina	8	750	38
	Fused silica	0.5	800	2
	Glass	8	800	1

Table 1.6 gives typical values of the linear expansivity, the specific heat capacity and the thermal conductivity for metals, polymers and ceramics.

Example

By how much will a 10 cm strip of (a) copper, (b) PVC expand when the temperature changes from 20 to 30°C? Use the data given in Table 1.4.

(a) For copper, using equation [15]: expansion = 18 × 10^{-6} × 0.10 × 10 = 18 × 10^{-6} m = 0.018 mm.
(b) For the PVC, using equation [15]: expansion = 75 × 10^{-6} × 0.10 × 10 = 75 × 10^{-6} m = 0.075 mm.
The amount of expansion with the PVC is some four times greater than that of the copper.

Example

The heating element for an electric fire is wound on an electrical insulator. What thermal considerations will affect the choice of insulator material?

The insulator will need to have a low heat capacity so that little heat is used to raise the material to temperature. This means using a material with as low a density, and hence low mass, and specific heat capacity as possible. It also will need to be able to withstand the temperatures realised without deformation or melting. A ceramic is indicated.

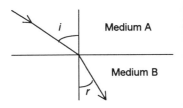

Figure 1.24 *Refraction with a light ray bending on passing from one medium to another*

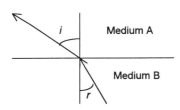

Figure 1.24 *Refraction for light ray passing from medium B to medium A*

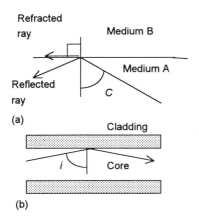

Figure 1.25 *(a) Critical angle is when a ray is reflected through ninety degrees and larger angles of incidence result in only reflection, (b) optical fibre using this property and having light transmitted through a fibre by internal reflection*

1.1.6 Optical properties

An important optical property of a material is its refractive index. When a ray of light passes from one medium to another, e.g. air into glass, reflection and refraction occur at the interface (Figure 1.23). With reflection the angle of incidence equals the angle of reflection. With refraction the ray of light bends from its straight-line path in passing across the interface. The *refractive index* in going from medium A to medium B, $_An_B$, is given by:

$$\text{refractive index } _An_B = \frac{\sin i}{\sin r} \qquad [18]$$

where i is the angle of incidence, i.e. the angle between the incident ray in medium A and the normal, and r the angle of refraction, i.e. the angle between the refracted ray in medium B and the normal. This is known as *Snell's law*.

In Figure 1.23 the ray of light is shown as starting in medium A and moving into medium B, bending towards the normal. This occurs because the velocity of light in medium A is greater than that in medium B. Suppose we reverse the path and have the ray of light passing from medium B into medium A (Figure 1.24). The same path is followed, but in the reverse direction, the ray now bending away from the normal. This is because the light is passing from a medium where the speed of light is lower to one where it is higher. Thus we have, when using the same notation for the angles, a refractive index in this case of:

$$_Bn_A = \frac{\sin r}{\sin i} = \frac{1}{_An_B} \qquad [19]$$

When a ray of light travels from a material into one in which it has a lower speed, it bends towards the normal. When a ray of light travels from a material into one in which it has a greater speed it bends away from the normal. When we have this condition of the ray bending away from the normal then we can have a particular incident angle which results in the refracted ray of light bending through 90° and thus not being transmitted across the interface (Figure 1.25(a)). The angle of incidence in such a case is termed the *critical angle C*. We then have:

$$_Bn_A = \frac{\sin C}{\sin 90°} = \sin C \qquad [20]$$

For angles of incidence greater than the critical angle, the ray of light is totally reflected at the interface, there being no refracted ray.

As an illustration of the significance of the critical angle in the choice of optical materials, consider the material used for *fibre optics*. The basic optical fibre consists of a central core of material in which the velocity of light is higher than in the surrounding cladding. Light for which the angle of incidence is greater than the critical angle is transmitted along such a fibre by total internal reflection, none of such light being lost from the fibre by being refracted through the cladding (Figure 1.25(b)).

The refractive index used above is that for light travelling from one material to another and is referred to as the *relative refractive index*. For example, we thus have $_An_B$ for light travelling from medium A to medium B. The refractive index is in fact the ratio of the velocities of light in the two media:

$$_An_B = \frac{\text{velocity of light in A}}{\text{velocity of light in B}} \qquad [21]$$

It is convenient to define an *absolute refractive index* of a medium as being that given when light travels from a vacuum into that medium, i.e.:

$$n_A = \frac{\text{velocity of light in a vacuum}}{\text{velocity of light in A}} \qquad [22]$$

To arrive at the relationship between the absolute refractive indices of two media A and B and the relative refractive index in going from A to B, consider the situation shown in Figure 1.26 when light travels from a vacuum into medium A and then into B. At the interface between the vacuum and medium A we can write:

$$n_A = \frac{c}{c_A}$$

where c is the velocity of light in a vacuum and c_A that in medium A. For the interface between medium A and medium B we can write:

$$_An_B = \frac{c_A}{c_B} = \frac{c_A}{c} \times \frac{c}{c_B} = \frac{n_B}{n_A} \qquad [23]$$

Thus knowing the absolute refractive indices for two media enables us to calculate the relative refractive index for the two media. But we have $_An_B = \sin\theta_A/\sin\theta_B$, thus we can write Snell's law as

$$n_A \sin\theta_A = n_B \sin\theta_B \qquad [24]$$

Table 1.7 shows values of the absolute refractive index for light of wavelength 589 nm (yellow light), this being taken as the value of refractive index which is typically used for white light.

Table 1.7 *Refractive index values at wavelength 589 nm*

Material	Refractive index
Borosilicate crown glass	1.51
Diamond	2.42
Acrylic	1.49
Polystyrene	1.59
Polyethylene	1.52

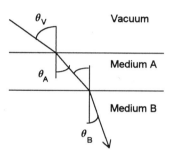

Figure 1.26 *Refraction from vacuum to A to B*

Light when incident on a material can be reflected, absorbed and transmitted. The transparency of a material, such as a plastic, depends on its light-absorbing and light-scattering properties. The term *total transmission factor* is used for the ratio of the total transmitted light intensity and the incident light intensity, assuming it is concentrated in a parallel beam perpendicular to the surface of the sample. For comparison purposes the values are usually quoted for a thickness of 1 mm. The *reflection factor* is the ratio of the light intensity reflected at an angle equal to the angle of incidence and the intensity of the incident beam, assuming it is concentrated in a parallel beam.

The *clarity* with which detail in an object can be seen when viewed through a sample of the material depends on the amount of light scattered in the material. It is perfect only when no light is scattered.

Polyethylene typically has a refractive index of about 1.52 and a direct transmission factor, for low density polyethylene, of about 40–45%. Polyvinyl chloride has a refractive index of about 1.54 and a direct transmission factor of about 90%. This high transparency means it is frequently used as a glass substitute, it having the advantage of not breaking so readily.

Example

Determine the critical angle for a glass–air interface if the glass has a refractive index of 1.5.

The refractive index of the glass is for light going from air to glass and so 1.5 = sin 90°/sin C. Hence the critical angle C is 41.8°.

Example

An optical fibre consists of a glass core clad with another material. The core has an absolute refractive index of 1.40 and the cladding an absolute refractive index of 1.42. What is the critical angle for light incident on the glass–cladding interface?

The refractive index for light passing from glass to the cladding is $_gn_c = n_c/n_g = 1.40/1.42 = 0.99$. Thus the critical angle C is given by 0.99 = sin C.sin 90° and so the critical angle is 81.9°.

1.1.7 Chemical properties

Attack of materials by the environment in which they are situated can be a major problem. The rusting of iron in air is an obvious example of such an attack. Tables are available giving the comparative resistance to attack of materials in various environments, e.g. in aerated water, in salt water, to strong acids, strong alkalis, organic solvents and ultraviolet radiation.

While some polymers are highly resistant to chemical attack, others are liable to stain, craze, soften, swell or dissolve completely. For example, nylon shows little degradation with weak acids but is attacked by strong acids; it is resistant to alkalis and organic solvents. Polymers have generally high resistance to attack in water and thus are widely used for containers and pipes. Polymers are generally affected by exposure to sunlight. Ultraviolet light, present in sunlight, can cause a breakdown of the bonds in the polymer molecular chains and result in surface cracking. For this reason, plastics often have an ultraviolet inhibitor mixed with the polymer when the material is produced.

1.1.8 Magnetic properties

In the vicinity of permanent magnets and current-carrying conductors, a magnetic field is said to exist. The magnetic field pattern can be plotted using a compass needle or demonstrated by scattering iron filings in the vicinity (Figure 1.27). The term *magnetic line of force* is used for a line traced out by such plotting or the iron filings. A useful way of considering magnetic fields is in terms of *magnetic flux*, this being something that is considered to flow along these lines of force like water through pipes. The term *magnetic flux density B* is used for the amount of flux passing through unit area. If flux Φ passes through an area A, then:

$$B = \frac{\Phi}{A} \qquad [25]$$

Magnetic flux is produced within magnetic materials when electrical currents pass through coils of wire wrapped round cores of such materials (Figure 1.28). The *magnetising field strength H* is NI/L, where N is the number of turns on the coil, I the current and L the length of the coil. The flux density produced for a given magnetising field depends on the magnetic material used for the core. With a vacuum for the core, the flux density B_0 is $\mu_0 H$, where μ_0 is a constant called the permeability of free space. The term *relative permeability μ_r* is used for the unitless factor by which the flux density in a material B compares with that which would have been produced with a vacuum as the core B_0:

$$\mu_r = \frac{B}{B_0} = \frac{B}{\mu_0 H} \qquad [26]$$

μ_r for air is about 1, since air is virtually the same as a vacuum.

Materials can be grouped into three general categories according to their permeabilities:

- ***Diamagnetic materials***
  These materials, e.g. copper, have relative permeabilities slightly below 1.

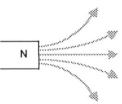

Figure 1.27 *Lines of force near the north pole of a permanent magnet*

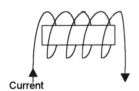

Current

Figure 1.28 *Magnetising a material*

- ***Paramagnetic materials***
  These materials, e.g. aluminium, have relative permeabilities slightly greater than 1.

- ***Ferromagnetic and ferrimagnetic materials***
  These have relative permeabilities considerably greater than 1, ferro-magnetic materials being metals and ferrimagnetic materials being ceramics. Iron, cobalt and nickel are examples of ferromagnetic materials, iron oxide Fe_3O_4 and nickel ferrite $NiFe_2O_3$ examples of ferri- magnetic materials. For iron the relative permeability is typically about 2000 to 10 000, though special steels can have values of the order of 60 000 to 90 000.

The relative permeability for a ferromagnetic or ferrimagnetic material is not constant, depending on the size of magnetising field used. When an initially unmagnetised material is placed in an increasing magnetising field, the flux density within the material increases in the way shown in Figure 1.29. The gradient of the graph, i.e. B/H, is not constant and so the relative permeability $B/\mu_0 H = B/B_0$ is not a constant. After a particular magnetising field is reached, the magnetic flux reaches a constant value, this being termed *saturation*. If the magnetic field is then reduced back to zero, the material may not simply just retrace its path back down the same graph line and may retain some magnetism when the applied magnetic field is zero. The retained flux density is termed the *remanent flux density* or *remanence*. To demagnetise the material, i.e. bring B to zero, a reverse field called the *coercive field* or *coercivity* must be applied. Figure 1.30 shows how the flux density B within the material might vary when the magnetising field is increased to saturation, then decreased to zero, then increased to saturation in the opposite direction, then decreased to zero, etc. The resulting graph is called a *hysteresis loop*.

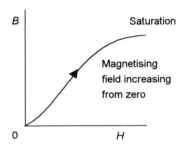

Figure 1.29 *Initial magnetiisation graph*

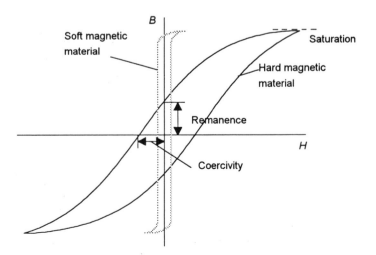

Figure 1.30 *Hysteresis loops for a soft and a hard magnetic material*

In Figure 1.30 the hysteresis loops are shown for two materials, termed *hard* and *soft magnetic materials*. A hard magnetic material has high remanence so that a high degree of magnetism is retained in the absence of a magnetic field, a high coercivity so that it is difficult to demagnetise and a large area enclosed by the hysteresis loop. The area of the loop is related to the energy dissipated in the material during each cycle of magnetisation. A soft material is very easily demagnetised, having low coercivity and the hysteresis loop only enclosing a small area. Hard magnetic materials are used for such applications as permanent magnets while soft magnetic materials are used for transformers where the magnetic material needs to be easily demagnetised and little energy dissipated in magnetising it. A typical soft magnetic material used for a transformer core is an iron–3% silicon alloy. The main materials used for permanent magnets are the iron–cobalt–nickel–aluminium alloys, ferrites and rare earth alloys.

Table 1.8 gives properties of typical soft magnetic materials and Table 1.9 gives details for hard magnetic materials. For hard magnetic materials an important parameter is the demagnetisation quadrant (Figure 1.31) of the hysteresis loop, it indicating how well a permanent magnet is able to retain its magnetism. The bigger the area the greater the amount of energy needed to demagnetise the material. A measure of this area is given by the largest rectangle which can be drawn in the area, this being the maximum value of the product *BH*.

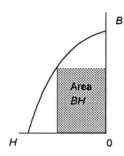

Figure 1.31   *The demagnetisation quadrant*

Example

Which of the following applications (a) a compass needle, (b) the core of an electromagnet, requires a soft and which a hard magnetic material?

(a) A hard magnetic material is required since the compass needle is required to be a permanent magnet.
(b) A soft magnetic material is required since the electromagnet has to lose its magnetism when the energising current is switched off.

Table 1.8 *Soft magnetic materials*

Material	Relative permeability		Remanence	Coercive field
	Initial value	Max. value	T	A/m
Fe–3% Si, silicon steel		90 000		6
Fe–70/80% Ni, Mumetal	60 000		0.5	1
Fe–24% Co, Permendur	300	2 000	1.7	950
Ni–Zn ferrite	20–600			
Mn–Zn ferrite	600–5000			

Table 1.9 *Hard magnetic materials*

Material	Remanence T	Coercive field kA/m	Max. *BH* T A/m
Fe–Co–Ni–Al alloys			
Alni	0.56	46	10
Alnico	0.72	45	14
Alcomax 2	1.30	46	43
Columax	1.35	59	60
Precipitation hardening alloys			
Comalloy	1.0	20	10
Vicalloy 1	0.9	24	8
Cunife 1	0.57	36	12
Steels			
6% tungsten	1.05	5.2	2.4
6% chromium	0.95	5.2	2.4
3% cobalt	0.72	10	2.8
9% cobalt	0.79	12.7	4.0
15% cobalt	0.83	14.3	4.9
Rare earth			
Samarium cobalt	0.9	670	160
Neodymium iron boron	1.1	890	220
Ferrites			
Feroba 1	0.22	135	8
Feroba 2	0.39	150	29
Feroba 3	0.37	240	26

1.1.9 Comparisons of properties

Table 1.10 summarises the relative properties of metals, polymers and ceramics, indicating the typical range of values likely to be encountered at about 20°C.

Example

Which type of material, metal, polymer or ceramic would be the most likely to give materials with each of the following properties: (a) high density, (b) high melting point, (c) high electrical conductivity, (d) low specific heat capacity, (e) low tensile modulus of elasticity.

(a) Metals tend to have the highest densities.
(b) The highest melting points are given by the ceramics.
(c) The highest electrical conductivities are from metals.
(d) The lowest specific heat capacities are from metals.
(e) Polymers give the lowest modulus of elasticity.

Table 1.8 *The range of properties*

Property	Metals	Polymers	Ceramics
Density Mg/m³	Medium–high, 2–16	Low, 1–2	Generally medium, 2–4
Melting point °C	Medium–high, 200–3500	Low, 70–200	High, 2000–4000
Thermal conductivity	High	Low	Medium–low
Thermal expansion	Medium	High	Low
Specific heat capacity	Low	Medium	High
Electrical conductivity	High	Very low	Very low
Optical properties	Opaque	Some transparent, some opaque	Some transparent, some opaque
Tensile strength MPa	Medium–high, 100–2500	Generally low*, 30–80	Generally low, 10–400
Compressive strength MPa	Medium–high, as tensile	Generally low*, as tensile	High, 1000–5000
Tensile modulus GPa	Medium–high, 40–400	Low*, 0.1–4	High, 150–450
Toughness	Good	Some good, some poor*	Poor
Hardness	Medium	Low	High
Wear resistance	Medium	Low–moderate	High
Resistance to corrosion	Medium–poor	Good–medium	Good

Note: 1 Mg/m³ = 1000 kg/m³. * Polymers are widely used with fillers, such as fibres and particles, and these can markedly change their properties, in particular making them stiffer, stronger and tougher.

Costs

Costs can be considered in relation to the basic costs of the raw materials, the costs of manufacturing products, and the life and maintenance costs of the finished product. In comparing the basic costs of materials, the comparison is often on the basis of the cost per unit weight or cost per unit volume. Table 1.11 shows the relative costs of some materials.

Often a more important comparison is on the basis of the cost per unit strength or cost per unit stiffness for the same volume of material. This enables the cost of, say, a beam to be considered in terms of what it will cost to have a beam of a certain strength or stiffness. Hence if, for comparison purposes, we consider a beam of volume 1 m³ then, if the tensile strength of the material is 500 MPa and the cost per cubic metre £800, the cost per MPa of strength will be 800/500 = £1.6.

The costs of manufacturing will depend on the processes used. Some processes require a large capital outlay and then can be used to produce large numbers of the product at a relatively low cost per item. Other processes may have little in the way of setting-up costs but a large cost per unit product.

The cost of maintaining a material during its life can often be a significant factor in the selection of materials; many metals need a surface coating to protect them from corrosion by the atmosphere. The rusting of steels is an obvious example of this and dictates the need for the continuous repainting of the Forth Railway Bridge

Key point

The cost per unit property is a useful basis for comparing the merits of materials.

Table 1.11 *Relative costs of materials relative to a mild steel ingot*

Material	Relative cost/kg	Relative cost/m^3
Cobalt	100	112
PTFE	39	11
Nickel	28	32
Chromium	26	24
Tin	19	18
Titanium	17	10
Brass sheet	16	17
Al–Cu alloy sheet	14	5.3
Nylon 66	12	1.8
Phosphor bronze ingot	10	12
Magnesium ingot	9.2	2.1
Acrylic	8.9	1.4
Copper tubing	8.7	10
ABS	8.3	1.1
Aluminium ingot	4.3	1.5
Polystyrene	3.6	0.50
Zinc ingot	3.6	3.3
Polyethylene (HDPE)	3.4	0.43
Polypropylene	3.2	0.34
Natural rubber	3.1	0.50
Polyethylene (LDPE)	2.3	0.29
PVC, rigid	2.3	0.43
Mild steel sheet	1.9	1.9
Mild steel ingot	1.0	1.0
Cast iron	0.8	0.79

Example

Compare the costs per unit strength of the following two materials for the same volume of material: low-carbon steel, cost per kg £0.1, density 7800 kg/m^3, strength 1000 MPa; aluminium–manganese alloy, cost per kg £0.22, density 2700 kg/m^3, strength 200 MPa.

For the steel, the volume of 1 kg is 1/7800 = 0.000 128 m^3 and so the cost per m^3 is 0.1/0.000 128 = £780. The cost per MPa of strength is thus 780/1000 = £0.78. For the aluminium alloy, the volume of 1 kg is 1/2700 = 0.000 37 m^3 and so the cost per m^3 is 0.22/0.000 37 = £590. Thus although the cost per kg for the aluminium alloy is greater than that of the steel, because of the lower density, the cost per cubic metre is less. For the aluminium alloy, the cost per MPa of strength is thus 590/200 = £2.95. Hence on a comparison of the strengths of equal volumes, it is cheaper to use the steel.

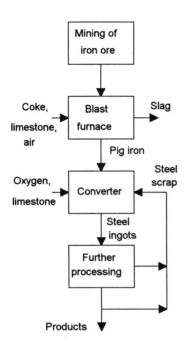

Figure 1.32 *Steel making*

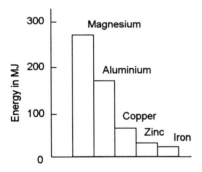

Figure 1.33 *Energy to produce 1 kg of metals from ores*

1.1.10 Environmental factors

Consider the resources involved in the production of steel (Figure 1.32). To start with there is the mining of the iron ore. This is then used in a blast furnace with coke and limestone, also mined, and air to produce pig iron and slag, a waste product. The molten pig iron is then mixed with steel scrap, oxygen and limestone in a converter to produce steel ingots. These may then be further processed by rolling, forging, etc. in order to produce products which might still require some machining and finishing.

But Figure 1.32 gives only part of the picture. There will be other outcomes at each stage, such as mine waste, pollution, waste heat, etc., and other inputs, in particular energy. It has been reckoned that energy currently accounts for about 11% of the costs in steel making. The energy needed to produce 1 kg of liquid steel amounts to about 23 million joules. Figure 1.33 shows an estimate of the relative amounts of energy needed to produce 1 kg of metals from their ores. Though iron requires the least, the great amount of steel products used means that the total amount of energy devoted to steel making is large.

The above figures can often, however, be very markedly reduced if waste materials are recycled. Thus the recycling of aluminium can reduce the energy requirements for 1 kg of aluminium by about 80%; in the recycling of copper the energy requirements can be reduced also by about 80%. Steel produced from scrap requires only about 25% of the energy required with ore (this recycling using scrap is illustrated in Figure 1.33). Thus considerable energy savings can be made by recycling metals.

The term *recycling* is used for the feeding of waste material, i.e. scrap, back into the processing cycle so that it can be reused in the production of fresh metal for fabrication into products. In the case of steel making an obvious example is the crushing of old car bodies to form steel cubes which can be fed into the steel furnaces, along with the iron ore, to make fresh steel. Another example is the recycling of aluminium cans in the production of aluminium. Figure 1.34(a) illustrates the stages involved in the production of aluminium cans when recycled material is used. It has been estimated that of the order of 40% of aluminium in Britain is recycled. Because aluminium requires a large amount of energy to produce it from its ore, the energy savings, as well as the saving in aluminium, can be very significant. Another example is the recycling of glass bottles. Typically of the order of 30% of the glass in a glass bottle comes from recycled glass (Figure 1.35(b)). There is, however, no great energy saving advantage in using scrap glass compared to using the raw materials. There is, however, a big saving if bottles are reused.

1.1.11 Case studies

The following case studies illustrate how properties dictate constraints of a sensor and the interplay between the properties required of materials and the design of components made of them.

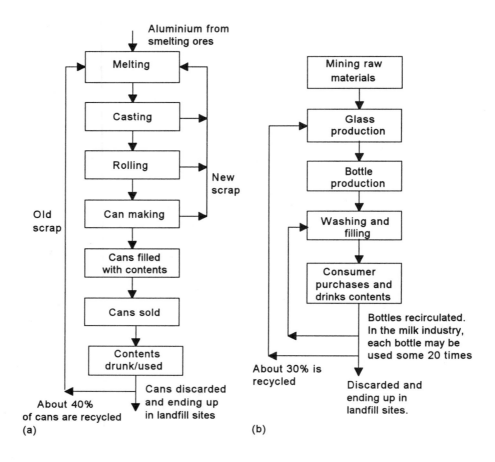

Figure 1.34 *Using (a) recycled aluminium in the production of aluminium cans, (b) recycled glass bottles*

Electrical resistance strain gauges

The electrical resistance strain gauge is used to measure the strain on the surface of a component under strain. It consists of a resistance element in the form of a either a thin wire of a high resistivity metal alloy which is in the form of a flat coil on a paper backing or a metal foil conducting track on a polymer film (Figure 1.35). The gauge is stuck to the surface of the component, like a postage stamp being stuck on an envelope, with its long elements in the direction of loading. When the component is put in tension, the long elements of the resistance element are stretched and increase in resistance as a consequence of the length increasing and also the width decreasing (note that these are not the only changes, the resistivity is also increased by strain). The fractional change in resistance is proportional to the strain, the constant of proportionality being termed the gauge factor. Typical values are about 2.1. This change in resistance can then be used to enable the strain to be determined.

However, a change in temperature will also cause a change in resistance of the strain gauge as a result of the resistivity of the

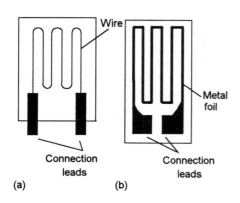

Figure 1.35 *Strain gauges: (a) metal wire, (b) metal foil*

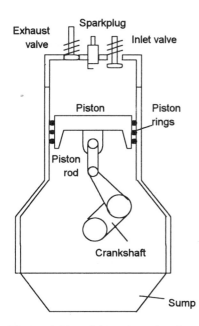

Figure 1.36 *Internal combustion engine*

gauge material changing and also as a result of expansion. Thus, when using strain gauges, methods have to be used to eliminate the effects of temperature changes. This can be done by using a 'dummy gauge' which is subject to the same temperature changes that the 'active gauge' is but not to the strain that is to be measured. By incorporating the two gauges in opposite arms of a Wheatstone bridge, the effects of temperature changes can be eliminated.

Pistons

Consider the materials used for the pistons in internal combustion engines. Figure 1.36 shows the basic form of an internal combustion engine and the piston. Fuel enters the cylinder and is ignited by a spark at the spark plug. The piston then has the function of retaining the hot expanding gas on one side, moving as a result of the expansion and communicating the movement via the piston rod to the crankshaft; it then moves back up the cylinder and expels the products of the combustion through the exhaust valve.

Thus, pistons are subject to high temperatures on one side and relatively cold on the other and have the problem of accommodating thermal expansion since they must be a reasonably tight fit in the cylinder but not expand so much that they become a tight fit. The upper part of a piston is fitted with piston rings to help retain the combustion gases in the upper part of the cylinder and also to take the wear resulting from sliding up and down the cylinder wall and avoid the piston rubbing against the wall and wearing.

Earlier low speed engines had pistons of cast iron to match the material used for the cylinder. With increasing engine speeds, modern pistons are made from an aluminium alloy, e.g. the aluminium casting alloy LM13. Such an alloy has a high thermal conductivity and also has the great benefit of a low density, so allowing lighter weight pistons. The high thermal conductivity enables heat to be more rapidly conducted away and so results in the piston running at a lower temperature than otherwise would be the case. The material used must retain its properties at temperatures up to about 300°C. Thermal expansion is, however, a problem. This is particularly the case when they are in a cylinder made of cast iron since cast iron has a different coefficient of thermal expansion to that of an aluminium alloy; Table 1.12 gives thermal data for both materials. In the absence of special design features, the difference in expansion could result in seizure when hot (leaving enough room for the expansion would mean too loose a fitting piston when cold and result in 'piston slap'). Such design features include the piston being tapered (Figure 1.37) so that there is additional clearance when cold near the top; it is this part which attains the highest temperature when hot and thus expands more than the cooler lower part of the piston. Casting or forging can be used to manufacture the pistons.

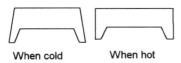

When cold   When hot

Figure 1.37 *Piston design*

Table 1.12 *Thermal properties*

Material	Density Mg/m^3	Thermal conductivity W m^{-1}K^{-1}	Coefficient of thermal expansion 10^{-6}/°K
Grey cast iron	7.2	44–53	11
Aluminium alloy LM13	2.7	117	19

Problems 1.1

1 Two wires of the same material are stretched by the same size force. Wire X has twice the cross-sectional area of the other wire, Y. Which of the materials will (a) stretch the most, (b) be subject to the highest stress?

2 Explain how you would expect a material to behave if it has:

(a) high tensile strength with a low percentage elongation,

(b) high tensile modulus of elasticity and good fracture toughness.

3 What is the tensile stress acting on a strip of material of cross-sectional area 50 mm^2 when subject to tensile forces of 1000 N?

4 Tensile forces act on a rod of length 300 mm and cause it to extend by 2 mm. What is the strain?

5 An aluminium alloy has a tensile strength of 200 MPa. What force is needed to break a bar of this material with a cross-sectional area of 250 mm^2?

6 A test piece of a material is measured as having a length of 100 mm before any forces are applied to it. After being subject to tensile forces, it breaks and the broken pieces are found to have a combined length of 112 mm. What is the percentage elongation?

7 A material has a yield stress of 250 MPa. What tensile forces will be needed to cause yielding if the material has a cross-sectional area of 200 mm^2?

8 A sample of high tensile brass is quoted as having a tensile strength of 480 MPa and a percentage elongation of 20%. An aluminium–bronze is quoted as having a tensile strength of 600 MPa and a percentage elongation of 25%. Explain the significance of this data in relation to the mechanical behaviour of the materials.

9 A grey cast iron is quoted as having a tensile strength of 150 MPa, a compressive strength of 600 MPa and a percentage elongation of 0.6%. Explain the significance of the data in relation to the mechanical behaviour of the material.

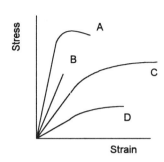

Figure 1.38   *Problem 10*

10 Figure 1.38 shows the stress-strain graph for four materials. Which of the materials is (a) the most ductile, (b) the most brittle, (c) the strongest, (d) the stiffest?

11 A number of parallel plate capacitors are to be made by inserting sheets of the following materials, between and in contact with two metal plates. Which one of the following capacitors will have the highest breakdown voltage? One made with:

(a) a mica sheet of thickness 0.5 mm,
(b) a polyethylene sheet of thickness 0.5 mm,
(c) a sheet of dry paper of thickness 2.0 mm,
(d) a sheet of electrical ceramic of thickness 3.0 mm.

Material	Dielectric strength 10^6 V/m
Mica	60
Polyethylene	20
Dry paper	16
Electrical ceramic	12

12 Aluminium has a resistivity of 2.5×10^{-8} W m. What will be the resistance of an aluminium wire with a length of 1 m and a cross-sectional area of 2 mm²?

13 X and Y are two identical volume blocks of different materials and different densities, X having a higher specific heat capacity and a higher density than Y. When there is the same heat input to both blocks, for which block will the greatest change in temperature occur?

14 A sheet of a polymer of thickness 1 mm transmits 60% of the light intensity incident on it. What percentage of the incident light will be transmitted by a sheet of the same material 3 mm thick?

15 For three different materials X, Y and Z, X is weakly attracted by a magnet, Y is weakly repelled by a magnet and Z is strongly attracted by a magnet. Which of the materials is paramagnetic, which diamagnetic and which ferromagnetic?

16 Figure 1.39 shows the parts of the hysteresis loops involved in demagnetisation for three materials. On the basis of just this evidence, which of the materials will be most suited to being a permanent magnet in, say, a permanent magnet motor?

17 A magnetic material is required for use as a transformer core. This material is required to be easily demagnetised and requires little energy to become magnetised. What can be said about the requirements for coercivity and area of the hysteresis loop?

18 What are the main types of properties required for the following products?

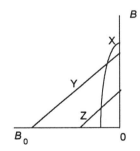

Figure 1.39   *Problem 16*

(a) a domestic kitchen sink,
(b) a shelf on a bookcase,
(c) a cup,
(d) an electrical cable,
(e) a coin,
(f) a car axle,
(g) the casing of a telephone.

19 For each of the products listed in Problem 18, identify a material that is commonly used and explain why its properties justify its choice for that purpose.

20 State the properties of a material you need to consider if you require materials which:

(a) are stiff,
(b) capable of being bent into a fixed shape,
(c) do not fracture when small cracks are present,
(d) do not easily break,
(e) act as an electrical insulator,
(f) are a good conductor of heat,
(g) can be used as the lining for a tank storing acid.

21 You read in a textbook that:

'Designing with ceramics presents problems that do not occur with metals because of the almost complete absence of ductility with ceramics.'

Explain the significance of the comment in relation to the exposure of ceramics to forces.

22 Compare the specific strengths, and costs per unit strength for equal volumes, for the materials giving the following data:

Low-carbon steel:
cost per kg £0.1,
density 7800 kg/m^3,
strength 1000 MPa

Polypropylene:
cost per kg £0.2,
density 900 kg/m^3,
strength 30 MPa

23 In a scrap recycling plant for domestic refuse a physical method is required to separate polymers, aluminium alloys and steel from each other. Suggest the properties of the materials on which the separation methods might be based.

1.2 Properties data

This section is about where data can be found for the various properties of materials, Section 1.3 detailing methods used for obtaining such data.

Standards

A *standard* is a technical specification drawn up with the co-operation and general approval of all interests affected by it with the aims of benefiting all concerned by the promotion of consistency in quality, rationalisation of processes and methods of operation, promoting economic production methods, providing a means of communication, protecting consumer interests, promoting safe practices, and helping confidence in manufacturers and users. Standards, both national and international, are used in the case of materials to ensure such things as consistency of quality, consistency in the use of terms, rationalisation of testing methods, and provide an efficient means of communication between interested parties. Thus if a material is stated by its producer to be to a certain standard, tested by the methods laid down by certain standards, then a customer need not have all the details written out of what properties and tests have been carried out by the producer in order to know what properties to expect of the material. There is, for example, the standard for tensile testing of metals (BSEN 10002, a European standard adopted as the British Standard and replacing BS 18). This lays down such things as the sizes of the test pieces to be used (see Chapter 3 for such details). There are standards for materials such as copper and copper alloy plate (BS 2875), steel plate, sheet and strip (BS 1449), the plastic polypropylene (BS 5139), etc. which lay down the composition and properties for such materials. There are many thousands of standards laid down by national standards bodies such the *British Standards Association* (BS) and international bodies such as the *International Organisation for Standardisation* (ISO).

1.2.1 Data sources

Data on the properties of materials is available from a range of sources. These include:

- Specifications issued by bodies responsible for standards, e.g. the British Standards Institution, the American Society for Metals, the International Organisation for Standardisation, etc. The standards operating in Britain are those issued by the British Standards Institution and listed in their catalogue. This lists all the British Standards and also whether they are also European and/or international standards.

- Data books, e.g. the *ASM Metals Reference Book* (American Society for Metals 1983), *ASM Engineered Materials Reference Book* (American Society for Metals 1989), *Smithells Metals Reference Book* by E. A. Brandes and G. B.

Brook (Butterworth-Heinemann 1992), *Metals Databook* by C. Robb (The Institute of Metals 1987), *Handbook of Plastics and Elastomers* edited by C. A. Harper (McGraw-Hill 1975), *Handbook of Properties of Technical and Engineering Ceramics*, parts 1 and 2 by R. Morrell (National Physical Laboratory 1985), *Engineers Guide to Composite Materials* edited by J. W. Weeton, D. M. Peters and K. L. Thomas (Society for Metals 1987), and, for a cheap, concise source, the *Newnes Engineering Materials Pocket Book* by W. Bolton (Newnes, Butterworth- Heinemann 1989, 1996, 2000).

- Computerised databases which give materials and their properties with means to rapidly access particular materials or find materials with particular properties, e.g. *Cambridge Materials Selector* (Cambridge University Engineering Department 1992), *MAT.DB* (American Society for Metals 1990).

- Trade associations, e.g. the Copper Development Association, which issues brochures giving technical details of compositions and properties of copper and copper alloy materials.

- Data sheets supplied by suppliers of materials.

- In-company tests. This is more often than not used to check samples of a bought-in material to ensure that it conforms to the standards specified by the supplier from which the material was bought.

1.2.2 Classification systems

Classification systems using codes are used to refer to particular metals. Such codes can relate to the chemical composition or the type of properties it has and are a concise way of specifying a particular material without having to write out in full its chemical composition or properties. The following are some of the classification systems commonly used:

- *Steels: British*
  The system specified by the British Standards Association uses three digits followed by a letter and two further digits. The first three digits of the code represent the type of steel:

000 to 199	Carbon and carbon–manganese types, the number being 100 times the manganese contents
200 to 240	Free-cutting steels, the second and third numbers being 100 times the mean sulphur content
250	Silicon–manganese spring steels
300 to 499	Stainless and heat resistant valve steels, these numbers corresponding with the AISI-SAE numbers (see below)

500 to 999   Alloy steels with groups of numbers allocated to different alloys according to their main alloying element: 500–519 nickel, 520–539 chromium, 540–549 molybdenum, etc.

The fourth symbol is a letter:

A   The steel is supplied to a composition determined by chemical analysis

H   The steel is supplied to a hardenability specification

M   The steel is supplied to mechanical property specifications

S   The steel is stainless

The fifth and sixth digits correspond to 100 times the mean percentage carbon content of the steel.

As an illustration of the above coding system, consider a steel with a code 070M20. The first three digits are 070 and since they are between 000 and 199 the steel is a carbon or carbon–manganese type. The 070 indicates that the steel has 0.70% manganese. The fourth symbol is M and so the steel is supplied to mechanical property specifications. The fifth and sixth digits are 20 and so the steel has 0.20% carbon.

•   *Steels: American*

The AISI-SAE (American Iron and Steel Institute, Society of Automotive Engineers) use a four digit code. The first two digits indicate the type of steel. The third and fourth digits indicate 100 times the percentage carbon content.

1000   Carbon steel, manganese steels 1300

2000   Nickel steels

3000   Nickel–chromium steels

4000   Molybdenum steels 4000, 4400; chromium–molybdenum steels 4100; nickel–chromium–molybdenum steels 4300, 4700; nickel–molybdenum steels 4600, 4800

5000   Chromium steels

6000   Chromium–vanadium steels

7000   Tungsten–chromium steels

8000   Nickel–chromium–molybdenum steels

9000   Silicon–manganese steels 9200, nickel–chromium–molybdenum steels 9300, 9400

For example, 1010 is a carbon steel with about 0.10% carbon; 5120 is a chromium steel with 0.20% carbon.

•   *Aluminium*

Aluminium cast alloys use the code of LM followed by a number, the number being used to indicate a specific alloy. The coding system for wrought aluminium alloys is that of the

Aluminium Association. This uses four digits with the first digit representing the principal alloying element, the second digit modifications to impurity limits, and the last two digits for the 1XXX alloys the aluminium content above 99.00% in hundredths and for others as identification of specific alloys. The 1XXX alloys have the principal alloying element of 99.00% minimum aluminium, the 2XXX copper as the principal alloying element, 3XXX manganese, 4XXX silicon, 5XXX magnesium, 6XXX magnesium and silicon, 7XXX zinc, 8XXX other elements.

• *Copper*

A commonly used system is that of the Copper Development Association (CDA). This uses the letter C followed by three digits. The first digit indicates the group of alloys concerned and the remaining two digits alloys within the group. The groups C1XX to C7XX are used for wrought alloys and C8XX and C9XX for cast alloys.

C1XX	Copper with a minimum copper content of 99.3%. High copper alloys, with more than 96% copper
C2XX	Copper–zinc alloys (brasses)
C3XX	Copper–zinc–lead alloys (leaded brasses)
C4XX	Copper–zinc–tin alloys (tin brasses)
C5XX	Copper–tin alloys (phosphor bronzes)
C6XX	Copper–aluminium alloys (aluminium bronzes), copper–silicon alloys (silicon bronzes), miscellaneous copper–zinc alloys
C7XX	Copper–nickel, copper–nickel–zinc alloys (nickel silvers)
C8XX	Cast coppers, high copper alloys, brasses, manganese bronze and copper–zinc–silicon alloys
C9XX	Cast copper–tin, copper–tin–lead, copper–tin–nickel, copper– aluminium–iron, copper–nickel–iron and copper– nickel–zinc alloys

The British Standards Association coding for wrought copper and copper alloys consists of two letters followed by three digits. The two letters indicate the alloy group and the three digits the alloy within that group.

C	Copper and alloys containing a very high percentage of copper
CA	Copper–aluminium alloys, i.e. aluminium bronzes
CB	Copper–beryllium alloys, i.e. beryllium bronzes
CN	Copper–nickel alloys, i.e. cupro-nickels
CS	Copper–silicon alloys, i.e. silicon bronzes
CZ	Copper–zinc alloys, i.e. brasses
NS	Copper–zinc–nickel alloys, i.e. nickel silvers
PB	Copper–tin–phosphorus alloys, i.e. phosphor bronzes

For cast copper alloys, letters followed by a digit are used. For example, AB1, AB2 and AB3 are aluminium bronze, LB1, LB2, LB4, and LB5 are leaded bronzes, SCB1 and SCB2 are general purpose brasses.

The above gives an indication of some of the codes developed by standards bodies to assist communication between suppliers and users of materials. In addition, since the properties of materials depend on the heat treatment and degree of working they have undergone, there are codes to specify such treatments. The conditions being generally referred to as the *temper* of the material. Thus, for example:

- *Aluminium*
  With aluminium alloys M indicates as manufactured (F in American code for as fabricated), O annealed, H followed by a number between 1 and 8 the degree of hardening resulting from cold working, T followed by various letters different forms of heat treatment.

- *Copper*
  Copper alloys are designated as temper O if the material is supplied in the annealed condition, H if supplied in the hard condition resulting from cold working with ¼H and ½H to indicate quarter and half-hard conditions, M for as manufactured, W(H) for the heat treatment of solution treatment to the hardened condition, W(P) for precipitation hardening to the hardened condition.

1.2.3 Case studies

The following case studies illustrate the searches for a material with the properties required for a particular product.

Electrical conductor

Consider the search for a material for use as a conductor of electricity where high conductivity is the main property required. Since metals are in general good conductors and polymers and ceramics very poor conductors then the choice would seem to be among metals. When tables are consulted the following information can be found for electrical conductivities, at 20°C:

Aluminium 40×10^6 S/m,
Copper 64×10^6 S/m,
Gold 50×10^6 S/m,
Iron 11×10^6 S/m,
Silver 67×10^6 S/m

If cost is also a factor then gold and silver are likely to be ruled out. Thus copper looks to be the optimum choice. If we now consider what form of copper, then tables are likely to yield data in the following form:

C101	Electrolytic tough-pitch h.c. copper	101.5–100
C103	Oxygen-free h.c. copper	101.5–100
C105	Phosphorus deoxidised arsenical copper	50–35
C108	Copper–cadmium	92–80

The conductivities are not expressed in S/m but in units called IACS (International Annealed Copper Standard) units and written as a percentage. This scale is based on 100% being the conductivity (or resistivity) of annealed copper at 20°C. This is a resistivity of 1.7241×10^{-8} Ω m or a conductivity of 58.00×10^6 S/m. Thus, if there are no other considerations C101 or C103 would appear to be the choice. Often, however, there are other factors to be taken into account, such as strength.

A cast iron with high tensile strength and ductility

Consider interpreting the data in Table 1.13, taken from *Newnes Engineering Materials Pocket Book* by W. Bolton (Heinemann Newnes 1989, 1996, 2000), for a number of cast irons.

Table 1.13 *Mechanical properties of cast irons*

Material	Tensile strength MPa	Yield stress MPa	Percentage elongation
Grey irons			
BS 150	160	98	0.6
BS 180	180	117	0.5
BS 220	220	143	0.5
BS 260	260	170	0.4
BS 300	300	195	0.3
BS 350	350	228	0.3
BS 400	400	260	0.2
Malleable irons			
Blackheart B32-10	320	190	10
Blackheart B35-12	350	200	12
Whiteheart W38-12	380	200	12
Whiteheart W40-05	400	220	5
Whiteheart W45-07	450	260	7

A high ductility means a high percentage elongation. Grey irons have very low percentage elongations and so are brittle. The malleable irons show greater percentage elongations and thus the selection needs to be one of these irons. The malleable iron with the greatest tensile strength Whiteheart W45-07, does not, however, have so high a percentage elongation as Whiteheart W38-12. Thus if ductility is more important than strength the choice might be Whiteheart W38-12.

Metal for die casting

Consider a requirement for a metal with a low melting point to use in die casting for the production of small components for toys, e.g. toy car steering wheels and drive shafts. The following are some of the melting points for metals from tables:

Aluminium 600°C,
Lead 320°C,
Magnesium 520°C,
Zinc 380°C.

Lead and zinc have the lowest melting points. If we add another requirement of reasonable strength in the as-cast condition then tables give for tensile strengths:

Lead 30 MPa,
Zinc 280 MPa.

Thus, taking strength into account, zinc would appear to be the choice. There are, however, often other considerations that have to be taken into account before a choice can be made.

Problems 1.2

1   Determine from tables the following data for materials at about 20°C:
   (a) the tensile strength of the carbon steel 1030 in the as-rolled condition,
   (b) the electrical conductivity on the IACS scale of the unalloyed aluminium 1060 in the annealed state,
   (c) the percentage elongation of the brass SCB1,
   (d) the yield stress of the manganese steel 120M19 in the quenched in tempered state,
   (e) the tensile strength of the stainless steel 302S31 in the soft state,
   (f) the density of the plastic ABS,
   (g) the plane strain fracture toughness of the plastic polypropylene,
   (h) the tensile strength of the elastomer natural rubber,
   (i) the thermal expansivity of the plastic high density polythene,
   (j) the tensile modulus of the plastic ABS.
 2 The rain water guttering used for buildings is required to have a high stiffness per unit weight so that it does not sag under its own weight. Use tables to either obtain the specific modulus or values of the modulus and density and hence compare cast iron, aluminium alloys, and the plastic PVC as possible materials.
 3 The panels used for car bodywork need to be in sheet form and stiff. Use tables to obtain modulus of elasticity values and hence compare carbon steel, an aluminium alloy, polypropylene, and a composite formed by polyester with 65% glass fibre cloth.

4 The fan in a vacuum cleaner needs to be made of a low-density material and a high tensile strength, i.e. a high specific strength. The aluminium alloy LM6 has been suggested because the fan could then be die cast. Use tables to obtain the specific strength of the material.

5 The plastic ABS has been suggested for use as the casing for a radio. The properties required include high stiffness. Determine from tables the modulus of elasticity and compare it with other plastics.

6 The material high tensile brass HTB1 has been suggested as a material for use as a marine propeller. Use tables to obtain values of its tensile strength, 0.1% proof stress and percentage elongation.

7 The alloy steels 150M36 and 530M40 have been suggested as the material for a car axle. Determine from tables the tensile strengths, yield stresses and percentage elongations for both materials, in the quenched and tempered state, so that a comparison can be made.

8 Table 1.14 is taken from *Newnes Engineering Materials Pocket Book* by W. Bolton (Heinemann-Newnes 1989, 1996, 2000) and gives data for cast aluminium alloys when sand cast and in the as-manufactured condition. Select a material which is likely to be tough.

Table 1.14 *Mechanical properties of cast aluminium alloys*

Material	Tensile strength MPa	Percentage elongation
LM4	140	2
LM5	140	3
LM6	160	5

9 Table 1.15 is taken from *Newnes Engineering Materials Pocket Book* by W. Bolton (Heinemann-Newnes 1989, 1996, 2000) and gives data for polymers. Select a material which will be stiff and not too brittle.

Table 1.15 *Mechanical properties of polymers*

Polymer	Tensile strength MPa	Tensile modulus GPa	Percent. elong.
ABS	17–58	1.4–3.1	10–140
Acrylic	50–70	2.7–3.5	5–8
Cellulose acetate	24–65	1.0–2.0	5–55
Cellulose acetate butyrate	18–48	0.5–1.4	40–90
Polyacetal, homopolymer	70	3.6	15–75
Polyamide, Nylon 66	80	2.8–3.3	60–300

1.3 Materials testing

This section is a discussion of standard tests that are used for the determination of the properties of materials. The standard tests used in Britain are those specified by the British Standards Association and European standards organisations; standards which are specified as BSEN are European standards which have been adopted as British Standards. In the United States, standards organisations are The American Society for Testing and Materials (ASTM), the American Iron and Steel Institute (AISI) and the Society of Automotive Engineers (SAE).

Why are standard tests necessary? Well they mean that a customer can buy material and be able to rely on the testing procedure used without requiring details of the tests used in order to interpret the results; it also makes it easier to compare data from different manufacturers.

1.3.1 Tensile test

In tensile test, measurements are made of the force required to extend a standard test piece at a constant rate, the elongation of a specified gauge length of the test piece being measured by some form of extensometer. British and European standards (BSEN 10002 Part 1) state the rate at which the stresses are applied should be between 2 and 10 MPa/s if the tensile modulus is less than 150 GPa and between 6 and 30 MPa/s if the tensile modulus is equal to or greater than 150 GPa. In order to eliminate any variations in tensile test data due to differences in the shapes of test pieces, standard shapes and sizes are adopted.

The test piece

Test pieces are said to be *proportional test pieces* if the relationship between the gauge length L_0 and the cross-sectional area A of the gauge length is:

$$L_0 = k\sqrt{A} \tag{27}$$

British and European standards specify the constant k should have the value 5.65 and the gauge length should be 20 mm or greater. With circular cross- sections of diameter d, $A = \frac{1}{4}\pi d^2$ and thus to a reasonable approximation this value of k gives:

$$L_0 = 5d \tag{28}$$

With circular cross-sectional areas which are too small for this value of k, a higher value may be used, preferably 11.3. With proportional test pieces, the same test results are given for the same test material when different size test pieces are used.

Figure 1.40 shows the standard size test pieces for round and flat samples of metals with Table 1.16 showing the standard dimensions that can be used. An important feature of the dimensions is the radius given for the shoulders of the test pieces.

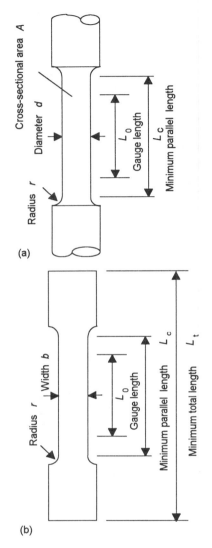

Figure 1.40 *Standard test pieces:*
(a) round, (b) flat

Very small radii can cause localised stress concentrations which
may result in the test piece failing prematurely.

For the tensile test data for the same material to give essentially
the same results, regardless of the length of the test piece used, it is
vital that the standard dimensions are adhered to.

Table 1.16 *Dimensions of standard test pieces:*
Flat test pieces

b mm	L_0 mm	L_c mm	L_f mm
20	80	120	140
12.5	50	75	87.5

Round test pieces (proportional)

d mm	A mm^2	L_0 mm	L_c mm
20	314.2	100	110
10	78.5	50	55
5	19.6	25	28

Note: $k = 5.85$.

Validity of tensile test data

The purpose of taking tensile test pieces and carrying out the tests
is to obtain data which enables judgements to be made about the
material from which the test piece was cut. The samples of a
material have to be taken in such a way that the properties deduced
from the tensile test are representative of the material as a whole.
There may, however, be problems in assuming this. The following
outlines some of these problems:

- *The properties may not be the same in all parts of a product*
  With a casting there may be different cooling rates in different
  parts of a casting, e.g. the surface compared with the core, or
  thin sections compared with thick sections. As a result, the
  internal structure of the material may differ in different parts
  of the casting. A tensile test piece cut from one part may not
  thus represent the properties of the entire casting. For the
  same reason, the properties of a separately cast test piece may
  not be the same as those of the cast product because the
  different sizes of the two lead to different cooling rates.

- *The size of an item affects its properties after heat treatment*
  If the mechanical properties of metals are looked up in tables
  you will often find that different values of the properties are
  quoted for different limiting ruling sections. The *limiting
  ruling section* is the maximum diameter of a round bar at the
  centre of which the specified properties may be obtained. The
  reason for the difference of mechanical properties of the same
  material for the different diameter bars is that during the heat
  treatment different rates of cooling occur at the centres of such
  bars due to their differences in sizes. Consequently there are

differences in microstructure and hence differences in mechanical properties, e.g. steel 070M55 with a limiting ruling section of 19 mm may have tensile strengths of 850 to 1000 MPa, with a limiting ruling section of 63 mm strength 777 to 930 MPa and for a limiting ruling section of 100 mm strength 700 to 830 MPa. The *limiting ruling section* is the maximum diameter of round bar at the centre of which the specified properties may be obtained.

• *The properties of a product may not be the same in all directions*
For example, with rolled sheet there is a directionality of properties with the tensile properties in the longitudinal, transverse and through the thickness of the sheet differing, e.g. with rolled brass strip we might have tensile strengths of 740 MPa in the direction of the rolling and 850 MPa at right angles to it.

• *The temperature in service of the product may not be the same as that of the test piece when the tensile test data was obtained*
The tensile properties of metals depend on temperature. In general, the tensile modulus and tensile strength both decrease with an increase in temperature; the percentage elongation tends to increase.

• *The rate of loading of a product may differ from that used with the test piece*
The data obtained from a tensile test is affected by the rate at which the test piece is stretched, so in order to give a standardised result the tests are carried out at a constant stress rate, between 2 and 20 MPa/s if the tensile modulus is less than 150 GPa and between 6 and 30 MPa/s if equal to or greater than 150 GPa.

Interpreting tensile test data

The results from tensile tests can be used to determine the safe stresses to which a material can be subject. Thus, the higher the yield stress of a metal the higher the stresses that it can be exposed to in service without yielding. Another important deduction that can be made is whether the material is brittle or ductile. A brittle material will show little plastic behaviour and have a low percentage elongation. A ductile material will show considerable plastic behaviour and have a high percentage elongation.

A consequence of the heat treatment and working of a material during the fabrication of products is a change in mechanical properties. Thus, tensile test data enable the effectiveness of heat treatments and the effects of working to be monitored.

Tensile tests for plastics

Tensile tests can be used with plastic test pieces to obtain stress–strain data. The stress–strain properties of plastics are much

Key point

The stress–strain properties of plastics are much more dependent than metals on the rate at which the strain is applied and temperature.

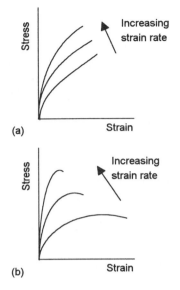

(a)

(b)

Figure 1.41 *Stress–strain graphs for plastics: (a) a brittle, (b) a ductile plastic*

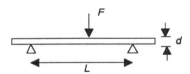

Figure 1.42 *A bend test that can be used with brittle materials to determine the flexural strength*

more dependent than metals on the rate at which the strain is applied. Thus, for example, the tensile test may indicate a yield stress of 62 MPa when the rate of elongation is 12.5 mm/min but 74 MPa when it is 50 mm/min. Also the form of the stress–strain graph may change with a ductile material at low strain rates becoming a brittle one at high strain rates. Figure 1.41 shows the general forms of stress–strain graphs for plastics at different strain rates. Another factor that is more marked than with metals is the effect of temperature on the properties of plastics.

Verification of tensile test equipment

The British and European standard BSEN 10002 Part 2 describes how the force readings given by such a machine can be verified. A given force indicated by the machine is compared with the true force indicated by a force-proving instrument or exerted by weights. Three series of measurements should be taken with increasing force, each series having at least five steps at regular intervals from 20% of the maximum range of the scale.

1.3.2 Bend tests

With many brittle materials, such as ceramics or glass, the conventional tensile test cannot be used because of the problems of preparing suitable test pieces and effectively holding them in the test machines. The presence of flaws at the surface, e.g. produced by the act of clamping them in the test machine, can so easily lead to failure. For such materials a *bend test* is used. BS 2782: Part 3 gives details of this test for rigid plastics. The materials are in the forms of beams and bent by three-point bending (Figure 1.42). The term *flexural strength* or *modulus of rupture* is used for the surface stress in the beam when breaking occurs. For a beam, the stress σ at a distance y from the neutral axis is related to the bending moment M and the second moment of area I of the beam section by $\sigma/y = M/I$. With three-point loading $M = FL/4$. For a rectangular cross-section beam, $I = bd^3/12$, where b is the breadth of the section and d its depth. The maximum stress will occur at the surface when $y = d/2$. Thus, for a rectangular cross-section beam:

$$\text{flexural strength} = \frac{3FL}{2bd^2} \qquad [29]$$

Flexural strength values for materials tend to be about twice their tensile strength values. For example, alumina (99% purity) has a flexural strength of 350 MPa, a tensile strength of 210 MPa and a compressive strength of 2625 MPa. Because cracks and flaws tend to close up in compression, brittle materials tend to be much stronger in compression than tension.

Ductility test

A simple test that is often quoted by suppliers of materials as a measure of ductility is the *bend test*. The test involves bending a

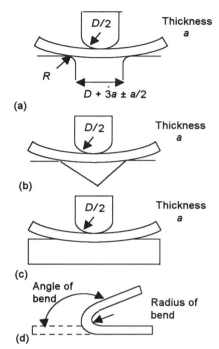

Figure 1.43 *Bend tests: (a) mandrel test, (b) on a vee block, (c) on a block of soft material, (d) the angle of bend*

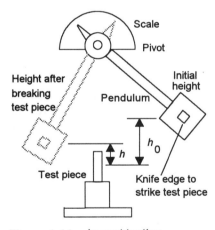

Figure 1.44 *Impact testing*

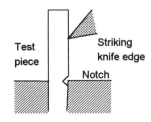

Figure 1.45 *Izod form of test*

sample of the material through some angle and determining whether the material is unbroken and free from cracks after the bending. There are a number of ways that can be used to carry out such a test, BS 1639 listing the British standards.

The simplest method is the mandrel form of test shown in Figure 1.43(a), this being suitable for medium and thin thickness sheet for angles of bend up to 120°. Figure 1.43(b) shows how the test can be conducted on a vee block, this being suitable for medium thickness sheet with bend angles up to 90°. Figure 1.43(c) shows the form of test possible for thin sheet with bend angles up to 90°, the material being bent on a block of soft material. Other methods can also be used, e.g. bending round a mandrel, free bending and pressure bending (see the British Standard for more details). The results of a bend test are quoted in terms of the angle of bend that can be withstood without breaking or cracking, as illustrated in Figure 1.43(d).

1.3.3 Impact tests

Impact tests are designed to simulate the response of a material to a high rate of loading and involve a test piece being struck a sudden blow. There are two main forms of test, the *Izod* and *Charpy* tests. Both tests involve the same type of measurement but differ in the form of the test pieces. Both involve a pendulum (Figure 1.44) swinging down from a specified height h_0 to hit the test piece and fracture it. The height h to which the pendulum rises after striking and breaking the test piece is a measure of the energy used in the breaking. If no energy were used the pendulum would swing up to the same height h_0 it started from, i.e. the potential energy mgh_0 at the top of the pendulum swing before and after the collision would be the same. The greater the energy used in the breaking, the greater the 'loss' of energy and so the lower the height to which the pendulum rises. If the pendulum swings up to a height h after breaking the test piece then the energy used to break it is $mgh_0 - mgh$.

Izod test pieces

With the Izod test the energy absorbed in breaking a cantilevered test piece is measured, as illustrated by Figure 1.45. The test piece has a notch and the blow is struck on the same face as the notch and at a fixed height above it. In the case of metals, the test pieces used are generally either 10 mm square or 11.4 mm in diameter if round. Figure 1.46 shows details of one form of the square test piece. With the 70 mm length the notch is 28 mm from the top of the piece. If a longer length is used then more than one notch is used. With a length of 96 mm there are two notches on opposite faces, one 28 mm from the top and the other twice that distance from the top. With a longer length test piece of 126 mm there are three notches, on three of the faces. The first notch is 28 mm from the top, the second twice that distance and third three times that distance from the top.

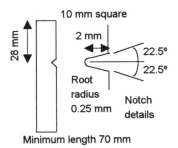

Figure 1.46 *Izod metal test piece*

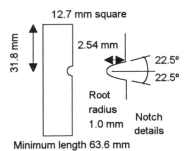

Figure 1.47 *Izod plastic test piece*

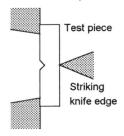

Figure 1.48 *Charpy test*

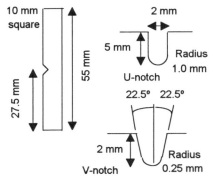

Figure 1.49 *Charpy metal test pieces*

In the case of plastics, the test pieces are 12.7 mm square or 12.7 mm by 6.4 to 12.7 mm depending on the thickness of the material concerned. (Figure 1.47). With metals the pendulum strikes the test piece with a speed of between 3 and 4 m/s, with plastics a lower speed of 2.44 m/s is used.

Charpy test pieces

With the Charpy test, the energy absorbed in breaking a test piece in the form of a beam is measured (Figure 1.48). The standard machine has the pendulum hitting the test piece with an energy of 300 ± 10 J. The test piece is supported at each end and notched at the midpoint between the two supports. The notch is on the face directly opposite to where the pendulum strikes the test piece. The British and European Standard is BSEN 10045.

For metals, the test piece generally has a square cross-section of side 10 mm and length 55 mm with there being 40 mm between the supports. Figure 1.49 shows details of such a test piece and the forms of notch commonly used. With the V-notch, reduced width specimens of 7.5 mm and 5 mm can be used. For plastics, the test pieces may be unnotched or notched. A standard test piece is 120 mm long, 15 mm wide and 10 mm thick for moulded plastics. For sheet plastics the width can be the thickness of the sheet with a U-shaped notch of width 2 mm and a radius of 0.2 mm at its base. For moulded plastics the depth below the notch is 6.7 mm, for sheet plastics either 10 mm or two-thirds of the sheet thickness.

Impact test results

In stating the results of impact tests it is vital that the form of test is specified. There is no reliable relationship between the values obtained by the two forms of test and so values from one test cannot be compared with those from the other. In addition there is no reliable relationship between the impact energies given for breaking test pieces of different sizes or different notches with the same test method. The impact energy value obtained for a material is influenced by such factors as the temperature, the speed of impact, any degree of directionality in the properties of the material from which the test piece was cut, and the thickness of the test piece.

For both the Izod and Charpy tests, the impact strengths for metals are expressed in the form of the energy absorbed, i.e. as, for example, 30 J. For plastics, with the Izod test the results are expressed as the energy absorbed in breaking the test piece divided by the width of notch and with the Charpy test as the energy absorbed divided by either the cross-sectional area of the specimen for unnotched test pieces or by the cross-sectional area behind the notch for notched test pieces, e.g. 2 kJ/m^2.

Interpreting impact test results

When a material is stretched energy is stored in the material. Think of stretching a spring or a rubber band. When the stretching

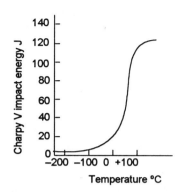

Figure 1.50   *Effect of temperature for a 0.2% carbon steel on the results of the Charpy test*

force is released the material springs back and the energy is released. However, if the material suffers a permanent deformation then all the energy is not released. The greater the amount of such plastic deformation the greater the amount of energy not released. Thus when a ductile material is broken, more energy is 'lost'.

The fracture of materials can be classified roughly as either brittle or ductile fracture. With brittle fracture there is little plastic deformation prior to fracture and so little energy is required to break the test piece. With ductile fracture the fracture is preceded by a considerable amount of plastic deformation and so more energy is required to break the test piece. Thus the impact test can be used to give information about the type of fracture that occurs. For example, Figure 1.50 shows the effect of temperature on the Charpy V-notch impact energies obtained for test pieces of a 0.2% carbon steel. Above about 0°C the material gives ductile failures, below that temperature, brittle failures. Such graphs have a great bearing on the use that can be made of the material, since at low temperatures the steel can be easily shattered by impact. Table 1.17 shows some typical impact strengths for metals.

Table 1.17   *Impact strengths at room temperature for metals*

Materials	Charpy V strength J
Aluminium, commercially pure, annealed	30
Aluminium–1.5% Mn alloy, annealed	80
hard	34
Copper, oxygen-free HC, annealed	70
Cartridge brass (70% Cu, 30% Zn), annealed	88
¾ hard	21
Cupronickel (70% Cu, 30% Ni), annealed	157
Magnesium–3% Al, 1% Zn alloy, annealed	8
Nickel alloy, Monel, annealed	290
Titanium–5% Al, 2.5% Sn, annealed	24
Grey cast iron	3
Malleable cast iron, Blackheart, annealed	15
Austenitic stainless steel, annealed	217
Carbon steel, 0.2% carbon, as rolled	50

The appearance of the fractured surfaces after an impact test also gives information about the type of fracture that has occurred. With a brittle fracture of metals, the surfaces are crystalline in appearance. With a ductile fracture, the surfaces are rough and fibrous in appearance. Also with ductile failure there is a significant reduction in the cross-sectional area of the test piece, but with brittle fracture there is virtually no such change.

With plastics, a brittle failure gives fracture surfaces which are smooth and glassy or somewhat splintered, with a ductile failure the surfaces often have a whitened appearance. Also, the change in cross-sectional area can be considerable with a ductile failure but negligible with brittle failure. At room temperature, plastics can be grouped with regard to their impact properties as:

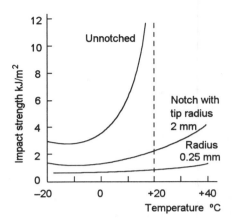

Figure 1.51 *Impact strength for propylene homopolymer*

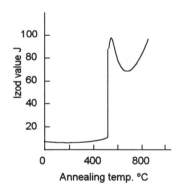

Figure 1.52 *Effect of annealing temperature on Izod values*

1 Essentially brittle materials even when unnotched, e.g. polystyrene and dry glass-filled nylon 6.6.

2 (a) Tough when unnotched and brittle when bluntly notched, e.g. polyvinyl chloride and acetal copolymer; (b) tough unnotched and brittle when sharply notched, e.g. high-density polyethylene and dry nylon 6.6.

3 Essentially tough under all conditions, e.g. low-density polyethylene and wet nylon 6.6.

Table 1.18 gives some typical values of impact strengths for plastics at 20°C for the Izod test with a notch tip having a radius of 0.25 mm and a depth of 2.75 mm. The impact properties of plastics vary quite significantly with temperature, changing in many cases from brittle to tough at some particular transition temperature. Figure 1.51 shows the impact strength as a function of temperature for propylene homopolymer (at 20°C, category 2(b) from above).

Table 1.18 *Impact strengths for plastics*

Material	Impact strength kJ/m^2
Polythene, high density	30
ABS	25
Nylon 6.6, dry	5
Polyvinyl chloride, unplasticised	3
Polystyrene	2

One use of impact tests is to determine whether heat treatment has been successfully carried out. A comparatively small change in heat treatment can lead to quite noticeable changes in impact test results. The changes can be more pronounced than changes in other mechanical properties such as percentage elongation or tensile strength. Figure 1.52 shows the effect on the Izod impact test results for cold-worked mild steel of annealing to different temperatures. The impact test can thus be used to indicate whether annealing has been carried out to the required temperature.

Example

A sample of unplasticised PVC has an impact strength of 3 kJ/m^2 at 20°C and 10 kJ/m^2 at 40°C. Is the material becoming more or less brittle as the temperature is increased?

Because there is an increase in the impact energy the material is becoming more tough.

1.3.4 Toughness test

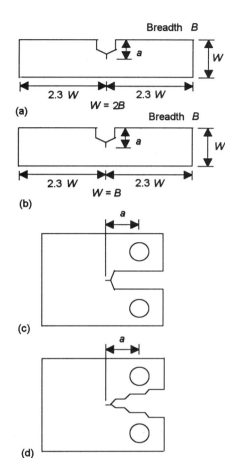

(a)

(b)

(c)

(d)

Figure 1.53 *Fracture toughness test pieces: (a) and (b) for three-point bending, (c) and (d) for tensile loading*

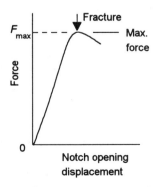

Figure 1.54 *Force–displacement graph*

Fracture toughness testing involves test pieces with sharp notches being strained until a crack propagates and the test piece fails. The problem in obtaining test pieces is producing the sharp notches. This is done by taking a test piece with a machined notch and then using a standardised pre-cracking procedure involving loading with an alternating stress (fatigue loading) in order to obtain a sharp crack at the base of the machined notch. Figure 1.53 shows forms of test piece, as specified by BS 7448.

The stress occurring in the region of a crack is greater than the applied stress and can be described by a factor termed the *stress intensity factor K*. For the various forms of test piece:

$$K = \sigma_n Y \sqrt{\pi a} \qquad [30]$$

where σ_n is the nominal stress, Y depends on the test piece geometry and is usually expressed in the form of a polynomial in a/W with a being the crack length and W the specimen width. For an infinite plate, $Y = 1$. Other Y values are correction factors because a plate is not infinite.

For the test, a steadily increasing force is applied to a test piece, the notch opening as a result, and the maximum value of the force recorded before breaking occurs as a result of the crack propagating. Figure 1.54 shows a typical graph of force plotted against notch opening displacement. For the standard test pieces, the critical value of K at which crack propagation occurs, i.e. K_c, is given by:

$$K_c = \frac{F_{max}}{BW^{0.5}} \times f\left(\frac{a}{W}\right) \qquad [31]$$

where the term for the function of (a/W), i.e. $f(a/W)$, is given by:

$$\frac{\left[2 + \frac{a}{W}\right]\left[0.886 + 4.64\frac{a}{W} - 13.32\frac{a^2}{W^2} + 14.72\frac{a^3}{W^3} - 5.6\frac{a^4}{W^4}\right]}{\left[1 - \frac{a}{W}\right]^{1.5}} \qquad [32]$$

The value of K_c depends on the thickness of the test plate. As thickness increases, K_c decreases to eventually become constant. This constant value is called the *plane-strain fracture toughness* K_{Ic}. With the test pieces, if $2.5(K_c/\sigma_{ys})^2$ is less than a, B or $W - a$, then the value quoted for K_c is the plane-strain fracture toughness K_{Ic}. If these conditions are not realised, then the test is considered to be not valid as a measure of the plane-strain fracture toughness. There are three different modes of testing. These are referred to as mode I, mode II or mode III (the first three Roman numerals). In mode I, the specimen is pulled at right angles to the plane of the crack; the arrangements shown in Figure 1.53 are all mode I tests. The plane strain fracture toughness measured from this mode is referred to a K_{Ic}. This is also normally referred to just as fracture toughness. Mode II and mode III tests are shear tests where the parts above and below the crack in, for example, Figure 1.53(c) are moved in opposing directions but parallel to the plane of the crack. Table 1.19 shows some typical values.

Table 1.19 *Fracture toughness values*

Material	Plane-strain fracture toughness (MPa m$^{1/2}$)
Metals	
Steels	80–170
Cast irons	6–20
Aluminium alloys	5–70
Copper alloys	30–120
Nickel alloys	100–150
Titanium alloys	50–100
Polymers	
Polyethylene, low density	1–2
Polyethylene, high density	1–5
Polypropylene	3–4
Nylons	3–5
Ceramics	
Alumina	3–5
Silicon nitride	4–5
Soda glass	0.7

1.3.5 Hardness tests

The *hardness* of a material may be specified in terms of some standard test involving indenting or scratching of the surface of the material, the harder a material the more difficult it is to make an indentation or scratch. There is no absolute scale for hardness, each hardness form of test having its own scale. Though some relationships exist between results on one scale and those on another, care has to be taken in making comparisons because the different types of test are measuring different things.

The most common form of hardness tests for metals involves standard indentors being pressed into the surface of the material concerned. Measurements associated with the indentation are then taken as a measure of the hardness of the surface. The Brinell test, the Vickers test and the Rockwell test are the main forms of such tests.

The Brinell test

With the Brinell test, a hardened steel ball is pressed for a time of 10 to 15 s into the surface of the material by a standard force (Figure 1.54). After the load and ball have been removed, the diameter of the indentation is measured. The Brinell hardness number, signified by HB, is obtained by dividing the size of the force applied by the surface area of the spherical indentation:

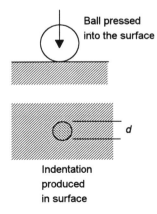

Ball pressed into the surface

Indentation produced in surface

Figure 1.54   *Brinell test: a hardened steel ball is pressed into the surface and the diameter of the indentation measured*

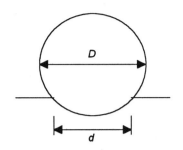

Figure 1.55 *Brinell test*

$$\text{Brinell hardness} = \frac{\text{applied force}}{\text{surface area of indentation}} \qquad [32]$$

The units used for the area are mm² and for the force kgf (1 kgf = 9.8 N and is the gravitational force exerted by 1 kg). The area can be obtained, from the measured diameter of the indentation and ball diameter (Figure 1.55), either by calculation or the use of tables:

$$\text{area} = \tfrac{1}{2}\pi D\left[D - \sqrt{(D^2 - d^2)} \;\right] \qquad [33]$$

where D is the diameter of the ball and d that of the indentation.

The diameter D (mm) of the ball used and the size of the applied force F (kgf units) are chosen, for the British Standard, to give F/D^2 values of 1, 5, 10 or 30 with the diameters of the balls being 1, 2, 5 or 10 mm. In principle, the same value of F/D^2 should give the same hardness value, regardless of the diameter of the ball used. It is necessary for the impression to have a diameter of between $0.25D$ and $0.50D$ if accurate values of the hardness are to be obtained. The F/D^2 value is thus chosen to fit the materials concerned, the harder the material the higher the value used. For steels and cast iron the value used for F/D^2 is 30, for copper alloys and aluminium alloys 10, for pure copper and aluminium 5 and for lead, tin and tin alloys 1.

The Brinell test cannot be used with very soft or very hard materials. In the one case the indentation becomes equal to the diameter of the ball and in the other there is either no or little indentation on which measurements can be based. Also, with very hard materials the material deforms the indenter. The Brinell test is thus limited to materials with hardnesses up to about 450 HB with a steel ball and 600 HB with a tungsten carbide ball.

The thickness of the material being tested should be at least ten times the depth of the indentation if the results are not to be affected by the thickness of the material. Also, because of the large depth of the indentation, it cannot be used on plated or surface hardened materials since the result will be affected by the underlying material.

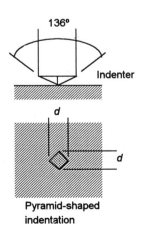

Figure 1.56 *Vickers hardness test: a diamond indenter is pressed into the surface and the diagonals of the indentation measured*

Example

For a Brinell test with a steel, what load should be used with a 10 mm diameter ball?

F/D^2 is taken as 30 and so $F = 30 \times 10^2 = 3000$ kg.

The Vickers test

The Vickers hardness test involves a diamond indenter, in the form of a square-based pyramid with an apex angle of 136°, being pressed under load for 10 to 15 s into the surface of the material under test (Figure 1.56). The result is a square-shaped impression.

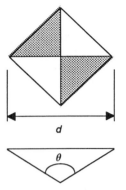

Figure 1.57 *Vicker's test*

After the load and indenter are removed the diagonals d of the indentation are measured. The Vickers hardness number (HV) is obtained by dividing the size of the force F, in units of kgf, applied by the surface area, in mm², of the indentation:

$$\text{Vickers hardness} = \frac{\text{applied force}}{\text{surface area of indentation}} \qquad [34]$$

The surface area can be calculated from the mean diagonal value, the indentation being assumed to be a right pyramid with a square base and an apex angle θ of 136° (Figure 1.57), or obtained by using tables:

$$\text{Area} = \frac{d^2}{2 \sin \theta/2} = \frac{d^2}{1.854} \qquad [35]$$

Thus the Vickers hardness HV is given by:

$$\text{HV} = \frac{1.854F}{d^2} \qquad [36]$$

The Vickers test has the advantage over the Brinell test of the increased accuracy that is possible in determining the diagonals of a square as opposed to the diameter of a circle. The square indentation produced for a particular material depends on the force used, but because the indentation is always the same square shape regardless of how big the force and the surface area is proportional to the force, the hardness value obtained is independent of the size of the force used. Typically a load of 30 kg is used for steels and cast irons, 10 kg for copper alloys, 5 kg for pure copper and aluminium alloys, 2.5 kg for pure aluminium and 1 kg for lead, tin and tin alloys. Up to a hardness value of about 300, the hardness value number given by the Vickers test is the same as that given by the Brinell test.

Microhardness test

This form of hardness testing is used when it is necessary to monitor the hardness distributions within a material on a microscale or determine the hardness of thin films or small objects. A diamond indenter is pressed against the surface by a load of between 1 and 1000 gf (gram-force, 1 gf = 0.009 81 N) to give a micro-indentation, the diagonals of which are measured by means of a microscope. The *Vickers microhardness test* uses a square-based pyramidal diamond with face angles of 136°, as with the Vickers hardness test outlined in Section 3.6.2 but with much smaller forces. The surface area is given, as before, by equation [35] and the hardness by equation [36]. The *Knoop test* uses a diamond pyramid indenter which is designed to give a long thin impression, the length being seven times greater than the width (Figure 1.58). The surface area of the indentation is given by:

Figure 1.58 *Knoop test: a diamond indenter is pressed into the surface and the length L measured*

$$\text{surface area} = \frac{L^2}{14.23} \qquad [37]$$

and thus the Knoop hardness HK by:

$$HK = \frac{14.23F}{L^2}$$ [38]

The shape of the Knoop indentation has the advantage over the Vickers microhardness test of giving a length which is more easily measured, being for the same hardness material and indenter load about three times greater than the diagonal length given by the Vickers test. The range of loads used with the Knoop test is similar to that used with the Vickers test.

The Rockwell test

The Rockwell hardness test differs from the Brinell and Vickers hardness tests in not obtaining a value for the hardness in terms of the area of an indentation but using the depth of indentation, this depth being directly indicated by a pointer on a calibrated scale. The test uses either a diamond cone or a hardened steel ball as the indenter (Figure 1.59). A preliminary force is applied to press the indenter into contact with the surface. A further force is then applied and causes the indenter to penetrate into the material. The additional force is then removed and there is some reduction in the depth of the indenter due to the deformation of the material not being entirely plastic. The difference in the final depth of the indenter and the initial depth, before the additional force was applied, is determined. This is the permanent increase in penetration e due to the additional force. The Rockwell hardness number is then given by:

Rockwell hardness number (HR) = $E - e$ [39]

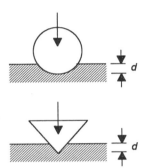

Figure 1.59 *Rockwell test: a hardened steel ball or diamond iindenter s pressed against the surface and the depth of indentation measured*

where E is a constant determined by the form of the indenter. For the diamond cone indenter E is 100, for the steel ball 130. There are a number of Rockwell scales (Table 1.20), the scale being determined by the indenter and the additional force used. In any reference to the results of a Rockwell test the scale letter must be quoted. For metals the B and C scales are probably the most commonly used ones.

For the most commonly used indenters with the Rockwell test the size of the indentation is rather small. Localised variations of structure, composition and roughness can thus affect the results. The Rockwell test is, however, more suitable for workshop or 'on-site' use as it less affected by surface conditions than the Brinell or Vickers tests which require flat and polished surfaces to permit accurate measurements. A variation of the Rockwell test has to be used for thin sheet, this test being referred to as the *Rockwell superficial hardness test*. Smaller forces are used and the depth of indentation which is correspondingly smaller is measured with a more sensitive device. The initial force used is 29.4 N. Table 1.21 indicates the scales given by this test; the N scales are for materials that would have been tested on the C scale, the T scales for those on the B scale.

Table 1.20 *Rockwell hardness scales*

Scale	Indenter	Additional load kg	Typical applications
A	Diamond	60	Extremely hard materials, e.g. tool steels
B	Ball 1.588 mm dia.	100	Softer materials, e.g. Cu alloys, Al alloys, mild steel
C	Diamond	150	Hard materials, e.g. steels, hard cast irons, alloy steels
D	Diamond	100	Medium case hardened materials
E	Ball 3.175 mm dia.	100	Soft materials, e.g. Al alloys, Mg alloys, bearing metals
F	Ball 1.588 mm dia.	60	As E, the smaller ball being more appropriate where inhomogeneities exist
G	Ball 1.588 mm dia.	150	Malleable irons, gun metals, bronzes
H	Ball 3.175 mm dia.	60	Soft aluminium, lead, zinc, thermoplastics
K	Ball 3.175 mm dia.	150	Aluminium and magnesium alloys
L	Ball 6.350 mm dia.	60	Soft thermoplastics
M	Ball 6.350 mm dia.	100	Thermoplastics
R	Ball 12.70 mm dia.	60	Very soft thermoplastics

Note: the diameter of the balls arise from standard sizes in inches, 1.588 mm being 1/16 in, 3.175 mm being 1/8 in, 6.350 mm being 1/4 in, and 12.70 mm being 1/2 in.

Table 1.21 *Rockwell superficial hardness scales*

Scale	Indenter	Additional load in kg
15-N	Diamond	15
30-N	Diamond	30
45-N	Diamond	45
15-T	Ball 1.588 mm dia.	15
30-T	Ball 1.588 mm dia.	30
45-T	Ball 1.588 mm dia.	45

Key points

Because the different forms of hardness test are concerned with different measurements, the values given by the different methods can differ for the same material. There are no simple theoretical relationships between the various hardness scales.

Comparison of the different hardness scales

The Brinell and Vickers tests both involve measurements of the surface area of indentations, the forms of the indenters used being different. The Rockwell test involves measurements of the depth of penetration of indenters. Because the tests are concerned with different measurements as an indication of hardness, the values given can differ for the same material.

There are no simple theoretical relationships between the various hardness scales, though some simple approximate, experimentally derived, relationships have been obtained. Different relationships, however, hold for different metals. The relationships are often presented in the form of tables. Table 1.22 shows part of a table for steels. Up to a hardness value of 300 the Vickers and Brinell values are almost identical.

There is an approximate relationship between hardness values and tensile strengths:

$$\text{tensile strength} = k \times \text{hardness} \qquad [40]$$

where k is a constant for a particular material. Thus for annealed steels the tensile strength in MPa is about 3.54 times the Brinell hardness value, and for quenched and tempered steels 3.24 times the Brinell hardness value. For brass the factor is about 5.6 and for aluminium alloys about 4.2.

Table 1.22 *Comparison of hardness scales for steels*

Brinell value	Vickers value	Rockwell B	Rockwell C
112	114	66	
121	121	70	
131	137	74	
140	148	78	
153	162	82	
166	175	86	4
174	182	88	7
183	192	90	9
192	202	92	12
202	213	94	14
210	222	96	17
228	240	98	20
248	248	102	24
262	263	103	26
285	287	105	30
302	305	107	32
321	327	108	34
341	350	109	36
370	392		40
390	412		42
410	435		44
431	459		46
452	485		48
475	510		50
500	545		52

Example

An aluminium alloy has a hardness of 45 HB when annealed and 100 HB when solution treated and precipitation hardened. Estimate the tensile strengths of the alloy in these conditions if a factor of 4.2 is assumed.

Using a factor of 4.2, then the tensile strength in the annealed condition is 4.2 × 45 = 189 MPa. For the heat-treated condition it is 4.2 × 100 = 420 MPa. The measured values were 180 MPa and 430 MPa.

Hardness measurements with plastics

The Brinell, Vickers and Rockwell tests can be used with plastics. The Rockwell test with its measurement of penetration depth, rather than surface area, is more widely used. Scale R is a commonly used scale.

Another test that is used with plastics involves an indenter, a ball of diameter 2.38 mm, being pressed against the plastic by an initial force of 0.294 N for 5 s and then an additional force of 5.25 N being applied for 30 s. The difference between the two penetration depths is measured and expressed as a *softness number*. This is just the depth expressed in units of 0.01 mm. Thus a difference in penetration of 0.05 mm is a softness number of 5. The test is carried out at a temperature of 23 ± 1°C.

Another form of test that is used is the *Shore durometer*. This is a hand-held device which involves a rounded indenter being pressed into the surface of the material under the action of a spring or weight, a pointer then registering the hardness value on a scale. A number of scales are used, ranging from Shore A for the very soft to Shore D for the very hard.

The Moh scale of hardness

A completely different form of hardness test, called the Moh scale, is based on assessing the resistance of a material to being scratched. Ten styli with points made of different materials are used. The styli materials are arranged in a scale so that each one will scratch the one preceding it in the scale but not the one that follows it. The scale and materials are:

1	Talc
2	Gypsum
3	Calcspar
4	Fluorspar
5	Apatite
6	Felspar
7	Quartz
8	Topas
9	Corundum
10	Diamond

Thus, for example, felspar will scratch apatite but not quartz. Diamond will scratch all the materials while talc will scratch none of them. In this test the various styli are used until the lowest number stylus is found that will just scratch it. The hardness number is then one less since it is the number of the stylus that just fails to scratch the material. For example, glass can just be scratched by felspar but not by apatite. The glass thus has a hardness number of 5.

1.3.6 Electrical tests

The *electrical resistivity* or *conductivity* of a material requires a measurement of the resistance of a strip or block of the material. The British Standard for resistivity measurements with metals is BS 5714. In the case of metals the resistivity is very low and so the resistance to be measured can be low. For example, the resistance of a 1 m length of copper wire with a diameter of 1 mm is about 0.03 Ω at 20ºC. Such a resistance is not easy to measure, since the means by which it is connected to the measurement system can have resistances of the same order of size or even larger. A smaller gauge wire of 0.1 mm gives a resistance of about 2.1 Ω and is easier to measure. For routine measurements with resistances greater than 1 Ω the test piece can be what is termed a *two-terminal device*, i.e. there is just a single terminal at each end of the test piece for connections. For resistances less than 1 Ω the test piece should be a *four-terminal device*, i.e. there are two terminals at each end. This means that the circuit connections to each end give less ambiguity as to between which points measurements are being made. For routine resistance measurements the method used should be capable of an accuracy of at least ±0.30%. The method used for such resistance measurements is likely to be a form of Wheatstone resistance bridge, with possibly a Kelvin double bridge for small resistances.

In addition to measuring the resistance, the length and cross-sectional area of the test piece is required. The area can be obtained by direct measurement; however, an alternative method which is often used for small cross-sections is to weigh the test piece and calculate the area from a knowledge of the density and length, the area being mass/(density × length).

Since resistivity changes with temperature and the dimensions of the test piece also change with temperature, it is important that the temperature t at which a measurement is made is noted. The following equation can then be used to correct the result to the reference temperature t_0 at which the result is required:

$$\rho_{t_0} = \frac{\rho_t}{1 + (a + \gamma)(t - t_0)}$$ [41]

where a is the temperature coefficient of resistance at the reference temperature and γ the coefficient of linear expansion.

1.3.7 Chemical tests

Metallic materials corrode in moist air with some metals corroding at a faster rate than others. Corrosion testing can involve field trials or suitable conditions simulated in the laboratory. Essentially the tests are the observation of what happens to the metals over a period of time. Metals exposed to corrosive environments are often protected by being coated with a material such as paint. Tests are then used to investigate the weathering characteristics of the painted material. An accelerated weathering process is often used with exposure to radiation from an electric arc, with intermittent exposure to a spray of water to simulate rain. BS 6917 gives details of corrosion testing in artificial atmospheres, indicating the requirements for specimens, apparatus and procedures. BS 3900: Part G gives details of environmental tests on paint films.

The use of metals at high temperatures is often restricted by surface attack or scaling which gradually reduces the

---

Key points

For materials such as plastics or ceramics, the problem in making electrical measurements is that they have very high resistivities. This can present the problem that the surface layers, perhaps as a result of the absorption of moisture, might have a significantly lower resistivity than the bulk of the material and so the value indicated by the measurement is not that of the bulk material. Polymers also present the problem that when a voltage is applied across a sample that the current through the material slowly decreases with time. Thus resistivity measurements need to have a time quoted with them, e.g. the value one minute after the application of a voltage.

cross-sectional area and hence the stress-bearing ability of the item. The build-up of oxide layers at high temperatures is very much influenced by the environment, e.g. metal pipes exposed to superheated steam or hot gases from furnaces. Materials are tested by exposing them to such situations and measuring the reduction in the metal thickness as a consequence of the corrosive attack.

Plastic materials may dissolve in some liquids or absorb sufficient of the liquid to have their properties changed. When absorption occurs the plastic becomes permeable to the liquid, i.e. liquid can leak through it. This permeability is of vital concern if the plastic is being considered for used as a container for liquids, e.g. a soft drink bottle.

Plastics are not generally subject to corrosion in the same way as metals but they can be adversely affected by weathering, i.e. exposure to light, heat, rain, sun. This can show itself as a fading of the colour of the plastic and/or a loss of flexibility. Tests are used to determine the weathering resistance of plastics with them being subject to an accelerated weathering process involving exposure to radiation from an electric arc, with intermittent exposure to a spray of water to simulate rain. The tests tend to be comparative ones with standard colours/materials being simultaneously exposed and performances compared. There are a number of standard tests (BS 2782: Part 5) for such factors as colour fastness in water, effects of exposure to damp, heat, water spray and salt mist, changes in colour and variations in properties after exposure to daylight under glass, natural weathering or artificial light.

Example

The following are corrosion test results for different metals suspended in the hot fumes from the combustion of fuel oils.

Material	Corrosion rate in mm/year
Steel: 25% Cr, 20% Ni	Completely corroded
35% Cr–65% Ni alloy	Completely corroded
50% Cr–50% Ni alloy	4
60% Cr–40% Ni alloy	2

The 50% Cr–50% Ni alloy has a tensile strength of 550 MPa, yield stress 340 MPa and Charpy impact strength 37 J. The 60% Cr–40% Ni alloy has a tensile strength of 760 MPa, yield stress 590 MPa and Charpy impact strength 7 J. On the basis of the above data discuss the possible choice of a metal for pipes which would be exposed to the fumes.

On the basis of the corrosion tests the choice is between the 50% Cr–50% Ni and the 60% Cr–40% Ni alloys, with the latter having better corrosion properties. The mechanical properties indicate that the 60% Cr–40% Ni alloy is stronger but considerably less tough. The more tough and shock resistant properties are likely to mean that the 50% Cr–50 % Ni alloy is the choice.

Example

Tables indicate that a weight loss of 1 mg per exposed area of 0.01 m^2 (1 dm^2) per day for cast iron is a penetration of corrosion into the cast iron surface of 4.65 μm per year. What will be the penetration of corrosion into a cast iron product if it suffers a weight loss of 0.5 mg/dm^2 in a day?

Since 1 mg/dm^2 in a day is 4.65 μm per year then 0.5 mg/dm^2 is 2.325 μm per year.

1.3.8 Magnetic tests

BS 6404: Part 4 gives details of the standard test methods used for the determination of magnetic properties of materials. For magnetic field strengths below 5 kA/m the *ring method* can be used to determine the magnetic properties of a material. The material is in the form of an anchor ring (Figure 1.60) with a magnetising winding of N_1 turns wrapped round it. A measured current I is passed through the coil and varied by means of a resistor and also reversed by a reversing switch in order to give the complete hysteresis cycle.

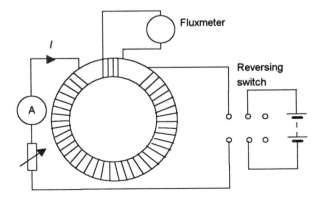

Figure 1.60 *Ring method of determining magnetic properties*

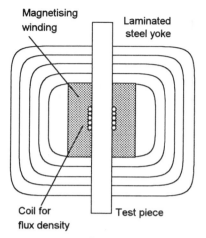

Figure 1.61 *Permeameter method for magnetic properties*

The magnetic field strength H is then $N_1 I/L$, where L is the mean circumference of the ring. The magnetic flux density in the material is determined by means of a coil of N_2 turns. This is connected to a ballistic galvanometer, fluxmeter or flux integrator to give a reading of the magnetic flux density B. The results can then be plotted as a graph of magnetic flux density against the magnetic field strength (as in Figure 1.30)

There are a number of forms of *permeameter methods* that can be used to determine magnetic properties. The National Physical Laboratory (NPL) form of this type of test can be used for magnetic field strengths between 2 kA/m and 200 kA/m. With this method (Figure 1.61), the test specimen is a bar at least 250 mm long and generally of square cross-section 10 mm by 10 mm. It is clamped between two massive laminated steel yokes to give a closed path for the magnetic flux. The magnetising field is provided by means of a winding formed on a bobbin surrounding the test piece. The magnetic flux produced in the test piece is determined by a coil wrapped round it and connected to a ballistic galvanometer, fluxmeter or flux integrator.

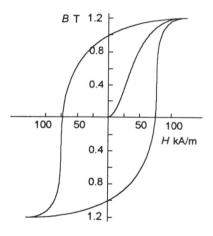

Figure 1.62 *Example*

Example

Determine the remanence, coercive field, initial permeability, maximum permeability and saturation magnetisation for the material giving the hysteresis loop shown in Figure 1.62.

The remanence is the value of the flux density remaining in the material when the magnetising field has been reduced to zero. It is thus 1.0 T. The coercive field is the value of the field strength needed to give zero flux density in the material. It is thus 75 kA/m. The initial permeability is obtained from the initial slope of the graph line when the unmagnetised material is first magnetised. μ is thus about 0.4/25 000 = 0.000 004 T/A m^{-1} and so the initial relative permeability μ_r is about 0.000 004/μ_0 = 0.000 004/4$\pi \times 10^{-7}$ = 3.2. Maximum permeability is the maximum gradient of the initial graph line, about 1.2/50 000 = 0.000 012 T/A m^{-1} and a relative permeability of 19.2. The saturation magnetisation for the material is the maximum value of the magnetic flux density and is thus 1.2 T.

1.3.9 Case study

In 1985 a resident in one of the apartments in Newnham House in Putney, London, noticed a smell of gas. A few minutes later there was a huge explosion which demolished the six apartments in Newnham House and broke windows in surrounding buildings. Only one of the nine people in the apartments survived. Investigation shows that the origin of the leak was the gas supply main under the access road at the back of Newnham House. This was grey cast-iron pipe buried at a depth of 0.8 m below the road. The road had originally been reinforced concrete panels but it had been dug up on a number of occasions and when the surface had been reinstated it had not been reinforced. As a consequence the road had settled to leave depression which was as much as 200 mm in one place. The ground below the gas main was also found to be poorly consolidated. As a result, the gas pipe was subject to a bending moment. Tensile test pieces were cut from the pipe material and gave the result of tensile strengths between 131 and 138 MPa. For a beam which is subject to bending, the maximum stress is My/I, where M is the bending moment, y the maximum distance from the neutral axis and I the second moment of area. For the pipe we have $y = R$, the external radius, and $I = \pi/4(R^4 - r^4)$. Thus the maximum stress is $4MR/[\pi(R^4 - r^4)]$. The pipe had $R = 86$ mm and $r = 75$ mm and so a maximum stress of 138 MPa would be obtained with a bending moment of 29 kN m. Consideration of the loading acting on the pipe, particularly when lorries pass along the road, indicates that the bending moment acting on the pipe could be greater than this and so this could be responsible for the pipe breaking.

Problems 1.3

1 The following results were obtained from a tensile test of an aluminium alloy before it broke. The test piece had a diameter of 11.28 mm and a gauge length of 56 mm. Plot the stress–strain graph and determine (a) the tensile modulus, (b) the 0.1% proof stress.

Load/kN	0	2.5	5.0	7.5	10.0	12.5	15.0	17.5
Ext./mm	0	1.8	4.0	6.2	8.4	10.0	12.5	14.6

Load/kN	20.0	22.5	25.0	27.5	30.0	32.5	35.0
Ext./mm	16.3	19.0	21.2	23.5	25.7	28.1	31.5

Load/kN	37.5	38.5	39.0	39.0
Ext./mm	35.0	40.0	61.0	86

2 The following results were obtained from a tensile test of a polymer. The test piece had a width of 20 mm, a thickness of 3 mm and a gauge length of 80 mm. Plot the stress–strain graph and determine (a) the tensile strength, (b) the secant modulus at 0.2% strain.

Load/N	0	100	200	300	400	500	600	650	630
Ext./mm	0	0.08	0.17	0.35	0.59	0.88	1.33	2.00	2.40

3 The following results were obtained from a tensile test of a steel. The test piece had a diameter of 10 mm and a gauge length of 50 mm. Plot the stress–strain graph and determine (a) the tensile strength, (b) the yield stress, (c) the tensile modulus.

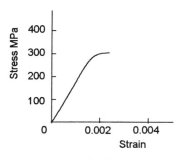

Figure 1.63   *Problem 6*

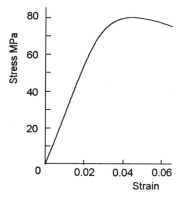

Figure 1.64   *Problem 7*

Load/kN	0	5	10	15	20	25
Ext./mm	0	0.016	0.033	0.049	0.065	0.081

Load/kN	30	32.5	35.8
Ext./mm	0.097	0.106	0.250

4 A flat tensile test piece of steel has a gauge length of 100.0 mm. After fracture, the gauge length was 131.1 mm. What is the percentage elongation?

5 The following data was obtained from a tensile test on a stainless steel test piece. Determine (a) the limit of proportionality stress, (b) the tensile modulus, (c) the 0.2% proof stress.

Stress/MPa	0	90	170	255	345	495	605	700
Strain/$\times 10^{-4}$	0	5	10	15	20	30	40	50

Stress/MPa	760	805	845	880	895
Strain/$\times 10^{-4}$	60	70	80	90	100

6 Estimate from the stress–strain graph for cast iron in Figure 1.63 the tensile strength and the limit of proportionality.

7 Estimate from the stress–strain graph for a sample of nylon 6 given in Figure 1.64 the tensile modulus and the tensile strength.

8 A round test piece of cast iron has a gauge length of 100.00 mm. After fracture the gauge length is 118.55 mm. What is the percentage elongation?

9 The effect of working an aluminium alloy (1.25% Mn) is to change the tensile strength from 110 MPa to 180 MPa and the percentage elongation from 30% to 3%. What is the effect of the working on the properties of the material?

10 An annealed titanium alloy has a tensile strength of 880 MPa and a percentage elongation of 16%. An annealed nickel alloy has a tensile strength of 700 MPa and a percentage elongation of 35%. Which alloy is (a) the stronger, (b) the more ductile in the annealed condition?

11 Cellulose acetate has a tensile modulus of 1.5 GPa and polythene a modulus of 0.6 GPa. Which of the two plastics will be the stiffer?

12 The following are Izod impact energies at different temperatures for samples of annealed cartridge brass (70% Cu–30% Zn). What can be deduced from the results?

Temperature °C	+27	−78	−197
Impact energy J	88	92	108

13 The following are Charpy V-notch impact energies for annealed titanium at different temperatures. What can be deduced from the results?

Temperature °C	+27	−78	−196
Impact energy J	24	19	15

14 The following are Charpy impact strengths for nylon 6.6 at different temperatures. What can be deduced from the results?

Temperature °C	−23	−33	−43	−63
Impact strength kJ/m²	24	13	11	8

15 The impact strengths of samples of nylon 6, at a temperature of 22°C, are found to be 3 kJ/m² in the as-moulded condition but 25 kJ/m² when the sample has gained 2.5% in weight through water absorption. What can be deduced from the results?

16 With the Vickers hardness test a 30 kg load gave for a sample of steel an indentation with diagonals having mean lengths of 0.530 mm. What is the hardness?

17 With the Vickers hardness test a 30 kg load gave for a sample of steel an indention with diagonals having mean lengths of 0.450 mm. What is the hardness?

18 With the Vickers hardness test a 10 kg load gave for a sample of brass an indentation with diagonals having mean lengths of 0.510 mm. What is the hardness?

19 With the Brinell hardness test a 10 mm diameter ball and 3000 kg load gave an indentation with a diameter of 4.10 mm. What is the hardness?

20 With the Brinell hardness test a sample of cold-worked copper with a 1 mm diameter ball and 20 kg load gave an indentation of diameter 0.630 mm. What is the hardness?

21 After 4000 hours' exposure to the fumes in an oil-fired furnace samples of metals were found to show the following corrosion rate: Steel 25% Cr–12% Ni 0.11 mm/year, Steel 25% Cr–20% Ni 0.28 mm/year, 65% Ni–35% Cr alloy 0.02 mm/year. Explain the significance of the data?

22 Oxide penetration on steels exposed for 5000 hours to steam at about 600°C was found to be as follows: 0.11% C steel 0.40 mm, 0.34% C steel 0.25 mm, 1.24% Cr–0.5% Mo–1.4% Si steel 0.12 mm, 2.25% Cr–0.5% Mo–0.75% Si steel 0.09 mm. Discuss the significance of the data.

23 The corrosion rate for mild steel test plates was found to give averages of 0.050 mm per year in rural surroundings, 0.070 mm per year in marine surroundings and 0.150 mm per year in heavy industrial surroundings. Discuss the significance of the data.

24 Specify the type of test that can be used in the following instances:
   (a) A storekeeper has mixed up two batches of steel, one batch having been surface hardened and the other not. How could the two be distinguished?
   (b) What test could be used to check whether tempering has been correctly carried out for a steel?

(c) A plastic is modified by the inclusion of glass fibres. What test can be used to determine whether this has made the plastic stiffer?

(d) What test could be used to determine whether a metal has been correctly heat treated?

(e) What test could be used to determine whether a metal is in a suitable condition for forming by bending?

25 Determine the remanence, coercive field, initial relative permeability, maximum relative permeability and saturation magnetisation for the material giving the hysteresis loop shown in Figure 1.65.

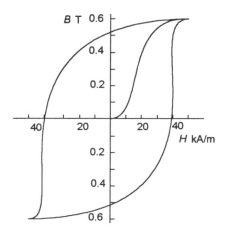

Figure 1.65   *Problem 25*

26 Determine the remanence, coercive field, initial relative permeability, and saturation magnetisation for the material giving the hysteresis loop shown in Figure 1.66.

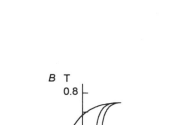

Figure 1.66   *Problem 26*

2 Structure of materials

Summary

Why are metals good conductors of electricity? Why are polymers weaker and less stiff than metals? Why are ceramics so stiff? How can we make metals more ductile? In order to obtain an understanding of why materials behave as they do it is necessary to have an appreciation of the structure of materials. This chapter considers the basic structural elements of materials and how atoms and molecules can be arranged in solids, hence leading to the structure of metals, alloys, polymers and ceramics. Consideration of the charge carriers in materials leads to the development of models to describe electrical and magnetic properties.

Objectives

By the end of this part, the reader should be able to:

- appreciate the basic features of the Bohr and quantum number model of atoms, explaining the metallic, ionic, covalent, van der Waals and hydrogen bonds;
- describe the form of the basic close-packed crystalline structures and interstitial sites;
- identify the mechanism of diffusion;
- explain the behaviour of metals in terms of slip and the movement of dislocations;
- know the forms alloys can take and use phase diagrams to explain how alloys solidify;
- explain the behaviour of polymers in terms of their chain structures and ceramics in terms of their network structures;
- explain the electrical conductivity of conductors, semiconductors and insulators and use the energy band model;
- explain the behaviour of dielectrics;
- explain the atomic origins of magnetism.

2.1 Basic structure of materials

To gain an understanding of the properties of materials, we need to consider their structure. All elements have their particular properties by virtue of how their atoms are made up from the basic building blocks of protons, neutrons and electrons. Solids contain vast numbers of atoms and we need to consider the forces between the constituent particles and the ways in which they are packed together in solids. This can lead us to an understanding of mechanical properties and so how materials behave when stretched or compressed.

This section follows a brief discussion of atomic structure with a consideration of the bonding between particles in solids and their arrangement in crystalline structures, hence leading to a consideration of the basic structure of metals, polymers and ceramics and how the properties of such materials can be explained in terms of structure and modified by structural changes.

2.1.1 The structure of atoms

The atom consists of a positively charged nucleus surrounded by negatively charged electrons. The nucleus contains neutrons, particles with no charge, and positively charged protons. The electric charge carried by each electron and proton is 1.6×10^{-19} C, there being equal numbers of electrons and protons in a neutral atom. The *atomic number* of an element is equal to the number of protons in each atom, each element having its own characteristic atomic number. A material that is made up of just atoms all with the same atomic number is called an *element*, e.g. the element oxygen has eight protons and eight neutrons and so has an atomic number of 8. All the atoms of a particular element do not necessarily contain the same number of neutrons, though they have the same number of protons; atoms of the same element with different numbers of neutrons are termed *isotopes*.

The *atomic mass* of an atom is concentrated in the nucleus, the mass of each proton and each neutron being 1.67×10^{-24} kg and that of each electron 9.11×10^{-28} kg; neutrons and protons thus having masses which are nearly 2000 times that of the electron. The total number of protons and neutrons in the nucleus is termed the *mass number*. The values quoted for atomic masses are in terms of a unit called the *atomic mass unit (a.m.u.)*, this being one-twelfth of the mass of an atom of carbon which contains six protons and six neutrons. The a.m.u. is 1.6598×10^{-27} kg.

The term *mole* is used for the amount of a substance which contains as many elementary entities as there are atoms in 0.012 kg of the form of carbon having six protons and six neutrons in each atom. Equal molar masses of any substances always contain the same number of elementary entities, this number being termed the *Avogadro number* 6.023×10^{23}.

Example

A neon atom has an atomic number of 10 and an atomic mass number of 21. How many (a) protons, (b) neutrons, (c) electrons does a neutral atom of neon contain?

(a) The number of protons is equal to the atomic number and so is 10.

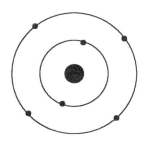

Figure 2.1 *Bohr model for a carbon atom: a positively charged nucleus around which 6 electrons orbit with 2 in one fixed orbit and 4 in another*

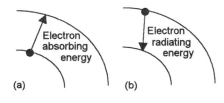

Figure 2.2 *Bohr model of an atom: (a) absorbing energy by an electron jumping to a more outer orbit, (b) radiating energy by falling back to a more inner orbit*

Key points

- Atomic electrons are in their ground state when in the fixed orbits and can only absorb energy if it is enough to enable them to move to a vacant space in a further out orbit, emitting energy when they fall back to a more inner orbit.
- The *valence* of an atom is equal to the number of electrons in the outermost shell or the number of electrons needed to fill the outermost shell.

(b) The atomic mass number is equal to the number of protons plus the number of neutrons. Thus the number of neutrons is 21 – 10 = 11.
(c) The number of electrons in a neutral atom is equal to the number of protons and is thus 10.

Electrons in atoms – Bohr model

The negatively charged electrons are held to the positively charged nucleus by an electrostatic force of attraction (opposite charges attract, like charges repel). For such forces to result in stable atoms, Bohr proposed a model for atomic structure in which:

- Electrons move in fixed orbits round the nucleus, like planets orbiting the sun. Only certain orbits are possible and only when electrons are in these orbits is there stability (Figure 2.1). The term shell is often used for an orbit.

- Electrons in orbits close to the nucleus have stronger forces of attraction holding them in orbit than electrons which are in orbits further out and thus each orbital electron possesses a specific amount of energy. We can thus consider each orbit as representing for electrons a fixed *energy level*. The *ground state* is defined as being the energy situation existing when electrons are not excited, i.e. have not received an external input of energy.

- An atomic electron can only emit or absorb energy when making a transition from one possible orbit to another (Figure 2.2).

- To jump between energy levels an electron must collect or emit exactly the right amount of energy. When an electron is excited by an input of the right amount of energy, it moves to an orbit of greater radius. When the electron subsequently moves back, it emits the energy as a packet of energy. Energy is thus absorbed or emitted in discrete packets termed *quanta*. In terms of the frequency f of the radiation emitted or received, one quantum of energy is hf, where h is called Planck's constant and has a value of 6.62×10^{-34} J s.

With the Bohr model the number of electrons that can be accommodated in each shell is limited. The innermost shell, termed the K shell, can accommodate 2, the next shell L can accommodate 8, the next shell M can accommodate 18. Thus hydrogen which has just one electron has, when in the ground state, the electron in the K shell. Helium which has two electrons has both the electrons in the K shell. Lithium which has three electrons has two filling its K shell and one in the L shell.

The *valence* of an atom is a measure of the ability of the atom to enter into chemical combination with other elements. In terms of

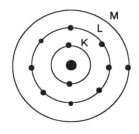

(a) One electron in
     outer shell

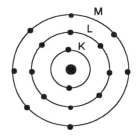

(b) Seven electrons in
     outer shell

(c) Six electrons in
     outer shell

(d) Four electrons in
     outer shell

Figure 2.3 *Bohr models:*
(a) sodium, (b) chlorine, (c) oxygen,
(d) carbon

the Bohr model it is equal to the number of electrons in the outermost shell or the number of electrons needed to fill the shell, the aim of chemical combinations being to end up with full shells of electrons. Thus sodium (Figure 2.3(a)) has a valence of one since it has just one electron in its outermost shell. Chlorine (Figure 2.3(b)) is just one electron short of a complete shell and so has a valence of one. Oxygen (Figure 2.3(c)) has two electrons in the inner shell and six in the outer shell. As it needs two further electrons to complete its outermost shell it has a valence of two. Carbon (Figure 2.3(d)) has two electrons in its inner shell and four electrons in its outer shell. This means it has a valence of four.

Most of the volume of an atom is occupied by the orbiting electrons and it is the electron orbits which define the size of an atom. When two atoms are brought close together it is the electrons of the atoms which interact.

Electrons in atoms – quantum number model

In arriving at his model, Bohr used a so-called *quantum number n*. This number could only take integer values and was needed to ensure that only certain orbits were possible. With $n = 1$ he obtained the innermost orbit K, with $n = 2$ the next orbit outwards L, with $n = 3$ the next M, and so on. The Bohr model for the atomic electron systems has nowadays been replaced by one based on the idea of the electrons having four quantum numbers, these being denoted by n, l, m_l and m_s.

- **Principal quantum number n**
  This can have an integer value 1, 2, 3, 4, etc. and is the number of the shell, roughly defining the energy of an electron.

- **Angular momentum quantum number l**
  This defines the angular moment of an electron and gives the sub-shells. It can have an integer value in the range 0 to $(n - 1)$. Values of l are denoted by the letter symbols s, p, d, f, where $s = 0$, $p = 1$, $d = 2$, $f = 3$.

- **Magnetic quantum number m_l**
  This describes the magnetic moment of an electron. It can have integral values from $-l$ to $+l$, including 0.

- **Spin quantum number m_s**
  This describes the concept of an electron spinning in either a clockwise or anticlockwise direction and is $+\frac{1}{2}$ or $-\frac{1}{2}$.

There is an important principle called *Pauli's exclusion principle* governing the set of quantum numbers that atomic electrons have: *no two electrons in an atom can ever have the same set of quantum numbers*.

Consider the situation with $n = 1$. A consequence of this value is, since $(n - 1)$ is zero, that l is zero. Because l is zero, then m_l is zero. Since m_s can be either $+\frac{1}{2}$ or $-\frac{1}{2}$, there are just two possible

$n = 1$

l	m_l	m_s	
0	0	$+\frac{1}{2}$	1s
0	0	$-\frac{1}{2}$	

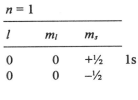

Figure 2.4 *Box form of notation*

$n = 2$

l	m_l	m_s	
0	0	$+\frac{1}{2}$	2s
0	0	$-\frac{1}{2}$	
1	0	$+\frac{1}{2}$	2p
1	0	$-\frac{1}{2}$	
1	1	$+\frac{1}{2}$	
1	1	$-\frac{1}{2}$	
1	-1	$+\frac{1}{2}$	
1	-1	$-\frac{1}{2}$	

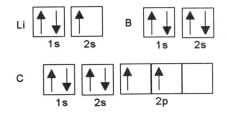

Figure 2.5 *Box diagrams for lithium, boron and carbon*

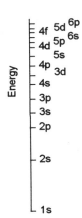

Figure 2.6 *Energy levels of the various electronic states*

combinations of quantum numbers that can occur with n equal to 1. Thus for $n = 1$, there are just two possible electron systems and so we have:

Hydrogen as the atom containing just one electron and the set of quantum numbers $n = 1$, $l = 0$, and m_s either $+\frac{1}{2}$ or $-\frac{1}{2}$.

Helium as the atom containing two electrons with one having the quantum numbers $n = 1$, $l = 0$, and $m_s = +\frac{1}{2}$ and the other $n = 1$, $l = 0$, and $m_s = -\frac{1}{2}$.

The electronic structure of a helium atom can be written as $1s^2$, with the number 1 denoting the value of n, the s the value of l as 0, and the 2 indicating that two electrons both occupy 1s quantum states. An alternative notation is shown in Figure 2.4, this being a box labelled as 1s with arrows indicating that the spins of the two electrons are opposite.

Now consider the situation with $n = 2$. A consequence of this value is that l can have the values 0 or 1, m_l the values 0, −1 or +1 and m_s either $+\frac{1}{2}$ or $-\frac{1}{2}$. The possible combinations of quantum numbers are thus: as shown opposite. There are eight possible sets of distinct quantum numbers for $n = 2$ and thus eight possible electrons. Thus for $n = 2$ we have:

Lithium as the atom containing three electrons, two occupying the two 1s states and the third one in the 2s state, i.e. $1s^2 2s^1$.

Beryllium as the atom containing four electrons, two occupying the two 1s states and the other two occupying the two 2s states, i.e. $1s^2 2s^2$.

Boron as the atom containing five electrons, two occupying the two 1s states, two occupying the two 2s states and one in the 3p state (p is used to indicate $l = 1$), i.e. $1s^2 2s^2 2p^1$.

Carbon as the atom containing six electrons, two occupying the two 1s states, two occupying the two 2s states and two in the 3p states, i.e. $1s^2 2s^2 2p^2$.

And so on for nitrogen, oxygen, fluorine and neon to complete the filling of the 2p states.

Figure 2.5 shows the box diagrams for the above elements. With the carbon we might have considered filling the first of the 2p boxes so that we have the two 2p spins paired in opposite directions. However, the state is as shown in the figure. A rule we can use in such circumstances is called *Hund's rule*, this states that in the ground state of an atom, the total spin of the electrons always has its maximum possible value.

Table 2.1 gives the electronic structures of atoms for atomic numbers up to 32. The electrons occupy the quantum states which give the lowest possible value of energy. For this reason 1s is first filled, then 2s, then 2p, then 3s, then 3p, then 4s, then 3d, then 4p, and so on. Figure 2.6 gives an approximate indication of the relative energies of the various states.

Table 2.1 *Electronic structures of atoms*

Atomic number	Element	1s	2s	2p	3s	3p	Atomic number	Element	1s	2s	2p	3s	3p	3d	4s	4p
1	Hydrogen	1					17	Chlorine	2	2	6	2	5			
2	Helium	2					18	Argon	2	2	6	2	6			
3	Lithium	2	1				19	Potassium	2	2	6	2	6		1	
4	Beryllium	2	2				20	Calcium	2	2	6	2	6		2	
5	Boron	2	2	1			21	Scandium	2	2	6	2	6	1	2	
6	Carbon	2	2	2			22	Titanium	2	2	6	2	6	2	2	
7	Nitrogen	2	2	3			23	Vanadium	2	2	6	2	6	3	2	
8	Oxygen	2	2	4			24	Chromium	2	2	6	2	6	4	2	
9	Fluorine	2	2	5			25	Manganese	2	2	6	2	6	5	2	
10	Neon	2	2	6			26	Iron	2	2	6	2	6	6	2	
11	Sodium	2	2	6	1		27	Cobalt	2	2	6	2	6	7	2	
12	Magnesium	2	2	6	2		28	Nickel	2	2	6	2	6	8	2	
13	Aluminium	2	2	6	2	1	29	Copper	2	2	6	2	6	10	1	
14	Silicon	2	2	6	2	2	30	Zinc	2	2	6	2	6	10	2	
15	Phosphorus	2	2	6	2	3	31	Gallium	2	2	6	2	6	10	2	1
16	Sulphur	2	2	6	2	4	32	Germanium	2	2	6	2	6	10	2	2

The valence of an atom is determined by the number of electrons in the outermost sp level. For example, lithium has the electronic structure $1s^2 2s^1$. The outermost sp level has just one electron and so there is a valence of 1. Chlorine has the electronic structure $1s^2 2s^2 2p^6 3s^2 3p^5$. The outermost sp level is one electron short of being completed and so there is a valence of 1. Germanium has the electronic structure $1s^2 2s^2 2p^6 3s^2 2p^6 3d^{10} 4s^2 4p^2$. The outermost sp level is four electrons short of being completed and so there is a valence of 4.

The periodic table

The periodic table (Table 2.2) can be considered to be a grouping of the elements in ascending atomic number sequence, so that elements in the same vertical column have similar chemical properties. It is also a grouping according to the atomic electron structures since these determine the chemical properties.

Example

With reference to Table 2.2, the Periodic Table, determine in terms of its quantum numbers, the electronic structure of an atom with atomic number of 47.

This is the element silver (Ag). It has the electronic structure of $1s^2 2s^2 2p^6 3s^2 3p^6 4s^2 3d^{10} 4p^6 5s^1 4d^{10}$.

Table 2.2 *Periodic table*

Group

n	1	2											3	4	5	6	7	0
1	1 H																	2 He
2	3 Li	4 Be											5 B	6 C	7 N	8 O	9 F	10 Ne
3	11 Na	12 Mg					Transition elements						13 Al	14 Si	15 P	16 S	17 Cl	18 Ar
4	19 K	20 Ca	21 Sc	22 Ti	23 V	24 Cr	25 Mn	26 Fe	27 Co	28 Ni	29 Cu	30 Zn	31 Ga	32 Ge	33 As	34 Se	35 Br	36 Kr
5	37 Rb	38 Sr	39 Y	40 Zr	41 Nb	42 Mo	43 Tc	44 Ru	45 Rh	46 Pd	47 Ag	48 Cd	49 In	50 Sn	51 Sb	52 Te	53 I	54 Xe
6	55 Sb	56 Ba	71 Lu	72 Hf	73 Ta	74 W	75 Re	76 Os	77 Ir	78 Pt	79 Au	80 Hg	81 Tl	82 Pb	83 Bi	84 Po	85 At	86 Rn
7	87 Fr	88 Ra	103 Lr	104 Rf	105 Ha													

Lanthanides

57 La	58 Ce	59 Pr	60 Nd	61 Pm	62 Sm	63 Eu	64 Gd	65 Tb	66 Dy	67 Ho	68 Er	69 Tm	70 Yb
89 Ac	90 Th	91 Pa	92 U	93 Np	94 Pu	95 Am	96 Cm	97 Bk	98 Cf	99 Es	100 Fm	101 Md	102 No

Actinides

| s filling | f filling | d filling | p filling |

2.1.2 Bonds

Atoms are held together in solids and molecules by electric forces. An electric force of attraction occurs between positively and negatively charged particles, a force of repulsion between two positively or two negatively charged particles. There are various ways in which such electric force bonds can be produced between particles in solids. Bonds can be classified in two groups:

- *Primary bonds*

  These are *metallic*, *ionic* and *covalent* bonds, and all are relatively strong bonds.

- *Secondary bonds*

  These are *van der Waals* and *hydrogen* bonds. Both are relatively weak.

Metals and ceramics are held together by primary bonds, mainly the metallic bond for metals and the ionic and covalent bonds for ceramics. These strong bonds are responsible for the high strength and stiffness of these materials. Secondary bonds provide the bonds responsible for holding together solid polymer materials. Their relatively weak nature means that such materials are less strong and less stiff.

The metallic bond

The term *metal* is used for elements, such as copper, which have atoms which so readily lose electrons that in the solid state at room temperature there are many free electrons. Thus in the solid state, copper consists of an array of atoms each of which has lost one electron (Figure 2.7). This leaves each copper atom as having a net positive charge and it is termed a positive ion. The electrons that have been lost do not combine with any one ion but remain as a cloud of negative charge floating between the ions. The result is rather like a glue in that the cloud of electrons holds the positive ions together, positive ions being attracted to electrons which in

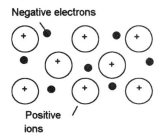

Negative electrons

Positive ions

Figure 2.7 *Metallic bond: positive ions with free electrons 'floating' between them*

turn attract other positive ions. This is what is termed the *metallic bond*. It is the dominant, though not the only, bond in metals.

In a metal we will have the attractive forces between positive ions and the electrons and repulsive forces between the positive ions. At the normal separation of the metal ions, the forces of attraction are just balanced by the forces of repulsion. If we try to compress a metal and so move the positive ions closer together, the repulsive force predominates. If we try to stretch a metal, the attractive force predominates. Since the bonds formed between the positive ions can be formed in any direction without any restrictions, a simple model we can use to describe the structure of metals is to think of the ions in a metal being like spheres, a sphere being a shape which imposes no directionality rules on how they are packed together.

The free electrons explain why metals are good conductors of electricity, since they have free charge carriers which are easily moved through the solid by the application of a voltage. Insulators have no free electrons and the atoms in the solid are bonded together in a different way.

The ionic bond

An individual atom is electrically neutral, having as much positive charge in the nucleus as negative charge in its electrons. However, if an atom loses an electron, it must then have a net positive charge and is a *positive ion*. If an atom gains an extra electron, it ends up with a net negative charge, becoming a *negative ion*. Sodium chloride, common salt, is an example of an ionic bonded material. In order for the sodium and the chlorine atoms to assume the 'full shell' configuration for their atoms, the sodium has to lose an electron and the chlorine gain one. The sodium thus transfers an electron to the chlorine. The result is that we have a positive ion and a negative ion. Unlike-charged particles attract each other. Thus there is a force of attraction and this is what is referred to as the *ionic bond* (Figure 2.8).

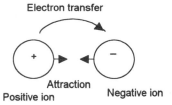

Figure 2.8 *Ionic bond: attractive force betwen oppositely charged ions*

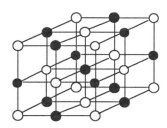

- ● Positive sodium ion
- ○ Negative chlorine ion

Figure 2.9 *Sodium chloride crystalline structure*

The solid sodium chloride exists not just as a pair of ions but a vast structure of sodium and chlorine ions (Figure 2.9). In such an array, unlike-charged ions attract each other and like-charged ions repel each other. Thus sodium ions repel sodium ions, chlorine ions repel chlorine ions, but sodium and chlorine ions attract each other. A stable structure exists because the arrangement of the sodium and chlorine ions are such that the attractive forces are just balanced by the repulsive forces.

Because the electrons are all tied up in bonds, there are no free electrons to act as the charge carriers for electrical currents. Thus materials with ionic bonds are electrical insulators. Because ionic bonds are strong bonds, such materials tend to have high melting points, a lot of thermal energy being needed to break the points.

The covalent bond

Covalent bonding can be considered to be where neighbouring atoms share electrons in order to realise the 'full shell'

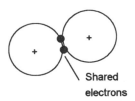

Figure 2.10 *Covalent bond: shared electrons between positive ions*

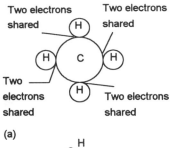

(a)

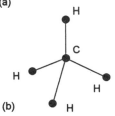

(b)

Figure 2.11 *Methane with covalent bonding*

configuration. Figure 2.10 illustrates this in a simplistic manner. We can think of a shared electron as being in orbit about the pair of atoms, rather than just one of the atoms. As a result there is a greater chance of the shared electrons being between the atoms and so acting as a 'glue' holding together the atoms. We have one positive ion attracted towards the shared electrons which in turn attracts the other positive ion.

An atom can share electrons with more than one other atom, if by doing so it can achieve the 'full shell' configuration. For example, a carbon atom needs four electrons to obtain a 'full shell' and hydrogen needs one electron for a 'full shell'. This may be achieved by carbon sharing electrons with four hydrogen atoms, the result being a methane molecule. Each hydrogen atom shares one of the carbon electrons and the carbon atom obtains a share in one electron from each hydrogen atom. There are thus four covalent bonds. Figure 2.11(a) gives a two-dimensional picture of this methane molecule. However, the covalent bond is highly directional and so a more realistic representation of the methane molecule is given by the three-dimensional figure shown in Figure 2.11(b) where the angle between any pair of CH bonds is the same.

Diamond is another example of a material formed as a result of covalent bonds, the bonds being between carbon atoms with each atom forming bonds with four other atoms.

Covalent bonding is the dominant type of bond in silicate ceramics. It is a strong bond and thus such materials have high melting points, i.e. a lot of thermal energy is needed to break the bonds. Because the electrons are all tied up in bonds, there are no free electrons to act as the charge carriers for electrical currents and so materials with covalent bonds are electrical insulators.

Van der Waals bonds

With an atom having no net charge, the charge on the nucleus is just balanced by the charge carried by the electrons. We might thus expect that, when we have two atoms with no net charge, there will be no bonding forces between them. However, the distribution of the negative charge about an atom can vary with time. Thus, at some instant of time, we might have an atom with more negative charge on one side than the other. Some molecules might have a permanent polarisation of charge in this way with some parts with a net negative charge and some with a net positive charge, the molecules being then said to be *dipoles*. We can thus have electrostatic forces of attraction between such polarised atoms or molecules. The situation becomes something like that shown in Figure 2.12. This is the force responsible for what is termed *van der Waals bonding* and occurs as the bonding between the polymer molecules in a polymer. It is a much weaker force than the covalent bonds that are usual among the atoms in the polymer molecules.

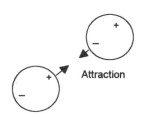

Figure 2.12 *Van der Waals bonding provided by the distribution of charge giving rise to dipoles*

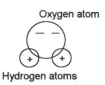

Figure 2.13 *Hydrogen bonding*

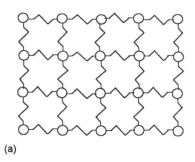

(a)

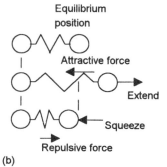

(b)

Figure 2.14 *A simple model of a solid*

Hydrogen bonding

The water molecule consists of two oxygen atoms and a hydrogen atom. When water freezes to form ice, we must have bonds formed to hold the material together. The covalent bonding in the water molecule results in the shared electrons being closer to the oxygen atom than the hydrogen atoms. This results in the two hydrogen ends of the molecule being slightly positive relative to the oxygen atom. Each water molecule is said to be a *dipole* (Figure 2.13). Hydrogen bonding occurs as a result of electrostatic attraction between these molecular dipoles. The cell walls of plants, and hence wood, are made of a naturally occurring polymer called cellulose. The cellulose molecules are bound to each other by hydrogen bonds.

2.1.3 Forces between particles

Consider what forces there must be between atoms and molecules in a solid. Since a solid has a fixed shape and does not spontaneously change its shape, there must be no resultant force acting on the particles constituting the solid. Only if there is no resultant force will the particles not move. The particles all stick together in the solid form, so there must be forces of attraction that hold the particles together and resist them being pulled apart. But if we squeeze a solid, it resists being squashed. There must, therefore, be forces of repulsion opposing the particles being forced further together.

We can think of the situation in a solid as being like that of an array of particles with each particle linked to its neighbours by springs (Figure 2.14(a)). The particles are held together by attractive forces exerted on each of them by the spring. If you try to pull the particles further apart, then the spring exerts attractive forces on the particles to pull them back to their normal separation (Figure 2.14(b)). However, if we apply forces on the particles to push them closer together, the spring exerts repulsive forces (Figure 2.14(b)). The more we try to push the particles together, the greater the repulsive force. We can thus think of there being attractive forces and repulsive forces which vary with separation between particles. At the normal separation of the particles, they neither move further in nor out and so the attractive forces must be just cancelled by the repulsive forces. At greater separations, the attractive force predominates and at closer separations the repulsive force predominates.

Figure 2.15 shows how we might expect these forces to vary with the separation distance between particles in a solid. The force of attraction decreases as the separation increases. The force of repulsion also decreases as the separation increases, but at a faster rate. At the normal separation of the particles in a solid, the repulsive force and attractive force will cancel so there is no resultant force acting on the particles. They are in equilibrium.

The graph shows how the resultant force, i.e. the sum of the attractive and repulsive forces, varies with separation. For

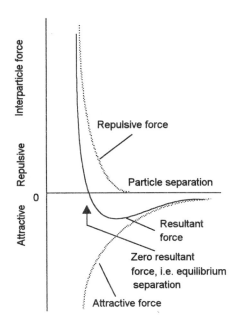

Figure 2.15 *Forces between particles*

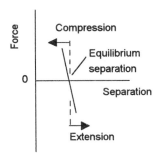

Figure 2.16 *Hooke's law*

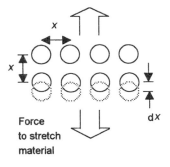

Figure 2.17 *Stretching a material*

separations greater than the normal separation, there is a resultant attractive force which acts to pull the particles back to the normal separation. With separations less than the normal separation, there is a resultant repulsive force which pushes the particles back out to the normal separation.

If we consider the resultant force on the particles for separations close to their equilibrium separation (Figure 2.16), then the force is reasonably proportional to the amount by which we extend or compress the particles together from their equilibrium position. In other words, the extension or compression is proportional to the applied force and we have *Hooke's law*.

Maths in action

Consider two neighbouring planes of particles (Figure 2.17) between which we have forces of the form described above. Each atom effectively 'occupies' a space of area x^2, where x is the separation of the particles. If the cross-sectional area of these planes is A, then the number of atoms in such an area is A/x^2. If we pull these planes further apart by an amount δx, then there will be an attractive force developed between the particles of δF. The total force resisting the extension is thus $(A/x^2)\delta F$. Hence the stress is this force divided by the area and so is $\delta F/x^2$. The strain is the extension per unit length and so is $\delta x/x$. Thus the tensile modulus of elasticity E is:

$$E = \frac{\text{stress}}{\text{strain}} = \frac{\delta F/x^2}{\delta x/x} = \frac{1}{x}\frac{\delta F}{\delta x} \qquad [1]$$

Thus, with this simple model of a solid, the tensile modulus of elasticity is inversely proportional to the separation of the particles and directly proportional to the gradient dF/dx of the force–separation graph at the equilibrium separation.

Potential energy

Consider two particles exerting attractive forces on each other. If we start to increase the separation of the particles then we have to do work. Think of it being like stretching a spring between them. The further we move one particle from the other, the more work we have to do. Thus when one of the particles has been moved to an infinite distance from the other, then the maximum amount of work would have been done. When a particle on the end of a tethered spring is pulled so that the spring stretches, then work is done and the particle gains potential energy. Thus, when two particles, between which there are attractive forces, are pulled apart, they gain potential energy and when one has been moved to

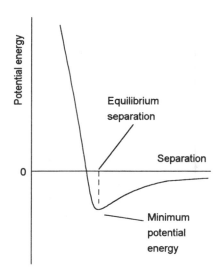

Figure 2.18   *Inter particle energy*

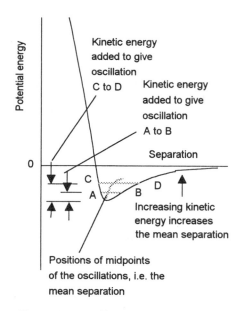

Figure 2.19   *Expansion*

an infinite distance from the other, then we would expect the potential energy to be a maximum. However, it is convenient to regard the potential energy of each particle to be zero when they are an infinite distance apart. This is because at such a separation they have no influence on each other. As a consequence of this, when there is a resultant force of attraction between two particles their potential energy is a negative quantity. Only with it negative can it be said to increase when the separation is increased to infinity.

Figure 2.18 shows how the potential energy between atoms or molecules in a solid varies with the separation between them (note that a potential energy–separation graph can be derived from the force–separation graph since the energy at a particular separation is the area under the force–separation graph up to that separation). At the equilibrium separation, i.e. the separation when there is no resultant force, the potential energy is a minimum. The deeper this 'potential energy well', the greater the energy that has to be supplied to separate the particles completely in the solid. Thus the greater the *binding energy*.

When a material is heated, its particles gain energy. In a gas where the molecules are so far apart that usually we can ignore intermolecular forces for all but the small amounts of time when they collide, the result of heat energy being supplied is to increase the kinetic energy of the molecules so that they move around faster. In a solid the inter particle forces are such that the particles are not free to move about as molecules can in a gas. What we can, however, consider to happen is that the influx of heat energy causes the particles to oscillate about their equilibrium positions. With a mass oscillating on the end of a tethered spring, when the mass is at the end of its travel and the maximum distance away from its equilibrium position and momentarily has zero velocity, all its energy is in the form of potential energy. When the oscillating mass is passing through its equilibrium position its velocity is a maximum and potential energy has been transformed into kinetic energy. In the case of a solid, when it is at the absolute zero of temperature, it has no kinetic energy and thus the particles are in their equilibrium positions with the minimum potential energy value. At temperatures above the absolute zero, there is an influx of heat energy to the solid and so the particles will have kinetic energy which results in them oscillating with this energy being transformed into potential energy and then back into kinetic energy, and so on repeatedly. The particle thus oscillates between the points A and B on the potential energy graph (Figure 2.19). If there is a bigger input of heat energy, i.e. the solid is raised to a higher temperature, then the particle oscillates between the points C and D, higher up the potential energy graph than points A and B. The midpoint of these oscillations moves to greater separations as the temperature increases. Thus as the temperature of the solid increases, the solid expands.

2.1.4 Particle arrangement in solids

The arrangement of particles, i.e. atoms, ions and molecules, in solids determines the microstructure and is an important aspect in determining the properties. For example, a change in microstructure can change a metal from being ductile to brittle. A material has short-range order if the arrangement of the particles extends only to their nearest neighbours. Such an arrangement is a characteristic of glasses. A solid having only short-range order in the arrangement of its constituent particles is said to be *amorphous*. Materials which display both short- and long-range order have their particles arranged in a regular repetitive arrangement which extends throughout the entire material. Such materials are termed *crystalline*.

Consider the stacking together of spheres in an orderly manner. Since spheres can be stacked in any way, a sphere can be considered to be a model for an atom, ion or molecule in a solid when there is no directionality to the bonding forces. One of the simplest arrangement of spheres is that of the *simple cubic structure*. Figure 2.20(a) shows the structure obtained by stacking four spheres with the centres of each sphere at the corners of a cube. The surfaces of each sphere touch the surfaces of each of its neighbours in such a way that the length of the side of the cube is equal to the diameter of the spheres. The line joining the centres of the spheres (Figure 2.20(b)) encloses what is termed the *unit cell*, this being the smallest arrangement of particles that when regularly repeated forms the crystal. The resulting solid would consist of a completely orderly array of spheres, i.e. particles, and we would expect the surfaces of such a solid to be smooth and flat with the angles between adjoining faces always 90°. Such a solid would when broken up always have the appearance of stacked cubes. This is a description of a *cubic crystal*.

The simple cubic crystal shape is arrived at by stacking spheres in one particular way. By stacking spheres in a closer manner other crystal shapes can be produced: body-centred cubic, face-centred cubic and hexagonal close-packed structures. These three close-packed structures represent the structures occurring with solid metals.

With the *body-centred cubic* unit cell (Figure 2.21), a square arrangement of spheres is used in each layer but successive layers are displaced so that spheres fit within the hollows of the layer underneath. As a consequence, the arrangement is slightly more complex than the simple cubic unit cell in having an extra sphere in the centre of the cell. With the *face-centred cubic* unit cell (Figure 2.22), the spheres in each layer are packed as close as possible and successive layers are displaced so that spheres fit within the hollows of the layer underneath. There is, when compared with the simple cubic unit cell, a sphere at the centre of each face of the cube. With the *hexagonal close-packed* unit cell (Figure 2.23), the spheres are packed in a close array which gives a hexagonal form of structure.

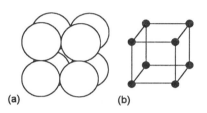

(a)   (b)

Figure 2.20 *Simple cubic structure*

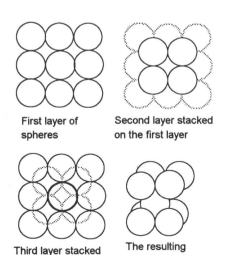

First layer of spheres

Second layer stacked on the first layer

Third layer stacked on the second layer

The resulting structure

Figure 2.21 *Body-centred cubic structure*

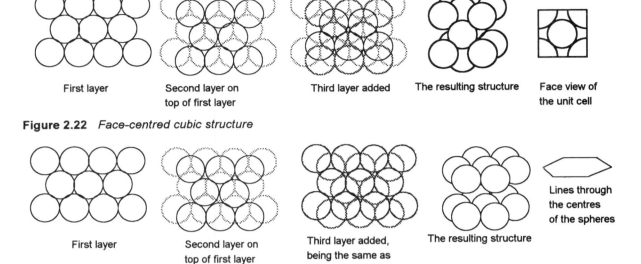

First layer — Second layer on top of first layer — Third layer added — The resulting structure — Face view of the unit cell

Figure 2.22 *Face-centred cubic structure*

First layer — Second layer on top of first layer — Third layer added, being the same as the first layer — The resulting structure — Lines through the centres of the spheres

Figure 2.23 *Hexagonal close-packed structure*

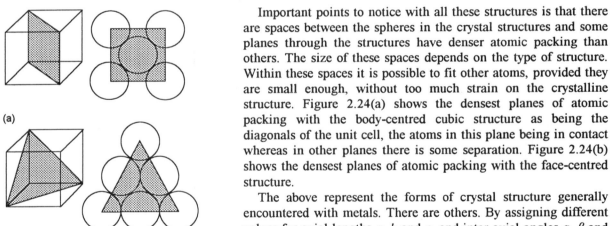

(a)

(b)

Figure 2.24 *Planes with densest atomic packing*

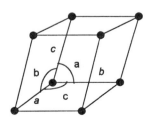

Figure 2.25 *Axes and inter axial angles*

Important points to notice with all these structures is that there are spaces between the spheres in the crystal structures and some planes through the structures have denser atomic packing than others. The size of these spaces depends on the type of structure. Within these spaces it is possible to fit other atoms, provided they are small enough, without too much strain on the crystalline structure. Figure 2.24(a) shows the densest planes of atomic packing with the body-centred cubic structure as being the diagonals of the unit cell, the atoms in this plane being in contact whereas in other planes there is some separation. Figure 2.24(b) shows the densest planes of atomic packing with the face-centred structure.

The above represent the forms of crystal structure generally encountered with metals. There are others. By assigning different values for axial lengths a, b and c, and inter axial angles a, β and γ (Figure 2.25), seven different types of crystal structure have been identified:

- **Cubic**
  For the three axial lengths of the unit cell, $a = b = c$ and all inter axial angles are 90°.

- **Tetragonal**
  For the three axial lengths of the unit cell, $a = b \neq c$ and all inter axial angles are 90°.

- **Orthorhombic**
  For the three axial lengths of the unit cell, $a \neq b \neq c$ and all inter axial angles are 90°.

- *Hexagonal*

  For the three axial lengths of the unit cell, $a = b \neq c$ and two inter axial angles are 90° and one 120°.

- *Rhombohedral*

  For the three axial lengths of the unit cell, $a = b = c$ and all inter axial angles are equal with none being 90°.

- *Monoclinic*

  For the three axial lengths of the unit cell, $a \neq b \neq c$ and two inter axial angles are 90° and one not 90°.

- *Triclinic*

  For the three axial lengths of the unit cell, $a \neq b \neq c$ and all inter axial angles are different and none are 90°.

Figure 2.26 shows the fourteen types of unit cells that can describe all possible lattices.

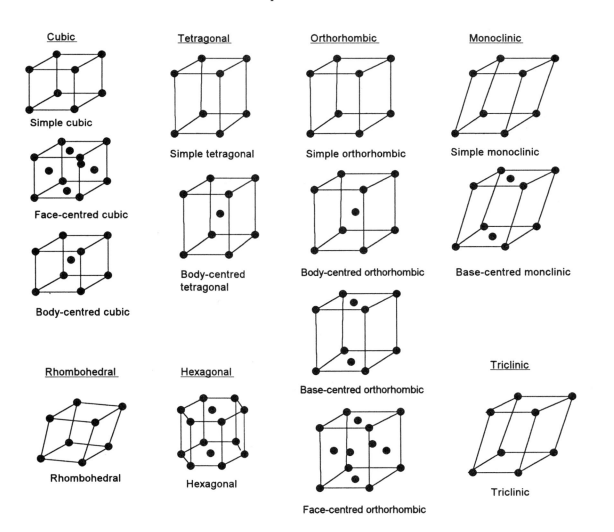

Figure 2.26 *Unit cells*

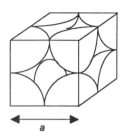

Figure 2.27 *Simple cubic unit cell*

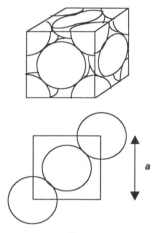

Figure 2.28 *Face-centred cubic unit cell*

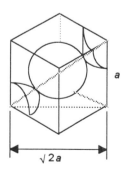

Figure 2.29 *Body-centred cubic unit cell*

Unit cell size and particle radii

Consider the simple cubic structure unit cell (Figure 2.27). The unit cell has sides a with lengths equal to the diameter of the particle. Thus, if r is the radius:

$$a = 2r \qquad [2]$$

Within the unit cell there is effectively one particle, being made up of eight one-eighth segments of particles. The total volume occupied by the particle in the unit cell is $4\pi r^3/3$. The length of a side of the unit cell is $2r$ and its total volume is $(2r)^3$. Thus the fraction of the unit cell occupied, termed the *packing factor*, is:

$$\text{packing factor} = \frac{\frac{4}{3}\pi r^3}{(2r)^3} = 0.52 \qquad [3]$$

This means that 52% of the unit cell is occupied, the remaining 48% being free space.

Consider the face-centred cubic structure (Figure 2.28). The particles contact each other across the diagonal of the face. Thus, using Pythagoras, $(4r)^2 = 2a^2$ and:

$$a = \frac{4r}{\sqrt{2}} \qquad [4]$$

In the unit cell, the sum of the parts of particles adds up to four complete particles. Thus the occupied volume is $4 \times 4\pi r^3/3$ and:

$$\text{packing fraction} = \frac{4 \times \frac{4}{3}\pi r^3}{\left(\frac{4r}{\sqrt{2}}\right)^3} = 0.74 \qquad [5]$$

Thus, compared with the simple cubic structure, the face-centred cubic structure is more closely packed.

For the body-centred cubic unit cell, the particles are in contact with each other along the diagonal from one corner of the cube to the opposite corner (Figure 2.29). Thus, using Pythagoras, $a^2 + (\sqrt{2}a)^2 = (4r)^2$. Hence:

$$a = \frac{4r}{\sqrt{3}} \qquad [6]$$

One complete particle is located at the centre of the cell and an eighth of a particle at each corner and thus there is a total of 2 particles in a unit cell. Hence:

$$\text{packing fraction} = \frac{2 \times \frac{4}{3}\pi r^3}{\left(\frac{4r}{\sqrt{3}}\right)^3} = 0.68 \qquad [7]$$

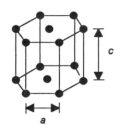

Figure 2.30 *Hexagonal close-packed unit cell*

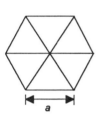

Figure 2.31 *Base of unit cell*

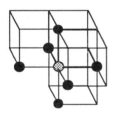

Figure 2.32 *Simple cubic structure*

For the hexagonal close-packed unit cell there are six one-sixth particles at the corners and half a particle in the centre in the top and bottom layers and three complete particles in the middle layer. Thus there are the equivalent of six complete particles in the unit cell. The ratio of the height c of the unit cell to the basal side a is 1.633 (Figure 2.30) when the particles are packed as close as possible. The base of the unit cell can be considered to be composed of six equilateral triangles (Figure 2.31), each having an area of ½ base × height = $\frac{1}{2}a \times a \sin 60° = 0.43a^2$. Thus the volume of the unit cell is $6 \times 0.433a^2 \times c = 6 \times 0.433a^2 \times 1.633a = 4.24a^3$. Since $a = 2r$, the packing factor is:

$$\text{packing factor} = \frac{6 \times \frac{4}{3}\pi r^3}{4.24(2r)^3} = 0.74 \qquad [8]$$

Note that with the hexagonal close-packed structure, the c/a ratio of 1.633 is an ideal value, most metals with this type of structure having ratio values that differ slightly from 1.633. For example, magnesium has a ratio value of 1.623, titanium 1.587 and zinc 1.856. The lower values for magnesium and titanium would suggest that the particles in their structure are slightly compressed, while for zinc the particles are elongated.

Another indication of how tightly particles are packed together is given by the *coordination number*. This is the number of particles touching a particular particle. Thus in the simple cubic structure (Figure 2.32), each particle is in contact with six other particles and thus the coordination number is 6. With the body-centred cubic structure (look at Figure 2.21) the coordination number is 8, with the face-centred cubic structure (look at Figure 2.22) 12 and with the hexagonal close-packed structure (look at Figure 2.23) 12.

Table 2.3 summarises the packing factors and coordination numbers for the simple cubic, face-centred cubic, body-centred cubic and hexagonal close-packed structures. The hexagonal close-packed and the face-centred cubic structures are the most close-packed structures. Table 2.4 lists metals with their crystalline structures and lattice parameters at 20°C. No metals have simple cubic structures. Some materials can have more than one crystalline structure, such materials being termed *allotropic* or *polymorphic*. For example, iron at low temperatures has a body-centred cubic structure but at higher temperatures a face-centred cubic structure.

Table 2.3 *Packing factors*

Unit cell structure	Coordination number	Packing factor
Simple cubic	6	0.52
Face-centred cubic	12	0.74
Body-centred cubic	8	0.68
Hexagonal close packed	12	0.74

Table 2.4 *Crystal structures for metals at 20°C*

Metal	Lattice size a nm	Lattice size c nm	c/a ratio	Atomic radius r nm
Body-centred cubic				
Chromium	0.289			0.125
Iron	0.287			0.124
Molybdenum	0.315			0.136
Potassium	0.533			0.231
Sodium	0.429			0.186
Tantalum	0.330			0.143
Tungsten	0.316			0.137
Vanadium	0.304			0.132
Face-centred cubic				
Aluminium	0.405			0.143
Copper	0.362			0.128
Gold	0.408			0.144
Lead	0.495			0.175
Nickel	0.352			0.125
Platinum	0.393			0.139
Silver	0.409			0.144
Hexagonal close-packed				
Cadmium	0.2973	0.5618	1.890	0.149
Beryllium	0.2286	0.3584	1.568	0.113
Cobalt	0.2507	0.4069	1.623	0.125
Magnesium	0.3209	0.5209	1.623	0.160
Titanium	0.2950	0.4683	1.587	0.147
Zinc	0.2665	0.4947	1.856	0.133
Zirconium	0.3231	0.5148	1.593	0.160

Note: the atomic radii were obtained from the lattice constants.

Example

Copper has a face-centred structure and an atomic radius of 0.1278 nm. The atomic mass of copper is 63.5 g/mol and Avogadro's number is 6.02×10^{23}. Calculate the density of copper from this data.

The face-centred structure has four atoms per unit cell. The mass of each copper atom is $63.5/6.02 \times 10^{23} = 1.055 \times 10^{-22}$ g. Thus the mass of a unit cell is 4.219×10^{-25} kg. The volume of a unit cell is a^3 with $a = 4r/\sqrt{2}$. Hence the estimated density is:

$$\text{density} = \frac{4.219 \times 10^{-25}}{\left(4 \times 0.1278 \times 10^{-9}/\sqrt{2}\right)^3} = 8933 \text{ kg/m}^3$$

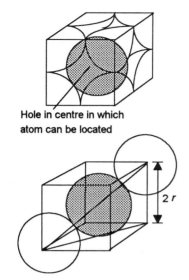

Hole in centre in which
atom can be located

Figure 2.33   *Simple cubic*

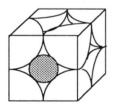

Figure 2.34   *Body-centred cubic*

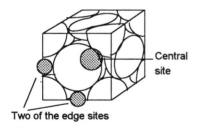

Central
site

Two of the edge sites

Figure 2.35   *Face-centred cubic*

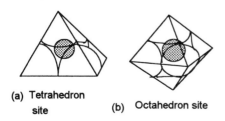

(a) Tetrahedron
      site

(b) Octahedron site

Figure 2.36   *Interstitial sites*

Example

Determine the percentage volume change that occurs when iron changes from a body-centred cubic structure to a face-centred cubic structure.

With a body-centred cubic structure, there are two atoms per unit cell and equation [6] gives $a = 4r/\sqrt{3}$. Thus the volume per atom is $a^3/2 = (4r/\sqrt{3})^3/2 = 6.16r^3$. With a face-centred cubic structure, there are four atoms per unit cell and equation [4] gives $a = 4r/\sqrt{2}$. Thus the volume per atom in this structure is $(4r/\sqrt{2})^3/4 = 5.66r^3$. Assuming there is no change in the atomic radius, the percentage volume change during the change is thus:

$$\% \text{ change} = \frac{5.66r^3 - 6.16r^3}{6.16r^3} \times 100 = -8.1\%$$

The iron thus contracts when the transformation occurs.

Interstitial sites

In crystal structures there are holes, the packing fraction being less than 100%. Atoms can be placed in these holes, such locations for atoms being termed *interstitial sites*.

With a cubic structure there is an interstitial site at the centre of the unit cell (Figure 2.33). An atom placed at the centre and touching the atoms around it will have a coordination number of eight. If R is its radius, then Pythagoras gives $(2R + 2r)^2 = (2r)^2 + [(2r)^2 + (2r)^2]$ and so a radius R of $0.73r$, where r is the radius of the atoms forming the cubic structure.

The body-centred cubic structure has holes in the centre of the faces (Figure 2.34). An atom placed in one of these sites and touching the atoms around it will have a coordination number of 6 and a radius of $0.41r$.

The face-centred cubic structure (Figure 2.35) has holes at the centre of each edge of the cube, as well as in the centre of the unit cell. An atom placed in the central site and touching the atoms around it will have a coordination number of 6 and a radius of $0.41r$ while the atoms placed at the centres of each edge will have coordination numbers of four and a radius of $0.22r$.

An interstitial site which is formed by four spheres in contact, i.e. a coordination number of four, is termed a *tetrahedron site* since the cluster of atoms in contact with the interstitial atom form a tetrahedron (Figure 2.36(a)). An interstitial site which is formed by six spheres in contact, i.e. coordination number six, is termed an *octahedron site* since the atoms in contact with the interstitial atom form an octahedron (Figure 2.36(b)). An interstitial site which is formed by eight spheres in contact, i.e. coordination number eight, is termed a *cube site* since the atoms in contact with the interstitial atom form a cube (as in Figure 2.33). An interstitial

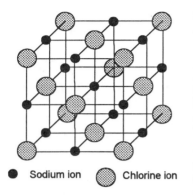

Sodium ion ⚫ **Chlorine ion** ◯

Figure 2.37 *Sodium chloride crystal*

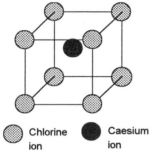

Chlorine ion ◯ **Caesium ion** ⚫

Figure 2.38 *Caesium chloride crystal*

site formed by twelve spheres in contact, i.e. coordination number 12, is termed a *cuboctahedron site*.

2.1.5 Ionic crystals

Ionic crystals are composed of positive and negative ions strongly held together by electrostatic forces. With the sodium chloride crystal, we have positive sodium ions and negative chlorine ions. One way of considering the resulting crystalline structure is to think of there being a a structure formed by the larger chlorine ions with the smaller sodium ions as interstitial particles. The size of the ions are such that a cubic structure is formed (Figure 2.37) with each sodium ion being surrounded by six chlorine ions. When the radius ratio of the ions is between 0.414 and 0.732, the coordination number is 6 and the interstitial ion occupies an octahedral site.

With a caesium chloride crystal we have positive caesium ions and negative chlorine ions. The sizes of the two ions in this case are similar and the cubic structure has each caesium ion surrounded by eight chlorine ions (Figure 2.38). When the radius ratio is between 0.732 and 1.000, the coordination number is 8 and the interstitial ion occupies a cubic site.

If the radius ratio is between 0.225 and 0.414, the coordination number is 4 and the interstitial ion occupies a tetrahedronal site. Zinc blende ZnS has zinc ions with radii of 0.074 nm and sulphur ions with radii 0.184 nm, giving a ratio of 0.402. Zinc blende thus gives a cubic structure with the zinc ions occupying tetrahedronal interstitial sites.

Table 2.5 gives ionic radii for some commonly encountered ions.

Example

Determine the packing factor for a sodium chloride crystal with sodium ions radius of 0.098 nm and chlorine ions 0.181 nm.

Table 2.5 *Ionic radii*

Element	State of ionisation	Radius nm	Element	State of ionisation	Radius nm
Aluminium	+3	0.053	Magnesium	+2	0.066
Barium	+2	0.136	Molybdenum	+4	0.070
Bromine	−1	0.196	Nickel	+2	0.070
Cadmium	+2	0.095	Oxygen	−2	0.140
Caesium	+1	0.167	Potassium	+1	0.133
Calcium	+2	0.099	Silicon	+4	0.040
Chlorine	−1	0.181	Selenium	−2	0.191
Cobalt	+2	0.074	Sodium	+1	0.098
Copper	+2	0.073	Sulphur	−2	0.184
Iron	+3	0.065	Uranium	+4	0.097
Lithium	+1	0.078	Zinc	+2	0.074

In the unit cell there are segments amounting to four sodium ions and four chlorine ions. Thus the total volume of a unit cell which is occupied is $4(4\pi 0.098^3/3) + 4(4\pi 0.181^3/3) = 0.1151$ nm^3. The cubic unit cell has a side of length $2 \times 0.098 + 2 \times 0.181 = 0.558$ nm. The volume of the unit cell is thus 0.1737 nm^3. Thus the packing factor is $0.1151/0.1737 = 0.662$.

Example

Magnesium oxide MgO has magnesium ions with radii 0.066 nm and oxygen ions with radii 0.132 nm. Determine the type of crystalline structure for magnesium oxide and its density. Magnesium has an atomic mass of 24.312 g/mol and oxygen 16 g/mol. The Avogadro number is 6.02×10^{23}.

The ratio of the radii is $0.066/0.140 = 0.47$ and thus there is a coordination number of 6 for the interstitial magnesium ion and it will occupy an octahedral site. The structure will thus be like that of sodium chloride (Figure 4.37). With this type of structure, the ions touch along the edge of the cube and thus the structure shown is a cube with sides of $2 \times 0.066 + 2 \times 0.140 = 0.412$ nm. The structure has segments totalling four magnesium ions and four oxygen ions. Thus the mass is $(4 \times 24.312 + 4 \times 16)/6.02 \times 10^{23} = 2.679 \times 10^{-22}$ g and the density is 3831 kg/m^3.

The Perovskite structure

In the above discussion, ionic structures were considered which had only two types of ions. There are, however, ionic structures with three or more types of ions. An important form of structure with three ions is the *Perovskite structure*. This is found in several electrical ceramics, e.g. barium titanate $Ba^{2+}Ti^{4+}O_3^{2-}$, and is named after a mineral which has the same structural form. Figure 2.39 shows the basic form of the structure.

For barium titanate, the Ba^{2+} ion has a radius of 0.136 nm, the Ti^{4+} ion a radius of 0.068 nm and the O^{2-} ion a radius of 0.140 nm. The ratios of ion sizes are titanium/oxygen $0.068/0.140 = 0.49$ and barium/oxygen $0.136/0.140 = 0.97$. The barium/oxygen ratio is virtually that which would give a cuboctahedron interstitial site (1.00 ratio), the titanium/oxygen an octahedron. Thus for the barium/oxygen we need the barium to have twelve oxygen atoms as neighbours and for the titanium/oxygen we require the titanium to have six oxygen neighbours. The Perovskite structure is indeed one which provides such a combination of interstitial sites. Figure 2.40 shows these elements occurring in barium titanate.

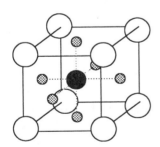

Figure 2.39   *Perovskite structure*

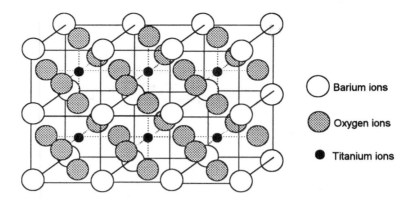

Figure 2.40 *Barium titanate*

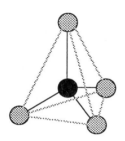

Figure 2.41 *Tetrahedron*

2.1.6 Covalent structures

In covalent structures, each atom is linked to its neighbours by covalent bonds. The coordination number depends on the number of valence bonds. Elements such as silicon, germanium and carbon have a valence of four, each atom thus having a coordination number of four, and can form covalent bonds to produce a bonding arrangement in the form of a tetrahedron (Figure 2.41). Figure 2.42 shows how this arrangement is used to build up a diamond structure. There are four atoms completely within the unit cell and fourteen partly. There are eight segments within the unit cell.

The strength of the covalent bonds and the way in which each atom is held in place in the structure makes this diamond structure a very strong and hard material with a high melting point. Table 2.6 gives some examples.

Table 2.6 *Diamond structures*

Element	Lattice size nm	Covalent radius nm	Melting point °C
Carbon	0.357	0.077	3550
Silicon	0.543	0.117	1414
Germanium	0.568	0.122	942

Diamonds are covalently bonded structures in the tetrahedron structure which are formed with just carbon atoms and are very hard. Another structure formed by carbon atoms is graphite, the lead in pencils that so easily makes a mark on paper. This is, by contrast, a very soft material with layers of the material easily removed to make marks on paper. The way in which the carbon atoms are arranged in graphite (Figure 2.43) is, however, different from the diamond structure. It can be considered to be a 'layered' structure since the atoms are strongly bonded together with covalent bonds in two-dimensional layers, with only very weak van der Waals bonds between the layers. The van der Waals bonds are easily broken and so one layer of carbon atoms easily slides over another.

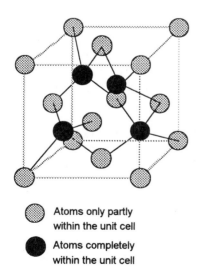

● Atoms only partly within the unit cell

● Atoms completely within the unit cell

Figure 2.42 *Diamond structure unit cell*

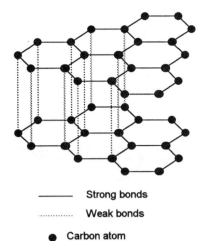

——— Strong bonds

············· Weak bonds

● Carbon atom

Figure 2.43 *Graphite*

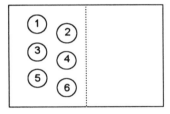

Figure 2.44 *Activity*

Key points

The mechanism responsible for diffusion, whether in a gas, liquid or solid, is the random motion of particles. This random motion results in a net flow of particles from a high concentration region to a low concentration region.

2.1.7 Diffusion

The term *diffusion* is used for the process whereby particles spontaneously spread out with time. For example, with gases we have the spread of cooking odours from the kitchen to other parts of the house. For liquids we have the spread of a blob of ink dropped into water until the entire water is coloured. For solids, there are the spread of electrons across semiconductor junctions or the spread of atoms within metals. The mechanism responsible for diffusion, whether in a gas, liquid or solid, is the random motion of particles.

Activity

For this activity, you need paper, a pencil/pen, scissors, and a six-sided die.

Divide a piece of paper into two compartments, as shown in Figure 2.44. Cut from another sheet of paper, six discs and label them 1 to 6. In the first compartment place the 6 discs. Now throw the die. Each throw will identify one of the six numbers. Now move the corresponding disc from its compartment into the other compartment.

1. After six throws of the die, how many discs are there in each compartment?
2. After another six throws, how many discs are there in each compartment?
3. Now repeat step 2 several times, noting the number of discs in each compartment after each set of throws.
4. How does the distribution of the 'particles' between the two compartments change with number of throws?.

Each disc will undergo 'random motion' between the zones. Initially we have a high concentration of particles in the left-hand compartment. After playing the 'game' for a while, an equilibrium state is reached when, although particles still move from left to right and from right to left, there are roughly constant numbers of particles in the compartments. When equilibrium occurs we end up with roughly the same number of particles in each compartment. Thus the particles have diffused from the high concentration region to the low concentration region.

The rate at which particles diffuse in steady state conditions, i.e. when there are no changes in the system with time, is proportional to the concentration gradient dC/dx, this being known as *Fick's first law*:

$$\text{rate of diffusion} = -D\frac{dC}{dx}$$

[9]

D is called the *diffusion coefficient* (in units of m²/s), its value depending on the temperature and system concerned. The minus sign is to show that a positive flow of particles goes in the direction of falling concentration, i.e. a negative value of dC/dx. The rate of diffusion is in atoms moved per square metre per second or kilograms per square metre per second, the concentration in atoms per cubic metre or kilograms per cubic metre, x in metres and hence D in m²/s.

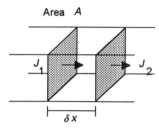

Area A

Figure 2.45 *Diffusion into a small region*

Maths in action

For most engineering applications, such as the diffusion of carbon into the surface of iron to harden its surface, the concentration of carbon atoms in the surface will change with time as more and more carbon moves into the surface. Thus if the rate of diffusion into a small region (Figure 2.45) is J_1 and the rate out of the region is J_2, then in a time dt the number of particles in the region will increase by $(J_2 - J_1)A dt$. Thus the change in concentration dC is given by:

$$dCAdx = (J_2 - J_1)Adt$$

Hence dC/dt = $(J_2 - J_1)$/dx = −dJ/dx, where −dJ is the reduction in rate of diffusion. Thus, using equation [9]:

$$\frac{dC}{dt} = -\frac{dJ}{dx} = \frac{d}{dx}\left(D\frac{dC}{dx}\right)$$

and, if D is constant, for cases of such non-steady state diffusion where the concentration changes with time:

$$\frac{dC}{dt} = D\frac{d^2C}{dx^2} \tag{10}$$

This is termed *Fick's second law*. In the case of diffusion where the concentration of the supply atoms remains constant, the solution of the above differential equation is:

$$\frac{C_s - C_x}{C_s - C_0} = \text{erf}\left(\frac{x}{2\sqrt{Dt}}\right) \tag{11}$$

where C_s is the concentration of the diffusing atoms and is constant, C_0 is the concentration of the diffused atoms at the start of the process, i.e. $t = 0$, C_x is the concentration at some distance x from the source of diffusing atoms at time t, and erf is a special function called the *Gaussian error function*. Table 2.7 shows values erf y for different values of the bracketed term y. Up to about 0.6, erf y is approximately the same as y.

Table 2.7 *erf y values*

y	erf y	y	erf y
0	0	0.85	0.7707
0.05	0.0564	0.90	0.7970
0.10	0.1125	0.95	0.8209
0.15	0.1680	1.00	0.8427
0.20	0.2227	1.1	0.8802
0.25	0.2763	1.2	0.9103
0.30	0.3286	1.3	0.9340
0.35	0.3794	1.4	0.9523
0.40	0.4284	1.5	0.9661
0.45	0.4755	1.6	0.9763
0.50	0.5205	1.7	0.9838
0.55	0.5633	1.8	0.9891
0.60	0.6039	1.9	0.9928
0.65	0.6420	2.0	0.9953
0.70	0.6778	2.2	0.9981
0.75	0.7112	2.4	0.9993
0.80	0.7421		

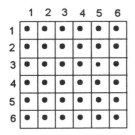

(a) Initially

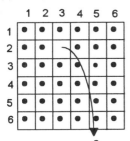

(b) Dice give 3, 2

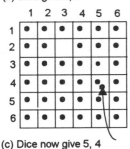

(c) Dice now give 5, 4

Figure 2.46 *Distributing quanta*

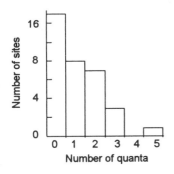

Figure 2.47 *Distribution*

Diffusion and temperature

To move, say, an atom from one interstitial site in a solid to another vacant site requires energy. Energy is needed to break bonds, this being termed the *activation energy*. For this to spontaneously occur, the energy must come from the solid being at some temperature. The thermal energy will not be uniformly distributed among the atoms in a solid but randomly distributed with some at some instant of time having more than others. At any temperature, only a fraction of the atoms might have sufficient energy to reach the activation energy value and so be able to move.

Suppose we have n packets of energy among N atoms, the term quantum is used for such a packet or energy and quanta when we have more than one. How will the quanta be arranged among the atoms? Consider a simple model of a solid as a board divided into 36 squares, six squares by six, and having quanta moved purely by chance among the board squares. Initially, suppose we have one packet of energy per square (Figure 2.46(a)). We use the chance throw of two six-sided dice to determine which square to select to move a packet of energy from (Figure 2.46(b)) and then a further two throws to select the square to move it to (Figure 2.46(c)). After 100 moves the type of result obtained is 17 squares with no quanta, 8 squares with one, 7 squares with two, 3 squares with three, 0 squares with four and 1 square with five. Figure 2.47 shows a histogram of these results. The result with just this number of sites and quanta is fairly crude, but the general picture shows a distribution which approximates to an exponential. With more quanta to spread around, i.e. a higher temperature, we obtain a less steep exponential. This leads to the probability of an atom having the activation energy E_a at a particular temperature as:

$$\text{probability} \propto \exp\left(-E_a/kT\right) \qquad [10]$$

where k is Boltzmann's constant and T the temperature on the kelvin scale. The probability is the number of such atoms divided by the total number present in the system. The rate R at which atoms diffuse into the vacancies is proportional to the number of atoms able to diffuse and so:

$$R = A \exp\left(-E_a/kT\right) \qquad [11]$$

This equation is termed *Arrhenius's rate equation*. Thus the greater the activation energy the slower the diffusion rate, the greater the temperature the faster the diffusion rate.

We can thus express the diffusion coefficient in terms of equation [11] as:

$$D = D_0 \exp\left(-E_a/kT\right) \qquad [12]$$

where D_0 is the proportionality constant and is a property of the material concerned.

Example

For the movement of nickel atoms in face-centred cubic iron, D_0 is 7.7×10^{-5} m^2/s and E_a is 280 kJ/mol. $k = 1.38 \times 10^{-23}$ J/K. Determine the diffusion coefficient at (a) 500°C, (b) 1000°C.

Using equation [12]:

(a)$D = 7.7 \times 10^{-5}\ e^{-(280 \times 10^3 / 6.02 \times 10^{23})/(1.38 \times 10^{-23} \times 773)}$

$= 8.9 \times 10^{-24}$ m^2/s

(b)$D = 7.7 \times 10^{-5}\ e^{-(280 \times 10^3 / 6.02 \times 10^{23})/(1.38 \times 10^{-23} \times 1273)}$

$= 2.4 \times 10^{-16}$ m^2/s

2.1.8 Case studies

An example of a process involving diffusion are the carburising treatment for the surface hardening of steels. In carburising, the steel is placed in an atmosphere which is rich in carbon; in pack carburising the steel is placed in a container filled with charcoal and heated so that the charcoal forms carbon monoxide. Diffusion of carbon then occurs into the steel surface.

Example

A low-carbon steel which initially contained just 0.20% carbon is to be surface hardened by diffusing more carbon into its surface. Determine the length of time needed to obtain a concentration of carbon at 0.35% at a distance 1.00 mm below the surface if the carbon concentration at its surface remains constant at 0.95% and the diffusion coefficient at the temperature of the process is 1.74×10^{-11} m^2 s^{-1}.

Using equation [11]:

$$\frac{0.95 - 0.35}{0.95 - 0.2} = 0.80 = \mathrm{erf}\left(\frac{1.0 \times 10^{-3}}{2\sqrt{1.74 \times 10^{-11}t}}\right)$$

Table 4.6 indicates that when erf $y = 0.8$, $y = 0.9$. Thus

$$0.9 = \frac{1.0 \times 10^{-3}}{2\sqrt{1.74 \times 10^{-11}t}}$$

Hence $t = 17\ 738$ s or nearly 5 hours.

Production of transistors

In the production of semiconductor devices, e.g. transistors, a process that is used to replace some of the atoms in a semiconductor by atoms of another element, i.e. change its doping, involves a gas containing the dopant in contact with the hot semiconductor so the dopant then diffuses into the surface.

Example

In the production of a transistor, phosphorus is to be diffused into silicon. Determine the time needed to give a concentration of 10^{21} atoms/m^3 at a depth of 1μm if initially there was zero concentration, the surface concentration of phosphorus remains constant at 10^{24} atoms/m^3 and the diffusion coefficient at the temperature of the process is 3.0×10^{-17} m^2/s.

Using equation [11]:

$$\frac{10^{24} - 10^{21}}{10^{24}} = 1 - 0.001 = \text{erf}\left(\frac{1 \times 10^{-6}}{2\sqrt{3.0 \times 10^{-17}t}}\right)$$

When erf $y = 0.999$, y is about 2.4. Thus t is about 1447 s or 0.40 h.

2.1.9 Viscosity of liquids

A characteristic of liquids is that they can be made to flow. Some liquids flow more easily than others and the property used to describe the ease with which a liquid flows is called *viscosity*. The lower the viscosity the more easily a liquid will flow. For example, a heavy machine oil has a higher viscosity than water and does not, therefore, pour out of a can as quickly as water. Viscosity depends on temperature; the higher the temperature the lower the viscosity and so the more easily the liquid flows. Warm machine oil flows more readily than cold machine oil.

The simplest form of flow is called *streamline* or *laminar* flow in that the flow can be considered in terms of layers of liquid sliding over each other. If liquid flow over a stationary surface is considered (Figure 2.48), the layer of liquid in contact with the surface is stationary and successive layers, as distance x from the surface increases, have increasing velocities so that a velocity gradient exists near the surface. This velocity gradient (dv/dx) depends on the shearing force per unit area (F/A) responsible for sliding the layers over the surface and the viscosity η of the liquid:

$$\frac{F}{A} = \eta \frac{dv}{dx} \tag{13}$$

Velocities in successive layers

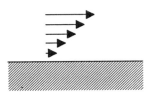

Figure 2.48 *Streamline flow*

Key points

Viscosity is an important property when it comes to pouring liquid metals, glasses or polymers into moulds. Liquid metals tend to have, at their melting points, viscosities similar to that of water at room temperature and they flow fairly easily. Glasses and polymers, however, can have much higher viscosities and so flow less readily into all parts of a mould.

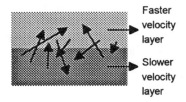

Figure 2.49 *Molecules moving between layers*

A liquid consists of a large number of molecules in random motion, the higher the temperature the greater the average molecular speed. When there is laminar flow there is an orderly velocity imposed in a particular direction and superimposed on top of the random molecular velocity. The molecules immediately adjacent to the stationary surface experience short-range attractive forces to the atoms in that surface and so give a slower moving layer adjacent to the surface. The random velocity element of the velocity of molecules means that they will move between layers (Figure 2.49). We have a situation comparable to that of diffusion discussed earlier in this chapter. This means that the faster velocity layer will lose some of its molecules and gain some of the molecules from a slower moving layer. The effect of lower velocity molecules entering the faster moving layer will be to lower its average velocity and so exert a drag on it. This drag is what we have termed viscosity.

The viscosity of many liquids is independent of the velocity gradient in a liquid. Such liquids are termed *Newtonian liquids*. There are, however, liquids, notably polymers, which decrease in viscosity when the velocity gradient increases. Non-drip paints are such liquids, flowing more easily when painted onto a surface and then less easily in the tin and when the painting action ceases. Polymers are long chain molecules. These long chains become easily tangled with each other. The paint in the tin consists of tangled molecules and because of this does not flow easily. When the paint is brushed on a surface, the act of brushing aligns many molecules so that their chains point in the same direction. This allows them to slide more easily over one another and so the viscosity decreases.

2.1.10 Surfaces

Molecules in the bulk of a liquid are surrounded by other molecules and are subject to attractive forces which are roughly the same in all directions (Figure 2.50). However, at the surface of the liquid, the molecules have no liquid molecules above them so there is just a net attractive force downwards from intermolecular attraction. If the surface of the liquid is to be increased, more molecules have to be moved into the surface against this attractive force. Energy is thus required to increase surface area; the term *free surface energy* is defined as the energy required to produce unit area of surface. Water at room temperature has a free surface energy of about 0.070 J/m^2 and liquid metals much higher values, e.g. molten aluminium 0.50 J/m^2 and molten iron 1.50 J/m^2.

In the same way that liquids have free surface energies, so also do solids. Atoms in the bulk of a solid are subject to attractive forces in all directions, while those in the surface are subject only to inward-directed forces. Thus to increase the surface area of a solid requires energy. For a crack to propagate, new solid surfaces have to be produced and this requires energy.

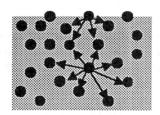

Figure 2.50 *Intermolecular force in liquids*

Activity

When a fine mist of water is sprayed onto a horizontal pane of glass, the fine droplets of liquid form an approximately hemispherical shape. Why is this? If the water is sprayed onto a vertical pane of glass, what is the shape taken by the droplets? Would you expect it to be different?

The force of attraction between the molecules in a small droplet is greater than the force of gravity acting on them. Hence the droplet clings to itself and forms the hemispherical shape. If the pane of glass were vertical, then we would expect the force of gravity to pull the water molecules apart. If the droplet is small enough, then the attraction between the molecules is still greater than the force of gravity and so the droplet remains hemispherical. As the droplet becomes larger, the force of gravity will eventually overcome the force of attraction between molecules and cause the water to run down the vertical pane of glass. But what happens to a large droplet on a horizontal pane of glass?

Problems 2.1

1 An isotope of germanium has an atomic number of 32 and an atomic mass number of 74. How many (a) protons, (b) neutrons, (c) electrons, have neutral atoms of this isotope?
2 Give the notation for the electronic structure of (a) Fe, (b) Fe^{2+}, (c) Li^+, (d) Cl^-.
3 Describe the types of bonds between atoms, ions or molecules.
4 Figure 2.51 shows how the resultant force between two ions in two solids X and Y vary with their separation. Which of the two has (a) the largest ions, (b) the largest tensile modulus?
5 How does a crystalline material differ in structure from an amorphous material?
6 Explain how differences in the way ions are packed in a solid lead to the structures of simple cubic, face-centred cubic, body-centred cubic and close-packed hexagonal.
7 Determine the density of iron with a body-centred structure if the iron has an atomic radius of 0.124 nm. The atomic mass of iron is 55.85 g/mol and Avogadro's number is 6.02×10^{23}.
8 Determine the lattice parameter a for iron when in a body-centred cubic structure the atom has a radius of 0.1238 nm.
9 Determine the density of nickel with a face-centred structure if the nickel has an atomic radius of 0.124 nm. The atomic mass of nickel is 58.69 g/mol and Avogadro's number 6.02×10^{23}.

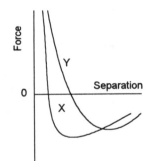

Figure 2.51   *Problem 4*

10 Determine the density of aluminium with a face-centred structure if the aluminium has an atomic radius of 0.143 nm. The atomic mass of aluminium is 26.97 g/mol and Avogadro's number 6.02×10^{23}.

11 Determine the atomic radius of molybdenum atoms if molybdenum has a body-centred cubic structure and a density of 10.2 kg/m³. The atomic mass of molybdenum is 95.94 g/mol and Avogadro's number 6.02×10^{23}.

12 Determine the volume of the unit cell for beryllium if it has a hexagonal close-packed structure with $a = 0.228$ 56 nm and $c = 0.358$ 32 nm.

13 Determine the percentage volume change that occurs when titanium changes from a body-centred cubic structure to a hexagonal close-packed structure. With the body-centred cubic structure $a = 0.332$ nm and with hexagonal close-packed structure $a = 0.2956$ nm and $c = 0.4683$ nm.

14 Gallium has an orthorhombic crystalline structure with the lattice parameters $a = 0.451$ 86 nm, $b = 0.451$ 86 nm and $c = 0.765$ 70 nm. If the atomic radius is 0.1218 nm, the atomic mass 69.72 g/mol, the density 5904 kg/m³ and Avogadro's number is 6.02×10^{23}, determine the form of orthorhombic structure and the packing factor.

15 Determine the crystal form and the packing factor for a potassium chloride crystal if the potassium ions have a radius of 0.133 nm and the chlorine ions 0.181 nm.

16 Determine the diameter of the atom that could fit in the octahedral interstitial site of body-centred cubic iron if the iron ions have radii of 0.1258 nm.

17 Determine the diameter of the atom that could fit in the tetrahedral interstitial site of face-centred cubic nickel if the nickel ions have radii of 0.069 nm.

18 A low-carbon steel which initially contained just 0.20% carbon is to be surface hardened by diffusing more carbon into its surface. Determine the length of time needed to obtain a concentration of carbon at 0.40% at a distance 0.50 mm below the surface if the carbon concentration at its surface remains constant at 0.90% and the diffusion coefficient at the temperature of the process is 1.74×10^{-11} m² s⁻¹.

19 A low-carbon steel which initially contains just 0.10% carbon is to be surface hardened by diffusing more carbon into its surface. Determine the length of time needed to obtain a concentration of carbon at 0.45% at a distance 2.00 mm below the surface if the carbon concentration at its surface remains constant at 1.2% and the diffusion coefficient at the temperature of the process is 1.69×10^{-11} m² s⁻¹.

20 In the production of a transistor, arsenic is to be diffused into silicon. Determine the time needed to give a concentration of 1.5×10^{22} atoms/m³ at a depth of 1.2 mm if initially there was zero concentration, the surface concentration of arsenic remains constant at 5×10^{24} atoms/m³ and the diffusion coefficient at the temperature of the process is 3.0×10^{-18} m²/s.

21 Determine the diffusion coefficient for the movement of carbon atoms in face-centred cubic iron at 1200 K if D_0 is 2.0×10^{-5} m²/s and E_a is 142 kJ/mol. $k = 1.38 \times 10^{-23}$ J/K.

22 A low-carbon steel which initially contains just 0.10% carbon is to be surface hardened by diffusing more carbon into its surface. Determine the time needed to obtain a concentration of carbon of 0.45% at 2.00 mm below the surface at 1000ºC if the carbon concentration at its surface remains constant at 1.2%, $D_0 = 0.23 \times 10^{-4}$ m²/s and $E_a = 137.7$ kJ/mol.

2.2 Structure of metals

Figure 2.52 *Grain structure in a metal, i.e. regions of crystallinity*

Metals are crystalline substances. This may seem a strange statement in that metals do not generally look like crystals, with their geometrically regular shapes. However, if we consider a metal in solidifying from the liquid as not growing as a single crystal but having crystals starting to grow at a number of points within the liquid, then the result is a mass of crystals. Each crystal is prevented from reaching geometrically regular shapes by neighbouring crystals growing and restricting its growth (Figure 2.52). The result is said to be *polycrystalline* and the term *grain* is used to describe regions in the metal for which there are orderly arrangements of particles, the arrangement following the crystalline form for that metal.

Activity

A simple model of a metal with grains is given if a raft of bubbles is produced on the surface of a soapy liquid by bubbling a gas through a jet (Figure 2.53). The bubbles pack together in an orderly and repetitive manner, but if 'growth' is started at a number of centres then 'grains' are produced. At the boundaries between the 'grains' the regular pattern breaks down as the pattern changes from the orderly arrangement in one 'grain' to the orderly arrangement in the next 'grain' and gives a good representation of a metal. You can try this for yourself at home with a fine tube and some soapy water.

Exposing the grains in metal surface

The grains in the surface of a metal are not generally visible, though an exception is the very large grains which are readily visible in the surface of galvanised steel objects. Grains can, however, be made visible by careful etching of the polished surface of the metal with a suitable chemical. The chemical preferentially attacks the grain boundaries. For example, in the case of copper and its alloys, concentrated nitric acid can be used. In the case of carbon and alloy steels of medium carbon content, an etchant called nital can be used. Nital is a mixture of nitric acid and alcohol, typically 5 ml of acid to 95 ml of alcohol.

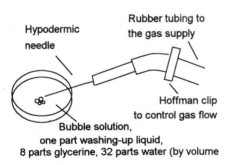

Hypodermic needle

Rubber tubing to the gas supply

Hoffman clip to control gas flow

Bubble solution, one part washing-up liquid, 8 parts glycerine, 32 parts water (by volume

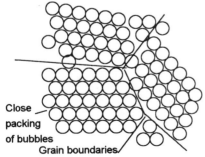

Close packing of bubbles

Grain boundaries

Figure 2.53 *(a) Simple arrangement for producing bubbles, (b) a raft of bubbles acting as a model for grains*

Key points

A grain within a metal is a regular, ordered three-dimensional arrangement of atoms. This is also called a crystal. Metals are polycrystalline and adjacent grains or crystals will have the same arrangement of atoms but a different orientation relative to each other.

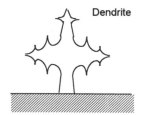

Growth out from a cold surface

(a)

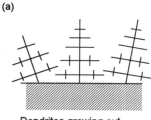

Dendrites growing out from cold surface

(b)

Figure 2.54 *Dendrite growth*

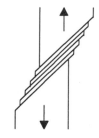

Figure 2.55 *Slip: planes of atoms sliding over each other*

Growth of metal crystals

As indicated in Table 2.4, chromium and iron can form body-centred cubic lattices, aluminium and copper face-centred cubic and magnesium and titanium hexagonal close-packed lattices. If we made a crystal model from spheres, we could extend the model by adding spheres equally to all faces or perhaps preferentially add them to one face. However, when crystals grow, they tend to grow in a 'tree like' form of growth, adding branches and then branches on branches.

Energy is needed to change a solid at its melting point to a liquid without any change in temperature occurring; this energy is called *latent heat*. Similarly, when a liquid at the fusion point (i.e. the melting point) changes to a solid, energy has to be removed, no change in temperature occurring during the change of state. This is the latent heat. Thus when the liquid metal in the immediate vicinity of a metal crystal face solidifies, energy is released which warms up the liquid in front of the advancing crystal face. This slows, or stops, further growth in that direction. The result of this action is that spikes develop as the crystal grows in the directions in which the liquid is coolest (Figure 2.54(a)). As these warm up the liquid in turn, so secondary, and then tertiary, spikes develop as the growth continues in the directions in which the liquid is coolest. This type of 'tree-like' growth is termed a *dendrite*. It is the pattern formed by water vapour freezing on a cold window.

With face-centred and body-centred cubic structures, the cube faces are the preferred planes of growth and thus the dendrites grow in the direction of the cube edges. The dendrites grow (Figure 2.54(b)) until they meet other dendrites. The space between the dendrites then solidifies. When solidification is complete, there is little evidence of the dendrite structure remaining.

2.2.1 Slip

If we think of a crystal as an orderly arrangement of stacked spheres, then it turns out that, in certain directions, the arrangement behaves as though is has layers of spheres and we can slide one layer over another. Likewise, within a metal single crystal movement occurs along *slip planes* and the result is rather like Figure 2.55. The following are terms used to describe slip:

- *Slip plane*
  This is the plane between the planes of atoms on which slip occurs.

- *Slip direction*
  This is the direction in which slip occurs.

- *Slip step*
  This is the step produced on the surface of a crystal surface as a result of slip.

- *Slip vector*

  This is the direction and magnitude of an increment of slip.

- *Slip system*

  This is the combination of a slip plane and a slip direction.

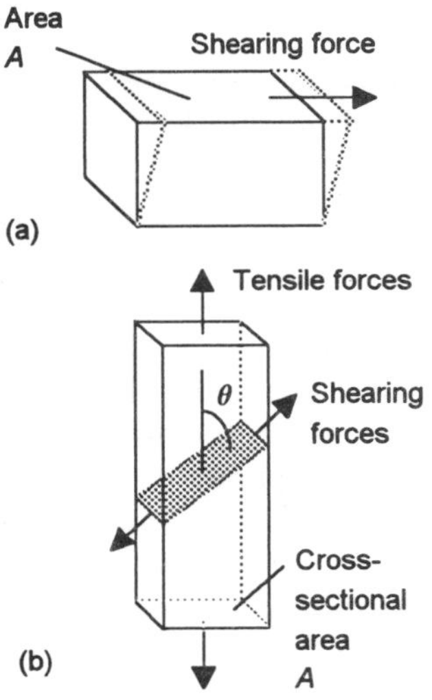

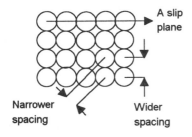

Figure 2.56 *Shearing*

Maths in action

The slip plane is at an angle to the stretching forces in Figure 2.55 because we are concerned with shearing. If you think of a pack of cards, we are sliding one card over its neighbour (Figure 2.56(a)). Thus, for Figure 2.56(b), the shear stress τ is the shearing force divided by the area being sheared. Since the area is $A/\sin\theta$ and the shearing force is the component $F\cos\theta$ of the stretching force F, we have $\tau = (F\cos\theta \sin\theta)/A$. Since the tensile stress is $\sigma = F/A$ then:

$$\tau = \sigma\cos\theta \sin\theta = \tfrac{1}{2}\sin 2\theta \qquad [14]$$

The maximum value of the shear stress is thus when $\theta = 45°$ and is $\tfrac{1}{2}\sigma$.

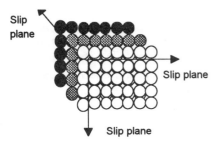

Figure 2.57 *Slip with a simple cubic crystal*

Slip planes

Crystals do not slip on any arbitrary plane or in any arbitrary direction. The favoured slip plane is always that in which the atoms are closest packed. This is because the most closely packed planes are wider apart than others and 'smoother'. Figure 2.57 illustrates this for a simple cubic crystal, there being two directions of possible slip for the plane indicated, to the left and to the right. Because crystals are generally highly symmetrical, they have a number of equivalent slip planes and directions. With the simple cubic crystal, there are identical planes at right angles to the slip plane indicated in Figure 2.57. Figure 2.58 illustrates this. The system has thus three equivalent sets of slip planes. Slip in metallic crystalline structures occurs on a number of slip systems, these being characteristic of the crystal structure concerned. Figure 2.59 shows the slip systems for face-centred cubic, body-centred cubic and hexagonal close-packed crystals.

With the face-centred cubic structure, no matter from which direction a stress is applied, there will be resolved shearing stresses acting on several slip planes. For this reason, such structures give materials which are comparatively soft and ductile. With the body-centred cubic, though there are many slip directions possible, metals with this structure tend to be harder and less ductile because the planes are less densely packed than those of the face-centred structure. With the hexagonal close-packed system, the limited number of slip systems restricts the ductility.

Figure 2.58 *Equivalent slip planes*

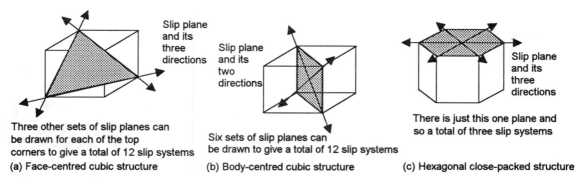

Three other sets of slip planes can be drawn for each of the top corners to give a total of 12 slip systems
(a) Face-centred cubic structure

Six sets of slip planes can be drawn to give a total of 12 slip systems
(b) Body-centred cubic structure

There is just this one plane and so a total of three slip systems
(c) Hexagonal close-packed structure

Figure 2.59 *Slip systems*

Slip does not extend from one grain to another in a poly-crystalline material, the grain boundaries restricting the slip to within a grain. As a consequence, the bigger the grains the more slip that can occur and hence the greater the plastic deformation. A fine grain structure should therefore have less slip and be less ductile. A brittle material is thus one in which each little slip process is confined to a short run in the metal and not allowed to spread, a ductile material being when the slip is not confined to a short run and can spread over a large part of the metal.

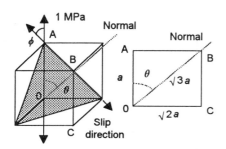

Figure 2.60 *Example*

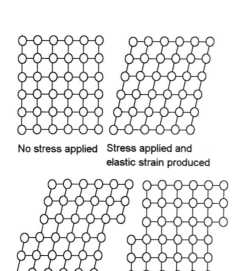

No stress applied   Stress applied and elastic strain produced

Stress applied and yielding occurs   Stress removed showing permanent deformation

Figure 2.61 *Block slip model*

Example

Determine the component of the shear stress on the slip system of the unit cell of the face-centred copper crystal shown in Figure 2.60 if a tensile stress of 1 MPa is applied in the direction shown.

The area of the slip plane is $A/\cos\theta$, where A is the surface area of the unit cell and θ is the angle between the normal to the slip plane and the direction of the applied tensile stress. The shear force is the component of the applied force F along the slip plane and is thus $F\cos\phi$, where ϕ is the angle between the slip direction and the applied force. Thus the shear stress τ is $F\cos\phi/(A/\cos\theta) = (F/A)\cos\phi\cos\theta$ and so:

$$\tau = \sigma\cos\phi\cos\theta \qquad [15]$$

where σ is the tensile stress. This is known as *Schmid's law*. The angle ϕ between the applied stress and the slip direction is 45° and $\cos\theta = a/\sqrt{3}a = 1/\sqrt{3}$, where a is the lattice constant. Thus $\theta = 54.7°$ and the shear stress = 1 cos 45° sin 54.7° = 0.58 MPa.

Block slip model

A simple theory to explain the elastic and plastic behaviour of metals is the *block slip model* (Figure 2.61). A metal is considered to be made of blocks of atoms which can be made to move relative

Key points

The bigger the grains the more slip that can occur and hence the greater the plastic deformation. Thus, a fine grain structure should be less ductile. A brittle material has slip confined to short runs in the metal, a ductile material has slip allowed to spread over a large part of the metal.

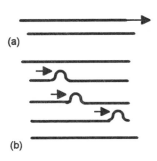

Figure 2.62 *Moving a carpet: (a) as a whole, (b) by moving a ruck along*

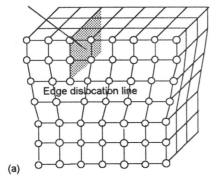

to each other by the application of shear stress. With this model, slip occurs in a single movement involving one whole plane of atoms sliding bodily over another. The shear stress needed to achieve this can be calculated and gives a yield stress of the order of 10^{10} Pa for a pure metal. This compares with the values obtained in practice of about 10^6 Pa to 10^7 Pa. The model is clearly inadequate.

2.2.2 Dislocations

The block slip model has all the atoms arranged in a perfectly orderly manner. If, however, we consider the arrangement to be imperfect then permanent deformations can be produced with much less stress. When you have a large carpet which is perfectly flat on the floor, it requires quite an effort to slide the entire carpet and make it move across the floor (Figure 2.62(a)). But if there is a ruck in the carpet (Figure 2.62(b)), then the carpet can be slid over the floor by pushing the ruck along a bit at a time and considerably less effort is required. This is the type of movement which is considered to take place within a metal, the 'ruck' in the lattice being a *dislocation* of atoms due to imperfect packing of the atoms within the lattice.

Figure 2.63(a) shows the type of arrangement of atoms that might be considered to occur with what is called an *edge dislocation*. With such a dislocation, the line of dislocation is at right angles to the slip plane. Figure 2.63(a)(b)(c) shows the sequence of movements that occur when stress is applied and permanent deformation occurs. The dislocation moves through the array of atoms without wholesale movement of planes of atoms past each other; it is a bit-by-bit process like the ruck in the carpet. The movement of dislocations is similar to that given by the block slip model but requires smaller stresses as only a few bonds are being altered or broken at any one time.

The magnitude and direction of the slip resulting from the motion of a single dislocation is called the *Burgers vector*. Thus in Figure 2.63(c), **b** is the Burgers vector for the edge dislocation. The vector is perpendicular to the line of the dislocation.

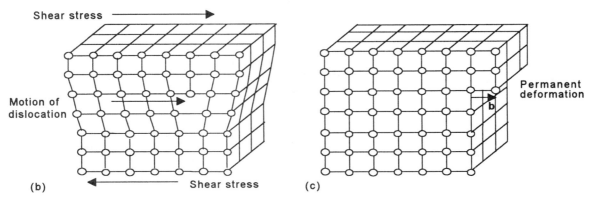

Figure 2.63 *Shear stress causing deformation when an edge location is present*

Figure 2.64 shows the form of a *screw dislocation* and its movement through the lattice of atoms under the action of shearing. With a screw dislocation, the line of dislocation is parallel to the slip plane (with the edge location it was at right angles to the slip plane). The direction of motion of the dislocation is perpendicular to the Burgers vector, this being in the direction of the slip.

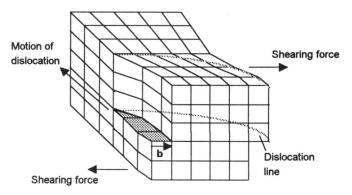

Figure 2.64 *A screw dislocation*

In practice, dislocations are often considered to be a combination of edge and screw dislocations and are termed *mixed dislocations*. For the curved dislocation line in Figure 2.65, the dislocation is of the purely screw type at the left where it enters the crystal and of the edge type where it leaves the crystal on the right. Within the crystal the dislocation is a mixture of edge and screw components.

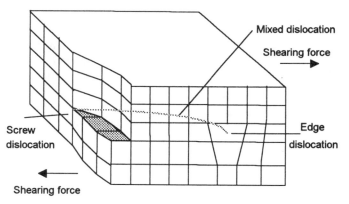

Figure 2.65 *Screw dislocation at the left, becoming mixed and finally edge at the right*

Screw dislocation and crystal growth

Screw dislocations were first postulated to explain how crystals grow, rather than the yielding of metals. Suppose we have a material crystallising in the simple cubic crystal lattice. If at some instant the crystal has grown to the shape shown in Figure 2.66(a), then it is easier if the next atoms are deposited as the steps (Figure 2.66(b)). A consequence of this is that the infilling of the steps

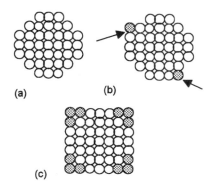

Figure 2.66 *Crystal growth*

results in the emergence of a cubic shape (Figure 2.66(c)) with no steps for further infilling. However, if a screw dislocation is present at the surface (Figure 2.67), it creates a perpetual step since the surface becomes like a spiral ramp so atoms can be added indefinitely to the step.

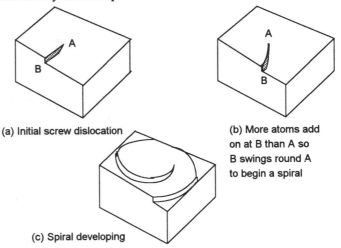

(a) Initial screw dislocation

(b) More atoms add on at B than A so B swings round A to begin a spiral

(c) Spiral developing

Figure 2.67 *Growth spiral on crystal surface*

Point defects

Edge and screw locations are *line defects*, being long in one direction while measuring only a few atomic diameters at right angles to their length. *Point defects* can occur in crystal lattices, these being highly localised disruptions of the lattice involving one or just a few atoms, and can be:

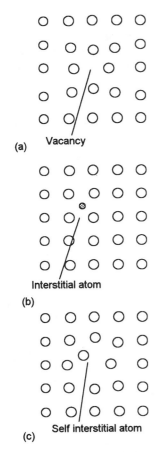

(a) Vacancy

(b) Interstitial atom

(c) Self interstitial atom

Figure 2.68 *Point defects*

- ### Vacancies

  A vacancy occurs as a result of an atom missing from a normal site (Figure 2.68(a)). The distribution of heat quanta among atoms can result in a particular atom acquiring sufficient energy to escape, the higher the temperature the more likely this is to occur (equation [10]). Thus the number of vacancies per cubic metre n_v is given by:

  $$n_v = n\,e^{-E_a/kT}$$

  where n is the number of lattice points per cubic metre, E_a the energy needed to produce a vacancy, T the temperature on the kelvin scale and k is Boltzmann's constant. The concentration of vacancies is important in determining the amount of diffusion that can occur in solids with atoms moving from lattice sites into vacancies.

- ### Interstitial defects

  This type of defect is formed when an extra atom is inserted into the lattice at a normally unoccupied site (Figure 2.68(b)). Such atoms may be impurities or deliberate alloying additions.

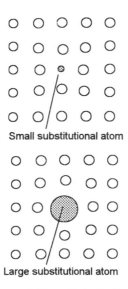

Small substitutional atom

Large substitutional atom

Figure 2.69 *Substitutional point defects*

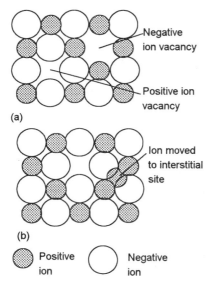

(a)

(b)

Positive ion   Negative ion

Figure 2.70 *(a) Schottky defect, (b) Frenkel defect*

• *Self-interstitial defect*

This is when an atom is displaced from its normal position within the lattice to an interstitial position (Figure 2.68(c)).

• *Substitutional defects*

This is when one atom in a lattice is replaced by one of a different type (Figure 2.69). Such substitutional atoms may be smaller or larger than the atoms being replaced. Such atoms may be impurities or deliberate alloying additions.

All the point defects produce a local distortion in the lattice, the atoms neighbouring the point defect being either pulled closer to the point or moved further away.

Additional effects can occur when the removal or addition of atoms changes the local charge on the lattice. Thus with a sodium chloride crystal, the removal of a positive sodium ion to create a vacancy must be counterbalanced by the removal of a negative chlorine ion in order to maintain neutrality. The pair of vacancies (Figure 2.70(a)) is termed a *Schottky defect*. If a positive ion moves from a normal lattice point to an interstitial point, then the defect is termed a *Frenkel defect* (Figure 2.70b)).

Example

Determine the number of vacancies per cubic metre occurring in copper at 100°C if 83.7 kJ/mol are required to produce a vacancy. Copper has a face-centred lattice with a lattice parameter of 0.3615 nm. Avogadro number $= 6.02 \times 10^{23}$, Boltzmann's constant $k = 1.38 \times 10^{-23}$ J/K.

There are four atoms per unit cell and so the number of lattice points per cubic metre is $4/(0.3615 \times 10^{-9})^3 = 8.47 \times 10^{28}$ atoms/m³. Thus, using equation [15]:

$$n_v = 8.47 \times 10^{28} \; e^{-(83\,700/6.02\times10^{23})/(1.38\times10^{-23}\times373)}$$

$$= 1.57 \times 10^{17}$$

Planar defects

Planar defects are atomic in one dimension and large in the others. Grain boundaries, i.e. the interfaces between crystals in a polycrystalline material, are planar defects. There are other planar defects, a common one being the *twin interface*. A crystal is twinned when one portion of its lattice is a mirror image of the neighbouring portion (Figure 2.71). Twins can be produced during the growth of the crystal or by shear stress causing one portion of the crystal to become deformed. The twinning process differs from slip in that with twinning there is rotation of part of the lattice.

The stress to twin a crystal tends to be higher than that required for slip and thus generally slip is the normal deformation process. However, in hexagonal close-packed polycrystalline structures,

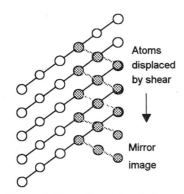

Figure 2.71 *Atoms in a twin*

twinning may play a significant part in deformation because such structures only have three slip planes and thus slip is restricted. Twinning is also found in body-centred cubic crystals, e.g. iron and chromium, which were deformed at very low temperatures. Face-centred cubic metals show the least tendency to twin.

Movement of dislocations

What happens when two dislocations come close to each other during their movement through a metal? With a dislocation, the atoms on one side of the slip plane are in compression and on the other side in tension. When two dislocations come together, as in Figure 2.72, the regions of compression can impinge on each other and so hinder the movement of the dislocations. If the movement of the dislocations is such as to bring the compression region of one dislocation against the tension region of another dislocation (Figure 2.73) then it is possible for the two dislocations to annihilate each other.

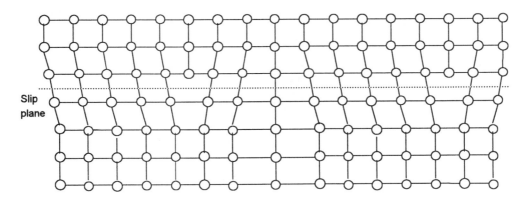

Figure 2.72 *Two dislocations of the same sign on the same slip plane 'repelling' each other*

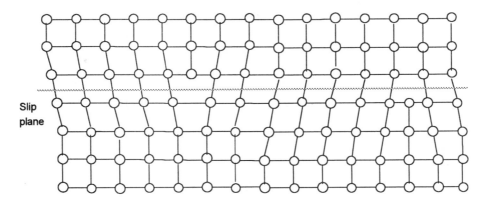

Figure 5.22 *Two dislocations of opposite sign on the same slip plane can move together and annihilate each other*

In general, the more dislocations a metal has, the more they get in the way of each other and so the more difficult it is for the dislocations to move through the metal. More stress is needed to cause yielding. *Work hardening* occurs as a result of a material being plastically deformed, this increasing the dislocation density.

The movement of dislocations through a metal is also hindered by the grain boundaries. The more grain boundaries there are in a metal, the more difficult it is to produce yielding. More grain boundaries occur when the grain size is small, thus a treatment which reduces grain size makes a metal stronger while one which increases grain size makes it weaker.

The movement of dislocations is hindered by anything that destroys the continuity of the atomic array. The presence of 'foreign' atoms can distort the atomic array of a metal and so hinder the movement of dislocations. *Dispersion hardening* increases the yield stress of a metal by producing a dispersion of fine particles throughout the material, these hindering the movement of dislocations. *Alloying* involves the introduction of foreign atoms into a crystal lattice, producing interstitial and substitutional point defects which hinder the movement of dislocations. Hence alloys tend to have a higher yield stress than the parent metal alone, indeed pure metals like copper and iron are very soft and no use as an engineering material but when alloyed become much stronger. This is referred to as *solution hardening*.

Dislocations and temperature

Dislocations can annihilate each other if they are of opposite sign and moving along the same slip plane. However, this does not occur too frequently at room temperature. At higher temperatures another annihilation mechanism can occur. This is because at higher temperatures diffusion of atoms can become significant and leads to what is called *dislocation climb* when an edge dislocation moves in a direction at right angles to its slip plane to annihilate another dislocation (Figure 2.74). A consequence of this type of action is that more dislocations can annihilate each other.

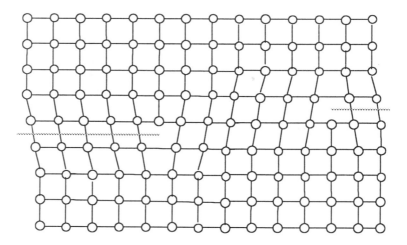

Figure 2.74 *Dislocation climb*

If a work-hardened material is heated to about 0.3 to 0.4 times its melting temperature (in degrees kelvin), sufficient diffusion occurs for dislocation climb to occur and the number of dislocations become reduced. At this temperature there is no change in grain size. The result of such changes is that residual stresses are released, *recovery* being said to occur, and there is a slight reduction in yield stress. At higher temperatures *recrystallisation* occurs and new grains of low dislocation density are produced. The result is a marked decrease in yield stress. The heat treatment which allows recrystallisation to occur with the consequent decrease in yield stress is called *annealing*.

The effects of diffusion and consequent annihilation of dislocations as a result of dislocation climb is particularly evident in the behaviour of metals when subjected to a load for a long period of time at a high temperature. The strain increases steadily with time. This effect is called *creep*.

Dislocation multiplication

The number of dislocations passing through a unit area of a polycrystalline material is of the order of 10 to 100 million per square centimetre. These occur as a result of the fabrication process. However, if the material is subject to a plastic deformation, this dislocation density increases to as much as a million million per square centimetre. A possible mechanism for this multiplication is called the *Frank-Read source*, these being the names of those who postulated such a source.

Consider the dislocation shown in Figure 2.75(a). The movement of the ends of the dislocation are hindered by other dislocations or possibly point defects. Thus, when a stress is applied, the ends of the dislocation cannot move. The result is that the dislocation bows out, as in Figure 2.75(b). Initially the bowed dislocation has a large radius. However, as the stress increases the radius decreases until the minimum radius is reached when the dislocation forms a semicircle (Figure 2.75(c)). This is the condition of maximum stress. Beyond that point the radius decreases and the stress to keep the dislocation expanding decreases. The dislocation forms a loop which grows by sweeping round the fixed ends (Figure 2.75(d), (e)) until eventually the two sides meet to form a complete loop (Figure 2.75(f)). When this occurs, the portions of the dislocation loop A and B annihilate each other and the final result is an expanding dislocation loop which is free of the points anchoring the ends of the initial dislocation and the original pinned location with ends A and B. This pinned dislocation can continue multiplying.

2.2.3 Case study

The term *hot rolling* is used when the rolling takes place at a temperature greater than the recrystallisation temperature of the material being rolled. The rollers plastically deform the grains in the material. However, because of the temperature at which the rolling takes place, the grains are able to recrystallise and the result is undistorted grains and a reduction in dislocation density. The material has thus a low yield stress. With cold rolling, no such

Key point

Recrystallisation results in new grains of low dislocation density and thus a marked decrease in yield stress.

(a) Dislocation with anchored ends

(b) Dislocation bows as stress increases

(c) Semi-circle, maximum stress

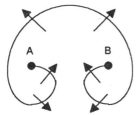

(d) Radius decreases

(e) Loop grows by sweeping round the ends

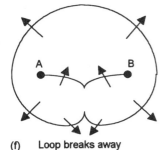

(f) Loop breaks away

Figure 2.75 *Frank-Read source*

recystallisation occurs and the resulting sheet has distorted grains, elongated along the direction of rolling, with an increased density of dislocations. Thus, cold rolling results in work hardening and a higher yield stress. In rolling to obtain thin sheet, the rolling has generally to take place in a number of stages with the sheet being annealed between the stages since the work hardening makes the material too hard.

Problems 2.2

1 Explain what is meant by the term grain.
2 Explain why slip only occurs in particular directions in a crystal.
3 Determine the component of the shear stress on the slip system of the unit cell shown in Figure 2.60 if a tensile stress of 2 MPa is applied in the direction shown.
4 The shear stress at which slip occurs for a face-centred cubic metal, in the direction shown in Figure 2.60, is found to be 2 MPa. What direct stress applied at right angles to the unit cell will produce this slip.
5 Give a simple explanation of elastic and plastic deformation of metals in terms of the block slip model.
6 Distinguish between edge and screw dislocations and explain how slip occurs with both types of dislocation.
7 Why do dislocations give a better explanation of slip than the block slip model?
8 Determine the number of vacancies per cubic metre occurring in copper at just below its melting point of 1085°C if 83.7 kJ/mol are required to produce a vacancy. Copper has a face-centred lattice with a lattice parameter of 0.3615 nm. Avogadro number = 6.02×10^{23}, Boltzmann's constant $k = 1.38 \times 10^{-23}$ J/K.
9 Determine the number of vacancies per cubic metre occurring in aluminium at 500°C if 73.2 kJ/mol are required to produce a vacancy. Aluminium has a face-centred lattice with a lattice parameter of 0.4050 nm. Avogadro number = 6.02×10^{23}, Boltzmann's constant $k = 1.38 \times 10^{-23}$ J/K.
10 Carbon atoms in the face-centred cubic lattice of iron, lattice parameter 0.3571 nm, are located at octahedral sites at the centre of each edge of the unit cell and at the centre of the unit cell. Carbon atoms in the body-centred lattice of iron, lattice parameter 0.2866 nm, are located in tetrahedral sites. If the carbon atoms have radii of 0.071 nm, in which form of iron will they produce the greatest distortion of the lattice?
11 Describe how dislocation density, grain size and foreign atoms affect the mobility of dislocation within a metal and hence the yield stress of the metal.
12 Explain in terms of dislocations (a) work hardening, (b) dispersion hardening, (c) solution hardening, (d) recovery, (e) annealing.
13 Explain how the Frank-Read source of dislocations is able to multiply dislocations.

2.3 Structure of alloys

Most engineering materials are not single component systems with just one type of atom but consist or more than one, e.g. brass is composed of atoms of copper and zinc, steel of iron and carbon. The term *alloy* is used generally for a metallic material consisting of two or more elements; a *binary alloy* has two components, a *ternary alloy* three. Pure metals do not always have the appropriate combination of properties needed; alloys can, however, be designed to have them.

If you put sand in water, the sand and water retain their individual identities. The sand is *insoluble* in water and the result is said to be a mixture. With a mixture we can vary the percentages of the constituents. In such a *physical mixture*, each component retains its own physical structure and properties. We can say that this mixture has two *phases*. A *phase* is a physical entity which is identifiable by its physical state and internal structure, having the same chemical composition and structure throughout. Thus we have the phase of the sand and the phase of the water. However, if we put a pinch of salt in water we can obtain a *solution*. The salt and the water have interacted to form a new single phase. With a solution we can, within limits, vary the percentages of the constituents.

Sodium is a very reactive element, which has to be stored under oil to stop it interacting with the oxygen in the air, and chlorine is a poisonous gas. Yet when these two elements interact, the product, sodium chloride (common salt), is eaten by you and me every day. Sodium chloride NaCl is a compound. In a *compound* the components have interacted and the product does not have the properties of its constituents. With a compound the percentages of the constituent elements are fixed. Thus, in sodium chloride there has to be equal numbers of sodium and chlorine atoms.

This section is about the structure of alloys and how they can be described by *phase diagrams*. Such diagrams map the range of composition, usually given as the percentage by weight of each component of an alloy, e.g. 70% copper–30% nickel, and temperatures over which particular phases are stable for a system of materials and thus summarise the data on innumerable different mixtures of those materials.

2.3.1 Solid solutions

When two liquids are mixed the result can be:

- One liquid completely dissolves in the other, e.g. alcohol in water.

- Each liquid is partially soluble in the other. Thus if a small amount of liquid A is mixed with liquid B a solution might be formed, but if more is added a limit of solubility is reached and the end result is a solution of A and B and undissolved B. It is like mixing salt with water, a small amount of salt will dissolve but if you add too much you end up with a solution and undissolved salt.

- Each liquid is completely insoluble in the other. Such a mixture of liquids will always separate out into two layers. Oil and water behaves in this way.

When liquid metals are mixed the result is that in most cases one liquid is completely dissolved in the other, a homogeneous solution being produced.

When a liquid mixture solidifies the result can be:

- The components separate out with each in the solid state maintaining its own separate identity and structure. The components are said to be *insoluble* in each other in the solid state.

- The components remain completely mixed in the solid state. The components are then said to be soluble in each other in the solid state, the components forming a *solid solution*.

- On solidifying the components show *limited solubility* in each other.

- On solidifying, elements combine to form a compound.

Substitutional solid solution

When liquid copper and liquid nickel are mixed, the result is complete solubility in the liquid state. When the mixture solidifies, a solid solution is produced. This solid solution has a face-centred cubic lattice. The copper and nickel atoms are virtually the same size and so the resulting solid solution is formed by atoms of one substituting for atoms of the other in the crystal lattice (Figure 2.76). Copper atoms have a radius of 0.128 nm and nickel atoms 0.125 nm, a difference of about 2%; substitutional solid solution solubility being limited if the substituted atom has a radius which is significantly bigger than the solute atoms. Silver atoms have a radius of 0.144 nm. Since this is about 13% greater than the radius of copper atoms, only limited substitutional solid solution is feasible for a copper–silver alloy. Generally the arrangement of solute atoms in a substitutional solid solution is random.

It is not only the size of the atoms which determines whether a substitutional solid solution can be formed, other factors are involved. The conditions, known as the *Hume-Rothery rules*, are:

- Atomic sizes within about 15%.

- Similar crystal structure for the two metals.

- Similar valence. Generally a metal of high valence solute can dissolve very little of a metal of low valence, though a metal of low valence can dissolve an appreciable amount of a high valence metal.

- Similar chemical affinity. When two metals have a high chemical affinity for each other, compounds are likely to be produced instead of solid solutions.

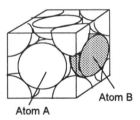

Figure 2.76 *Substitutional solid solution*

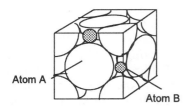

Figure 2.77 *Interstitial solid solution*

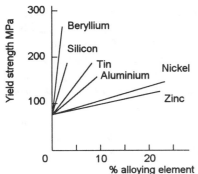

Atomic radius nm		
Copper	0.128	Least effect
Zinc	0.133	because atoms
Nickel	0.125	almost same
Aluminium	0.143	size as copper
Tin	0.151	Most effect
Silicon	0.118	because atoms
Beryllium	0.113	differ most in
		size from copper

Figure 2.78 *Solid solution strengthening*

Copper and nickel fit all the above rules: sizes within 15%, both having a face-centred cubic structure, both having a valence of 2. The solubility is high. Copper and zinc have sizes within 15%, both have a valence of 2 but copper has a face-centred cubic structure and zinc a hexagonal close-packed structure. As a consequence the solubility is reasonably good but limited. If the liquid solution contains more than about 30% zinc, some of the excess zinc atoms combine with some of the copper atoms to form a copper–zinc compound. Copper and tin have sizes which differ by more than 15% (tin, a radius of 0.158 nm), different valence (tin, 4) and different crystal structures (tin, tetragonal). Thus solubility is limited.

Interstitial solid solution

Another form of solid solution can occur when the added atoms are small enough to fit in the interstitial spaces in the crystal lattice (Figure 2.77), though this tends to be rarer than the substitutional solid solution. This requires a radius of about 0.41 or less of the radius of the solvent atom for face-centred cubic lattices and body-centred cubic lattices for perfect fits. Carbon can form an interstitial solid solution with the face-centred cubic form of iron, carbon atoms having a radius of 0.077 nm and iron atoms 0.128 nm.

Solid solution strengthening

Solid solutions have 'foreign' atoms embedded in the lattice of an element and hinder the movement of dislocations, so increasing strength and hardness. The extent to which the strength and hardness is increased depends on how big the 'foreign' atoms are in comparison with those of the lattice into which they are introduced and how many are introduced. The bigger the disruption of the lattice the greater the obstacle for dislocations and hence the greater the improvement in strength and hardness. Figure 2.78 illustrates this for copper alloys with different elements.

2.3.2 Phase diagrams

When pure water is cooled to 0°C it changes from liquid to solid, i.e. ice is formed. Figure 2.79 shows the type of graph produced if the temperature of the water is plotted against time during a temperature change from above 0°C to below 0°C. Down to 0°C the water only exists in the liquid state. At 0°C solidification starts to occur and while solidification is occurring the temperature remains constant, latent heat being extracted during this time. We have phase changes from the liquid phase to the liquid plus solid phase and then to the solid phase.

All pure substances show the same type of behaviour as water when they change state. Figure 2.80 shows the cooling graph for copper, the transition from liquid copper to solid copper taking place at 1084°C.

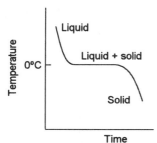

Figure 2.79 *Cooling curve for water during solidification*

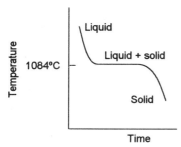

Figure 2.80 *Cooling curve for copper during solidification*

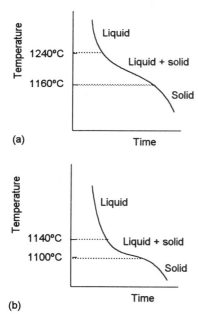

Figure 2.81 *Cooling curves: (a) 70% Cu–30% Ni, (b) 90% Cu–10% Ni*

The cooling curves for an alloy do not show a constant temperature occurring during the change of state. Figure 2.81 shows cooling curves for two copper–nickel alloys. With alloys, the temperature is not constant during solidification. The temperature range over which this solidification occurs depends on the relative proportions of the elements in the alloy. For the alloy with 70% copper and 30% nickel, the transition between liquid and solid starts at 1240°C and is completed at 1160°C when all the alloy is solid. For the alloy with 90% copper and 10% nickel, the transition between liquid and solid starts at 1140°C and is completed at 1100°C when all the alloy is solid.

If the cooling curves are obtained for the entire range of copper–nickel alloys, a composite diagram can be produced which shows the effect of the relative proportions of the constituents on the temperatures at which solidification starts and that at which it is complete. Figure 2.82 shows such a diagram for copper–nickel alloys. Such a diagram is called a *thermal equilibrium diagram*, or, a *phase diagram* since it shows the phases existing for each alloy composition and each temperature. The line drawn through the points at which each alloy starts to solidify from the liquid is called the *liquidus* and the line through the points at which they become completely solid the *solidus*.

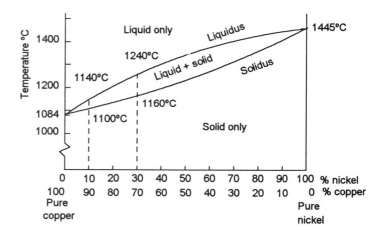

Figure 2.82 *Thermal equilibrium diagram for copper–nickel alloys*

Liquid copper and liquid nickel are completely miscible and the resulting solution of the two liquids is completely homogeneous and thus at the temperature at which they are liquid there is but one phase present. When the liquid alloy is cooled it solidifies. In the solid state the two metals are completely soluble in each other and so the solid state for this alloy has but one phase. In the case of 70% copper–30% nickel alloy, the liquid phase exists above 1240°C; between 1240°C and 1160°C there are two phases when both liquid and solid are present. Below 1160°C there is just one phase, that of the solid solution.

The thermal equilibrium/phase diagram is constructed from the results of a large number of experiments in which the cooling curves are determined for the whole range of alloys in the group. The diagram provides a forecast of the states that will be present when an alloy of a specific composition is heated or cooled to a specific temperature. The diagrams are obtained from cooling curves produced by very slow cooling of the alloys concerned. They are slow because time is required for equilibrium conditions to obtain at any particular temperature, hence the term thermal equilibrium diagram.

Example

When pure copper is heated it starts to melt at 1083°C. During melting, both solid and liquid forms are present. How many phases are present (a) prior to 1083°C, (b) during melting, (c) above 1083°C?

A *phase* is a physical entity which is identifiable by its physical state and internal structure, having the same chemical composition and structure throughout. Thus the number of phases are (a) one, since there is just one form present, (b) two, since there are the solid and liquid phases, (c) one, since there is just the liquid phase.

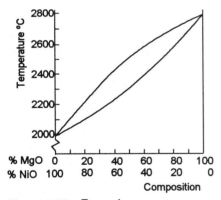

Figure 2.83 *Example*

Example

Figure 2.83 shows the phase diagram for a NiO–MgO refractory material. Select a composition that can be used at 2200°C without melting but can be melted and cast at 2600°C.

We require a liquidus below 2600°C and a solidus above 2200°C. For a liquidus above 2600°C there must be less than about 65% MgO. For a solidus above 2200°C there must be at least about 40% MgO. Thus a composition of between 40% and 65% MgO could be used.

Solidification

Consider the solidification of a copper–nickel alloy, e.g. 70% copper–30% nickel. Figure 2.84 shows the relevant phase diagram. When the liquid copper–nickel alloy cools to the liquidus temperature, small dendrites of copper–nickel alloy form. Each dendrite will have the composition of 53% copper–46% nickel. This is the composition of the solid that can be in equilibrium with the liquid at the temperature concerned, the composition being obtained from the phase diagram by drawing a constant temperature line at this temperature from the liquidus point and finding its intersection of the line with the solidus. Such a constant temperature line is termed a *tie line* and lies within a two-phase region, the ends of the tie line representing the compositions of the two phases in equilibrium.

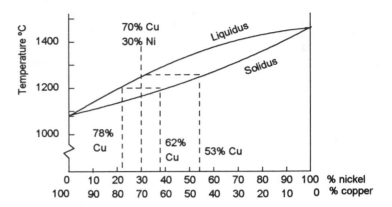

Figure 2.84 *Solidification of 70% copper–30% nickel alloy*

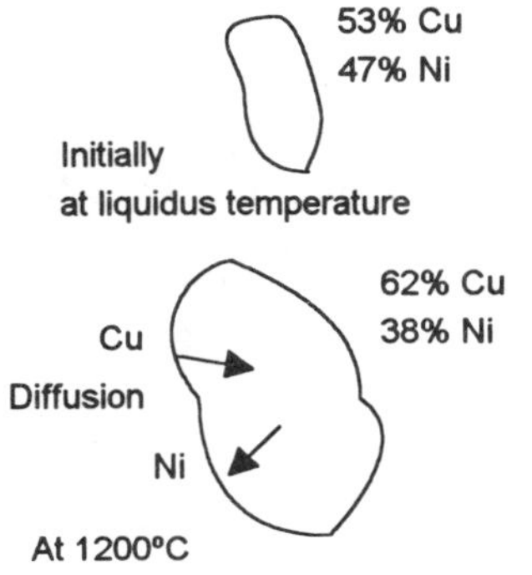

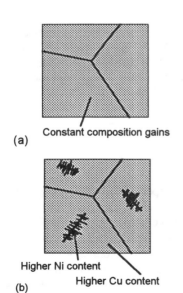

Figure 2.85 *Diffusion during crystal growth*

As the alloy cools further, so the dendrites grow. At 1200°C the expected composition of the solid material as indicated by the tie line at that temperature would be 62% copper–38% nickel, the liquid having the composition 78% copper–32% nickel. So the percentage of copper in the dendrite has increased from 53% at the liquidus temperature, while the percentage of nickel has decreased. If the dendrite is to have a constant composition, movement of the atoms within the solid by diffusion will have to occur. Nickel atoms have to move outwards from the initial dendrite core and copper atoms move inwards (Figure 2.85). This diffusion takes time.

As the alloy cools further, so the expected composition of the solid material changes until, at the solidus temperature, the composition becomes 70% copper–30% nickel. For this to happen, the entire process of cooling from the liquid must take place slowly (Figure 2.86(a)). In the normal cooling of an alloy, in perhaps the production of a casting, the time taken for the transition from liquid to solid is relatively short and inadequate for sufficient diffusion to have occurred for the constant composition solid to be achieved. The result is that the earlier parts of the crystal growth have a higher percentage nickel and lower percentage copper than the later growth parts (Figure 2.86(b)). This effect is called *coring*.

Coring can be eliminated after an alloy has been solidified by heating it to a temperature just below that of the solidus and then holding it at that temperature for sufficient time to allow diffusion to occur and a uniform composition to be achieved.

Lever rule

For a two-phase region, the *lever rule* enables the relative amounts to be calculated of each phase in equilibrium, at a particular temperature. Consider an alloy of two components A and B with a composition X% of B and (100 – X)% of A. If the phase diagram

Figure 2.86 *(a) No coring, (b) coring*

is as shown in Figure 2.87, then at a temperature T_1 the alloy will just be beginning to change from liquid to solid. At temperature T_2 the alloy will consist of a mixture of liquid and solid. To obtain the composition of this mixture, a horizontal, constant temperature (isothermal) line is drawn at temperature T_2. This line is called a *tie line*. The intercepts of the tie line with the solidus and liquidus give the composition of the liquid and solid in the mixture. Thus at temperature T_1 the composition of the liquid will be L' and that of the solid S'.

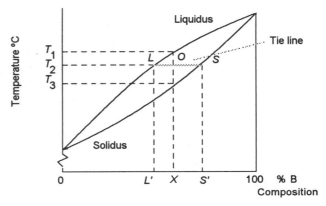

Figure 2.87 *The lever rule*

The mass of A in the alloy mixture at this temperature must be equal to the masses in the solid m_{sA} and in the liquid m_{lA}:

$$m_A = m_{sA} + m_{lA} \tag{16}$$

Likewise, the mass of B in the alloy must at this temperature be equal to the masses in the solid m_{sB} and in the liquid m_{lB}:

$$m_B = m_{sB} + m_{lB} \tag{17}$$

But $m_B = Xm_A/100$, $m_{sB} = S'm_{sA}/100$ and $m_{lB} = L'm_{lA}/100$. Thus:

$$Xm_A = S'm_{sA} + L'm_{lA} \tag{18}$$

Hence, substituting in equation [18] with m_A from equation [17]:

$$X(m_{sA} + m_{lA}) = S'm_{sA} + L'm_{lA}$$

$$m_{lA}(X - L) = m_{sA}(S' - X) \tag{19}$$

$(X - L)$ is the same as LO and $(S' - X)$ is the same as OS, thus:

amount of liquid phase × LO = amount of solid phase × OS [20]

If we think of the tie line as a simple beam resting on a pivot at O, then equation [20] describes the condition for balance according to the principle of moments. Hence the equation is referred to as the *lever rule*.

We can use this equation to determine the fraction of an alloy A that will be solid at a particular temperature. Thus, if the mass of A present in the alloy is m_A, then:

$$m_A = m_{sA} + m_{lA} = m_{sA} + m_{sA}\frac{OS}{LO}$$

and so:

$$\frac{m_{sA}}{m_A} = \frac{LO}{LO + OS} \tag{21}$$

LO + OS is the full length of the tie line. The fraction of the alloy that is solid is thus proportional to the lever arm LO. Similarly, the fraction of the alloy that is liquid is proportional to the length of the lever arm OS.

Thus if a series of tie lines are drawn for different temperatures between T_1 and T_3, the fraction of the alloy that is solid increases from zero at T_1 to 100% at T_3.

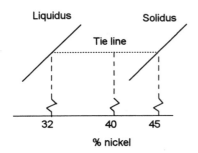

Figure 2.88 *Example*

Example

Determine the amounts of solid and liquid present in a 60% copper– 40% nickel alloy at a temperature of 1250ºC given the segment of the phase diagram in Figure 2.88.

Using equation [6], the fraction solid is 8/13 = 0.62 and the fraction that is liquid is 5/13 = 0.38.

Gibb's phase rule

How many phases can coexist in equilibrium in a particular system? The *Gibb's phase rule*, derived by thermodynamics, states that:

$$P + F = C + 2 \tag{22}$$

where P is the number of phases that can coexist at any one time in a given system, F the number of degrees of freedom, i.e. the number of variables such as pressure, temperature and composition that can be changed without changing the number of equilibrium phases present, and C the number of components in the system, i.e. the elements or compounds which make up the system. With a pure metal there is just one component, with a binary alloy, e.g. brass with copper and zinc, there are two components, with a ternary alloy, e.g. a stainless steel with chromium, nickel and iron, there are three components.

Consider a pure metal; we have $C = 1$. Suppose we are in the process of melting the metal and so have both solid and liquid present, i.e. two phases (Figure 2.89). Thus, with $P = 2$ we must

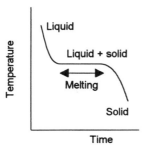

Figure 2.89 *Cooling curve for a pure metal during melting*

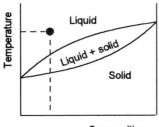

Figure 2.90   *Liquid phase*

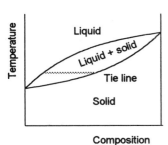

Figure 2.91   *Liquid + solid phase*

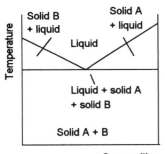

Figure 2.92   *Liquid + solid A + solid B phase*

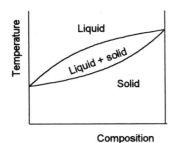

Figure 2.93   *Complete solubility in liquid and solid states*

have $F = 1$. Thus, we cannot have the temperature changing until the number of phases has changed.

If we apply the rule to the phase diagram shown in Figure 2.90, then if we consider the point shown when we have just the liquid phase, i.e. $P = 1$, the rule gives $F = 2 + 2 - 1 = 3$. If we have the pressure constant at atmospheric pressure then this is one of the degrees of freedom. This means to specify the characteristics of the liquid we need to specify the other two degrees of freedom, namely the composition and the temperature. Thus we can only specify the temperature of the liquid alloy if we specify the composition.

Now consider the situation for the phase diagram when the alloy exists as two phases, liquid plus solid. The rule gives $F = 2 + 2 - 2 = 2$. If we have the pressure constant at atmospheric pressure then this is one degree of freedom. This means to specify the characteristics of the alloy in this condition we need to specify just one variable (Figure 2.91). If we specify temperature, i.e. the tie line, we also fix the composition, this being given by the intersections of the tie line with the solidus and liquidus.

Suppose we have a binary alloy with three phases present, then the rule gives $F = 2 + 2 - 3 = 1$. If we have the pressure constant at atmospheric pressure then this is one degree of freedom. Thus there are no other degrees of freedom, all the other conditions are fixed and this thus represents a unique point on a phase diagram (Figure 2.92). With a binary alloy, three phases can only exist in equilibrium at a unique point. This is discussed in more detail in the next section.

If we have a ternary alloy with three components, e.g. a stainless steel with chromium, nickel and iron, then if we want four phases in equilibrium, the rule gives $F = 3 + 2 - 4 = 1$. If we fix the pressure at atmospheric pressure then there is no further degree of freedom to be specified. This thus represents a unique point on a ternary phase diagram.

2.3.3 Phase diagrams for binary alloys

Consider an alloy composed of just two elements, i.e. a binary alloy. The form of the phase diagram for a binary alloy depends on the solubility conditions pertaining in both the solid and liquid states and whether any reactions occur which result in the formation of compounds during the liquid to solid transition or during the cooling of the solid.

Complete solubility in both liquid and solid states

Figure 2.93 shows the form of phase diagram for an alloy involving two components which are soluble in each other in both the liquid and solid states. The phase diagram for the copper–nickel alloy is of this form.

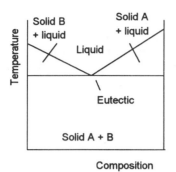

Figure 2.94   *Complete insolubility in solid phase*

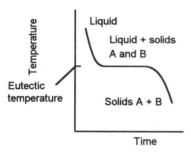

Figure 2.95   *Cooling curve for eutectic composition*

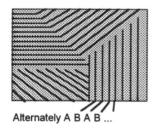

Alternately A B A B ...

Figure 2.96   *Grains in eutectic composition*

Key points

At the eutectic composition, the temperature at which solidification occurs is a minimum. The eutectic structure is generally a laminar arrangement of two phases.

Complete solubility in liquid state and complete insolubility in solid state

Figure 2.94 shows the type of phase diagram produced when the two alloy components A and B are completely soluble in each other in the liquid state but completely insoluble in each other in the solid state. Each of the two components in the solid alloy retains its independent identity. At one particular composition, called the *eutectic composition*, the temperature at which solidification occurs is a minimum. At this temperature, called the *eutectic temperature*, the liquid changes to the solid state without any change in temperature (Figure 2.95). Compare this with the type of cooling curves normally given by alloys, as in Figure 2.81. The solidification at the eutectic temperature, for the eutectic composition, has both the metals simultaneously coming out of the liquid and crystallising together. The resulting structure, known as the *eutectic structure*, is generally a laminar structure with layers of metal A alternating with layers of metal B (Figure 2.96).

The properties of the eutectic can be summarised as:

- Solidification takes place at a single fixed temperature.

- The solidification takes place at the lowest temperature in that group of alloys.

- The eutectic composition is a constant for that range of alloys.

- The eutectic structure is a mixture of the two solid phases.

- The solidified eutectic structure is generally a laminar structure of the two solid phases.

Consider the sequence of events and the resulting structure when different composition alloys are cooled from the liquid (Figure 2.97).

1   *For compositions prior to the eutectic composition*
     When the liquid alloy is cooled to the liquidus temperature, crystals of metal B start to grow. This means that as metal B is withdrawn from the liquid, the composition of the liquid must change to a lower concentration of B and a higher concentration of A. As the cooling proceeds and the crystals of B continue to grow, so the liquid further decreases in concentration of B and increases in concentration of A. This continues until the concentrations in the liquid reach that of the eutectic composition. When this happens, solidification of the liquid gives the eutectic structure. Alloys with less than the eutectic composition are termed *hypoeutectic*.

2   *For the eutectic composition*
     The transition from liquid to solid results in the eutectic structure. The resulting alloy has thus crystals of B embedded in a structure having the composition and structure of the eutectic.

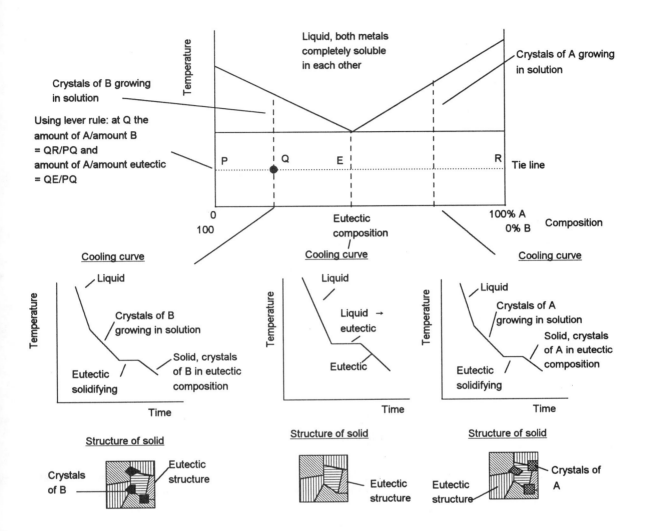

Figure 2.97 *A eutectic phase diagram and forms of structure occurring at different compositions*

3 ***For compositions after the eutectic composition***
when the liquid alloy is cooled to the liquidus temperature, crystals of metal A start to grow. This means that as metal A is withdrawn from the liquid, the composition of the liquid must change to a lower concentration of A and a higher concentration of B. As the cooling proceeds and the crystals of A continue to grow, so the liquid further decreases in concentration of A and increases in concentration of B. This continues until the concentrations in the liquid reach that of the eutectic composition. When this happens, solidification of the liquid gives the eutectic structure. The resulting alloy has thus crystals of A embedded in a structure having the composition and structure of the eutectic. Alloys with greater than the eutectic composition are termed *hypereutectic*.

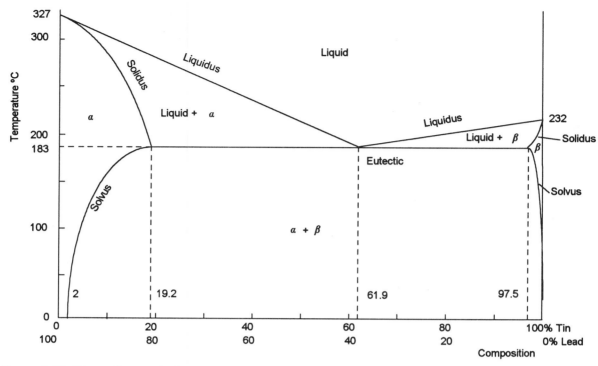

Figure 2.98 *Phase diagram for lead–tin alloys*

Complete solubility in the liquid state and limited solubility in the solid state

Few systems show complete solubility or complete insolubility in the solid state, in most cases one metal will dissolve in the other for only a limited range of concentrations, i.e. there is limited solubility. Lead–tin alloys are of this type, Figure 2.98 showing the phase diagram.

1 *Up to 2% tin*

Lead–tin alloys containing up to 2% tin show complete solubility in the solid phase, a single phase solid solution α forming during solidification. 2% is the limiting solubility at 0°C.

2 *Between 2% and 19.2% tin*

These lead-tin alloys exceed the solubility limit at 0°C of tin in lead. The *solvus* is the line on the phase diagram which gives the solubility of tin in lead at temperatures up to 183°C, this temperature being the one for which the maximum amount of tin can be dissolved in lead. These alloys solidify from the liquid to produce a single solid solution α. However, as the alloy continues to cool, a reaction takes place and when the solvus temperature is reached a second solid phase β is precipitated from the α phase. The result is a solid consisting of largely α phase with a low concentration of β phase. The α phase is a solid solution having a low concentration of tin in

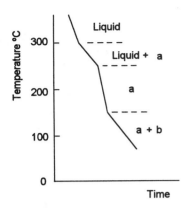

Figure 2.99 *Cooling curve for 10% tin alloy*

lead, the β phase a solid solution having a high concentration of tin in lead. Figure 2.99 shows a typical cooling curve.

Example

Determine the amount of α in comparison with β that forms if a 90% lead–10% tin alloy is cooled to 0°C.

Using the lever rule with a tie line drawn for 0°C, gives for the relative amounts:

$$\frac{\alpha}{\beta} = \frac{100 - 10}{10 - 2} = 11.25$$

The fraction of the solid which is α is $(100 - 10)/(100 - 2) = 0.918$ and the fraction which is β is $(10 - 2)/(100 - 2) = 0.082$.

Example

Determine the compositions of the α and β phases at (a) 0°C, (b) 170°C.

(a) The ends of the tie line for 0°C are at 2% and 100% tin. Thus the α phase is 2% tin–98% lead and the β phase 100% tin.
(b) The ends of the tie line for 170°C are at 15% and 98% tin. Thus the α phase is 15% tin–85% lead and the β phase 98% tin–2% lead.

---

3 *Between 19.2% and 61.9% tin*

The eutectic composition is 61.9% tin–31.8% lead. When an alloy with between 19.2% and 61.9% tin is cooled from the liquid, the behaviour is similar to that which occurs with two metals which are insoluble in the solid state. The result is a solid structure with crystals of α solid solution crystals in the eutectic structure. Figure 2.100 shows a typical cooling curve for one of these alloys.

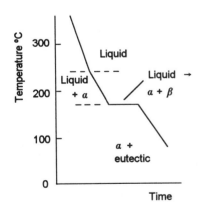

Figure 2.100 *Cooling curve for 40% tin alloy*

Example

Determine the amounts of α and β phase that forms if the 30% tin–70% lead alloy is cooled to 0°C.

Using the lever rule with the tie line drawn at 0°C, the amount which is $\alpha = (100 - 30)/(100 - 2) = 0.71$ and the amount $\beta = (30 - 2)/(100 - 2) = 0.29$.

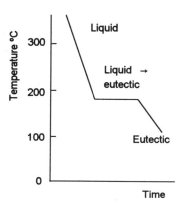

Figure 2.101 *Cooling curve for 61.9% tin alloy*

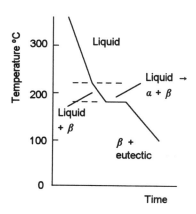

Figure 2.102 *Cooling curve for 80% tin alloy*

Example

Determine the amount of primary α relative to the amount of eutectic for a 30% tin–70% lead alloy when it has been cooled to 0°C.

The term primary is used for the α that is not 'locked up' as part of the eutectic structure. Using the lever rule with the tie line drawn at 0°C, the relative amount which is $\alpha = (61.9 - 30)/(30 - 2) = 1.14$.

4 *At 61.9% tin*

This is the eutectic composition. The structure consists of a laminar mixture of α and β phases. Figure 2.101 shows the cooling curve.

Example

Determine the amounts of α and β phase in the eutectic.

Using the lever rule with a tie line drawn at the eutectic temperature of 183°C, the fraction of the solid that is $\alpha = (97.5 - 61.9)/(97.5 - 19.2) = 0.45$ and the fraction that is $\beta = (61.9 - 19.2)/(97.5 - 19) = 0.55$.

Example

A lead–tin alloy contains 64% of primary α and 36% eutectic at a temperature marginally below the eutectic temperature. Determine its composition.

At the eutectic there is 0% primary α and all eutectic phase. Because there is primary α, the alloy must lie between 19.2% and 61.9% tin. If x is the percentage of tin in the alloy then, using the lever rule with a tie line marginally below the eutectic temperature, we have the fraction primary α/eutectic $= 0.64 = (61.9 - x)/(61.6 - 19.2)$ and so $x = 0.35$ and the alloy is 35% tin–65% lead.

5 *Between 61.9% and 97.5% tin*

When such an alloy is cooled from the liquid, the behaviour is similar to that which occurs with two metals which are insoluble in the solid state. The result is a solid structure with crystals of β solid solution crystals in the eutectic structure. Figure 2.102 shows a cooling curve.

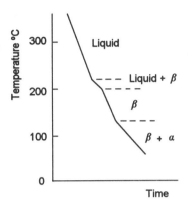

Figure 2.103 *Cooling curve for 98% tin alloy*

Key point

The solvus is the line on a phase diagram separating single- and two-phase regions. They always curve towards the two-phase region as the temperature is increased.

Key point

In the peritectic reaction, a single liquid phase combines with a single solid phase to form, on cooling, a second solid phase with a different composition.

Example

Determine the amount of α and β phase in an 80% tin–20% lead alloy at 0°C.

Using the lever rule with the tie line drawn at 0°C, the amount which is $\alpha = (100 - 70)/(100 - 2) = 0.31$ and the amount $\beta = (70 - 2)/(100 - 2) = 0.69$.

6   *Above 97.5% tin*

The limit of solubility of lead in tin is 2.5% at room temperature. The *solvus* is the line on the phase diagram which gives the solubility of tin in lead at temperatures up to 183°C, this temperature being the one for which the maximum amount of lead can be dissolved in tin. These alloys solidify from the liquid to produce a single solid solution β. However, as the alloy continues to cool, a reaction takes place and when the solvus temperature is reached a second solid phase α is precipitated from the β phase. The result is a solid consisting of largely β phase with a low concentration of α phase. Figure 2.103 shows a cooling curve.

Example

Determine the phase present in a 99% tin–1% lead alloy at a temperature of 170°C.

At 170°C the solvus line is at a composition of 98% tin–2% lead. Thus the composition is one of a single β solid solution.

The peritectic reaction

In many systems, during the cooling process from the liquid state, a reaction occurs between the solid that is first produced and the liquid in which it is forming so that, on cooling, a solid with a different composition is produced. We thus have:

liquid + solid phase 1 → solid phase 2

Such a reaction is termed a *peritectic reaction*. Figure 2.104 shows a phase diagram in which a peritectic reaction occurs. During the time we have the three phases coexisting, there is constant temperature (Gibb's rule: $P = 3$, $C = 2$ and so $F = C + 2 - P = 1$).

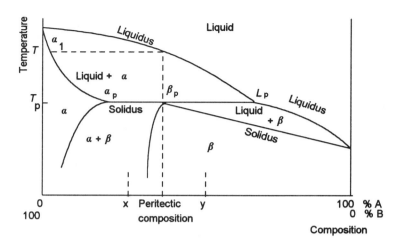

Figure 2.104 *Phase diagram with a peritectic reaction*

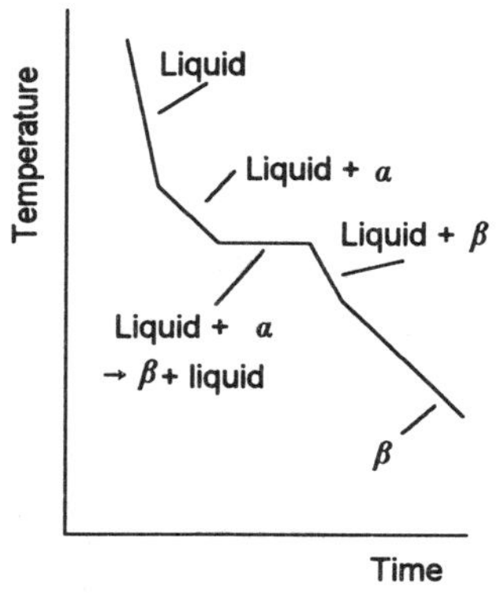

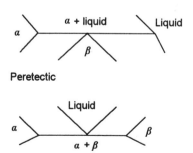

Figure 2.105 *Cooling curve*

Figure 2.106 *Points on a phase diagram*

Consider an alloy with the peritectic composition when cooling from the liquid. Solidification starts when the temperature falls to T with the formation of α phase with the composition α_1. As the temperature falls, solidification of the α phase continues until temperature T_p is reached. At this temperature the solid will be of composition α_p and the remaining liquid L_p. At this temperature the peritectic reaction occurs between the solid and the liquid and a solid phase β of composition β_p is produced:

$$\alpha_p + L_p \rightarrow \beta_p$$

The resulting structure is entirely β.

At composition x, when the temperature falls to T_p we have a solid of composition α_p with liquid L_p. There is, however, more α_p present than at the peritectic composition. Thus only some of the α_p reacts with the liquid to form β_p. The result is that the solid formed consists of a mixture of α and β phases.

At composition y, when the temperature falls to T_p we have a solid of composition α_p with liquid L_p. There is, however, more liquid present than at the peritectic composition. Thus all the α_p reacts with the liquid to form β_p and some liquid remains. After the reaction, the remaining liquid solidifies as β phase. The result is that the solid formed consists of just β phase. Figure 2.105 shows the form of the cooling curve.

A peritectic reaction appears on a phase diagram in the form of an inverted V at the peritectic composition; this differs from a eutectic reaction which is in the form of a V at the eutectic composition (Figure 2.106).

Example

Figure 2.107 shows part of the phase diagram for copper–zinc alloys. Identify the peritectic composition present on that diagram.

The peritectic composition shown on the figure is 37.5% zinc–62.5% copper. The full copper–zinc phase diagram shows five peritectic reactions.

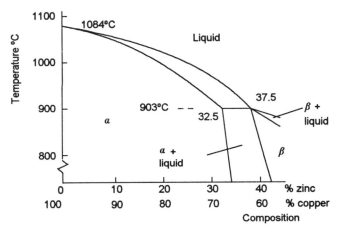

Figure 2.107 *Example*

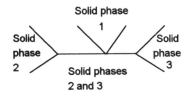

Figure 2.108 *Eutectoid*

Key point

The eutectoid reaction occurs when a cooling solid transforms into two other solid phases.

The eutectoid reaction

The eutectic reaction involves a liquid phase transforming into two solid phases. The *eutectoid reaction* occurs when a cooling solid transforms into two other solid phases at the same time.

solid phase 1 → solid phase 2 + solid phase 3

Figure 2.108 shows how the phase diagram appears for such a reaction. It has the appearance of a V resting on a line.

The compositions of the two new phases are given by the ends of the tie line through the eutectoid point. Eutectoid structures are like eutectic structures but finer in scale, the structure usually being fine parallel plates of the two phases.

2.3.4 Case studies

The following case studies are discussions of alloy compositions in relation to properties.

Soldering

For soldering electronic assemblies we need a solder with a low melting temperature to minimise the effect of the temperature on the electronic components. Most soft solders are based on lead–tin alloys; Figure 2.98 shows the phase diagram. The alloy with the lowest melting point is that of the eutectic composition, namely 61.9% tin–31.8% lead. Such an alloy also has the property of turning completely to liquid as soon it is heated up past 183°C and thus, as a liquid, flows easily into joints and so is generally used for electronic assemblies.

For soldering lead pipes, i.e. plumber's solder, an alloy is required which can be build up as a deposit and wiped into the required shape. The solder is thus required to be 'pasty' and not run in all directions. Eutectic solder would be too liquid and is not 'pasty'. An alloy is required which has a significant amount of liquid plus solid phase. A solder with about 35% tin–65 % lead gives such an alloy. At about 210°C the alloy with be about half liquid, half solid. The temperature is also not so high that the pipes themselves might melt.

Casting

Alloys with the eutectic composition have a lower melting point than alloys with other compositions and, since they change from liquid to solid without a change of temperature, have a well-defined melting point. Unlike the other alloys they do not have a 'pasty region' when they have a liquid plus solid phase and so when the melting temperature is reached they change directly to liquid and so have high fluidity. This makes them particularly suitable for casting.

In use, the eutectic alloys often have superior strengths to those with other compositions. Figure 2.109 shows how the tensile strengths for lead–tin alloys depend on the alloy composition. Up to 19.2% tin, there is strengthening of the alpha by the beta phase. Between 19.2% and the eutectic composition, the increasing amount of eutectic strengthens the alloy. The laminar structure of the eutectic is responsible for the increased strength. For alloys with higher compositions than the eutectic, the decreasing amount of eutectic results in lower strengths. Where, with a eutectic, both the phases are ductile or the main phase is ductile and the other brittle, then the result is a ductile eutectic with improved strength due to the obstruction of dislocations and the benefits gained from the composite structure having one phase reinforcing the other, like steel rods in reinforced concrete.

Zone refining

Ultra pure silicon is required for the manufacture of semiconductors with typically less than one atom impurity per 10^9 silicon atoms. The starting product might be silicon at 98% purity.

Zone refining can be used to obtain the required purity. It is based on the fact that the solubility of the impurities is much greater in the material when it is liquid than when it is solid. Figure 2.110 shows the relevant part of the silicon–impurity phase diagram. If we have an initial impurity concentration of C, then at the temperature T at which the liquid is in equilibrium with the

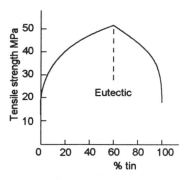

Figure 2.109 *Effect on strength of composition*

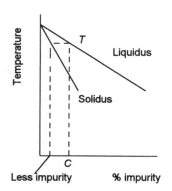

Figure 2.110 *Silicon–impurity phase diagram*

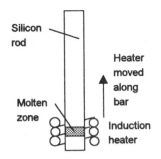

Figure 2.111 *Principle of zone refining*

solid we will have a solid with a smaller percentage of impurity. Impurity atoms diffuse across the solid–liquid interface to give the higher concentration of impurities in the liquid than in the solid.

The zone refining method involves a silicon ingot in the form of a rod (Figure 2.111). A short length of it is melted, usually by induction heating. The heating element is then slowly moved along the length of the rod. As the heating element moves, the freezing solid left behind contains less impurity than the liquid and the impurities become concentrated in the liquid. They can in this way be carried along to the end of the rod. Repeating this process a number of times can increase the purity of the body of the silicon rod and concentrate the impurities at one end which can then be cut off and discarded.

Gold alloys

Pure gold is very soft and malleable and can be beaten out to form extremely thin sheets. Because of these properties, jewellery formed from pure gold would be very impracticable. The strength and hardness, and hence wear resistance, of gold can be improved by forming a solid solution with another alloying element. The term *carat* is used to indicate the proportion that is the gold, a carat being a twenty-fourth part. Thus 18 carat gold has 18 parts by mass of gold and 6 parts alloying element. The addition of copper gives a red hue to the gold and the addition of silver a pale gold colour; to preserve the normal gold colour copper and silver are generally added in the ratio of three to one.

2.3.5 Precipitation

If a solution of sodium chloride in water is cooled sufficiently, sodium chloride precipitates out of the solution. This occurs because the solubility of sodium chloride in water decreases as the temperature decreases. Similar events can occur with solid solutions.

Figure 2.112 shows part of the copper–silver phase diagram. If a 5% copper–95% silver alloy is cooled from the liquid state to 800°C, a solid solution is produced. At this temperature the solution is not saturated but cooling to the solvus temperature makes the solution saturated. If the cooling is continued slowly, precipitation occurs. The result at room temperature is a solid solution containing a coarse precipitate.

The above discussion assumes that the cooling occurs very slowly. The formation of a precipitate requires the grouping together of atoms as a result of atoms diffusing through the solid solution. Diffusion in solids is a slow process. If, however, the solid solution is cooled rapidly from 800°C, i.e. quenched, the precipitation might not occur. The solution becomes *supersaturated*, i.e. it contains more of the *a* phase than the phase diagram predicts. The supersaturated solid solution may be retained in this form at room temperature, but the situation is not very stable and a very fine precipitate may occur with time. This

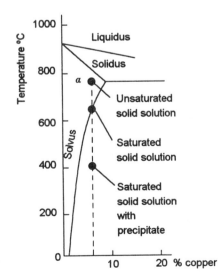

Figure 2.112 *Copper–silver phase diagram*

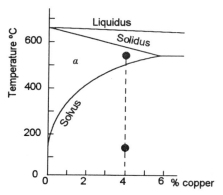

Figure 2.113 *Aluminium–copper phase diagram*

Slow cooling: coarse precipitate. Tensile strength 12 MPa, hardness 46 HB

Quenched: super saturated solution. Tensile strength 16 MPa, hardness 62 HB

Quenched and reheated to 165°C: fine precipitate. Tensile strength 24 MPa, hardness 110 HB

Figure 2.114 *Age hardening*

precipitation may be increased if the solid is heated for some time, the temperature being below the solvus temperature. The resulting precipitate tends to be very fine particles dispersed throughout the solid. Such a fine precipitate gives a much stronger and harder alloy than when the alloy is cooled slowly, dislocations being much more blocked. This hardening process is called *precipitation hardening*. The term *natural ageing* is sometimes used for the hardening process that occurs due to precipitation at room temperature and the term *artificial ageing* when precipitation occurs as a result of heating.

The treatment can thus be summarised as:

1   Heat to above the solvus and hold at that temperature until the formation of a solid solution is complete.

2   Quench to room temperature (sometimes hot water to minimise residual stresses) to prevent precipitation and form a supersaturated solid solution.

3   Age at room temperature for natural ageing or at a slightly elevated temperature, but less than the solvus temperature, for artificial ageing to obtain a fine precipitate.

Figure 2.113 shows part of the phase diagram for aluminium–copper alloys. If the alloy with 4% copper is heated to about 550°C and held at that temperature for a while, diffusion will occur and a homogeneous solid solution will form. If the alloy is then quenched to about room temperature, supersaturation occurs. This quenched alloy is relatively soft. If now the alloy is heated to a temperature of about 165°C and held at this temperature for about ten hours, a fine precipitate is formed. Figure 2.114 shows the effect on the alloy structure and properties of these processes being to give an alloy with a higher tensile strength and hardness.

Not all alloys can be treated in this way. Precipitation hardening can only occur if:

*   There is decreasing solubility with decreasing temperature.

*   The alloy forms a single phase on heating above the solvus line and a two phase region when cooled below it.

*   The alloy is quenchable so that the rapid cooling suppresses the forming of a precipitate.

*   The matrix is relatively soft and ductile and the precipitate hard and brittle.

Alloy systems that have some alloy compositions that can be treated in this way are mainly non-ferrous, e.g. copper–aluminium and magnesium–aluminium.

Problems 2.3

1 Explain what is meant by a solid solution, distinguishing between substitutional and interstitial solid solutions.

2 Based on the Hume-Rothery rules, which of the following systems might you expect to give unlimited solid solubility: (a) silver–gold, (b) molybdenum–tantalum, (c) magnesium–zinc?

3 Using Figure 2.84, estimate the liquidus and solidus temperatures for a 50% copper–50% nickel alloy.

4 Using the phase diagram for copper–nickel alloys in Figure 2.84, for a 70% nickel–30% copper alloy, what phases will be present at (a) 1350°C, (b) 1500°C?

5 For the phase diagram shown in Figure 2.115, determine:

(a) the liquidus temperature and solidus temperature for the ceramic with the composition 30% MgO–70% NiO.

(b) the phases present for a ceramic with the composition 40% MgO–60% NiO at 2500°C.

(c) a composition that can be used at 2400°C without melting but can be melted and cast at 2700°C

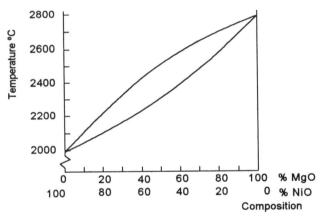

Figure 2.115 *Problem 5*

6 Describe the form of the phase diagrams that would be expected for alloys of two metals that are completely soluble in each other in the liquid state but in the solid state are (a) soluble, (b) completely insoluble, (c) partially soluble in each other.

7 Determine the amounts of solid and liquid present in a 74% copper–26% nickel alloy at a temperature of 1475°C given the segment of the phase diagram in Figure 2.116.

8 Determine the amounts of solid and liquid present in a 47% copper–53% nickel alloy at a temperature of 1300°C given the segment of the phase diagram in Figure 2.117.

9 Figure 2.118 shows the cooling curve for a 80% germanium–20% silicon alloy. Determine the liquidus and solidus temperatures.

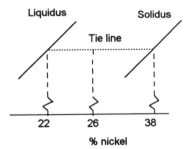

Figure 2.116 *Problem 7*

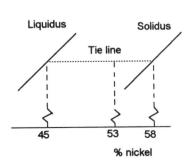

Figure 2.117 *Problem 8*

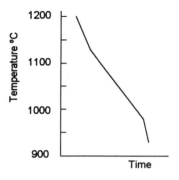

Figure 2.118 *Problem 9*

10 Using the lead–tin phase diagram given in Figure 2.98, determine the phases present and their amounts for a 40% tin–60% lead alloy at (a) 230°C, (b) 0°C.

11 Using the lead–tin phase diagram given in Figure 2.98, determine the solubility of tin in solid lead at 100°C.

12 Using the lead–tin phase diagram given in Figure 2.98, determine the phases present and their amounts for a 30% tin–70% lead alloy at (a) 300°C, (b) 200°C, (c) 0°C.

13 Using the lead–tin phase diagram given in Figure 2.98, determine the phases present and their amounts for a 70% tin–30% lead alloy at (a) 250°C, (b) 184°C, (c) 0°C.

14 Using the lead–tin phase diagram given in Figure 2.98, determine the composition of a lead–tin alloy having a structure with 23% primary α and 77% eutectic.

15 Figure 2.119 shows the phase diagram for aluminium–silicon alloys. What will be the phases present in (a) a 0.4% silicon alloy at 700°C, (b) a 0.4% silicon alloy at 550°C, (c) a 0.4% silicon alloy at 450°C, (d) a 10% silicon alloy at 700°C, (e) a 10% silicon alloy at 580°C, (f) a 10% silicon alloy at 450°C, (g) a 20% silicon alloy at 800°C, (h) a 20% silicon alloy at 600°C, (i) a 20% silicon alloy at 450°C?

16 Using the phase diagram for aluminium–silicon alloys shown in Figure 6.43, estimate the composition of a 3% silicon alloy at 600°C.

17 Using the phase diagram for aluminium–silicon alloys shown in Figure 2.119, estimate the composition of a 6% silicon alloy at 600°C.

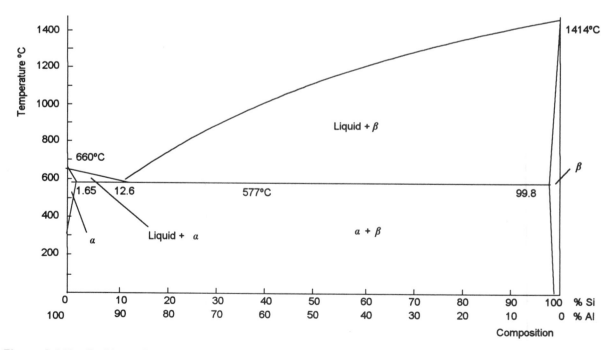

Figure 2.119 *Problem 15*

18 Using the phase diagram for aluminium–silicon alloys shown in Figure 2.119, what is the eutectic temperature for such alloys?

19 Using the phase diagram for aluminium–silicon alloys shown in Figure 2.119, what would be the optimum casting alloy in terms of its pouring properties?

20 Figure 2.120 shows the phase diagram for copper–silver alloys. What will be the phases present for a 94% copper–6% silver alloy at temperatures of (a) 1100°C, (b) 1000°C, (c) 800°C, (d) 400°C?

21 In Figure 2.120, a tie line has been drawn for a temperature of 500°C. Determine the percentages of a and b phases present in a 6% silver–94% copper alloy at this temperature.

22 In Figure 2.10, a tie line has been drawn for a temperature of 500°C. Determine the percentages of α and β phases present in a 94% silver–6% copper alloy at this temperature.

23 For the phase diagram shown in Figure 2.120, determine the eutectic composition.

24 For the copper–silver phase diagram shown in Figure 2.120, determine for the 75% copper–25% silver alloy the phases present, their chemical compositions and amounts at a temperature just below the eutectic temperature.

25 Explain how precipitation hardening is produced.

26 The relevant part of the aluminium–copper phase diagram is given in Figure 2.113. What type of microstructure would you expect for a 2% copper–98% aluminium alloy after it has been heated to 550°C, held at that temperature for a while, and then cooled (a) very slowly, (b) very rapidly to room temperature?

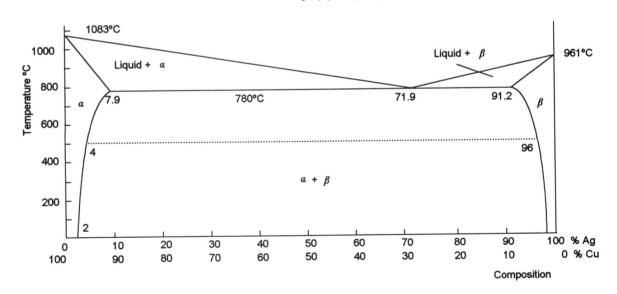

Figure 2.120 *Problem 20*

2.4 Structure of polymers

The plastic washing-up bowl, the plastic measuring rule and the plastic cup are all examples of materials that have polymer molecules as their basis. Polymers include such materials as plastics and rubbers and are used in a wide range of applications, e.g. home appliances, paints, car tyres, foams for packaging, toys, etc. and in composites. The properties of such materials which makes them so useful for so many articles are determined by their molecular structure and how such molecules are packed together in the solid. This section addresses this issue.

2.4.1 Polymer structure

The term *polymer* is used to indicate that a compound consists of many repeated structural units. The prefix 'poly' means many. Each repeating structural unit in the compound is called a *monomer* or *mer*. A polymer molecule in a plastic may have thousands of atoms all joined together in a long chain. The backbones of these long molecules are chains of atoms linked together ɔy covalent bonds. The chain backbone is usually predominantly carbon atoms and we have long molecules formed by the repetition of basic structural units formed by groups of atoms. *Polymerisation* is the process by which these smaller units are joined together to create these large chains.

For many plastics the monomer can be deduced by deleting the prefix 'poly' from its name. Thus the plastic called polyethylene is a polymer which has ethylene as its monomer base unit. The ethylene molecule C_2H_4 has its two carbon atoms joined by a double covalent bond, i.e. two electrons are shared between the carbon atoms. When the polyethylene is formed, the double bond is 'opened up' and as a result the ethylene molecule has an electron free for covalent bonding with another molecule. As a result, 'opened up' ethylene molecules $-CH_2-CH_2-$ can link together to form a long chain, i.e. the polyethylene molecule. Figure 7.1 shows the form of the ethylene molecule and part of the resulting polymer chain.

Figure 2.117 gives a two-dimensional representation of the polymer molecule. In fact the carbon atoms do not form a straight line because the carbon atoms have a directionality which means that the angle between the bonds between a carbon atom and its two neighbouring carbon atoms is about 109°. The result is rather a zig-zag form of molecule (Figure 2.118).

The *degree of polymerisation* of the polymer chain is the number of sub-units, i.e. mers, used to form the chain. With polyethylene this typically ranges from about 3500 to 25 000. If we disregard the contribution made to the molecular mass of a polymer by groups terminating the ends of the polymer molecule, then if M_0 is the molecular mass of the mer and the degree of polymerisation is n, the molecular mass of the polymer molecule is nM_0.

Any sample of polyethylene will contain individual chains of different lengths, the degree of polymerisation n not being constant. As a consequence, the value quoted for the degree of

Double covalent bond

Ethylene molecule

Single covalent bond

Part of the polymer chain

Figure 2.117 *The polymer, polyethylene*

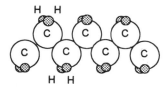

Figure 2.118 *Polyethylene molecular chain*

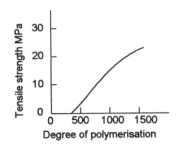

Figure 2.119 *Effect of degree of polymerisation on strength*

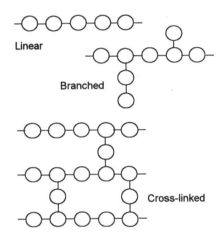

Figure 2.120 *Forms of chains*

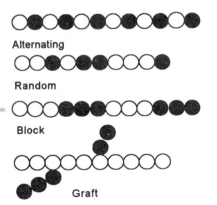

Figure 2.121 *Structures of copolymers made up of two monomers*

polymerisation and molecular mass of polyethylene is an average. Thus, since properties are affected by chain length, for polyethylene there can be a range of properties, these depending on the average size of molecular chain in the sample concerned. Figure 2.119 illustrates this for the tensile strength of polyethylene.

Example

The ethylene mer consists of **4** hydrogen atoms, each with an atomic mass 1 g/mol, and 2 carbon atoms, each with the an atomic mass of 12 g/mol. What will be its molecular mass and the mass of a polyethylene with 5000 mers in its chain?

The molecular mass of the mer is $2 \times 12 + 4 \times 1 = 28$ g/mol. Thus if we have 5000 mers in a chain, the molecular mass of the polyethylene will be $5000 \times 28 = 140\ 000$ g/mol.

Molecular chains

The molecular chains of polyethylene are said to be linear chains. Other possible forms of polymer chains are branched and cross-linked, as illustrated in Figure 2.120 (for simplicity, only carbon atoms are indicated).

The term *homopolymer* is used to describe those polymers that are made up of just one monomer: for instance, polyethylene is made up of only the monomer ethylene. Other types of polymers, *copolymers*, can be produced by combining two or more monomers in a single polymer chain. Figure 21.121 shows four possible types of structure of copolymers based on two monomers.

A solid polymer may thus consist of linear chains arranged in some way and held together in the solid by van der Waals bonding between chains and mechanical entanglement (Figure 2.122). The linear chains have no side branches or ionic or covalent cross-links with other chains and can thus move readily past each other, breaking and remaking van der Waals bonds. If, however, the chains have side branches, there is a reduction in the ease with which chains can move past each other and so the material is more rigid. If there are ionic or covalent cross-links, a much more rigid material is produced in that the chains cannot slide past each other at all and the solid may be considered to be almost just one large cross-linked chain.

Classification of polymers

Polymers can be classified as *thermoplastics*, *thermosets* or *elastomers*. A simple method by which thermoplastics and thermosets can be distinguished is when heat is applied. With a thermoplastic the material softens and removal of the heat results in hardening. With a thermoset, heat causes the material to char

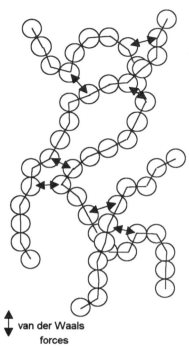

↕ van der Waals
forces

Figure 2.122 *Linear chains held together by van der Waals bonds and mechanical entanglement*

Key points

Thermoplastics have linear chains or branched chains for their structure. Thermosets have a cross-linked structure. Thermosetting polymers are stronger and stiffer than thermoplastics. Thermoplastics offer the possibility of being heated and then pressed into the required shapes. Thermosets cannot be so manipulated Elastomers are chain structures with some degree of cross-linking and can show very large, reversible strains when subject to stress.

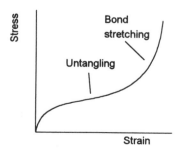

Figure 2.123 *Elastomers*

---

and decompose with no softening. An elastomer is a polymer that by its structure allows considerable extensions which are reversible. Thermoplastics have linear chains or branched chains for their structure. Thermosets have a cross-linked structure. Elastomers are chains with some degree of cross-linking.

The atoms in a thermoset form a three-dimensional structure of chains with frequent cross-links between chains. The bonds linking the chains are strong and not easily broken. Thus the chains cannot slide over one another. As a consequence, thermosetting polymers are stronger and stiffer than thermoplastics. Thermoplastics offer the possibility of being heated and then pressed into the required shapes. Thermosets cannot be so manipulated. The processes by which thermosetting polymers can be shaped are limited to those where the product is formed by the chemicals being mixed together in a mould so that the cross-linked chains are produced while the material is in the mould. The result is a polymer shaped to the form dictated by the mould. No further processes, other than possibly some machining, are likely to occur.

Elastomers are polymers that can show very large, reversible strains when subject to stress. The behaviour of the material is perfectly elastic up to considerable strains, e.g. you can stretch a rubber band up to more than five times its unstrained length and it is still elastic. Elastomers have a structure consisting of tangled polymer chains which are held together by occasional cross-linked bonds. The difference between thermosets and elastomers is that with thermosets, there are frequent cross-linking bonds between chains while with elastomers there are only occasional bonds. A simple model for the elastomer structure might be a piece of very open netting. In the unstretched state the netting is in a loose pile. In the elastomer, there will be some weak temporary bonds, van der Waals bonds, between chains in close proximity to each other, these being responsible for holding the tangled chains together. When the material begins to be stretched the netting just begins to untangle itself and large strains can be produced. The van der Waals bonds between the chains will cause the elastomer to spring back to its original tangled state when the stretching forces are removed. It is not until quite large strains are applied, when the netting has become fully untangled and the structure is orderly, that the bonds between atoms in the material begin to be significantly stretched. At this point the material becomes more stiff, i.e. stress–strain graph starts to become more steep, and much larger stresses are needed to give further extensions. Figure 2.123 illustrates the above points.

2.4.2 Molecular structures

Thermoplastics consist of polymers with long chain molecules that are either linear chains or long chains with small branches. Linear chains have no side branches or cross-links with other chains and can be regarded as 'smooth' and 'slippery', i.e. not 'sticking' to other molecules. Because of this they can easily move past each

Key point

Thermoplastics consist of polymers with long chain molecules that are either linear chains or long chains with small branches.

other. If, however, the chain has branches, then there is a reduction in the ease with which chains can be made to move past each other. This shows itself in the material being more stiff, i.e. less strain produced for a given stress. Making cross-links between chains makes the chains 'sticky' and it is even more difficult to stretch the material. Thermosets are polymers with considerable cross-links and consist of a network of linked atoms.

Polyethylene can be processed to have linear chains. The chains have a core of carbon atoms with hydrogen atoms attached, essentially an almost endless repetition of $-CH_2-$ units. The hydrogen atoms are small and bed into the carbon chain to give a very smooth, linear, chain. There is freedom for the chain to twist about any C–C bond and so the chain is flexible. Scaled up, the typical polyethylene chain is rather like a piece of string about 2 m long. The forces between the chains are due to the weak van der Waals bonding. The average length of the chains can be controlled and polyethylenes with different properties produced. As the length of the molecule increases so does the tensile strength (see Figure 2.119), the longer molecules becoming more easily entangled and so more stress is needed to stretch the material. However, we can add knobs and side branches to such a basic chain and so considerably alter the properties of the solid polymer.

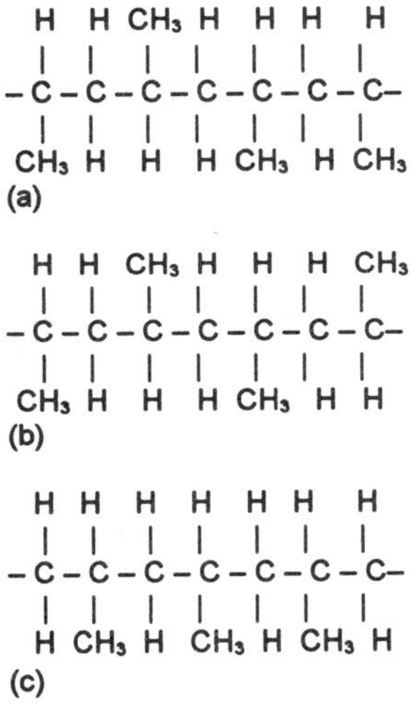

Polypropylene differs from polyethylene only to the extent that alternative carbon atoms have one of their hydrogen atoms replaced by CH_3 groups (Figure 2.124). This replacement can take a number of forms. With the *atactic* form it is random as to which side of the carbon atom the hydrogen atoms are replaced, with the *syndiatactic* form it alternates in a regular manner from one side to the other and with the *isotactic* form all the atoms replaced are on the same side. Commercial propylene is generally predominantly isotactic with small amounts of the other forms present. The result is knobbly, less slippery, chains, and a material more rigid and stronger than polyethylene in its linear form.

Polyvinyl chloride (PVC) has a linear chain, differing from polyethylene only to the extent that 'bulky' atoms, chlorine atoms, replace some hydrogen atoms on the chain (Figure 2.125). Commercial polyvinyl chloride is largely atactic. Because of this structure the chain is very knobbly and, when used without a plasticiser, it is a rigid and relatively hard material. Plasticisers are materials added to keep the knobbly chains apart and so permit more easily the sliding of one knobbly chain over another. Most PVC products are, however, made with a plasticiser incorporated with the polymer. The amount of plasticiser is likely to be between about 5 to 50% of the plastic, the more plasticiser added, the greater the degree of flexibility.

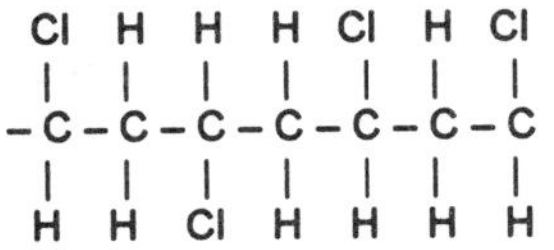

An alternative to putting knobs or branches on to a $-CH_2-$ chain is to make the chain stiffer by incorporating blocks in the backbone of the chain. An example of such a polymer is *polyethylene terephthalate (PET)*. This incorporates a six-carbon (benzene) ring in the backbone. This ring structure will not twist like the C–C bond and so the chain is stiffer. The polymer is widely used for the plastic bottles used for Coca-Cola and other

Figure 2.124 *Forms of polypropylene*

Figure 2.125 *PVC*

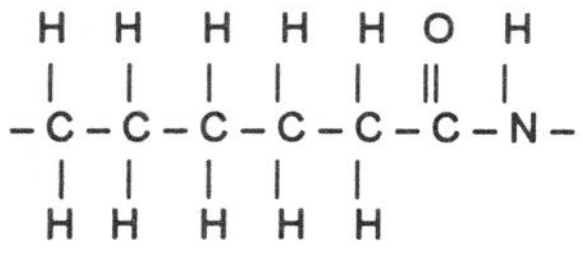

Figure 2.126 *The basic unit of nylon 6*

Key point

A copolymer is a combination of two or more mers making a polymer chain. A blend is a combination of two or more polymers to make a new polymer. Both are the equivalent of alloying.

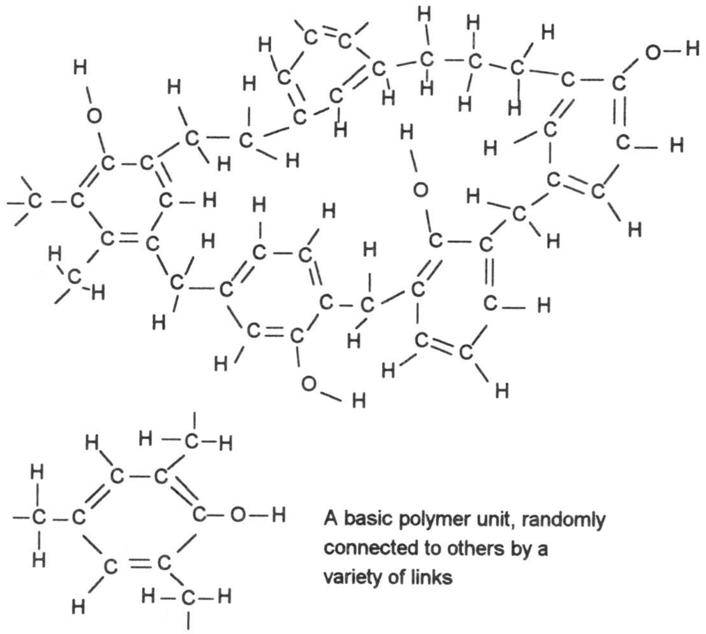

drinks. It is also used as fibres for clothing, being known then as Terylene or Dacron. Polyamides, i.e. nylons, consist of amide groups of atoms separated by lengths of (CH_2) chains. The lengths of these chains can be varied to give different forms of nylon. Figure 2.126 shows the form of nylon 6, the 6 referring to the number of carbon atoms in the chain before it repeats itself.

Another way of changing the chain structure is to combine two or more mers to give a copolymer. Ethylene and vinyl acetate can be combined to give a copolymer, the properties depending on the relative proportions of the two constituents. Increasing the vinyl acetate component increases the flexibility of the product, large amounts of vinyl acetate giving a polymer with properties more like those of a rubber than a thermoplastic. The copolymer is referred to as EVA and has the basic structure of a linear chain with short branches.

Yet another way of modifying the properties of a polymer is to blend two or more polymers. Polystyrene is mixed with rubbers to produce high impact polystyrene (HIPS). This overcomes the problem of brittleness that occurs with polystyrene alone.

The polymers discussed above are essentially linear polymer chains with the weak van der Waals bonding providing the forces between chains. In such materials, the individual chain molecules are distinct and capable of being separated. Such materials are termed *thermoplastics*. Another possibility with polymers is to have stronger cohesion between polymer chains than is provided by van der Waals bonds. Highly cross-linked chains are generally *thermosets* or *elastomers*. With thermosets, such a highly cross-linked structure gives a very stiff material. *Phenolics* (Figure 2.127) are an example of such a structure.

A basic polymer unit, randomly connected to others by a variety of links

Figure 2.127 *Cross-linked phenol formaldehyde*

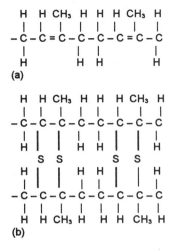

Figure 2.18 *(a) Isoprene chain,*
(b) isoprene chains linked by sulphur

Many elastomers consist of long linear chains linked by small molecules. *Natural rubber* is an example of such a material. The monomer from which natural rubber polymerises is isoprene C_5H_3. The monomer links up to form long chain molecules with some 20 000 carbon atoms. When rubber is *vulcanised*, molecules of sulphur form cross-links between chains (Figure 2.128). The sulphur breaks some of the double bonds between carbon atoms in the chains to give sulphur cross-links. The amount of sulphur added determines the amount of cross-links and hence the properties of the rubber. The greater the number of cross-links, the harder it is to stretch the rubber. The rubber of a rubber band has typically about one sulphur cross-link every few hundred carbon atoms.

The isoprene long chain molecule can exist in two different forms. In one form, referred to as the *cis structure*, the CH_3 groups are all on the same side of the chain (as in Figure 2.128). This concentration of the CH_3 groups all on one side of the chain allows the chain to bend easily and coil in a direction which puts the CH_3 groups on the outside of the bend. In the other form, the *trans structure*, the CH_3 groups alternate between opposite sides of the chain. This has the result that the chain cannot easily bend since the CH_3 groups get in the way. Cis-polyisoprene is natural rubber and shows a high degree of flexibility. Trans-polyisoprene is gutta-percha and is inflexible in comparison with the cis form.

Like thermoplastics, the properties of elastomers can be changed by copolymerisation and blending. Thus, randomly spaced units of styrene and butadiene give a linear chain with virtually no branches. This elastomer, styrene-butadiene rubber (SBR) can have its properties varied by varying the ratio of styrene to butadiene.

Example

What structural changes could be used to make a polymer consisting of linear chains of $-CH_2-$ groups a stiffer material?

The stiffness can be increased by: replacing some of the hydrogen atoms by bulky atoms or groups of atoms, introducing side branches to the chain, replacing some of the carbon atoms by groups of atoms, or introducing cross-links between polymer chains.

2.4.3 Crystalline and amorphous polymers

A *crystalline* structure is one in which there is an orderly arrangement of particles; a structure in which the arrangement is completely random is said to be *amorphous*. Many polymers are amorphous with the polymer chains being completely randomly

Figure 2.129 *Linear amorphous polymer*

Figure 2.130 *Folded linear chain*

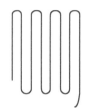

Figure 2.131 *A spherulite*

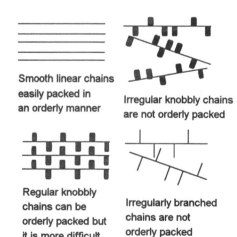

Smooth linear chains easily packed in an orderly manner

Irregular knobbly chains are not orderly packed

Regular knobbly chains can be orderly packed but it is more difficult

Irregularly branched chains are not orderly packed

Regularly branched chains can be orderly packed but it is more difficult

Figure 2.132 *Packing chains*

arranged in the material. Figure 2.129 illustrates this, the chains being shown as lines, individual atoms not being indicated. Linear polymer molecules can, however, assume an arrangement which is, at least partially, orderly. The long molecules fold backwards and forwards on themselves to give a concertina-like system, each loop being approximately 100 carbon atoms long (Figure 2.130). The folded chains produce thin platelets with a thickness of about 10 nm. The arrangement is said to be *crystalline*.

Often when a polymer is cooling from the melt, a number of such platelets grow with regions of amorphous material between them. As with the growth of metal and other crystals, growth from the melt occurs outwards from initiating nuclei. The platelets grow outwards, trapping amorphous material between them. This amorphous material wedges the platelets apart and they continue growing outwards until they bend back on themselves and touch, thus forming a sheaf-like sphere called a *spherulite* (Figure 2.131).

The tendency of a polymer to crystallise is determined by the form of the polymer chains, Figure 2.132 graphically illustrating the problem. Linear polymers can crystallise to quite an extent, complete crystallisation is not, however, obtained in that there are invariably some regions of disorder. For example, linear polyethylene chains can have some 95% of the material crystalline. PVC is essentially just the polyethylene molecule with some of its hydrogen atoms replaced by chlorine atoms to give a knobbly structure for the molecule. The molecule does not, however, give rise to a crystalline structure because the chlorine atoms are rather bulky and not regularly spaced along the chain and so get in the way of orderly packing. Polypropylene has a molecule rather like that of polyethylene but with some of the hydrogen atoms replaced by CH_3 groups. These are, however, regularly spaced along the molecular chain and thus some degree of orderly packing and hence crystallinity is possible. Table 2.8 shows the forms of molecular chains and degree of crystallinity possible for some common polymers.

Table 2.8 *Crystallinity of some common polymers*

Polymer	Form of chain	% possible crystallinity
Polyethylene	Linear	95
	Branched	60
Polypropylene	Regularly spaced groups on linear chain	60
Polyvinyl chloride	Irregularly spaced bulky chlorine atoms on linear chain	0
Polystyrene	Irregularly spaced bulky side groups on linear chain	0

Polymers with side branches show less tendency to crystallise since the branches get in the way of the orderly arrangement. If the branches are completely regularly spaced along the chain then some crystallinity is possible; irregularly spaced branches make crystallinity improbable.

Polyethylene can be produced in a linear form and a branched form, the linear form showing about 95% crystallinity while the branched form might show crystallinity in about 50% of the material. The greater the crystallinity of a polymer, the closer the polymer chains can be packed and so the greater the density of the solid polymer. The term *high density polyethylene* is used for the polyethylene with crystallinity of about 95%, *low density polyethylene* for that with crystallinity of about 50%. The high density polyethylene has a density of about 950 kg/m³, the low density about 920 kg/m³. The closer packing of the chains in the high density polyethylene means that there can be more inter-chain van der Waals bonds, hence a higher melting point (138°C) than the low density form (115°C). The more such van der Waals bonds the more stress has to be applied to produce a particular strain and so the greater the amount of crystallinity the higher the tensile modulus and tensile strength. Table 2.9 illustrates this.

Table 2.9 *Effect of crystallinity on polyethylene properties*

Polymer	% crystallinity	Modulus GPa	Strength MPa
Polyethylene	95	21 to 38	0.4 to 1.3
Polyethylene	60	7 to 16	0.1 to 0.3

When an amorphous polymer is heated, it shows no definite melting temperature but progressively becomes less rigid. This is because the arrangement of the chains in the solid is disorderly, just like in a liquid, and so there is no structural change occurring at melting. With a crystalline polymer, there is an abrupt change in structure at a particular temperature, termed the *melting point*, when the crystalline structure changes to a disorderly structure. If the density of the polymer were being monitored, there would be an abrupt change in density when this occurs as a result of a change in the way the chains are packed together.

Glass transition temperature

PVC, without any additives, at room temperature is a rather rigid material. It is often used in place of glass. But, if it is heated to a temperature of about 87°C a change occurs, it becomes flexible and rubbery. The PVC below this temperature gives only a moderate elongation before breaking, above this temperature it stretches a considerable amount. If amorphous polymers are heated, there is a temperature at which they change from being a stiff, brittle, glass-like material to a rubbery material. This is called the *glass transition temperature* T_g. Below this temperature, segments within the molecular chains are unable to move and the material is stiff with a high elastic modulus and generally rather brittle. Above this temperature, there is sufficient thermal energy for some motion of segments of the chains to occur. The material then

Key point

The *glass transition temperature* is a characteristic of an amorphous thermoplastic. Below this temperature, the polymer is said to be 'glassy', as these parts are rigid and cause failure in a brittle manner. If the polymer is heated above this temperature, it becomes flexible and is referred to as being 'rubbery'.

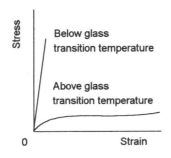

Figure 2.133 *Stress–strain graphs below and above the glass transition temperature*

becomes less stiff with a lower elastic modulus and more like an elastomer and rubbery (Figure 2.133).

Activity

A good model of the effect of the glass transition temperature can be seen with boiled spaghetti. Boil some spaghetti for a meal and then drain the water. Lay some strands on a flat tray. With your hands on either side of the spaghettis strands, try and squash the group. Notice how they will distort to take any shape that they may be pushed into. Now leave the strands on the tray overnight to completely dry. Then try pushing the strands together the next day. Describe what you see. Are there similarities between this behaviour and the description of the glass transition temperature?

The strands of spaghetti when dry, will show a high resistance to the pushing action and therefore a noticeable force is needed to cause any squashing, i.e. its rigidity increases. If the pushing force is increased sufficiently, then the dry strands will snap, i.e. fail in a brittle manner. The water in the freshly boiled spaghetti gives the strands the flexibility to bend and change orientation, just as polymer chains below their glass transition temperature can rotate to change their orientation. The dry spaghetti, however, loses this freedom, just like the polymer chains above the glass transition temperature.

Perspex is an example of a glassy polymer at room temperature. It has a glass transition temperature of 120°C. Thus when it is heated above this temperature it can be easily bent and twisted. Below that temperature it is much stiffer and fairly brittle. This property can be used to enable such polymers to be moulded into shapes required for products.

Below the glass transition temperature only a very limited molecular motion is possible; above the glass transition temperature quite a large amount of motion is possible. The extent to which motion is possible at any particular temperature depends on the structure of the polymer molecules and how well they can move past each other. Thus linear chain molecules tend to have lower glass transition temperatures than molecules with bulky side groups or branches and these, in turn, have lower values than cross-linked polymers. The greater the degree of linking the higher the glass transition temperature. Table 2.9 shows typical values.

In compounding a plastic, other materials are added to the polymer. These can affect the glass transition temperature. Thus an additive referred to as a plasticiser depresses the glass transition temperature by coming between the polymer molecules and weakening the forces between them.

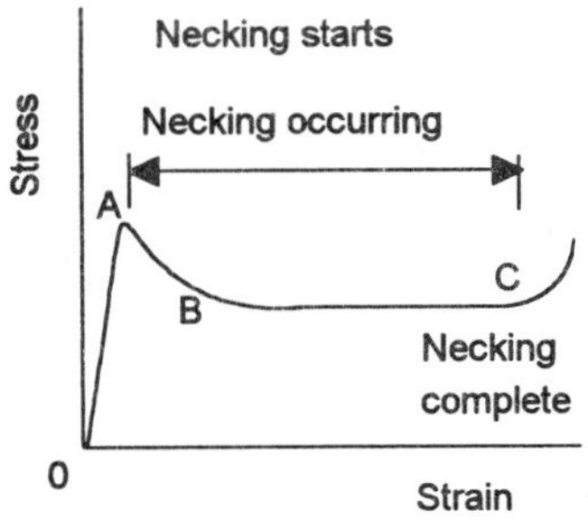

Table 2.9 *Glass transition temperatures*

Material		$T_g°C$
Thermoplastic:	Polyethylene, low density	−90
	Polypropylene	−27
	Polyvinyl chloride	+80
	Polystyrene	+100
Thermoset:	Phenol formaldehyde	Decomposes first
	Urea formaldehyde	Decomposes first
Elastomer:	Natural rubber	−73
	Butadiene styrene rubber	−58
	Polyurethane	−48

2.4.4 Mechanical properties

Consider what happens with a crystalline thermoplastic when it is stretched. Figure 2.134 shows the typical form of stress–strain graph. When stress is applied, the first thing that begins to happen is that there is some movement of folded chains past each other. However, when point A is reached the polymer chains start to unfold to give a material with the chains tending to line up along the direction of the forces stretching the material. The material shows this by starting to exhibit *necking* (Figure 2.135), i.e. a section of the material suddenly shows a marked contraction in its cross-section. As the stress is further increased, the necking spreads along the material with more and more chains unfolding. Eventually, when the entire material is at the necked stage, all the chains have lined up. The material is said to be *cold drawn*. Such a material has, as a result of the orientation of the molecular chains, different properties to the undrawn material. The material is stiffer, i.e. tensile modulus is higher. It is also stronger. Typically, with polyethylene, the tensile modulus increases from about 1 GPa to 10 GPa and the tensile strength from about 30 MPa to perhaps 200 MPa. However, the percentage elongation is reduced, typically from about a few hundred per cent to less than 10%.

The above sequence of events only tends to occur if the material is stretched slowly and sufficient time elapses for the molecular chains to unfold. If a high strain rate is used, the material is likely to break without the chains all reorientating. The plastic used for making polythene bags is a crystalline polymer.

Example

Cut a strip of polythene from such a bag and pull it between your hands and see the necking develop with low rates of strain. Try quickly breaking a strip of polyethylene before orientating the molecules and then another strip after it has been stretched and the molecules orientated, the difference in tensile strength should be apparent.

Figure 2.134 *Stress–strain graph for a crystalline polymer*

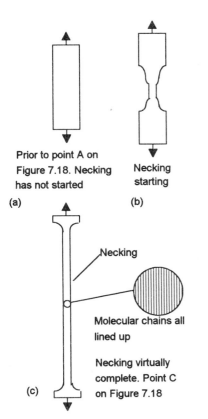

(a) Prior to point A on Figure 7.18. Necking has not started

(b) Necking starting

(c) Necking

Molecular chains all lined up

Necking virtually complete. Point C on Figure 7.18

Figure 2.135 *Necking with a polymer*

In order to improve the strength of polymer fibres, e.g. polyester fibres, they are put through a drawing operation to orientate the polymer chains. Stretching a polymer film causes orientation of the polymer chains in the direction of the stretching forces. The result is an increase in strength and stiffness in the stretching direction. Such stretching is referred to as *uniaxial orientation*. The material is, however, weak if forces are applied in directions other than the stretching forces and has a tendency to split. For the polyester fibres this does not matter as the forces will be applied along the length of the fibre; with film this could be a serious defect. The problem can be overcome by using a biaxial stretching process in which the film is stretched in two directions at right angles to each other. The film has then *biaxial orientation*. Rolling through compression rollers is similar in effect to the drawing operation and results in a uniaxial orientated product.

If orientated polymers are heated to above their glass transition temperatures, they lose their orientation. On cooling they are no longer orientated. This is made use of with *shrinkable films*. The polymer film is stretched and becomes longer and orientated. If it is then wrapped around some package and heated, the film loses its orientation and contracts back to its initially prestretched state. The result is a plastic film tightly fitting the package.

Stretching amorphous polymers

The above is a consideration of what happens when crystalline polymers are stretched. Now consider what happens with amorphous polymers.

Below the glass transition temperature, an amorphous polymer is glass-like and rather stiff and brittle. This is because, when so cold, no chains or parts of chains can move. If the temperature is increased to above the glass transition temperature, the material behaves in a rubbery fashion. This polymer is then very flexible, i.e. a much lower value for the elastic modulus, and is able to withstand large and recoverable strains, i.e. just like a rubber band. This is because there is now sufficient thermal energy being supplied for not only side groups on chains to be able to rotate but also entire segments of the chain also to rotate and move.

Figure 2.136 shows how the modulus of elasticity varies for such a material. Note that the modulus is plotted on a scale where each scale marking sees the value increase by a power of ten.

Amorphous polymers tend to be used below their glass transition temperature. They are, however, formed and shaped at temperatures above the glass transition temperature; they are then in a soft condition.

Elastomers are, under normal conditions, amorphous polymers that at room temperature are above their glass transition temperatures and so exhibit rubbery behaviour. However, if you cool a rubber sufficiently it becomes brittle and shows glassy behaviour. Most polymers become rubbery at some temperature, the exception being heavily cross-linked thermosets which decompose before they reach their glass transition temperatures.

Key points

Crystalline polymers are used up to their melting temperature. They can be hot formed and shaped at temperatures above the melting point or cold formed and shaped at temperatures between the glass transition temperature and the melting point.

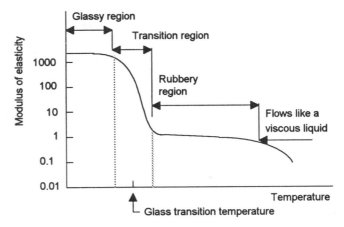

Figure 2.136 *Effect of temperature on the modulus of elasticity of an amorphous polymer*

2.4.5 Additives

The term *plastic* is commonly used to describe materials based on polymers. Such materials, however, invariably contain other substances that are added to the polymers to give the required properties. The following are some of the main types of additives:

- *Plasticisers*

  These are added to the polymers to make it more flexible. In one form this may be a liquid which is dispersed throughout the solid, filling the space between the polymer chains and acting like a lubricant and permitting the chains to more easily slide past each other. This is termed *external plasticisation*. The plasticiser decreases the crystallinity of polymers as it tends to hinder the formation of orderly arrays of polymer chains; it also reduces the glass transition temperature. Table 2.10 shows the effect of plasticiser on the properties of PVC. Unplasticised PVC (often referred to as uPVC) is rigid after processing and is used in making window and door frames, external guttering, drainpipes, etc. Plasticised PVC is flexible and is sued for electrical cable sheathing and tubing.

Table 2.10 *The effect of plasticiser on PVC properties*

	Tensile strength MPa	Percentage elongation
No plasticiser	52 to 58	2 to 40
Low amount of plasticiser	28 to 42	200 to 250
High amount of plasticiser	14 to 21	350 to 450

Internal plasticisation involves modifying the polymer chain by the introduction into it of bulky side groups. These force the polymer chains further apart, thus reducing the attractive forces between chains and so permitting easier flow of chains past each other.

- *Stabilisers*

  Since some polymers are damaged by ultraviolet radiation, protracted exposure to the sun can lead to a deterioration of mechanical properties. An ultraviolet absorber is thus often added to the polymer, such an additive being called a *stabiliser*. Carbon black is often used for this purpose. A heat stabiliser may be required to prevent the breakdown of chains during processing, e.g. lead carbonate with PVC.

- *Flame retardants*

  These may be added to improve fire-resistant properties. The *limiting oxygen index* (LOI) test measures the percentage by volume of oxygen in an oxygen–nitrogen mixture that will just support combustion. An LOI of at least 27% is required to achieve fire retardation. Flexible PVC, used for the insulation of electrical wiring, can have an LOI of 22% and thus a fire retardant, such as antimony oxide, is added.

- *Pigments and dyes*

  These are added to give colour to the material and enhance its appearance.

- *Fillers*

  The properties and cost of a plastic can be markedly affected by the addition of substances termed *fillers*. Since fillers are generally cheaper than the polymer, the overall cost of the plastic is reduced. Up to 80% of a plastic may be filler. Examples of fillers are glass fibres to increase the tensile strength and impact strength (see the section on composites in Chapter 3), mica to improve electrical resistance, graphite or molybdenum sulphide to reduce friction in such items as nylon gears, wood flour to increase tensile strength, carbon black to improve mechanical properties such as abrasion resistance in rubber tyres. One form of additive used is a gas to give foamed or expanded plastics (see the section on composites in Chapter 3). Expanded polystyrene is used as a lightweight packaging material, foamed polyurethane as a filling for upholstery.

- *Lubricants and heat stabilisers*

  These may be added to assist the processing of the material, examples being wax and calcium stearate.

2.4.6 Changing polymer properties

With metals, heat treatment can be used to change the properties. With polymeric materials there are few instances where heat treatments are used. The methods used to change the properties of polymeric materials involve changing the chemical reactions used to produce the material. The following are some of the outcomes of such changes (summarising points discussed earlier):

- ***Increasing the length of chain for a linear polymer***
  This increases the tensile strength and stiffness since longer chains more readily become tangled and so cannot easily be moved.

- ***Introducing large side groups into a linear chain***
  This increases the tensile strength and stiffness since the side groups inhibit chain motion.

- ***Producing branches on a linear chain***
  This increases the tensile strength and stiffness since the branches get in the way of chain movement.

- ***Introducing large groups into the chain***
  These reduce the ability of the chain to flex and so increase rigidity.

- ***Cross-linking chains***
  The greater the degree of cross-linking between chains the more chain motion is inhibited and so the more rigid the material.

- ***Introducing liquids between chains***
  The addition of external plasticisers, liquids which fill some of the space between polymer chains and keep chains apart, makes it easier for the chains to move and so increases flexibility.

- ***Making some of the material crystalline***
  With linear chains the degree of crystallinity can be controlled. The greater the degree of crystallinity the more dense the material and the higher its tensile strength and modulus of elasticity.

- ***Including fillers***
  The properties of polymeric materials can be affected by the introduction of fillers. Thus, for example, the tensile modulus and strength can be increased by incorporating glass fibres.

- ***Orientation***
  Stretching, or applying shear stresses during processing, can result in polymeric molecules becoming aligned in a particular direction and so give properties in that direction markedly different to those in the transverse direction.

- ***Copolymerisation***
  Combining two or more monomers in a single polymer chain will change polymer properties, the properties being determined by the ratio of the components.

- *Blending*

  Mixing two or more polymers to form a material will affect the properties, the result depending on the proportions of the polymers blended.

2.4.7 Case studies

The following case studies concern the use of plastics, showing how the structure determines the properties and hence their use.

Plastic bottles

Plastic bottles are widely used for packaging liquids, offering the advantages over glass bottles of being lighter and tougher; you can drop a plastic bottle and its does not shatter as a glass bottle is likely to do. A material widely used for plastic bottles is polyethylene terephthalate (PET). The process used for making the bottles is blow moulding and involves the hot plastic being moulded into the bottle shape. If the PET is cooled quickly it is amorphous; this gives a clear bottle but one that has loosely packed molecules and so is permeable to carbon dioxide (the 'fizz' of fizzy drinks) and also changes shape with time, i.e.. shows significant creep. If the PET is cooled slowly, crystallisation occurs and the resulting material is not permeable to carbon dioxide and has good creep resistance; however, it is opaque. To overcome this, the process involves the PET being stretched to give the bottle shape with alignment of the molecules; this is achieved by the initial shape being stretched by a mandrel being extended and expanded by air being blown into it. The result is to give a plastic bottle which has about 15 to 20% crystallisation and being impermeable to carbon dioxide, sufficient creep resistance and transparent.

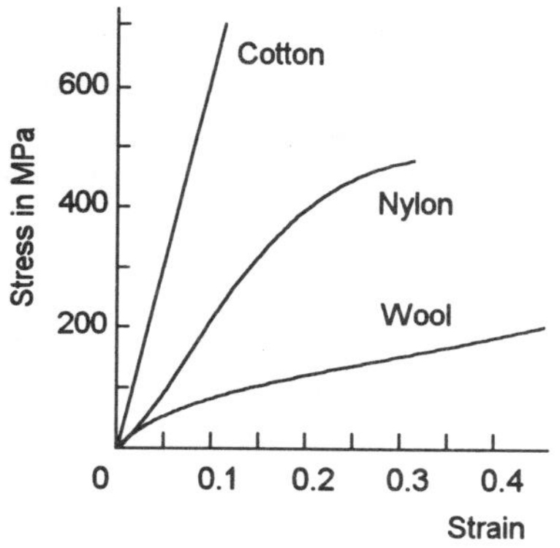

Clothing fibres

The fibres used for clothing include cotton, wool and polymers such as nylon and terylene. The properties required of the fibres include high tensile strength, able to be stretched and recover without deformation, flexibility, ability to be dyed, reasonable weathering properties, resistant to shrinkage.

Figure 2.137 shows the types of stress–strain graph that are found with cotton, wool and nylon fibres. Wool stretches much more than nylon or cotton. Cotton has a higher tensile strength than nylon, this in turn having a higher strength than wool. Thus, of the three fibres, wool is the weakest and the one that stretches the most. Linked with this stress–strain graph is the percentage recovery graph (Figure 2.138). A perfectly elastic material would have a 100% recovery, and spring back to its original length when the stress was removed. The recovery graph shows how, for each of the materials, the percentage recovery depends on the strain acting on the material. Nylon is thus a more elastic material than wool, with wool being more elastic than cotton. Thus if we

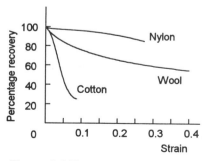

Figure 2.137 *Stress–strain graph for fibres*

Figure 2.138 *Recovery graph for fibres*

compare wool with nylon, perhaps when used to make socks, the nylon socks would be stronger than the wool and the material more elastic. Wool socks would stretch more and be more readily permanently stretched than the nylon. Thus we might expect nylon socks to last longer and maintain their original shape longer.

Wool, cotton and nylon are all polymeric fibre materials, wool and cotton being natural fibres while nylon is a synthetic fibre. Synthetic polymer fibres, such as nylon, are made by melting pieces of the polymer and then forcing the liquid through orifices to give long lengths of extruded fibres in which the nylon fibres are stretched to four or five times their initial length. This stretching orientates the polymer molecules along the length of the fibres and increases the strength, at the expense of the elongation possible.

Problems 2.4

1 What will be the average degree of polymerisation for a form of polyethylene with a molecular mass of 160 000?
2 Polytetrafluoroethylene is a polymer used to coat 'non-stick' pans. The molecular chain differs from that of polyethylene only to the extent that all the hydrogen atoms have been replaced by fluorine atoms. On the basis of this structure, explain the 'non-stick' property.
3 Increasing the amount of sulphur in a rubber increases the amount of cross-linking between the molecular chains. How does this change the properties of the rubber?
4 How does the form of the polymer molecular chain determine the degree of crystallinity possible with a polymer?
5 PVC has a glass transition temperature of 87°C. How would its properties below this temperature differ from those above it?
6 A polypropylene article was designed for use at room temperature. What difference might be expected in its behaviour if used at about −15°C if the glass transition temperature for polypropylene is −10°C?
7 What parameter needs to be considered in selecting a polymer for use in a situation where ductility is required at a low temperature?
8 The density of crystalline polyethylene at 20°C is 1000 kg/m^3 and that of amorphous polyethylene 860 kg/m^3. What will be the density of polyethylene which is 50% crystalline?
9 When a piece of polyethylene is pulled it starts necking at one point. Further pulling results in no further reduction of the cross-section of the material at the necked region but a spread of the necked region along the entire length of the material. Why doesn't the material just break at the initial necked section instead of the necking continuing?
10 Why are the properties of a cold drawn polymer different from those of the undrawn polymer?

11 What is the effect of a plasticiser on the mechanical properties of a polymer?
12 Explain how the elastic modulus of a polymer depends on (a) the degree of polymerisation, (b) crystallinity, (c) amount of covalent cross-links between chains.
13 The carbon–carbon bond in polyethylene has a length of 0.153 nm and the carbon–carbon–carbon bond angle is 112°. What would be the length of a polyethylene chain containing 10 000 carbon atoms?
14 Two polymers which show a high degree of crystallinity are mixed to produce a random copolymer. Would you expect the copolymer to show a similar degree of crystallinity?

2.5 Structure of ceramics

The term ceramic covers a wide range of materials, e.g. brick, earthenware pots, clay, glasses and refractory materials.

2.5.1 Bonds in ceramics

Ceramics are ionic or covalent bonded materials. The bonds between the atoms in ceramics are ionic or covalent. Because such bonds are strong, ceramics have high melting points. Ceramics which are predominantly ionic bonded are combinations of one or more metals with a non-metallic element, often oxygen. Examples are magnesium oxide MgO, alumina Al_2O_3 and zirconia ZrO_2. Ceramics which are predominantly covalent bonded are combinations of two non-metals like silica and oxygen or, sometimes, just pure elements like diamond with just carbon atoms.

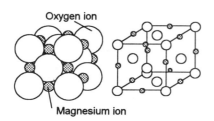

Figure 2.139 *Magnesium oxide*

Ionic crystalline ceramics

A large number of simple ionic bonded ceramics, e.g. MgO, FeO and CaO, have a crystalline structure like that of sodium chloride. The bonding is ionic. Figure 2.139 shows the type of structure occurring with MgO. Each magnesium atom loses two electrons to an oxygen atom to provide the ions and the consequential electric force of attraction between the resulting oppositely charged ions. The oxygen ions are comparatively large compared with magnesium ions and so the structure can be considered as a face-centred cubic structure of oxygen ions with the smaller magnesium ions tucked into the spaces between the oxygen ions. Magnesium oxide is an engineering ceramic which is used as a refractory in furnaces.

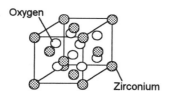

Figure 2.140 *Zirconium oxide*

Zirconium oxide ZrO_2 is another engineering ceramic which is ionic bonded. It is a face-centred cubic structure of zirconium ions with oxygen ions occupying the tetrahedral interstitial spaces (Figure 2.140). Alumina Al_2O_3 is an engineering ceramic which is used for cutting tools and grinding wheels and as a component in brick and pottery. It has oxygen ions arranged in a hexagonal close-packed structure with aluminium ions occupying two-thirds of the octahedral interstitial sites (Figure 2.141).

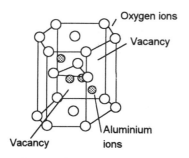

Figure 2.141 *Alumina*

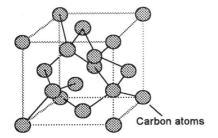

Figure 2.142 *Diamond: every atom is linked to four of its neighbours by covalent bonds in a tetrahedral arrangement*

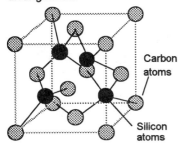

Figure 2.143 *Silicon carbide*

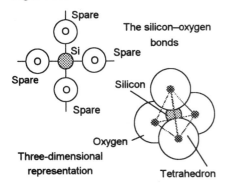

Figure 2.144 *Silicon-oxygen tetrahedron*

The above ceramics have crystal structures formed with just two ions, structures with more than two ions also occur. A common form with three ions is the Perovskite structure.

Simple covalent ceramics

Diamond (Figure 2.142) is an example of a simple covalent bonded ceramic, all the bonded atoms being carbon. The engineering ceramic silicon carbide SiC (Figure 2.143) has a structure similar to diamond with some of the carbon atoms being replaced by silicon.

2.5.2 Silicates

Silica forms the basis of a large variety of ceramics. A silicon atom forms covalent bonds with four oxygen atoms to give a tetrahedron-shaped structure (Figure 2.144). This form of structure leaves the oxygen atoms with 'spare' bonds with which to link up with other silicon–oxygen tetrahedra or metal ions. This bonding can be ionic.

Crystalline silica can be described as a number of the silicon–oxygen tetrahedra joined together in an orderly manner corner to corner to give three-dimensional structure. Quartz is such a crystalline silica structure.

Other silica structures can be produced by linking tetrahedra in chains and then linking adjacent chains by metallic ions (Figure 2/145(a)). The links in the chain are covalent bonds while the links between adjacent chains are ionic. Another chain structure involves a double chain of tetrahedra with links between these double chains being by metallic ions (Figure 2.145(b)). Asbestos is an example of such a material.

The tetrahedra can also link up to give sheets (Figure 2.146) instead of chains. Such a sheet is the basis of many minerals, e.g. clay. When water is added to clay, the water molecules attach by van der Waals bonds and so the sheets can glide easily over each other. As a consequence the clay is plastic and readily mouldable.

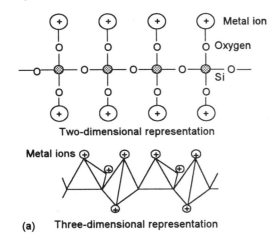

Two-dimensional representation

(a) Three-dimensional representation

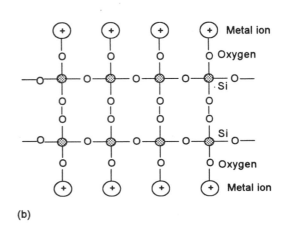

(b)

Figure 2.145 *(a) Single chain structure, (b) double chain structure*

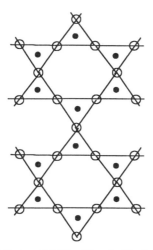

Figure 2.146 *Sheet silicate*

Key point

Liquid silica, when cooled rapidly, does not form into the ordered arrangement required of a crystal. Instead a disordered solid, a glass, is formed. The glass transition temperature is the temperature that marks the transition between the liquid and the glass form.

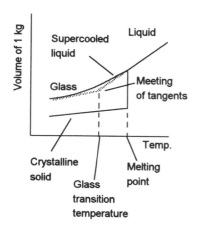

Figure 2.148 *Solidification of crystalline and glassy solids*

2.5.3 Glasses

If silica in the liquid state is cooled very slowly it crystallises at the freezing point. However, if the liquid silica is cooled more rapidly it is unable to get all its atoms into the orderly arrangement required of a crystal and the resulting solid is a disorderly arrangement which is called a *glass*. Figure 2.147 shows such a structure.

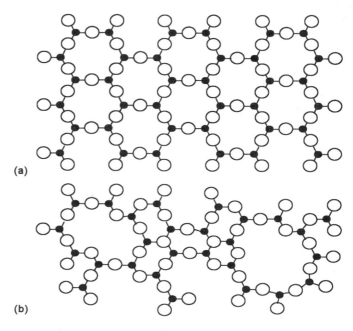

Figure 2.147 *Two-dimensional representations of (a) the orderly arrangement for a crystal, (b) the disorderly arrangement for a glass*

When liquids are cooled and solidify there is an abrupt change in density when the liquid changes state and the atoms become orderly packed into the smaller volume of the solid (Figure 2.148). This is, however, not the case with glass. There is no abrupt change in density but a gradual transition from a supercooled liquid to a glassy solid. The temperature at which the supercooled liquid silica is considered to have turned into a glass is called the *glass transition temperature* T_g.

The very viscous nature of melted silica means that silica crystals would grow only very slowly and so it is not too difficult a process to cool molten silica and form a glass. However, if some crystallisation does occur the resulting glass has a white translucent appearance due to the scattering of light at the surfaces of the crystals.

If a glass is heated to a temperature close to the glass transition temperature and held there for a while, some reorganisation of the atoms in the glassy structure can take place with the result that localised stresses in the glass are relieved. Such a process is called *annealing*.

Pure silica forms a glass with a high softening temperature of about 1200°C. It is difficult to work because it has a high viscosity but has high strength, low thermal expansion and hence high thermal shock resistance. Commercial glasses have their silica structure broken up by *network modifiers*. When an oxide such as sodium oxide Na_2O is added, the oxygen atoms of the oxide link into the silica structure and break up the network (Figure 2.149). The positive sodium ions do not enter the network but remain ionically bonded to their oxygen atoms and fill some of the interstices in the network. The resulting reduction in chain length and cross-linking softens the glass, reduces its glass transition temperature and reduces the viscosity of the molten glass. Such a glass is termed soda–silica glass and is widely used. There is, however, a limit to the oxide addition. This is because the reduction in chain length and cross-linking in the melt makes it easier for the liquid to crystallise and become opaque rather than form a glass. Other oxides, termed *intermediates*, are often added to silicate glasses. They cannot themselves form glasses but join in the silicate network. Such oxides enable special properties to be obtained, e.g. high refractive index as a result of adding lead oxide.

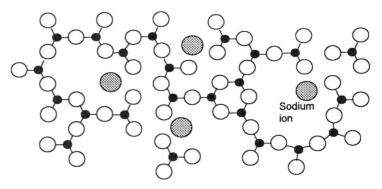

Figure 2.149 *Breaking up the network with a network modifier*

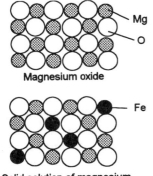

Magnesium oxide

Solid solution of magnesium oxide with substituted iron ions

Figure 2.150 *Forming a solid solution*

2.5.3 Phase diagrams for ceramics

As with metals, we can draw phase diagrams for ceramics and have diagrams showing complete solubility in the liquid with solidification as a solid solution or perhaps with limited solid solubility and some reaction occurring.

Substitutional solid solutions can occur with ionic-bonded materials in just the same way as with solid metals. Figure 2.150 illustrates this with the ceramic magnesium oxide in which some of the magnesium Mg^{2+} ions have been replaced by Fe^{2+} ions, the radii of the ions being similar. This substitution of ions does not involve any change in charge. Unless the two ions have the same valence, there will be a change in charge which has to be compensated for by some other change in charge elsewhere in the structure, e.g. by also substituting some of the oxide ions O^{2-} by ions such as F^-.

Figure 2.151 shows the phase diagram for alloys of silica with alumina. Materials based on silica and alumina are termed *refractories*, this term indicating that such materials are capable of withstanding high temperatures and can be used for furnace walls. As Figure 2.151 indicates, small amounts of alumina added to silica rapidly lower its melting point. A mixture of 5.5% alumina–94.5% silica gives the eutectic composition with a melting point of 1545°C. Thus, in general, refractories with between 3 and 8% alumina should be avoided because they are close to the eutectic composition and so have a lower melting point than other compositions. Higher amounts of alumina lead to higher melting points and the formation of a compound called mullite, this having the composition $(SiO_2)_2(Al_2O_3)_3$. The compositions from about 20 to 40% alumina are termed *fireclay refractories*, with the lower part of this range being low duty and the upper part high duty. Note that with a relatively slight change in alumina content at just over 70% alumina, there is a change in the solidus by about 255°C. For this reason mullite is an important refractory.

With more than about 80% alumina, the melting point is at its highest but the materials have poor resistance to *spalling*, i.e. the breaking of corners and flaking.

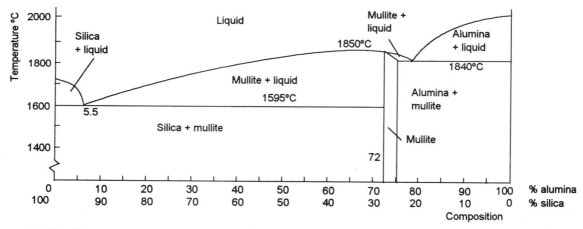

Figure 2.151 *Phase diagram for silica–alumina alloys*

2.5.4 Deforming ceramics

Ionic crystals have, like metal crystals, slip planes. In metals these are the densest packed planes with the slip direction being in the direction of the densest packed planes. This is also true of ionic crystals, but there are the additional provisos that the slip cannot bring ions of the same sign into juxtaposition and can only occur when it results in ions of the same charge replacing each other. Figure 2.152 shows a slip plane in an ionic crystal which is feasible under the above conditions. These conditions mean that very few planes in an ionic crystal permit slip. Because of this,

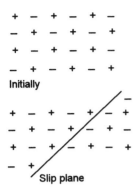

Figure 2.152 *Slip with an ionic crystal*

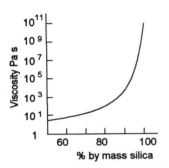

Figure 2.153 *Viscosity of silica with additive at 1300°C*

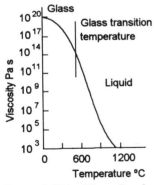

Figure 2.154 *Viscosity of soda-glass*

crystalline ceramics are brittle materials. There are insufficient slip planes for such materials to be ductile.

With covalent crystals, the bonding involves the sharing of electrons between specific atoms and is highly directional. Thus, when forces are applied to shear covalent crystals the bonds are liable to fracture and not reform. Thus such materials are brittle. However, some ceramic materials have a layered structure with ionic or covalent bonds between atoms in a layer but weaker van der Waals bonds between the layers. Because of this form of structure, one layer can be slid with relative ease over another.

Glasses have a disorderly arrangement of atoms and as a result do not have slip planes like crystalline materials. Glasses are three-dimensional bonded arrays of atoms and thus the movement of an atom or group of atoms through a glass is barely feasible. Because of this, glasses are brittle materials.

2.5.5 Case study

The glass used for bottles is based on silica. Molten silica is very viscous and has to be heated to about 2000°C to make a workable glass. The viscosity can, however, be lowered (Figure 2.153) to make a more easily workable glass at a temperature of about 700°C by the additions of oxides such as sodium oxide; these break up the silica network.

Typically, the sodium oxide modified silica, leaves the furnace at about 1000°C with a viscosity of 10^3 Pa s (Figure 2.154). During the bottle forming process its temperature has to drop to the glass transition temperature of about 500°C, the viscosity then being about 10^{14} Pa s. The usual working viscosity range is about 10^7 to 10^9 Pa s.

Problems 2.5

1 Alumina Al_2O_3 has a hexagonal close-packed lattice of oxygen ions O^{2-} with two-thirds of the octahedral interstitial sites occupied by aluminium ions Al^{3+}. Why are only two-thirds of the sites occupied?
2 Sketch the basic structural unit for silica and explain how such units can be combined to produce (a) a single chain structure, (b) a sheet structure, (c) a glass.
3 Explain the effect of the rate of cooling from the liquid state on the resulting structure of silica..
4 Suggest a reason why a sample of glass might have a white translucent appearance.
5 Using the silica–alumina phase diagram given in Figure 7.33, estimate for a 30% alumina–70% silica mixture the fraction of the total ceramic which would be liquid at 1600°C. Why are mixtures with about 5.5% alumina to be avoided for refractories?
6 What is the cause of the lack of plasticity in crystalline ceramics?

2.6 Structure of electric and magnetic materials

This section is a consideration of how the structure of materials determines their electric and magnetic properties. Why are some materials good conductors of electricity and others not? For example, the wires in an electrical circuit are likely to be made of copper because it has a high electrical conductivity. Why copper? Silicon chips are used for integrated circuits. Why silicon? The dielectric in a capacitor might be a ceramic. Why a ceramic? Why can some materials be used as permanent magnets? For example, what makes iron so magnetic a material?

2.6.1 Electrical conductivity

When an electrical field is applied across a material and a current flows, we have charge carriers moving. Unless the material was at a temperature of 0 K they will have been moving around in a random manner in the material before the electric field was applied but it gives them a drift velocity in a particular direction and hence the current. If the charge carriers have a charge q and a drift velocity v and move through a material with a cross-sectional area A (Figure 2.155), then all the charge carriers in a volume vtA will have moved through a cross-section in a time t. If there are n free charge carriers per unit volume able to participate in the current, then the amount of charge moved through the cross-section in time t is $vtAnq$. Since the current I is the rate of movement of charge:

$$I = vAnq \qquad [23]$$

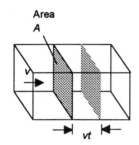

Figure 2.155 *Current as movement of charge carriers through a section of a material*

This equation is often written in terms of the current per unit cross-sectional area, this being called the *current density J*. Thus:

$$J = \frac{I}{A} = vnq \qquad [24]$$

The electrical conductivity σ of a material is L/RA, where L is the length of the material and R its resistance (equation [9], Chapter 1). Since $R = V/I$, where V is the potential difference applied across the material, then:

$$\sigma = \frac{L}{A}\frac{I}{V} = \frac{I/A}{V/L}$$

But I/A is the current density J and V/L is the potential gradient and equal to the electric field strength E, thus:

$$J = \sigma E \qquad [25]$$

Hence using equation [24] we can write:

$$\sigma = \frac{vnq}{E} = nq\mu \qquad [26]$$

where $\mu = v/E$ and is termed the *mobility*. It represents the ease of movement of charge carriers through the material under the action of an electrical field. Thus, for constant charges on the charge carriers, we can control the electrical conductivity by controlling the number of charge carriers per unit volume or controlling the mobility.

Example

What is the drift velocity of electrons in copper when a current of density 1.0×10^4 A/m^2 flows? Copper has 8.5×10^{28} electrons per cubic metre, with each electron having a charge of 1.6×10^{-19} C.

Using equation [2]:

$$v = \frac{J}{nq} = \frac{1.0 \times 10^4}{8.5 \times 10^{28} \times 1.6 \times 10^{-19}} = 7.4 \times 10^{-7} \text{ m/s}$$

Example

What is the mobility of electrons in copper if copper has a resistivity of 1.6×10^{-6} Ω m and 8.5×10^{28} electrons per cubic metre, each having a charge of 1.6×10^{-19} C?

Using equation [4], with conductivity as the reciprocal of resistivity:

$$\mu = \frac{1/\rho}{nq} = \frac{1}{1.6 \times 10^{-8} \times 8.5 \times 10^{28} \times 1.6 \times 10^{-19}}$$

$$= 4.6 \times 10^{-3} \text{ m}^2 \text{ V}^{-1} \text{ s}^{-1}$$

2.6.2 Electrical conductivity and metals

In terms of their electrical conductivity, materials can be grouped into three categories, namely conductors, semiconductors and insulators. Conductors have electrical conductivities of the order of 10^6 S/m, semiconductors about 1 S/m and insulators 10^{-10} S/m. Conductors are metals with insulators being polymers or ceramics. Semiconductors include silicon, germanium and compounds such as gallium arsenide.

Metals

If we consider the atomic structure of metals we find that they all have a structure of atoms with valence electrons which are loosely attached. For example, copper has the electronic structure

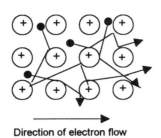

Direction of electron flow

Figure 2.156 *Electric current with a metal*

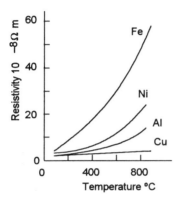

Figure 2.157 *Resistivity variation with temperature*

$1s^22s^22p^63s^23p^63d^{10}4s^1$ with the loosely attached $4s^1$ electron, aluminium the electronic structure $1s^22s^22p^63s^23p^1$ and the loose $3p^1$ electron. These electrons are so loosely attached that, when the atoms are packed close together in a solid, they come under the influence of other atoms and can drift off and move freely between the atoms. Typically a metal will have about 10^{28} free electrons per cubic metre. Thus, when a potential difference is applied across a metal, there are large numbers of free electrons able to respond and give rise to a current. We can think of the electrons pursuing a zig-zag path through the metal as they bounce back and forth between atoms (Figure 2.156).

An increase in the temperature of a metal results in a decrease in the conductivity, i.e. an increase in resistivity (Figure 2.157). This is because the temperature rise does not result in the release of any more electrons but causes the atoms to vibrate and scatter electrons more, so hindering their progress through the metal.

Example

Why is gold with the electronic structure $1s^22s^22p^6$ $3s^23p^63d^{10}4s^24p^64d^{10}4f^{14}5s^25p^65d^{10}6s^1$ a good conductor of electricity?

The atoms have a loosely attached $6s^1$ electron which is easily detached for conduction.

Insulators

Unlike metals, *insulators* have structures in which all the electrons are tightly bound to atoms. Ceramics with their atoms having all their electrons bound into covalent or ionic bonds are examples of insulators; likewise polymers with all their electrons tied up in covalent bonds. Thus there is no current when a potential difference is applied because there are no free electrons able to move through the material. To give a current, sufficient energy needs to be supplied to break the strong bonds which exist between electrons and insulator atoms. The bonds are too strong to be easily broken and hence normally there is no current. A very large temperature increase would be necessary to shake such electrons from the atoms.

Insulators while having very low conductivities do have some slight conductivity. This can be ascribed to the diffusion of ions in the applied electric field. For an ion to take place in diffusion it must break free of its bonds to neighbouring ions. When at some temperature, the energy quanta are distributed among the ions and the chance of an ion having enough quanta, i.e. activation energy E_a, to break free at some temperature T is given by (see earlier this chapter):

$$\text{probability} \propto \exp(-E_a/kT)$$

and so the rate R at which ions diffuse is given by Arrhenius's equation as:

$$R = A \exp(-E_a/kT) \tag{27}$$

where k is Boltzmann's constant. Thus high temperatures increase diffusion and hence conductivity.

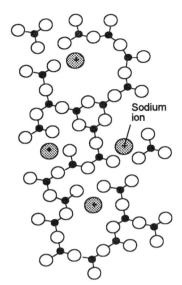

Figure 2.158 *Sodium–silica glass*

Example

Soda–silica glass consists of a network of silica which has been broken up to a limited extent by the inclusion of sodium oxide as a modifier (Figure 2.158). There are thus sodium ions Na^+ present in the glass. When heated up to about 300°C, the glass becomes quite a reasonable conductor of electricity, the resistivity dropping from about 10^{10} Ω m at 0°C to less than 10^4 Ω m at 300°C. Why does this occur and how do you think the resistivity of the glass can be improved?

The activation energy of the sodium ions E_a is relatively low and so raising the temperature leads to conduction occurring. To improve the resistivity of the glass, it is necessary to increase the activation energy so that higher temperatures are needed for a significant number of ions to break free and diffuse. This can be done by using bigger doubly charged ions such as Ba^{2+} instead of the sodium ions. The double charge means stronger bonding forces and the bigger size makes diffusion more difficult.

Semiconductors

Semiconductors can be regarded as insulators when at a temperature of absolute zero. However, the energy needed to remove an electron from an atom is not very high and at room temperature there has been sufficient energy supplied for some electrons to have broken free. Thus the application of a potential difference will result in a current. Increasing the temperature results in more electrons being shaken free and hence an increase in conductivity. At about room temperature, a typical semiconductor will have about 10^{16} free electrons per cubic metre and 10^{16} atoms per cubic metre with missing electrons.

Silicon, a semiconductor, is a covalently bonded solid with, at absolute zero, all the outer electrons of every atom involved in bonding with other atoms. Its atoms have the electronic structure $1s^2 2s^2 2p^6 3s^2 3p^2$, the two 3p electrons being shared with other silicon atoms (Figure 2.159). Germanium, another semiconductor, has a similar structure $1s^2 2s^2 2p^6 3s^2 3p^6 3d^{10} 4s^2 4p^2$. Thermal shaking of the atoms results in some of the bonds breaking and freeing electrons. When a silicon atom loses an electron, we can consider there to be a vacancy, i.e. hole, in its valence electrons. When

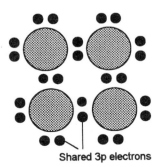

Figure 2.159 *Solid silicon*

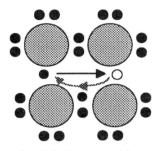

Freed electron moves into
a vacancy, the vacancy
moves in the opposite direction

Figure 2.160 *Movement of electrons
and holes*

electrons are made to move as a result of the application of a potential difference, i.e. an electric field, they can be thought of as hopping from valence site into a hole in a neighbouring atom, then to another hole, etc. (Figure 2.160). Not only do electrons move through the material but so do the holes, the holes moving in the opposite direction to the electrons. Thus when an electric field causes an electron to move in one direction, the hole moves in the opposite direction. The hole thus moves in the direction a positively charged particle would. It is thus useful to think of holes as mobile positive charges.

Activity

One way of picturing this behaviour of a hole is in terms of a queue of people at, say, a bus stop. Form a queue and consider what happens when a bus arrives.

When the first person gets on the bus, a hole appears in the queue at the position of the first person. Then the second person moves into the hole, which now moves to the second position. Thus as people move up the queue, the hole moves down the queue.

Key points

Doping is the deliberate introduction of controlled amounts of impurity into a semiconductor in order to change its electrical properties. Semiconductors which are doped to have more electrons available for conduction are called n-type; those with more holes available for conduction are called p-type.

The conductivity of a semiconductor can be very markedly changed by impurities. With the silicon used for the manufacture of semiconductor devices, the impurity level is routinely controlled to less than one atom in a thousand million silicon atoms. Foreign atoms can, however, be deliberately introduced in controlled amounts into a semiconductor in order to change its electrical properties. This is referred to as *doping*. Atoms such as phosphorus, arsenic or antimony when added to silicon add easily released electrons and so make more electrons available for conduction. Such dopants are called *donors*. Semiconductors with more electrons available for conduction than holes are called an *n-type semiconductor*. Atoms such as boron, gallium, indium or aluminium add holes into which electrons can move. They are thus referred to as *acceptors*. Semiconductors with an excess of holes are called a *p-type semiconductor*.

Since semiconductors can have both electrons and holes as charge carriers, the equation [26] used for electrical conductivity σ with a single type of charge carrier needs modification to:

$$\sigma = nq\mu_n + pq\mu_p \tag{28}$$

with n being the number of electrons per cubic metre with mobility μ_n and p being the number of holes per cubic metre with mobility μ_p. In a pure semiconductor, we have $n = p$ since each electron freed means a hole produced. Such semiconductors are said to be *intrinsic*. With doped materials, depending on the dopant used, we can have more electrons than holes or more holes than electrons. Such doped semiconductors are termed *extrinsic*. Generally, for n-type material the number of electrons per cubic metre is considerably in excess of the number of holes per cubic metre and we can neglect the contribution of the holes to the conductivity and so effectively $\sigma = nq\mu_n$. With p-type material the number of holes is likely to be considerably in excess of the number of electrons and we can neglect the contribution of the electrons to the conductivity and so effectively $\sigma = pq\mu_p$. Table 2.11 gives the number of conduction electrons and holes per cubic metre and their mobilities in intrinsic semiconductors at 300 K.

Table 2.11 *Electrons and holes in intrinsic semiconductors at 300 K*

Semi-conductor	Electrons/ m^3	Holes/m^3	Electron mobility m^2 V^{-1} s^{-1}	Hole mobility m^2 V^{-1} s^{-1}
Germanium	2.4×10^{19}	2.4×10^{19}	0.39	0.19
Silicon	1.4×10^{16}	1.4×10^{16}	0.15	0.048
Gallium arsenide	1.7×10^{12}	1.7×10^{12}	0.85	0.048

Example

Using the data given in Table 2.11, determine the electrical conductivity of intrinsic silicon at 300 K. The charges carried by the electrons and holes are both 1.6×10^{19} C.

$\sigma = nq\mu_n + pq\mu_p = 1.4 \times 10^{16} \times 1.6 \times 10^{-19} \times 0.15$
$+ 1.4 \times 10^{16} \times 1.6 \times 10^{-19} \times 0.048$

$= 4.4 \times 10^{-4}$ S/m

Example

What is the electrical conductivity at 300 K of silicon that has been doped with boron if there are 10^{21} boron atoms per cubic metre and the mobility of holes is 0.048 m^2 V^{-1} s^{-1}? The charge carried by a hole is 1.6×10^{-19} C.

Using equation [6], taking the number of holes to be equal to the number of boron atoms, and neglecting the contribution due to electrons: $\sigma = pq\mu_p = 10^{21} \times 1.6 \times 10^{-19} \times 0.048 = 7.7$ S/m.

2.6.3 Band theory

A simple theoretical model to explain conductivity and how it is affected by doping is called *band theory*. The theory identifies the

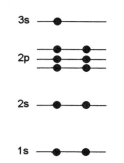

Figure 2.161 *Energy levels for single sodium atoms*

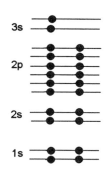

Figure 2.162 *Energy levels for two sodium atoms*

energy levels available for electrons in solids. In the model of an atom as consisting of electrons in orbit about a positively charged nucleus, the electrons can only exist in particular orbits and not at just any distance from the nucleus. Each orbit corresponds to a specific level of energy. Thus the electrons can only exist at particular energy levels. Figure 2.161 illustrates this for the energy levels for a single atom of sodium, this having the electronic structure $1s^2 2s^2 2p^6 3s^1$. No two electrons can have the same set of quantum numbers and thus each energy level can only accommodate two electrons, the spins of the two being in opposite directions (see Pauli's exclusion principle earlier in this chapter).

The electrons in isolated atoms occupy discrete energy levels. However, when atoms are packed close together to form a solid we have to consider that the atomic electrons can come under the influence of neighbouring atoms. Suppose we consider just two atoms coming together, each energy level can still only hold two electrons so each energy level doubles up to give the type of picture shown in Figure 2.162. Thus if we have N atoms brought together, each of the energy levels splits into N levels. For a typical solid N is about 10^{29} atoms per cubic metre. The separation between the levels becomes so small that we can regard levels merging to give bands of close packed energy levels. Figure 2.163 shows how the energy levels of sodium change as atoms are packed closer and closer together.

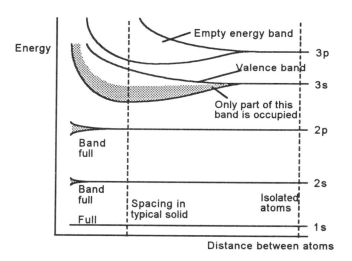

Figure 2.163 *Energy levels and atomic separation for sodium*

With sodium, the electrons involved in making bonds are those in the outer $3s^1$ level. With solid sodium, the electrons involved in the valence bonding occupy energy levels in the highest energy band containing occupied energy levels, this being termed the *valence band*.

For sodium, all the energy levels in the valence band are not occupied. Thus it is possible for electrons in that band to receive small amounts of energy and jump to empty energy levels. Electrons are not able to respond to an applied electric field, and

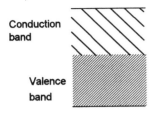

Figure 2.164 *Valence and conduction bands for sodium*

so not contribute to electrical conduction, unless they can move into empty energy levels. Electrical conduction is thus easy with sodium and the energy levels shown in Figure 2.163 for the spacing in the typical solid are characteristic of a good conductor. Figure 2.164 shows the energy level bands for solid sodium. Electrons that have gained enough energy to be released from being tied to atoms are said to be in the *conduction band*. For sodium the valence and conduction energy levels are both parts of the 3s band. Valence bands which are not full thus give good conductors of electricity. Note that the electrons in the filled lower energy bands of sodium do not contribute to electrical conductivity unless they are able to acquire sufficient energy to jump across the gap from the filled bands and into an empty energy level.

Figure 2.165 shows the energy levels for magnesium. Magnesium atoms have the electronic structure $1s^2 2s^2 2p^6 3s^2$. Magnesium has two electrons in its outermost s band and so that band in the solid is full. However, at the separation of the typical solid magnesium, the 3s band overlaps with the empty 3p band. As a result, electrical conduction is easy since valence electrons can acquire small amounts of energy and move into vacant energy levels. Figure 2.166 shows the energy level bands for solid magnesium. Good conductors of electricity are thus given by those elements which, in the solid state, have valence bands overlapping with empty bands.

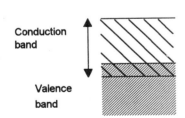

Figure 2.166 *Energy bands for magnesium*

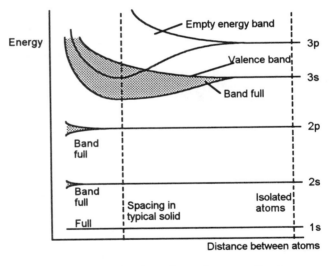

Figure 2.165 *Energy levels and atomic separation for magnesium*

It might be expected that solid carbon, silicon and germanium might be good conductors of electricity since they have electronic structures with just two electrons in their outer p shells. However, in the solid these atoms are covalently bonded. This produces a change in the band structure which leads to the s and p bands interacting to produce two bands separated by a gap, the lower band being full and the upper band empty. Figure 2.167 illustrates this energy band picture for carbon.

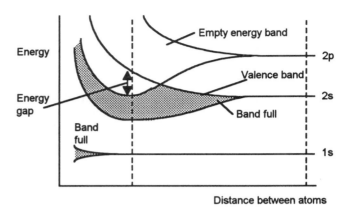

Figure 2.167 *Energy levels and atomic separation for carbon*

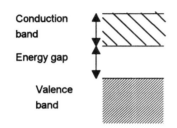

Figure 2.168 *Energy bands for an insulator*

Key points

Conductors have valence and conduction bands which overlap, semiconductors have a small gap between the valence and conduction bands, insulators have a large gap.

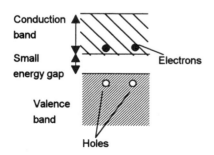

Figure 2.169 *Energy bands for an intrinsic semiconductor*

The differences between good conductors, insulators and semiconductors can be explained in terms of the size of the energy gap between the valence and conduction bands and hence the energy the valence electrons need to be able to participate in conduction. In the case of a metal, a good conductor, we have the valence electrons in an energy band which overlaps with a conduction energy band. Valence electrons are thus able to accept very small amounts of energy and move into vacant higher energy levels. Think of the electrons as being rather like a small number of people in a large square. Given energy they can move easily. With an insulator we have, at a temperature of absolute zero, the valence electrons in a full energy band that does not overlap with the conduction band and there is a large energy gap between them (Figure 2.168). The gap is too big for any electrons at room temperature to have received sufficient energy to have jumped from the valence band into the conduction band so there are no electrons in the conduction band. For electrons to move through the material and give electrical conduction, they must be able to move into vacant energy levels. Think of this as people packed together on one bank of a river. Before they can accept energy and move, they have to have enough energy to jump across the river to the empty bank opposite. With metals there is no gap between the valence and conduction bands, but with insulators the energy gap is high and too big for thermal energy at room temperature to get electrons across it. Thus at room temperature, an insulator has a very low conductivity.

With a pure semiconductor, i.e. an intrinsic semiconductor, we have a conduction band only a small distance above the valence band (Figure 2.169). At room temperature, some of the valence electrons have received sufficient energy to jump across the gap into the conduction band. We then have a few electrons in the conduction band that are capable of accepting small amounts of energy and so participating in the conduction process. But we also have holes in the valence band, i.e. the sites from which valence electrons jumped to the conduction band. This means that, when an electric field is applied, some movement of electrons can also occur in the valence band. We can think of the holes moving as an electron jumps into a hole, leaving another hole behind into which another electron can jump, and so on. There will be the same number of electrons in the conduction band as holes in the valence band.

Table 2.12 summarises the above points concerning conductors, semiconductors and insulators. Note that the energy is in units of eV (electron- volts), this being the energy acquired by an electron when accelerated through a potential difference of 1 V and thus being 1.6×10^{-19} J.

Table 2.12 *Conductors, semiconductors and insulators at 300 K*

Material	Resistivity Ω m	Energy gap in eV between valence and conduction bands	'Free' electrons for conduction per cubic metre
Good conductor	10^{-8} to 10^{-4}	0	10^{28}
Semiconductor	10^{-4} to 10^{7}	About 1	10^{16} to 10^{19}
Insulator	10^{12} to 10^{20}	More than 4 or 5	0

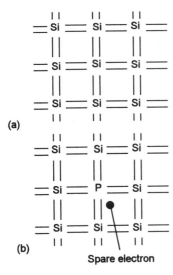

(a)

(b)

Spare electron

Figure 2.170 *(a) Silicon, (b) doped with a donor*

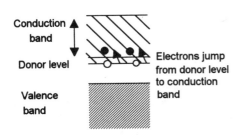

Conduction band

Donor level

Valence band

Electrons jump from donor level to conduction band

Figure 2.171 *Silicon doped with a donor*

Example

Which of the following solids might be expected at 300 K to be good conductors, semiconductors or insulators?

Aluminium, energy gap 0 eV
Aluminium nitride, energy gap 6.3 eV
Gallium antimonide, energy gap 1.95 eV

Aluminium is a good conductor because there is no energy gap. Aluminium nitride is an insulator because the energy gap is large. Gallium antimonide is a semiconductor because the energy gap is relatively small.

Doping

The balance between the number of electrons and holes can be changed by replacing some of the semiconductor atoms in the solid by atoms from other elements. This process is known as *doping* and typically about one atom in ten million might be replaced.

The silicon atom has four electrons which participate in bonding (Figure 2.170(a)). If atoms of an element having five electrons for participation in bonding are introduced into silicon, then there is a spare electron since only four electrons can participate in bonds within the silicon lattice (Figure 2.170(b)). This electron is thus easily made available for conduction. Elements which donate electrons in this way are termed *donors*. In terms of the energy band model, the extra electron is in an energy level that falls in the energy gap between the valence and conduction bands of silicon but very close to the conduction band (Figure 2.171). This energy gap between donor level and conduction band is so small that at room temperature virtually all

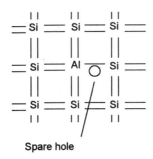

Spare hole

Figure 2.172 *Silicon doped with an acceptor*

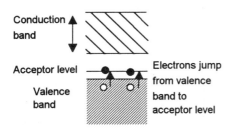

Figure 2.173 *Silicon doped with an acceptor*

the donor electrons have moved into the conduction band. Thus there are more electrons in the conduction band than holes in the valence band and so electrical conduction with such a doped material is more by electrons than holes. For this reason, this form of doped semiconductor is called *n-type*, the n indicating that the conduction is predominantly by negative charge carriers. Arsenic, antimony and phosphorus are examples of elements that are added to silicon to give n-type semiconductors.

If an element having atoms with just three electrons which participate in bonding is added to silicon, then all three of its electrons participate in bonds with silicon atoms. However, there is a deficiency of one electron and so one bond with a silicon atom is incomplete (Figure 2.172). A hole has been introduced. Elements that supply holes in this way are termed *acceptors*. With the energy band model, an energy level is introduced into the gap between the valence and conduction bands of the silicon but very close to the valence level (Figure 2.173). This gap between the acceptor energy level and the valence band is so small that at room temperature it will have been filled by the movement of electrons from the valence band. There are thus more holes introduced into the valence band so that we have more holes in the valence band than electrons in the conduction band. Electrical conduction with such a doped material is more by holes than electrons. For this reason, this form of doped semiconductor is called *p-type*, the p indicating that conduction is predominantly by positive charge carriers. Boron, aluminium, indium and gallium are examples of elements that are added to silicon to give p-type semiconductors.

Such doped semiconductors, n-type and p-type, are termed *extrinsic* meaning that there will be a majority charge carrier and a minority charge carrier. With n-type material, the majority charge carrier is electrons in the conduction band and for p-type it is holes in the valence band. Typically, doping replaces about one in every ten million atoms, i.e. 1 in 10^7. Since there are about 10^{28} atoms per cubic metre, about 10^{21} dopant atoms per cubic metre will be used. Each dopant atom will donate one electron or provide one hole. Thus there will be about 10^{21} electrons donated or holes provided per cubic metre. Since intrinsic silicon has about 10^{16} conduction electrons and holes per cubic metre, doping introduces considerably more charge carriers and swamps the intrinsic charge carriers. The majority charge carriers are thus considerably in excess of the minority charge carriers.

pn junction diode

The model used above can be used to give a simple explanation of the pn junction diode. Consider what happens if we have an n-type semiconductor in contact with a p-type semiconductor. Before contact we have two materials that are electrically neutral, i.e. in each the amount of positive charge equals the amount of negative charge. However, in the n-type semiconductor we have electrons available for conduction in the conduction band and in the p-type material we have holes available for conduction in the valence band. When the two materials are in contact then there is a higher

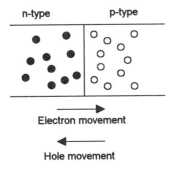

Figure 2.174 *The pn junction*

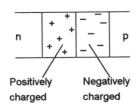

Figure 2.175 *Charge separation at junction*

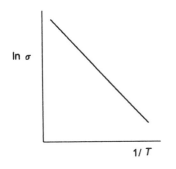

Figure 2.176 *Effect of temperature on conductivity of an intrinsic semiconductor*

concentration of electrons in the n material than in the p material and so a concentration gradient exists for electrons to diffuse from the n to p materials (Figure 2.174). There is also a higher concentration of holes in the p material than in the n material and so a concentration gradient exists which results in holes diffusing across the junction from the p to n materials. Because electrons leave the n-type semiconductor it is losing negative charge and so ends up with a net positive charge. Because the p-type material is gaining electrons it becomes negatively charged. Thus electrons and holes diffuse across the junction until the build-up of charge on each material is such as to prevent further charge movement. The result is shown in Figure 2.175. When an external potential difference is connected across a pn junction, we can easily get electrons to flow through the circuit when they flow in the direction from p to n, in the reverse direction they are opposed by the charges at the junction. Thus we have a device that allows current flow in just one direction.

Temperature and semiconductors

For an intrinsic semiconductor, the number of holes in the valence band equals the number of electrons in the conduction band. Hence the intrinsic carrier concentration per cubic metre $n_i = n = p$ and so equation [28] gives:

$$\sigma = nq\mu_n + pq\mu_p = n_i q(\mu_n + \mu_p) \qquad [29]$$

An increase in temperature for an intrinsic semiconductor results in more electrons having the energy to jump from the valence band to the conduction band. Thus the number of charge carriers n_i available for conduction increases. Though there is a decrease in mobility with an increase in temperature, the increase in n_i produces a much greater change in conductivity and is mainly responsible for the way the conductivity changes with temperature. Thus if we regard the other terms in the above equation as being constant, the electrical conductivity σ can be regarded as proportional to n_i. When at some temperature, the energy quanta are distributed among the charge carriers, the distribution will be exponential with only those in the 'tail' of the exponential having enough energy to be in the conduction band. The total number of charge carriers in this 'tail' will depend on the temperature. The number n_i is given by:

$$n_i = A\, e^{-E_g/2kT} \qquad [30]$$

where A is a constant and E_g the size of the energy gap between valence and conduction bands. Thus the electrical conductivity is given by:

$$\sigma = B\, e^{-E_g/2kT} \qquad [31]$$

where B is a constant. Thus, for an intrinsic semiconductor, a graph of ln σ against $1/T$ will be a straight line (Figure 2.176).

Experimental measurements of the electrical conductivity as a function of temperature enable the size of the energy gap E_g between the valence and conduction bands to be determined.

With an extrinsic semiconductor, at low temperatures the temperatures are not high enough to move many electrons from the valence band to the conduction band and thus, for an n-type semiconductor, most of the conduction is due to electrons moving from the donor levels into the conduction band. An increase in temperature increases the number of electrons jumping from the valence band to the conduction band. The conductivity thus increases. However, at the temperature when all the donor atoms have lost their electrons to the conduction band the numbers of electrons jumping from the valence band to the conduction band has not become dominant and so the number of charge carriers becomes essentially constant, being equal to the number of donor atoms. Because an increase in temperature decreases the mobility, for this temperature region the conductivity decreases as the temperature increases. At higher temperatures conduction becomes mainly due to electrons jumping from the valence band into the conduction band and the material behaves like an intrinsic semiconductor, the conductivity then increasing with increasing temperature.

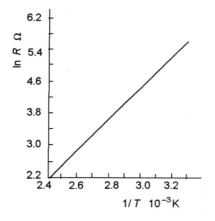

Figure 2.177 *Example*

Example

Figure 2.177 shows how the natural logarithm of the resistance of a sample of germanium varies with the reciprocal of the temperature on the kelvin scale. Determine the size of the energy gap for germanium. Boltzmann's constant $k = 1.38 \times 10^{-23}$ J/K.

Taking logs of equation [9] gives:

$\ln \sigma = \ln B - (E_g/2kT)$

Since $\sigma = L/RA$, where L is the length of the sample and A its cross-sectional area:

$\ln (L/A) - \ln R = \ln B - (E_g/2kT)$

$\ln R = (E_g/2kT) - $ a constant

Thus the slope of the graph is $(E_g/2k)$ and so:

$E_g = 2 \times 1.38 \times 10^{-23} \times (3.5/0.9 \times 10^{-3}) = 1.07 \times 10^{-19}$ J

$= (1.07 \times 10^{-19})/(1.6 \times 10^{-19}) = 0.67$ eV

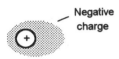

Figure 2.178 *A dipole*

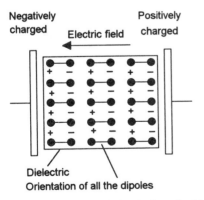

Figure 2.179 *Permanent dipole*

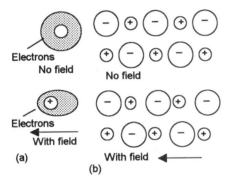

Figure 2.180 *(a) Electronic, (b) ionic polarisation*

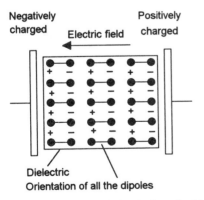

Figure 2.181 *Dipoles in the electric field of a parallel plate capacitor*

2.6.4 Dielectrics

An electric field is produced in a material when a potential difference is applied across it. Charged particles in electric fields experience forces, so if an electric field is produced in a conductor, the free electrons in it move and a current occurs. Insulators, however, have no free electrons and thus when an electric field is produced there is no movement of free electrons. But positive and negative charges within the particles of the material will be acted on by forces and can become displaced slightly.

The term *dipole* is used for atoms or molecules that effectively have a structure involving a positive charge and a negative charge separated by a distance (Figure 2.178). These may be *permanent dipoles* because of an uneven distribution of charge in a molecule (Figure 2.179). Such a material is said to show *molecular polarisation*.

With some materials, dipoles may also be temporarily created when an electric field is applied. This may occur as a result of the electric field distorting the arrangement of the electrons in orbit about the nucleus (Figure 2.180(a)). This is termed *electronic polarisation*. With an ionic bonded material, such as sodium chloride, since such materials consist of an orderly array of positive and negative ions and the arrangement is completely symmetrical there is, in the absence of an electric field, no permanent dipole. However, when an electric field is applied, forces act on the ions and pull the positive ions slightly in one direction and the negative ions slightly in the opposite direction. This resulting distortion results in temporary dipoles (Figure 2.180(b)). This is known as *ionic polarisation*. In a dielectric such as polystyrene, the predominant mode of polarisation is electronic polarisation. With crystalline ceramics such as alumina, the predominant mode of polarisation is ionic polarisation.

When an electric field is applied to a material containing dipoles, the dipoles become reasonably lined up with the field (just like compass needles line up with a magnetic field). Figure 2.181 illustrates this, the material being between the plates of a parallel plate capacitor and the electric field produced by a potential difference applied between the plates.

Relative permittivity

When there is no dielectric but just a vacuum between the plates of a capacitor, when a potential difference is applied across the plates they become charged (Figure 2.182(a)). The amount of charge Q_0 on the plates is determined by the potential difference V and is given by:

$$Q_0 = C_0 V \qquad [32]$$

where C_0 is the capacitance with a vacuum. The result of using a dielectric between the plates of the capacitor, whether the dipoles are permanent or temporary, is to give some alignment of the

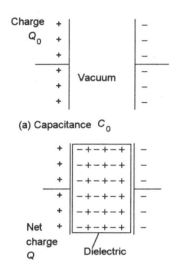

Charge Q_0

(a) Capacitance C_0

Net charge Q

Dielectric

(b) Capacitance C

Figure 2.182 *Capacitors*

dipoles with the electric field and the charge on the plates is partially cancelled by the charge on the dipoles adjacent to the plates (Figure 2.182(b)). Thus the applied voltage 'pushes' more charge onto the plates to maintain the same potential difference. Since:

$$Q = CV \qquad [33]$$

where C is the capacitance of the plates with the dielectric present, then a higher Q for the same V must mean a higher value of C. The factor by which the capacitance is increased is the *relative permittivity* ε_r:

$$C = \varepsilon_r C_0 \qquad [34]$$

The relative permittivity is thus a measure of the polarisation, the greater the polarisation, the greater the relative permittivity.

If a dielectric material behaved perfectly, when the electric field was reversed, the dipoles would all instantly follow the change and become realigned. However, with real materials the particles in the dielectric are held together by bonds which resist reorientation. Thus the rotation of dipoles is slowed down. With low frequency alternating voltages applied across the plates of a capacitor, the dipoles have sufficient time to become reoriented. However, with higher frequency alternating voltages, this may not be the case. As a consequence, the relative permittivity depends on the frequency.

When a dielectric material is used with an alternating electric field, a fraction of the energy supplied is 'lost' each time the field is reversed. This loss is because the motion of reorientation is opposed by frictional effects in the dielectric. The energy dissipated depends on the material and the frequency, the higher the frequency the greater the energy dissipated. This dissipated energy results in heating of the dielectric. This effect can be a problem in some instances but also can be put to industrial use for the heating of materials. Domestic microwave ovens utilise this effect and operate at a frequency of 2.4 GHz. At such frequencies, water molecules have higher energy losses and thus food containing water is heated up uniformly and rapidly. Ceramics, paper and glass are not affected at such frequencies and so do not heat up.

Barium titanate ceramics

When an electric field is applied to a dielectric, polarisation occurs and dipoles become aligned with the field. When the field is removed, the polarisation generally vanishes. This occurs even when there are permanent dipoles since in the absence of the field they become randomly orientated. There are, however, some materials for which this is not the case and they retain some polarisation when the field is removed. Such materials are called *ferroelectrics*. Barium titanate $BaTiO_3$ is such a material. It is a crystalline ceramic with a high value of relative permittivity, at

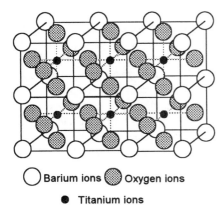

Barium ions ◯ Oxygen ions ◯

Titanium ions ●

Figure 2.183 *Barium titanate*

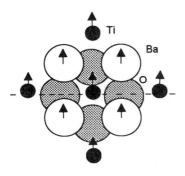

Figure 2.184 *Movement of the positive ions relative to the oxygen ions*

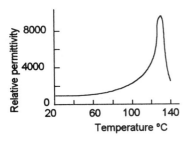

Figure 2.185 *Curie point*

room temperature about 2000. This high value arises from the form of its crystalline structure, it having a Perovskite structure (Figure 2.183).

The radius of the ion required to perfectly fit the cuboctahedron hole occupied by the Ba^{2+} ion is 0.154 nm; this ion has, however, a radius of 0.134 nm and so fits easily into the space. A consequence of this is that at room temperature, when an electric field is applied, the Ba^{2+} and Ti^{4+} ions move to give a distorted crystal structure (Figure 2.184) and this persists when the field is removed, thus resulting in a permanent electric dipole. The distortion increases as the temperature increases until at a temperature of 125°C the distortion is a maximum; above that temperature the distortion rapidly vanishes as the structure assumes the undistorted crystalline form. This temperature is called the *Curie point*. You can think of the ions vibrating back and forth as the temperature increases and needing 'more room' until at the Curie point they fill the spaces. Figure 2.185 shows the general form of the graph of relative permittivity with temperature.

The dielectrics used in commercial high permittivity ceramic capacitors are based on the use of ferroelectric materials like barium titanate. By substituting other ions for some of the barium and titanium ions, the characteristics of the material, e.g. the Curie point and the dependence of the relative permittivity on temperature, can be changed. For example, if some of the Ba^{2+} ions are replaced by Ca^{2+} ions, because these ions are smaller (0.099 nm), a higher Curie point occurs.

Codes (Table 2.13) are used to indicate the temperature range over which such capacitors can be used and the percentage capacitance change that occurs over that range. For example, Z5U has a temperature range of +10°C to +85°C and the capacitance changes by ±15% over the range. X7R has a temperature range −55°C to +125°C and the capacitance changes by ±3.3% over the range.

Table 2.13 *Coding for capacitors*

Code	Temperature range °C	Code	% capacitance change
X7	−55 to +125	D	±3.3
X5	−55 to +85	E	±4.7
Y5	−30 to +85	F	±7.5
Z5	+10 to +85	P	±10
		R	±15
		S	±22
		T	+22 to −33
		U	+22 to −56
		V	+22 to −82

Polymer dielectrics

Polymers such as polystyrene exhibit polarisation as a result of the displacement of electrons when an electric field is present, i.e.

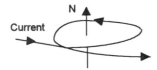

Figure 2.186 *Current in loop of wire giving a magnetic field*

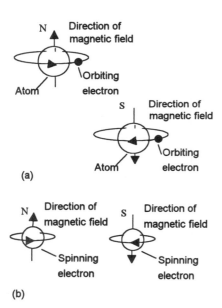

Figure 2.187 *Magnetic dipoles*

electronic polarisation. As a result, polystyrene has a relative permittivity of about 2.5. Because the polarisation is electronic, electrons with their very low mass are able to rapidly 'keep up' with alternating electric fields and so the relative permittivity is little affected by frequency. Some polymers have molecules which contain polar groups, e.g. C^+–Cl^-. As a result, they have higher relative permittivities, e.g. polyester gives a relative permittivity of 3.2. However, the greater mass of the polar groups means that they are not so capable of 'keeping up' with alternating electric fields and as a result the relative permittivity decreases with increasing frequency, e.g. the relative permittivity of polyester drops from 3.2 at 1 kHz to about 2 at 1 MHz.

2.6.5 Magnetism

A current through a loop of wire, i.e. charges moving in a circular path, produces a magnetic field (Figure 2.186). We can consider atoms to have electrons in orbit around a nucleus, thus such orbiting electrons can be considered to be like current moving in a loop of wire and so generate magnetic fields (Figure 2.187(a)). Likewise an electron spinning on its axis can be considered to produce a magnetic field (Figure 2.187(b)). Thus the electrons in atoms can be considered, by virtue of their movement, to make the atom behave like a tiny magnet. Such a tiny magnet is referred to as a *magnetic dipole*.

Electron spin is the most important source of magnetism in solids as the orbital effects tend to get cancelled out in the bonding. Each spinning electron in an atom thus behaves like a small magnet. However, when two oppositely spinning electrons are paired, their magnetism cancels out. In most elements the unpaired electron is a valence electron. However, because in solids the valence electrons have interacted to form bonds there is, on average, no net magnetism arising from these electrons. Certain elements do, however, have inner energy levels which are not completely filled. The transition elements scandium to zinc are such elements (see the periodic table). Table 2.14 shows the electron spin directions for these elements. Copper and zinc have completely filled 3d shells with all the electrons paired, thus they have no net magnetism from spin. The other elements have some unpaired 3d electrons and thus each of these elements have their atoms behaving as magnetic dipoles.

Types of magnetic materials

Materials can be grouped into three general categories (see Section 1.1.8 for a discussion of relative permeability):

- ## Diamagnetic materials
  These have relative permeabilities slightly less than 1. Bismuth, copper, mercury and water are examples of such materials.

Table 2.14 *Electron spin directions for transitional elements*

Element	3d electrons					4s electrons
Scandium	↑					↑↓
Titanium	↑	↑				↑↓
Vanadium	↑	↑	↑			↑↓
Chromium	↑	↑	↑	↑	↑	↑
Manganese	↑	↑	↑	↑	↑	↑↓
Iron	↑↓	↑	↑	↑	↑	↑↓
Cobalt	↑↓	↑↓	↑	↑	↑	↑↓
Nickel	↑↓	↑↓	↑↓	↑	↑	↑↓
Copper	↑↓	↑↓	↑↓	↑↓	↑↓	↑
Zinc	↑↓	↑↓	↑↓	↑↓	↑↓	↑↓

- ### *Paramagnetic materials*
  These have relative permeabilities slightly greater than 1. Aluminium and platinum are examples of such materials.

- ### *Ferromagnetic and ferrimagnetic materials*
  These have relative permeabilities considerably greater than 1. Ferromagnetic materials are metals, e.g. iron, cobalt and nickel, and ferrimagnetic materials are ceramics.

We can explain diamagnetism as occurring with materials that have atoms for which the movements of each of the electrons in their orbits tend to produce magnetic effects but the overall effect of all the electron orbits for an atom is to cancel each other out and there is thus no net magnetic field produced by an atom. When a magnetic field is applied to such a material, electromagnetic induction occurs for each orbiting electron; think of each orbiting electron being like a current-carrying loop of wire. The induced e.m.f. results in an induced current which is in such a direction as to oppose the change producing it and so the magnetic field produced by the induced current is in an opposite direction to the magnetic field responsible for its production. The result is that when a magnetic field is applied to a diamagnetic material, temporary magnetic dipoles are produced which oppose the field producing them and so the flux density in the material is less than would have been produced in a vacuum and the relative permeability is less than 1.

Paramagnetic materials have atoms that behave as magnetic dipoles but, in the absence of a magnetic field, the dipoles are all randomly orientated (Figure 2.188) and thus the material shows no permanent magnetism. When a magnetic field is applied, some small temporary degree of alignment occurs for the dipoles and thus the relative permeability is slightly greater than 1.

Ferromagnetic materials have very high relative permeabilities. The ferromagnetic properties of the transition elements iron, cobalt and nickel are due to them having atomic electrons with unpaired spins, the atoms in such materials thus behaving as permanent magnetic dipoles, and also the way in which the atoms

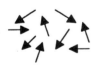

Figure 2.188 *Dipoles in paramagnetic materials*

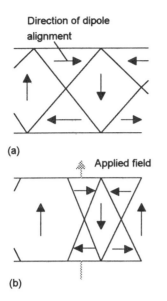

(a)

(b)

Figure 2.189 *Domains, i.e. regions with alignment of magnetic dipoles*

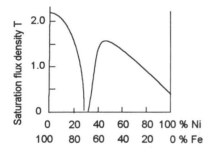

Figure 2.190 *Saturation flux density for iron–nickel alloys*

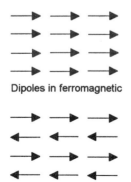

Dipoles in ferromagnetic

Dipoles in antiferromagnetic

Figure 2.191 *Dipoles in ferromagnetics and antiferromagnetics*

bond together in the solid results in neighbouring atomic dipoles aligning themselves all in the same direction. Electrons in one atom are thus 'aware' of the spin state of electrons in neighbouring atoms via the bonding electrons, this exchange of spin information being termed an *exchange interaction*. The magnetic contributions of neighbouring atoms thus add up.

The dipoles in such a material are not generally all aligned in the same direction but are in blocks within which they all are aligned in the same direction. Neighbouring blocks may have different directions. Such regions in a material where magnetic dipoles are aligned in this way are called *domains*. Figure 2.189(a) shows the type of domain structure a material might have in the absence of an external magnetic field. The direction of the magnetic dipoles varies from domain to domain and may be completely random with the result that the material shows no permanent magnetism. When a magnetic field is applied, those domains with magnetic dipoles most nearly in the direction of the field grow in size at the expense of neighbouring domains (Figure 2.189(b)). Increasing the applied field increases their growth until eventually all the domains have dipoles in the direction of the applied field. The material is then said to be *saturated* and showing its greatest amount of magnetism. When the applied field is removed, many of the domains may remain orientated in the same direction and thus the material retains some residual magnetism.

If the temperature of a ferromagnetic material is raised, the thermal motion of the atoms increases. This motion tends to disrupt the alignment of magnetic dipoles in the domains. At a certain temperature, called the *Curie temperature*, the thermal motion completely overcomes the magnetic alignment and the ferromagnetic material loses its ferromagnetic properties.

Ferromagnetism is not restricted to a few pure elements such as iron, cobalt and nickel but can also be shown by certain alloys and compounds. For example, alloys of iron and nickel form a range of magnetic materials. The composition determines the interactions between the dipoles of the two elements. Figure 2.190 shows how the saturation flux density depends on the composition of such an alloy. Alloys with about 27% nickel–73% iron are not ferromagnetic at room temperature. In general, commercial magnetic nickel–iron alloys tend to have about 50% nickel or 79% nickel. The 50% alloy has a high saturation flux density but moderate permeability, about 2500; the 79% alloy has a lower saturation flux density but a high permeability, about 100 000.

From an examination of Table 2.14 we might expect manganese, chromium, vanadium, titanium and scandium to be ferromagnetic because they have unpaired 3d electrons. However, they are not. They are *antiferromagnetic materials*. This is because, although each atom behaves as a magnetic dipole, in the solid the atoms bond together in such a way that although magnetic dipoles line up they do so in alternately different directions and so cancel each other out (Figure 2.191).

Ferromagnetic materials are metals with the bonds between atoms being the metallic bond. Ferrimagnetic materials are ceramics with ionic bonds. A consequence of this is that ferromagnetics are good conductors of electricity while ferrimagnetics are poor conductors. With a ferrimagnetic material we have two sets of magnetic dipoles. These are in opposite directions, as with antiferromagnetics, but the opposing dipoles are unequal and so a resultant magnetism is found.

Spinel ferrites have the general formula AB_2O_4 where A is a divalent metal ion and B a trivalent metal ion. The magnetic dipoles for the A and B ions are different strengths and these ions occupy interstitial sites in a cubic oxygen ion lattice. With a spinel structure there are 64 tetrahedral sites and 32 octahedral sites available per unit cell. However, these are not all occupied. With the normal form of the structure, eight of the tetrahedral sites are occupied by A^{2+} ions and sixteen of the octahedral sites are by B^{3+} ions. The ions in the tetrahedral sites give magnetic dipoles in the opposite direction to those in the octahedral sites. Thus the resultant magnetism is due to the differences between the ions in the two types of site. Because the A and B ions have different strength magnetic dipoles, there is a resultant magnetism.

Another type of occupancy of the spinel structure is possible, this being termed the *inverse spinel*. This has eight of the tetrahedral sites occupied by B^{3+} ions and the sixteen octahedral sites occupied by eight B^{3+} ions and eight A^{2+} ions. The result is that, since the magnetic dipoles of the B^{3+} ions will cancel each other out, the magnetism is due to just the eight A^{2+} ions.

An example of a ferrite is the naturally occurring *magnetite*, known as lodestone. This has Fe^{2+} and Fe^{3+} ions. If another divalent transition metal ion is substituted for the Fe^{2+} ion, e.g. Mn^{2+}, Co^{2+}, Ni^{2+}, Cu^{2+} or Zn^{2+}, then another ferrite is produced.

The above are examples of magnetically soft ferrites. Other ferrite structures can be used to give magnetically hard ferrites. One type has a hexagonal crystal structure, barium ferrite $BaFe_{12}O_3$ being of this type and known commercially as *Ferroxdure*. It is widely used for permanent magnets. Their magnetism can be explained in the same way as the soft ferrites with the spinel structure described above.

Dipole moments

For a coil of wire (Figure 2.192) of cross-sectional area A, number of turns N, carrying a current I and in a magnetic field of flux density B at right angles to the plane of the coil, the maximum torque acting on the coil is $NIAB$. The quantity NIA is called the *magnetic dipole moment* of the coil. Thus:

$$\text{maximum torque} = \text{magnetic dipole moment} \times B \qquad [35]$$

The bigger the dipole moment the larger the maximum torque for a given flux density.

We can similarly define a magnetic dipole moment for a spinning electron. For an isolated electron the fundamental magnetic dipole moment is called the *Bohr magneton* μ_B and is 9.27×10^{-24} A m².

Key point

The lode in the term lodestone has the same meaning as in the word lodestar for the North star. Both mean leading and indicate that they provide navigation aids.

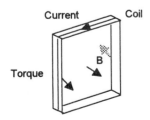

Figure 2.192 *Torque on a current-carrying coil*

Example

Iron saturates at a flux density of 2.18 T. What is the average number of electrons per atom in the solid state that appear to contribute to this magnetisation if the iron has a body-centred cubic structure with a lattice parameter of 0.287 nm?

If n electrons per atom contribute to the magnetisation and there are N atoms per cubic metre, then the amount of magnetism per cubic metre is $Nn\mu_B$. Note that this has units of $(/m^3)(A\ m^2) = A/m$. To find out the magnetism in tesla we need to multiply this by μ_0. Hence the amount of magnetism produced per cubic metre is $\mu_0 Nn\mu_B$ T. For a body-centred cubic structure there will be two complete atoms per unit cell and so:

$$N = 2/(0.287 \times 10^{-9})^3 = 8.46 \times 10^{28}\ \text{atoms/m}^3$$

Thus:

$$4\pi \times 10^{-7} \times 8.46 \times 10^{28} \times 9.27 \times 10^{-24}n = 2.18$$

and $n = 2.21$. This is less than the number, four, of unpaired electrons in the iron atom (see Table 2.14) and thus it appears that in the solid not all the electrons contribute to the magnetism.

Example

In magnetite $FeFe_2O_4$, when the iron atom is ionised to give the Fe^{2+} ion there will be four unpaired electrons left after the loss of the two 4s electrons. When the iron atom is ionised to give the Fe^{3+} ion there will be five unpaired electrons left after the loss of two 4s and one 3d electrons. In the magnetite unit cell, eight Fe^{2+} ions occupy eight octahedral sites and the sixteen Fe^{3+} ions are divided so that eight occupy octahedral sites and eight tetrahedral sites. Estimate the saturation magnetisation of magnetite if it has a unit cell with lattice parameter 0.839 nm, each unit cell containing eight molecules of $FeFe_2O_4$.

The magnetic dipoles from the electrons in the octahedral sites are in the opposite direction to that given by the electrons in the tetrahedral sites. Thus the magnetism is due entirely to the eight Fe^{2+} ions since the Fe^{3+} ions cancel each other out. Each Fe^{2+} ion will have a theoretical magnetic dipole moment of $4\mu_B$ and thus:

magnetic dipole moment for a unit cell
$$= 32\mu_B = 32 \times 9.27 \times 10^{-24} = 2.97 \times 10^{-22} \text{ A m}^2$$

The magnetism per cubic metre is thus

$$2.97 \times 10^{-22}/(0.839 \times 10^{-9})^3 \text{ A/m}$$
$$= 4\pi \times 10^{-7} \times 2.97 \times 10^{-22}/(0.839 \times 10^{-9})^3 = 0.63 \text{ T}$$

2.3.6 Case studies

The following case studies consider some of the practical applications of aspects considered in this chapter.

8.7.1 Resistors

Consider the requirements for resistors for general use in electrical and electronic circuits. They are required to obey Ohm's law and have resistances that do not markedly change when the temperature changes. This would suggest metals as the required material; the resistance of semi-conductors changes much more than metals when the temperature changes. The values of resistance frequently required are in the range 1 kΩ to 10 MΩ. Metals tend to have resistivities ρ of the order of 10^{-6} Ω m and so a resistance of 1 kΩ made with wire of diameter 1 mm would need a length of 785 m. This is not practical. One possible method by which we can use metals is to deposit very thin layers on an insulating substrate. This can then give a very thin thickness of metal. Thin films of thickness about 10 nm (1 nm = 10^{-9} m) are used. We can then make this into a reasonable length by etching a suitable pattern in the metal. Thus a spiral groove might be cut through the metal deposited on a cylindrical substrate (Figure 2.193). The resistance value can be adjusted to some required value by stopping the groove cutting when that value is obtained. Nickel–chromium alloys (nichrome) are widely used for resistors manufactured in this way.

Another alternative is to mix a conductive powder with an insulator and organic solvent, the resulting mixture then being spread over an insulating substrate as a film about 10 μm thick. The conductive powders used are highly conductive oxides such as PdO or RuO$_2$. They have resistivities of about 10^{-6} Ω m and behave as metals. The mixture is fired so that organic solvents evaporate and the insulator and conductive material are left bonded to the substrate. The dispersed conductive particles are considered to form convoluted chains through the insulator. Thus we effectively end up with a number of exceedingly small cross-section conductors. The resistance value for the resistor is then determined by the concentration of the conductive powder in the insulator.

There is an alternative to using a metal and that is to use carbon in the form of graphite. Diamond is a crystal structure based solely on carbon atoms, each atom being bound by strong covalent bonds

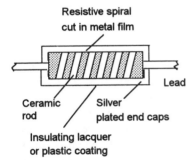

Resistive spiral
cut in metal film

Lead

Ceramic
rod

Silver
plated end caps

Insulating lacquer
or plastic coating

Figure 2.193 *Thin film resistor*

to four other carbon atoms. The result is a strong three-dimensional structure that is an electrical insulator because there are no free electrons. Diamond is also very hard. Graphite, however, is a very soft material. It is the material used as the lead in pencils. Graphite, like diamond, consists only of carbon atoms. However, the way in which the carbon atoms are arranged in the solid is quite different; it can be considered to be a 'layered' structure. The atoms are strongly bonded together with covalent bonds in two-dimensional layers, with only very weak bonds, van der Waals bonds, between the atoms in different layers. As a result, there are free electrons between the layers and the material conducts electricity. At room temperature, the resistivity of diamond is about 10^{10} Ω m, while that of graphite is about 10^{-6} Ω m.

Resistors using graphite all tend to be about the same size, but can have a wide range of values. The process used for making carbon resistors involves powdered graphite being mixed with naphthalene and an insulating filler such as china clay. The mix is then pressed into little cylinders and then fired. This gets rid of most of the naphthalene and leaves a porous graphite structure. The amount of naphthalene, and hence the degree of porosity, determine the resistance of the cylinder. Figure 2.194 shows the form such a resistor might take.

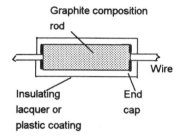

Figure 2.194 *Graphite composition resistor*

Thermistors

The oxide Mn_3O_4 is an electrical insulator with a structure consisting of Mn^{2+}, Mn^{3+} and oxygen ions. It is an insulator because there is no means by which electrons can move. However, if nickel oxide is added we have Ni^{2+} ions replacing some of the Mn^{3+} ions. This results in the gain of one electron for each such replacement and electron neutrality can only be maintained if a Mn^{3+} ion loses an electron to become Mn^{4+}. We now have Mn^{3+} and Mn^{4+} ions occupying similar places in the lattice and electrons are able to hop between them. The result is a semiconductor; this is the basis of thermistors with negative temperature coefficients.

Light-emitting diode (LED)

When a pn junction is biased one way (reverse bias) then virtually no current flows, while when it is biased the other way (forward bias) a current flows. The current is due to movement of holes from the p side of the junction to the n side and movement of electrons from the n side to the p side. When a hole reaches the n side it is readily filled by an electron; when an electron reaches the p side it readily fills a hole. We have electrons from the conduction band combining with holes in the valence band and this jumping between bands leads to the emission of light. Light-emitting diodes are made from semiconductors with band gaps in the range 1.8 to 3.0 eV. Pure gallium arsenide has an energy gap of 1.42 eV and so emits in the infrared, hence its use for TV remote controls. By alloying it with gallium phosphide the energy gap can be changed to give visible light and by adjusting the relative amounts of the two alloy constituents so the colour of the emitted light can be changed.

Magnetic recording media

Magnetic recording media in the form of tapes or discs, floppy and hard, consist of a substrate of plastic on which is the magnetic material. The substrate of tape is polyethylene terephthalate and that of discs polycarbonate. The most commonly used magnetic material is haematite (γ-Fe_2O_3). Particles of haematite are suspended in an organic liquid and applied to the substrate. A magnetic field is then applied and causes all the particles, which are needle-like, to orientate in the same direction. The liquid is then heated and evaporates and the tape/disc rolled or pressed to consolidate the coating. Other materials that are used are chromite (CrO_2) and cobalt modified haematite.

Problems 2.6

1 A solid has atoms with the electronic structure $1s^2 2s^2 2p^6 3s^2 3p^6 3d^{10} 4s^2\ 4p^6 4d^{10} 5s^1$. Would you expect it to be a good electrical conductor or an insulator?

2 What is the drift velocity of electrons in a copper wire of cross-sectional area 1.0×10^{-6} m^2 when a current of 1.0 A flows? Copper has 8.5×10^{28} electrons per cubic metre, each having a charge of 1.6×10^{-19} C.

3 What is the mobility of electrons in copper with an electrical conductivity of 5.5×10^7 S/m if it has a current density of 2.0×10^6 A/m? Copper has 8.5×10^{28} electrons per cubic metre, each having a charge of 1.6×10^{-19} C.

4 If copper forms a face-centred cubic structure with lattice parameter 0.361 51 nm and all the valence electrons contribute to current flow, determine the drift velocity of electrons in a 1 m length of copper wire when a potential difference of 1 V is applied between its ends. The conductivity of copper is 5.5×10^7 S/m and the charge on an electron is 1.6×10^{-19} C.

5 Determine the electrical conductivity of intrinsic germanium at 300 K if there are 2.4×10^{19} electrons per cubic metre, 2.4×10^{19} holes per cubic metre, the mobility of electrons is 0.39 m^2 V^{-1} s^{-1}, the mobility of holes is 0.19 m^2 V^{-1} s^{-1} and the charge carried by the electrons and holes is 1.6×10^{-19} C.

6 Determine the electrical conductivity of extrinsic germanium at 300 K if it is doped with 10^{22} donors per cubic metre, the mobility of electrons is 0.39 m^2 V^{-1} s^{-1} and the charge carried by the electrons and holes is 1.6×10^{-19} C.

7 Which of the following (a) silicon doped with gallium, (b) silicon doped with arsenic, (c) germanium doped with phosphorus, (d) germanium doped with indium, will be n-type and which p-type?

8 Silicon is to be doped with phosphorus to give an n-type semiconductor with an electrical conductivity of 200 S/m. Determine the number of phosphorus atoms required per cubic metre.

9 $KNbO_3$ has a Perovskite structure similar to that of barium titanate. The ion radii are K^+ 0.133 nm, Nb^{5+} 0.070 nm and O^{2-} 0.140 nm. Explain how you might expect the relative permittivity to vary with temperature.

10 With polymer dielectrics, how does the presence or otherwise of polar groups affect the relative permittivity?

11 Outline the reasons for the differences in relative permeability between diamagnetic, paramagnetic and ferromagnetic materials.

12 Explain how ferrimagnetism occurs.

13 A silicon steel is specified as having a Curie temperature of 745°C. How will its magnetic properties below that temperature differ from those above it.

14 Nickel has an average of 0.604 Bohr magnetons per atom. What is the saturation flux density?

15 Estimate the saturation magnetic flux density for the ferrite $NiFe_2O_4$ if the resultant moment is due to eight Ni^{2+} ions per unit cell and the unit cell has a lattice parameter of 0.837 nm.

16 Estimate the net magnetic moment per unit cell for the ferrite $CuFe_2O_4$ if Fe^{3+} ions occupy tetrahedral sites and equal numbers of Cu^{2+} and Fe^{3+} ions occupy octahedral sites in an inverse spinel structure.

17 Estimate the net magnetic moment per unit cell for the ferrite $CoFe_2O_4$ if Fe^{3+} ions occupy tetrahedral sites and equal numbers of Co^{2+} and Fe^{3+} ions occupy octahedral sites in an inverse spinel structure.

3 Alloys

Summary

The term alloy is used for metallic materials made by mixing two or more elements to form a material with a specific composition. The everyday metallic products around you will be made from alloys since pure metals tend not to have the properties required of them, generally being rather soft and weak. The term ferrous alloy is used for those alloys which contain iron as the main constituent and non-ferrous alloys when the main element is some other metal. This part is a consideration of metallic materials commonly used in engineering: steels, cast irons, and the non-ferrous alloys of aluminium, magnesium, titanium, zinc, copper, nickel and cobalt.

Objectives

By the end of this part, the reader should be able to:

- understand how the properties of carbon steels are related to the iron–carbon diagram;
- explain how the properties of steels are modified by alloying;
- explain the properties of cast irons;
- describe how the microstructure and properties of steels are modified by heat treatment;
- describe the range of non-ferrous alloys and how their properties are determined by their microstructure.

3.1 Ferrous alloys

The term *ferrous* alloys is used for those alloys that contain iron as the main constituent, non-ferrous alloys being those whose main element is any metal other than iron. Pure iron is a relatively soft material and is hardly of any commercial use in this state. Alloys of iron, particularly with carbon are, however, very widely used, the percentage of carbon alloyed with the iron having a profound effect on the properties of the alloy. This section is a discussion of the characteristics and properties of iron alloys:

- **Steel**
  These are ferrous alloys containing less than 2% carbon.

  Plain carbon steel
  Plain carbon steels are those steels having carbon contents from about 0.03% to about 1.2%. In addition, manganese and

silicon are almost always included to overcome the effects of the two main impurities of sulphur and oxygen which arise from the steel making process.

Alloy steel

The term *alloy steel* is used where significant amounts of elements such as nickel, chromium, molybdenum, manganese, silicon and vanadium are included. The term *low-alloy steel* is used when the total amount of alloying elements is less than about 2%, *medium alloy* for between 2% and 10% and *high alloy* for more than 10%.

• **Cast iron**
Cast irons have between about 2% and 4% carbon and are so called because they are cast to their finished shape from the melt.

3.1.1 The iron–carbon system

Pure iron exists at room temperature as a body-centred cubic structure, this being known as *alpha ferrite* or just *ferrite*. This continues to exist up to 912°C. At this temperature the structure changes to a face-centred cubic, known as *austenite* or *gamma iron*. At 1394°C this form changes to a body-centred cubic structure known as *delta ferrite*. At 1538°C, the iron melts. Figure 3.1 shows the cooling curve.

These transformations from a body-centred cubic, packing fraction 0.68, to a more closely packed face-centred cubic, packing fraction 0.74, and then back to the less well-packed body-centred cubic result in the volume of a sample of pure iron changing as it goes through these transformations. Figure 3.2 shows the changes.

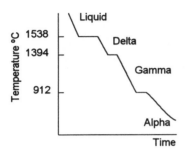

Figure 3.1 *Cooling curve for pure iron*

Key points

Pure iron exists at room temperature as a body-centred cubic structure known as *ferrite*. Above 912°C this changes to a face-centred cubic, known as *austenite*. Above 1394°C this changes to a body-centred cubic structure known as *delta ferrite*.

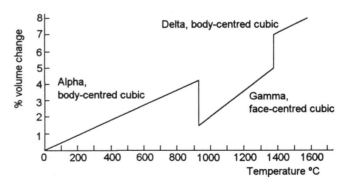

Figure 3.2 *Volume changes for pure iron*

The body-centred cubic structure of alpha ferrite is formed by iron atoms of diameter 0.256 nm. Between these atoms are interstitial sites which can accommodate atoms in an octahedral site up to 0.070 nm in diameter (Figure 3.3) and in a tetrahedral site of 0.038 nm in diameter. Carbon atoms have diameters of 0.154 nm. To accommodate carbon atoms in alpha ferrite in such

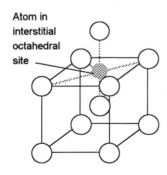

Figure 3.3 *Body-centred structure*

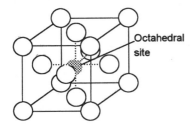

Figure 3.4 *Face-centred structure*

Key points

Cementite is hard and brittle. It has a composition of one atom of carbon to every three atoms of iron, i.e. Fe_3C. With carbon having an atomic mass of 12 g/mol and iron 56 g/mol:

percentage of carbon in cementite
= $12/(12 + 3 \times 56) \times 100\%$ = 6.67%

For this reason the iron–carbon phase diagram shows a vertical line at 6.67% carbon.

sites at room temperature would require a severe distortion of the lattice. Because of this, carbon has only very limited solubility in ferrite at defects, grain boundaries and dislocations. However, the interstitial sites increase in size as the temperature increases and so more carbon can be accommodated at higher temperatures.

The face-centred cubic structure of austenite has interstitial spaces which can accommodate atoms in an octahedral site (Figure 3.4) up to 0.104 nm in diameter and in a tetrahedral site 0.056 nm in diameter. Carbon atoms with their diameter of 0.154 nm can thus be accommodated with some slight distortion of the lattice and thus there is a higher solubility in austenite than in ferrite.

When iron containing carbon is cooled from the austenitic state to the ferrite state, the reduction in the solubility of carbon in iron means that some carbon must come out of solution. This occurs by the formation of a compound between iron and carbon called *cementite*. This has one carbon atom for every three iron atoms, hence it is often referred to as iron carbide Fe_3C. Cementite is hard and brittle.

Phase diagram

Figure 3.5 shows the iron–carbon system phase diagram for carbon content up to 6.67%.

Figure 3.6 shows an enlarged view of the phase diagram for very low carbon percentages. The alpha ferrite will accept up to about 0.02% carbon in solid solution. The solubility of alpha ferrite for carbon decreases as the temperature decreases. When an alloy with less than 0.02% carbon is cooled, carbon in excess of that which can be held by the alpha ferrite is precipitated as cementite.

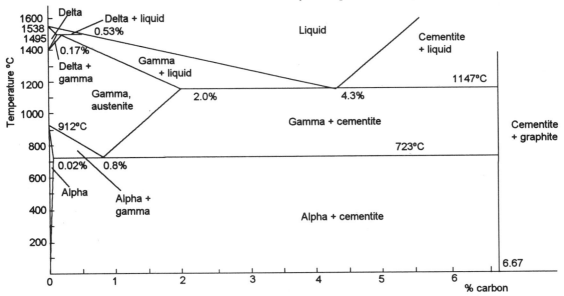

Figure 3.5 *The iron–carbon system. This figure describes the ideal microstructure that will form for any iron-carbon composition as the temperature is changed between melt temperature and ambient temperature. The boundary lines were obtained from cooling curves, such as Figure 3.1 for many compositions. The boundary lines represent temperatures at which the cooling rate is temporarily zero, the so-called 'arrest points'.*

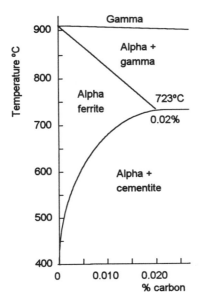

Figure 3.6 *Low-carbon part of phase diagram*

Key point

Pearlite is a laminated structure of alpha ferrite and cementite, i.e. it has alternating regions of soft alpha ferrite and hard cementite.

Consider the cooling from the liquid of an alloy with 0.8% carbon (Figure 3.7). For temperatures above 723°C, the solid formed is gamma ferrite, i.e. austenite. This is a solid solution of carbon in iron, a face-centred cubic structure. At 723°C there is a eutectoid and a sudden change to give a laminated structure of alpha ferrite plus cementite. This structure is called *pearlite* since it resembles mother-of-pearl. If we consider the tie line (a constant temperature line between two boundary lines on a phase diagram – see Chapter 2) in Figure 3.5 at a temperature marginally below 723°C and apply the lever rule:

$$\% \text{ of alpha ferrite} = \frac{6.67 - 0.80}{6.67 - 0.02} \times 100\% = 83.3\%$$

$$\% \text{ of cementite} = \frac{0.80 - 0.02}{6.67 - 0.02} \times 100\% = 11.7\%$$

Since the densities of alpha ferrite and cementite are similar, this means that the volume of pearlite that is alpha ferrite is about seven times that of the cementite.

Steels containing less than the eutectoid composition are called *hypoeutectoid steels*, those with between this composition and 2.0% carbon being called *hypereutectoid steels*.

Figure 3.8 shows the cooling of a 0.4% carbon steel, a hypoeutectoid steel, from the austenite phase to room temperature. When the alloy is cooled to below about 820°C, crystals of alpha ferrite start to grow in the austenite. The ferrite tends to grow at the grain boundaries of the austenite grains. At 723°C, the remaining austenite changes to the eutectoid structure, i.e. pearlite. The result is a network of ferrite along the grain boundaries surrounding areas of pearlite. At a temperature just above 723°C, if we consider a tie line and apply the lever rule:

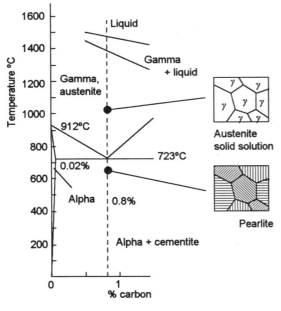

Figure 3.7 *The eutectoid composition*

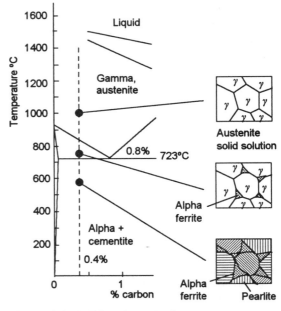

Figure 3.8 *0.4% carbon steel*

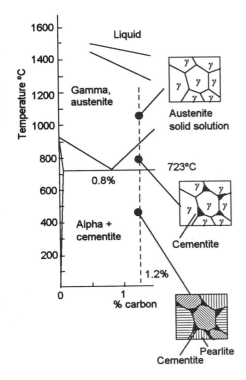

Figure 3.9 *1.2% carbon steel*

$$\% \text{ alpha ferrite} = \frac{0.8 - 0.4}{0.8 - 0.02} \times 100\% = 51.3\%$$

$$\% \text{ austenite} = \frac{0.4 - 0.02}{0.8 - 0.02} \times 100\% = 48.7\%$$

Since the austenite turns to pearlite when the temperature drops below 723°C, the composition becomes 51.3% alpha ferrite and 48.7% pearlite.

Figure 3.9 shows the cooling of a 1.2% carbon steel, a hypereutectoid steel, from the austenite phase to room temperature. When the alloy is cooled below about 900°C, cementite starts to grow at the grain boundaries of the austenite grains. At 723°C the remaining austenite changes to the eutectoid structure, i.e. pearlite. The result is a network of cementite along the grain boundaries surrounding areas of pearlite. At a temperature just above 723°C, if we consider a tie line and apply the lever rule:

$$\% \text{ cementite} = \frac{1.2 - 0.80}{6.67 - 0.8} \times 100\% = 6.8\%$$

$$\% \text{ austenite} = \frac{6.67 - 1.2}{6.67 - 0.8} \times 100\% = 93.2\%$$

Since the austenite turns to pearlite when the temperature drops below 723°C, the composition becomes 6.8% cementite and 93.2% pearlite.

In all the above discussion we have assumed that the cooling from the austenitic state is slow enough for diffusion to occur and all the changes to be completed. This is discussed later in this chapter.

The effect of carbon content on the properties

Ferrite is a comparatively soft and ductile material. Pearlite is a harder and much less ductile material. Thus the relative amounts of these two substances in a carbon steel will have a significant effect on the properties of that steel. Figure 3.10 shows how the percentages of ferrite and pearlite change with percentage carbon and how the mechanical properties are related to these changes.

For steels cooled slowly from the austenitic state up to the eutectoid composition, i.e. for hypoeutectoid steels, the decreasing percentage of ferrite and the increasing percentage of pearlite results in an increase in tensile strength and hardness. The ductility decreases, the percentage elongation being a measure of this. For hypereutectoid steels, increasing the amount of carbon decreases the percentage of pearlite and increases the percentage of cementite. This increases the hardness but has little effect on the tensile strength, the ductility also changes little.

Key points

Ferrite, being mostly pure iron, is a comparatively soft and ductile material. Pearlite, being a combination of ferrite and cementite, has properties intermediate between the two. Cementite is much harder and brittle. The relative amounts of these substances in a carbon steel has a significant effect on the properties of that steel.

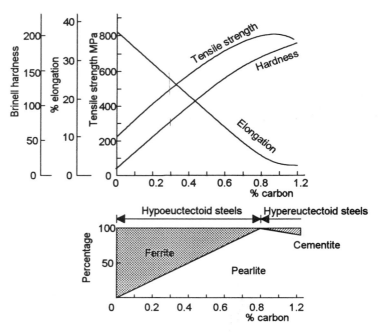

Figure 3.10 *The effect of carbon content on the structure and properties of carbon steels*

3.1.2 Heat treatment

Heat treatment can be used to change the microstructure and hence the properties of carbon steels. *Full annealing* involves heating the steel to a temperature at which it becomes austenitic, holding at that temperature until it is completely austenitic and then cooling very slowly in the furnace so that the structure has time to convert to an entirely new structure of ferrite and pearlite. The steel is then in its softest condition. *Normalising* is essentially the same as annealing but with the cooling from the austenitic state being in air rather than in the furnace, hence a more rapid rate of cooling. As the rate of cooling increases there is less time for the carbon to diffuse out of the iron matrix and so less ferrite is formed. This results in a slightly harder material.

For steels with more than about 0.3% carbon, rapid cooling by quenching the item in cold water can produce structural changes which result in the formation of *martensite*. This gives a much harder steel.

Martensite

Martensite is formed when there is not time for carbon atoms to diffuse out of the face-centred cubic austenite and produce the body-centred alpha ferrite with its lower solubility for carbon. The rate of cooling has to be sufficiently rapid not to give sufficient time for carbon atoms to diffuse out of the face-centred cubic austenite and produce the body-centred alpha ferrite. The minimum cooling rate that will give martensite is called the *critical cooling rate*. Cooling rates faster than this give a

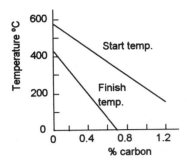

Figure 3.11 *Martensitic start and finish temperatures*

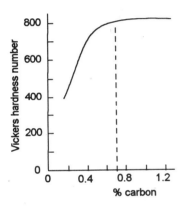

Figure 3.12 *Hardness of water-quenched plain carbon steels*

Key points

Martensite is formed when there is not time for carbon atoms to diffuse out of the face-centred cubic austenite and produce the body-centred alpha ferrite with its lower solubility for carbon. It has a low ductility because carbon atoms are locked in interstitial spaces which are too small for them and so high strains occur. Some ductility can be restored by tempering.

---

completely martensitic structure but rates slower than this will not. The value of the critical cooling rate depends on the percentage of carbon present, the smaller the percentage of carbon the higher the rate. Below about 0.3% carbon, the rate of cooling needed to prevent the diffusion of the relatively small numbers of carbon atoms is too high to be achievable by quenching the steel in cold water. Such steels are thus not subject to this form of heat treatment.

As the temperature falls from the austenitic state, the transformation of austenite to martensite starts at a temperature referred to as the *martensitic start temperature* M_s. The transformation of all the austenite to martensite is complete by the temperature referred to as the *martensitic finish temperature* M_f. The values of these temperatures depend on the composition of the alloy. Figure 3.11 shows the values for plain carbon steels. For a 0.4% carbon steel, the start temperature is about 450°C and the finish temperature about 200°C. Thus at room temperature, for such an alloy, all the austenite will be martensite. However, for a plain carbon steel with 1.0% carbon, the finish temperature is below room temperature and so all the austenite cannot be converted to martensite by cooling to room temperature. The martensitic finish temperature is below room temperature for plain carbon steels with more than about 0.7% carbon and so all these steels will contain both austenite and martensite.

The hardness of martensite increases with an increase in the carbon content. This means that plain carbon steels cooled to room temperature from the austenitic state will show an increase in hardness up to a carbon content of 0.7% carbon, since all the austenite is transformed into martensite. Above 0.7% carbon, not all the austenite is converted into martensite and the increasing hardness of the martensite is balanced by an increasing proportion of austenite, a relatively soft material. Figure 3.12 shows how the hardness of carbon steels, cooled at a rate greater than the minimum cooling rate, will vary with carbon content.

The rates of cooling at the centres of different diameter bars quenched to the same temperature will depend on their size, the bigger the diameter the slower will be the rate of cooling at the centre of the bar. There will, therefore, be a variation in properties across the section which will depend on the diameter. For this reason, the mechanical properties of steels are quoted for different size bars. The term *limiting ruling section* is used, this being the maximum diameter of round bar at the centre of which the specified properties may be obtained.

Martensite has a low ductility because in a plain carbon steel the carbon atoms are locked in interstitial spaces which are too small for them without lattice strains occurring. The result is that a large number of dislocations are produced and this high dislocation density hinders slip. Some ductility can be restored to a martensitic structure by *tempering*. This process involves heating the material to perhaps a few hundred degrees Celsius and then air cooling it. This allows carbon atoms to diffuse out of the martensite and so reduce its hardness.

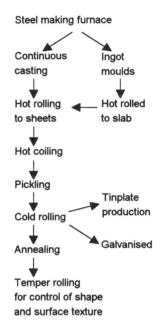

Steel making furnace

Continuous casting

Ingot moulds

Hot rolling to sheets ← Hot rolled to slab

Hot coiling

Pickling

Cold rolling → Tinplate production

Annealing → Galvanised

Temper rolling for control of shape and surface texture

Figure 3.13 *Processing route for mild steel strip*

3.1.3 Properties and uses of plain carbon steels

Table 3.1 shows typical mechanical properties of a range of plain carbon steels after different heat treatments. Carbon steels are grouped according to their carbon content.

1 *Less than 0.1% carbon*
   These steels are soft and ductile and cannot be usefully hardened by heat treatment. They have uses where high cold formability is required, e.g. car body work.

2 *0.1% to 0.25% carbon*
   These are called *mild steel*. Mild steel is a general purpose steel and, though reasonably strong, is used where hardness and tensile strength are not the most important requirements. It is ductile and cannot be usefully hardened by heat treatment. Typical applications are as joists in buildings, screws, nails, wire. The steel used for food and beverage cans, and referred to as tinplate, has 0.13% carbon and is coated with tin to give corrosion protection. Figure 3.13 shows the basic processing route followed by mild steel strip.

Table 3.1 *Properties of plain carbon steels*

% carbon	Condition	Tensile strength MPa	Percentage elongation	Brinell hardness	Izod impact strength J
0.2	Annealed	400	37	115	123
	Normalised	450	36	130	118
0.4	Annealed	520	30	150	44
	Normalised	600	28	170	65
	Quench, temper 200°C	910	16	260	
	Quench, temper 430°C	850	21	240	
	Quench, temper 540°C	780	23	210	
0.6	Annealed	635	23	180	11
	Normalised	790	18	230	13
	Quench, temper 200°C	1120	13	320	
	Quench, temper 430°C	1090	14	310	
	Quench, temper 540°C	970	17	280	
0.8	Annealed	620	25	170	6
	Normalised	1030	11	390	7
	Quench, temper 200°C	1330	12	390	
	Quench, temper 430°C	1300	13	375	
	Quench, temper 540°C	1130	16	320	
1.0	Annealed	660	13	190	3
	Normalised	1030	10	290	5
	Quench, temper 200°C	1300	10	400	
	Quench, temper 430°C	1200	12	360	
	Quench, temper 540°C	1090	15	320	

3 **0.25% to 0.50% carbon**

These are called *medium-carbon steel* and are strong and heat treatable to produce a wide range of properties in the quenched and tempered state. They are used where strength and toughness are required, e.g. agricultural tools, fasteners, dynamo and motor shafts, crankshafts, connecting rods, gears.

4 **More than 0.5% carbon**

These are called *high-carbon steel*. Such steels are very strong but ductility is low. High-carbon steel is used for withstanding wear and where hardness is a more necessary requirement than ductility. It is used for machine tools, saws, hammers, cold chisels, punches, axes, dies, taps, drills, razors. The main use of high-carbon steel is thus as a tool steel.

Plain carbon steels have limits on their engineering applications because:

- A high-strength steel can only be obtained by increasing the carbon content to such a level that the material becomes brittle. High strength cannot be obtained with good ductility and toughness.

- Hardness requires water quenching. The severity of this rapid rate of cooling often leads to distortion and cracking of the steel.

- Large sections cannot be hardened uniformly. The hardness depends on the rate of cooling and this will vary across a large section.

- Plain carbon steels have poor resistance to corrosion and oxidation at high temperatures.

3.1.4 Alloy steels

The term *alloy steel* is used to describe those steels to which one or more alloying elements, in addition to carbon, have been deliberately added in order to modify the properties of the steel. There are a number of ways in which the alloying elements can have an effect on the properties, the main one being:

- **Solution harden the steel**

  Most of the elements used in alloy steels form 'substitutional solid solutions' with the iron. This increases the tensile and impact strengths (Figure 3.14).

- **Form carbides**

  They may form stable, hard carbides which can, if in an appropriate form such as fine particles, increase the strength and hardness. Manganese, chromium and tungsten have this effect.

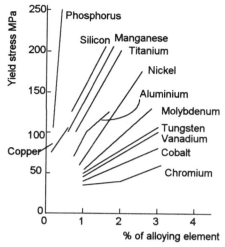

Figure 3.14 *Effect of alloying element on yield stress as a result of solid solution hardening*

- *Form graphite*

  They could cause the breakdown of cementite and lead to the presence of graphite in the alloy. Silicon and nickel have this effect. The result is a decrease in strength and hardness.

- *Stabilise austenite or ferrite*

  They may lower the temperature at which austenite is formed on heating the steel. Manganese, nickel, copper and cobalt have this effect. The lowering of this temperature means a reduction in the temperature to which the steel has to be heated for hardening by quenching. If a sufficiently high percentage of one of these elements is added, the transformation temperature to austenite may be decreased to such an extent that the austenite is retained at room temperature. With manganese, about 11 to 14% produces what is known as *austenitic steel*. Such steels have relatively good hardness combined with ductility, and so are tough.

  They may increase the temperature at which austenite is formed on heating the steel. This raises the temperature to which the steel has to be heated for hardening. Chromium, molybdenum, tungsten, vanadium, silicon and aluminium have this effect. If a sufficiently high percentage of one of these elements is added to the steel, on heating the transformation from ferrite to austenite may not take place before the steel reaches its melting point temperature. Thus such a steel cannot be hardened by quenching and is known as *ferritic steel*. With a 0.1% carbon steel, the addition of 12 to 25% chromium gives such a steel.

- *Change the critical cooling rate*

  Most alloying elements reduce the critical cooling rate. The effect of this is to make air or oil quenching possible, rather than water quenching. It also increases the hardenability.

- *Improve corrosion resistance*

  Some elements promote the production of adherent oxide layers on the surfaces of steel and so improve its corrosion resistance. Chromium is particularly useful in this respect. If it is present in a steel in excess of 12% the steel is known as *stainless steel* because of its corrosion resistance. Copper is also used to promote corrosion resistance.

- *Change grain growth*

  Some elements accelerate grain growth while others decrease grain growth. The faster grain growth leads to large grain structures and consequently to a degree of brittleness. The slower grain growth leads to smaller grain size and so to an improvement in ductility. Chromium accelerates grain growth and thus care is needed in the heat treatment of chromium steels to avoid excessive grain growth. Nickel and vanadium decrease grain growth.

- *Improve machinability*

  Sulphur and lead are elements added to improve the chip formation properties of steel and so lead to easier machining.

Table 3.2 indicates the main effects and functions of the various alloying elements. An alloying element generally affects the properties of the alloy in more than one way, some of the ways not always being beneficial. Alloying elements can thus be chosen to counteract the effects of impurities, counteract the effects of other alloying elements and improve the properties. The following sections outline some of the common forms of alloy steels.

Table 3.2 *Effects and functions of alloying elements*

Element	Main effects	Main functions
Aluminium	Ferrite stabiliser	Aids nitriding
Chromium	Carbide former	Improves corrosion resistance
	Ferrite stabiliser	Increases hardenability
	Forms surface oxide layers	Improves high-temperature properties
		Improves abrasion and wear resistance
Cobalt	Austenite stabiliser	Improves strength at high temperatures
Copper	Austenite stabiliser	Improves corrosion resistance
Lead	Improves chip formation	Improves machinability
Manganese	Solid solution hardening	Increases hardenability
	Carbide former	Combines with the sulphur in steel to reduce
	Austenite stabiliser	brittleness and so improve machinability
Molybdenum	Ferrite stabiliser	Improves hardenability
	Carbide former	Restricts austenitic grain growth
	Inhibits grain growth	Increases hot strength and hardness
		Improves corrosion resistance of stainless steels
		Prevents a form of embrittlement encountered during tempering of nickel–chromium steels
		Improves abrasion resistance with high carbon
Nickel	Austenite stabiliser	Improves strength and toughness
	Solid solution hardening	With high chromium content, makes steel austenitic
	Graphite former	
	Inhibits grain growth	
Phosphorus	Solid solution hardening	Strengthens low-carbon steels
	Improves chip formation	Improves machinability
Silicon	Ferrite stabiliser	Deoxidisation of liquid steel
	Solid solution hardening	Improves fluidity in casting
		Strengthens low-alloy steel
Sulphur	Improves chip formation	Improves machinability
Titanium	Carbide former	Forms compounds with carbon to improve chromium
	Solid solution hardening	steels
Tungsten	Carbide former	Improves hot hardness and strength
	Ferrite stabiliser	Gives hard, abrasion-resistant carbides in tool steels
Vanadium	Ferrite stabiliser	Restricts grain coarsening of austenite
	Inhibits grain growth	Increases hardenability
	Carbide former	Improves hot hardness

High-strength low-alloy steels

Mild steel or higher-carbon plain carbon steels have been widely used for structural work such as bridges, buildings and ships. The carbon content of the steels used is typically about 0.3%. However, the development of welding as a means of joining sections rather than riveting has led to problems. The relatively high carbon content of such plain carbon steels leads to a chance of failure by brittle fracture of welded structures. During welding, the metal nearest the weld changes to austenite and depending on the rate of cooling will produce martensite. Thus only steels with less than about 0.3% carbon are considered suitable for welding because of the increased chance of embrittlement and structural changes.

The properties generally required of a constructional steel are:

- High yield stress, this representing the maximum stress to which the material could be exposed in use.

- Tough, not brittle.

- Good weldability.

- Good corrosion resistance.

- At low temperatures steel becomes brittle, thus a value of this ductile to brittle transition temperature which is lower than will be encountered in service is required.

- Low cost.

High-strength low-alloy steels have been developed for use as structural steels. So that welding can be used, the carbon content of the alloy has to be kept low if brittle martensite is not to be formed in the heat-affected zones of welds. To maximise the yield strength, alloying elements can be added to give solution hardening of the ferrite and reduce grain size. However, most of the common alloying elements increase the temperature at which steel changes from being brittle to ductile and so makes it feasible that temperatures might occur which make the steel brittle. This is not desirable so the choice is restricted to those few elements which will give solution hardening without increasing this transition temperature. For this reason, nickel, manganese and titanium are the only options. Thus a carbon steel with about 0.2% carbon and 1.5% manganese gives a structural steel with a yield stress of about 250 MPa. The addition of small amounts of niobium or vanadium, coupled with controlled hot rolling, improves yield stress as a result of precipitation hardening occurring from the precipitation of carbides or nitrides and the production of fine grain size. One form of such a steel has 0.2% carbon, 1.5% manganese, between 0.003 and 0.1% niobium and between 0.003 and 0.1% vanadium and gives a yield stress of the order of 450 MPa and good toughness.

Table 3.3 lists the details of the properties of weldable structural steels, the grade numbers used being to the European specification BS EN 10025: 1993.

Key point

High-strength low-alloy steels are suitable for joining by welding since with only 0.3% carbon, the steel does not harden and become brittle on cooling after welding. The high strength is instead achieved by the addition of manganese, niobium and vanadium.

Table 3.3 *Weldable structural steels*

Grade	Tensile strength MPa	Min. yield stress MPa	Charpy V-notch energy J
S235JR	340–470	235	27 at 20°C
S235JO	340–470	235	27 at 0°C
S235J2G3	340–470	235	27 at –20°C
S275JR	410–560	275	27 at 20°C
S275JO	410–560	275	27 at 0°C
S275J2G3	410–560	275	27 at –20°C
S355JR	490–630	355	27 at 20°C
S355JO	490–630	355	27 at 0°C
S355J2G3	490–630	355	27 at –20°C
S355K2G3	490–630	355	40 at –20°C

Note: The number used with the specification code indicates the minimum yield stress in MPa. JR indicates the specification of the Charpy V-notch impact value as 27 J at 20°C, JO its value as 27 J at 0°C, J2 its value as 27 J at –20°C and K2 its value as 40 J at –20°C. G3 indicates the steel is supplied in the normalised condition. The tensile strength values are for thickness greater than 3 mm and less than 100 mm, the yield stress values for 16 mm thickness and the Charpy test values for thickness less than 150 mm.

Medium-alloy nickel–chromium steels

> **Key point**
>
> Medium-alloy nickel-chromium steels are widely used where a moderate to high tensile strength and hardness are required.

When chromium, of the order of 1%, is added to a steel, some of it dissolves in the ferrite to give solution hardening and the remainder forms chromium carbide. Chromium carbide is harder than cementite. The result is a harder steel. A problem with just the use of chromium is that it promotes the growth of grains and so gives a coarse grain structure steel and so brittleness may occur. The addition of nickel to a steel increases the strength by solution hardening and increases the toughness by limiting grain growth. Unfortunately it tends to cause cementite to break down to graphite. Thus nickel alone can only be used with steels having small amounts of carbon. However, if both chromium and nickel are added to a steel, the limiting grain size property of the nickel with the carbide-forming property of the chromium combine to give a very widely used alloy steel.

The combination of nickel and chromium changes the critical cooling rate so that martensite is formed at slower rates of cooling, thus larger sections do not run the risk of cracking through having to be cooled quickly to obtain the required hardness. With 4.25% nickel and 1.25% chromium, the critical cooling rate is so changed that the steel can be hardened to give a fully martensitic steel by quenching in an air blast rather than cold water. A slower rate of cooling also means that the differences in the rates of cooling of the core of a thick section compared with its surface are less and so there is less variation of hardness with depth.

Nickel–chromium steels tend to contain about 0.1 to 0.55% carbon, 1.0 to 4.75% nickel, 0.45 to 1.75% chromium and 0.3 to 0.8% manganese. Figure 3.15 shows the general form of the properties of such steels after quenching and tempering to different temperatures. A characteristic of nickel–chromium steels is that the impact strength shows a marked reduction if tempering at about 300°C is used and so tempering should not be used at such temperatures. This effect is termed *temper brittleness*. The impact strength is reduced by slow cooling from a higher tempering temperature, as it passes through the critical temperature region, and hence it must be cooled quickly by oil quenching. Temper brittleness can be overcome by adding a small amount of molybdenum, e.g. 0.3%. Table 3.4 shows the properties of common chromium and nickel–chromium–molybdenum steels.

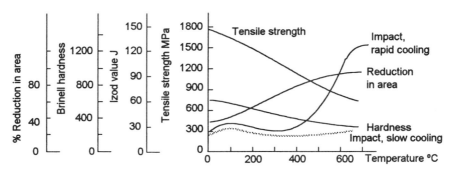

Figure 3.15 *Effect of tempering temperature on the properties of a typical nickel–chromium steel*

Table 3.4 *Properties of chromium and nickel–chromium–molybdenum steels*

BS ref.	Composition %	Condition OQ = oil quench, T = temper	Tensile strength MPa	Yield stress MPa	Percent. elong.	Typical uses
Chromium steel						
530M40	0.4 C, 0.6–0.9 Mn, 0.9–1.2 Cr	OQ 850–880°C, T 550–700°C	700–850	525	17	Widely used where strengths less than 700 MPa required, e.g. crankshafts, axles, connecting rods
Nickel–chromium–molybdenum steels						
817M40	0.4 C, 0.45–0.7 Mn, 1.0–1.4 Cr, 0.2–0.35 Mo, 1.3–1.7 Ni	OQ 820–850°C, T 660°C max.	850–1000	650	13	Widely used for high tensile engineering parts, e.g. connecting rods, bolts
826M40	0.4 C, 0.45–0.7 Mn, 0.5–0.8 Cr, 0.45–0.65 Mo, 2.3–2.8 Ni	OQ 820–850°C, T 660°C max	925–1075	740	12	Used where greater strengths than the above alloy required
853M30	0.3 C, 0.45–0.7 Mn, 1.4–1.8 Cr, 0.2–0.35 Mo, 3.9–4.9 Ni	OQ 810–840°C, T 200–280°C	1550 max.	1235	7	Can be air hardened. Used for gears, valve mechanisms

Manganese steels

About 0.25 to 1% manganese is added to steels to deoxidise them and combine with sulphur resulting from the steel making process. Higher percentages can, however, be used to increase the strength and toughness of steels. Steels with 1.6 to 1.9% manganese and 0.3 to 0.45% carbon have higher strengths than the plain carbon steels and are used for such items as gears, tie rods and axles. However, if the manganese content exceeds about 2%, the steels become brittle. With a manganese content of about 12% and a carbon content of 1.2% a steel, known as *Hadfield's manganese steel*, is produced which combines high surface hardness with toughness. If the steel is rapidly quenched from the austenitic state it remains austenitic at room temperature. However, as soon as it is deformed, the austenite transforms to martensite. Thus the surface becomes martensitic and hard while the interior remains austenitic and tough. Table 3.5 gives the properties of manganese steels.

Table 3.5 *Properties of manganese steels*

BS reference	Composition %	Condition OQ = oil quench, T = temper	Tensile strength MPa	Yield stress MPa	Percent. elong.	Typical uses
150M19	0.2 C, 1.3–1.7 Mn	OQ 860–900°C T 550–660°C	550–700	360	13	Weldable alloy used for large diameter shafts, haulage and lifting gear
150M36	0.32–0.4 C, 1.3–1.7 Mn	OQ 840–870°C, T 550–660°C	625–775	400	18	Axles, gun parts, levers, spindles
Hadfield steel	1.2 C, 13 Mn	Finish by quenching from 1050°C	Surface hardness 550 HB, interior 200 HB			Applications requiring wear resistance, e.g. railway crossings and points, dredging equipment

Maraging steels

Maraging steels are high-strength, high-alloy steels which can be precipitation hardened. The term maraging derives from the transformation which occurs after the steel is cooled from the austenitic state, the production of *mar*tensite followed by *ageing* (mar-ageing), i.e. precipitation hardening. The alloys have a high nickel content (18 to 22%) and a carbon content less than 0.3%. Other elements such as cobalt, titanium and molybdenum are present. These form intermetallic compounds with nickel. A typical maraging steel might contain 18% nickel, 7 or 8% cobalt, 5% molybdenum and about 0.5% titanium. The carbon content is kept low since otherwise the high nickel content could lead to the formation of graphite in the structure and a consequential drop in strength and hardness.

The steel is heated to about 830°C and then air cooled. This results in a martensitic structure being formed. Because of the low amounts of carbon in the steel, this martensitic structure is of low strength and the steel is easily machined and worked. Following machining and working, the steel is then precipitation hardened by heating to about 500°C for two or three hours. During this time, precipitates of intermetallic compounds are formed. The effect of this ageing process is to increase the tensile strength and hardness. Typically, prior to the ageing process the material might have a tensile strength of about 700 MPa and a hardness of 300 HV and afterwards 1700 MPa and 550 HV. Table 3.6 gives the properties of maraging steels with 18% nickel. The reference codes used for the steels quote the nominal percentage of nickel and the value for the nominal 0.2% proof stress. There are other groups of maraging steels based on 20% nickel and 25% nickel.

Maraging steels can give very high strengths with good fracture toughness but cost considerably more than conventional low-alloy steels. They are used where their special properties are essential, e.g. aircraft undercarriage parts, punches and dies for cold forging, die casting and extrusion dies for aluminium.

Table 3.6 *Properties of maraging steels*

Grade	Composition %	Condition	Tensile strength MPa	0.2% proof stress MPa	% elong.	Charpy impact value J
18Ni1400	17–19 Ni, 8–9 Co, 3–3.5 Mo, 0.15–0.25 Ti, 0.05–0.15 Al, 0.03 max. C	Solution treated 1 h at 820°C, air cooled, age harden 3 h at 480°C	1565	1400	9	52
18Ni1900	17–19 Ni, 8–9.5 Co, 4.6–5.2 Mo, 0.5–0.8 Ti, 0.05–0.15 Al, 0.03 max. C	Solution treated 1 h at 820°C, air cooled, age harden 3 h at 480°C	1965	1900	7.5	21
18Ni2400	17–18 Ni, 12–13 Co, 3.5–4 Mo, 1.6–2 Ti, 0.1–0.2 Al, 0.01 max. C	Solution treated 1 h at 820°C, air cooled, age harden 3 h at 480°C	2460	2400	8	11

3.1.5 Stainless steels

The addition of small percentages of chromium to a plain carbon steel results in the formation of hard carbides which, if dispersed as a fine precipitate, can increase strength and hardness and cause a reduction in the critical cooling rate. The percentages of chromium involved in these alloys would be less than 2%. If, however, much larger percentages are used, the steel is given exceptionally good corrosion resistance (Figure 3.16).

Ferritic steels

Ferritic steels contain between 12 and 25% chromium and less than 0.1% carbon. Figure 3.17 shows the phase diagram for iron–chromium alloys with this low percentage of carbon. When an alloy with this composition is cooled from the liquid, there is only one phase change and that is to ferrite. Because austenite cannot be formed with such a steel, hardening by quenching to give martensite cannot be used. They can, however, be work hardened. Since they work harden slowly, they are suitable for forming by deep drawing, spinning, etc.

Key point

The addition of more than 12% of chromium to steels gives exceptionally high corrosion resistance alloys.

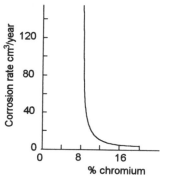

Figure 3.16 *Effect of percentage of chromium on corrosion rate*

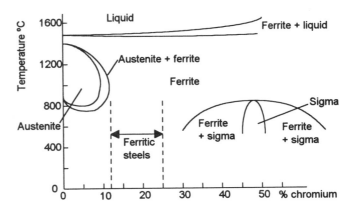

Figure 3.17 *Phase diagram for iron–chromium alloys with 0.1% carbon*

Ferritic steels are used for mouldings and trim for car bodies, gas and electric stoves and other domestic appliances; spoons and forks; car silencers; nuts, bolts, screws, etc. In general the applications are where good corrosion resistance is required without the need for high tensile strength. Table 3.7 shows the properties of typical ferritic stainless steels.

A problem that can occur with ferritic steels is the formation of *sigma phase*. This phase is hard and brittle and is formed from ferrite. In the formation, a considerable reduction in volume occurs and this can lead to the development of cracks. Ferritic steels containing high percentages of carbon can develop sigma phase if held at a temperature above 475°C for some time and so become brittle. It can be removed by heating to above 650°C.

Table 3.7 *Properties of ferritic stainless steels*

BS/AISI reference	Composition %	Condition AC = air cool, FC = furnace cool	Tensile strength MPa	Yield stress MPa	Percent. elong.	Typical uses
403S17	0.08 C, 1.0 Mn, 12–14 Cr	AC or FC from 700–780°C	420	280	20	Domestic articles such as forks, spoons
430S17	0.08C, 1.0 Mn, 16–18 Cr	AC from 750–820°C	430	280	20	General purpose, decorative trim, dishwashers, heaters, restaurant equipment

Martensitic steels

Martensitic steels contain between 12 and 18% chromium and between about 0.1 and 1.2% carbon. Increasing the carbon content increases the stability of the austenitic phase and thus the phase diagram shown in Figure 3.17 for 0.1% carbon steel becomes modified to basically that shown in Figure 3.18 for 1% carbon steel.

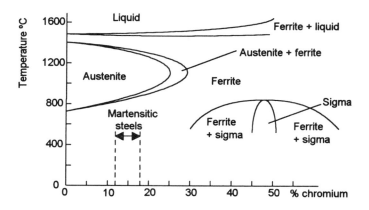

Figure 3.18 *Phase diagram for iron–chromium alloys with 1% carbon*

The loop on the phase diagram for the austenite phase increases as the carbon content increases, the loop being referred to as the *gamma loop*. With virtually zero carbon, the austenite phase can exist only out to a chromium content of about 12%, with 0.35% carbon the phase exists out to about 15% chromium, with higher percentages of carbon even further. Thus a suitable combination of chromium and carbon content for an alloy means that a steel will have an austenitic phase. This means the steel can be quenched to give martensite and so be hardened by heat treatment. The result is a steel with a much higher tensile strength and hardness than is possible with ferritic steel.

Martensitic steels are subdivided into three groups: stainless irons, stainless steels and high-chromium stainless steels. *Stainless irons* contain about 0.1% carbon and 12 to 13% chromium. Quenching from about 900 to 1000°C gives a fully martensitic structure. Tempering at about 659 to 750°C gives a structure consisting of ferrite with fine carbides dispersed through it. Such a material can be easily machined. *Stainless steels* contain about 0.25 to 0.30% carbon with 11 to 13% chromium. This form of steel is often referred to as cutlery stainless steel since it is widely used for cutlery. Quenching such a steel from about 850°C results in a martensitic structure containing carbides and consequently high hardness. *High-chromium stainless steels* contain 16 to 18% chromium, about 2% nickel and 0.05 to 0.15% carbon. Without the nickel such a low percentage of carbon with the high chromium content would lead to a ferritic steel. However, the small percentage of nickel enables austenite to form and hence, after quenching, martensite. The result is an alloy with a high corrosion resistance resulting from the high chromium content, with the mechanical properties of a lower chromium content alloy. Table 3.8 shows the properties of typical martensitic steels.

Table 3.8 *Properties of martensitic steels*

BS/AISI reference	Composition %	Condition OQ = oil quench, AC = air cool, T = temper	Tensile strength MPa	Yield stress MPa	Percent. elong.	Typical uses
410S21	0.09–0.15 C, 1.0 Mn, 11.5–13.5 Cr	OQ or AC from 950–1020°C, T 650–750°C	550–700	370	20	Machine parts, bolts, screws, cutlery, hardware
420S29	0.14–0.20 C, 1.0 Mn, 11.5–13.5 Cr	OQ or AC from 950–1020°C, T 650–750°C	700–850	525	15	Springs, machine parts, scissors, spindles, bolts

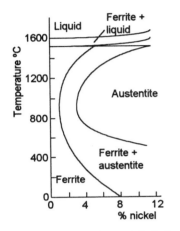

Figure 3.19 *Iron–nickel phases for 18% chromium alloy*

Austenitic steels

Austenitic steels contain 16 to 26% chromium, more than about 6% nickel and a very low percentage of carbon, about 0.1% or less. A common steel of this type has 0.05% carbon, 18% chromium and 8.5% nickel. Steels with this approximate ratio of chromium to nickel are referred to as *18/8 stainless steels*. Without the nickel, such a steel would be ferritic. However, the effect of the nickel is to give an austenitic phase. The stability of this austenitic phase increases as the percentage of nickel is increased, until the alloy becomes completely austenitic at room temperature (Figure 3.19). Such steels cannot be hardened by quenching. They are, however, usually quenched, not to produce martensite but to minimise the formation of chromium carbide as this causes a reduction in the corrosion resistance of the alloy. An increase in hardness can be produced by cold working. Table 3.9 shows the properties of typical austenitic steels.

Table 3.9 *Properties of austenitic steels*

BS/AISI reference	Composition %	Condition	Tensile strength MPa	0.2% proof stress MPa	% elong.	Typical uses
302S21	0.12C, 2.0 Mn, 17–19 Cr, 8–10 Ni	Softened 1000–1100°C	510	190	40	General purpose, food handling equipment, cook-ware, trim, springs
304S15	0.06 C, 2 Mn, 17.5–19 Cr, 8–11 Ni	Softened 1000–1100°C	480	195	40	Low carbon modification of 302S21 to restrict carbide precipitation during welding. Chemical and food processing equipment, gutters.
316S13	0.03 C, 2 Mn, 16.5– 18.5 Cr, 11.5–14.5 Ni, 2.5–3 Mo	Softened 1000–1100°C	490	190	40	Higher corrosion resistance than the 302 and 304 types. Chemical handling and photographic equipment
321S31	0.08 C, 2 Mn, 17–19 Cr, 9–12 Ni, Ti 5 times C	Softened 1000–1100°C	510	200	35	Stabilised for welding, aircraft exhaust manifolds, boiler shells, pressure vessels

Weld decay

During welding, when temperatures of 500 to 800°C are realised, stainless steels may undergo structural changes which are detrimental to the corrosion resistance of the material, the other properties being little affected. The effect is known as *weld decay* and results from the precipitation of chromium-rich carbides at grain boundaries. This removal of chromium from grains to the boundaries decreases the corrosion resistance. One way of overcoming this is to *stabilise* the steel by adding other elements such as niobium and titanium. These have greater affinity for the carbon than the chromium and so form carbides in preference to the chromium.

3.1.6 Steels at low and high temperatures

The effect of a reduction in temperature on the properties of the metals depends on the crystal structure. The strength, ductility and toughness of face-centred cubic metals improves as the temperature decreases. For this reason, austenitic stainless steels, along with aluminium, copper and nickel alloys, are widely used at low temperatures. Body-centred cubic metals, however, generally undergo a ductile to brittle transition at some temperature below room temperature. The temperature at which this transition occurs is markedly affected by small changes in the composition of the alloy, grain size, notches and flaws and the rate of loading. Body-centred cubic metals include all the plain carbon steels, low- and medium-alloy steels. Care is thus needed in the use of such steels at low temperatures. Hexagonal close-packed metals have properties intermediate between those of face-centred and body-centred cubic structures.

Heat-resisting steels

Steels for use at high temperatures must provide good resistance to creep and oxidation and suffer no detrimental changes in structure or properties. There must, therefore, be no such effects as temper embrittlement, sigma phase formation or carbide precipitation.

Carbon steels are limited to about 450°C, since at higher temperatures the rate of oxidation can become very high and there is a marked reduction in stress-bearing capabilities.

Low-alloy steels have been developed for use at higher temperatures. Such steels include small amounts of elements such as chromium, vanadium and molybdenum. These combine with the carbon to form carbides. Chromium and vanadium both contribute to high-temperature strength more effectively if molybdenum is also present and the chromium also reduces oxidation. For example, a 1% chromium–0.5% molybdenum steel has an oxidation limit of about 550°C and is more resistant to creep than a plain carbon steel and is used for steam pipes.

Activity

Look through sections 3.1.4 to 3.1.7 and list the benefits and disadvantages of the following as alloying elements in steel:
(a) nickel,
(b) manganese,
(c) chromium.

Key point

The limiting ruling section is the maximum diameter of round bar at the centre of which the specified properties may be obtained - see Section 3.2.3.

For temperatures in excess of 550°C, the oxidation problem can be reduced by increasing the chromium content. For example, a 12% chromium steel has an oxidation limit of about 575°C and is used for steam turbine rotors and blades.

For higher temperatures, austenitic steels can be used. Such steels do not transform to the ferritic structure on cooling to room temperature, but retain an austenitic structure. This can be obtained by adding 18% chromium and 8% nickel. Such a steel has an oxidation limit of about 650°C and superior creep resistance. A typical application is for superheater tubes. Enhanced creep resistance can be obtained by incorporating small amounts of niobium, titanium or molybdenum. Steels containing up to 25% chromium and 30% nickel have been developed for use up to about 759°C.

Nickel–chromium alloys can be used for higher temperatures. For example, an 80% chromium–20% nickel alloy has an oxidation limit of about 900°C and good creep resistance to quite high temperatures.

3.1.7 Steel selection

Table 3.10 gives commonly used steels for different levels of tensile strength, the relevant limiting ruling sections being quoted.

Table 3.10 *Steel selection*

Tensile strength (MPa)	BS steel code	Description of steel	Limiting ruling section (mm)
620 to 770	080M40	Medium carbon steel, hardened and tempered	63
	150M36	Carbon–Mn steel, hardened and tempered	150
	503M40	1% Ni steel, hardened and tempered	250
700 to 850	150M36	1.5% manganese steel, hardened and tempered	63
	708M40	1% Cr–Mo steel, hardened and tempered	150
	605M36	1.5% Mn–Mo steel, hardened and tempered	250
770 to 930	708M40	1% Cr–Mo steel, hardened and tempered	100
	817M40	1.5% Ni–Cr–Mo steel, hardened and tempered	250
850 to 1000	630M40	1% Cr steel, hardened and tempered	63
	709M40	1% Cr–Mo steel, hardened and tempered	100
	817M40	1.5% Ni–Cr–Mo steel, hardened and tempered	250
930 to 1080	709M40	1.5% Cr–Mo steel, hardened and tempered	63
	817M40	1.5% Ni–Cr–Mo steel, hardened and tempered	100
	826M31	2.5% Ni–Cr–Mo steel, hardened and tempered	250
1000 to 1150	817M40	1% Ni–Cr–Mo steel, hardened and tempered	63
	826M31	2.5% Ni–Cr–Mo steel, hardened and tempered	150
1080 to 1240	826M31	2.5% Ni–Cr–Mo steel, hardened and tempered	100
	826M40	2.5% Ni–Cr–Mo steel, hardened and tempered	250
1150 to 1300	826M40	2.5% Ni–Cr–Mo steel, hardened and tempered	150
1240 to 1400	826M40	2.5% Ni–Cr–Mo steel, hardened and tempered	150
>1540	835M30	4% Ni–Cr–Mo steel, hardened and tempered	150

3.1.8 Cast irons

The term *cast iron* arises from the method by which the iron is produced. Pig iron, the product of the blast furnace with about 3 to 4% carbon, is remelted in a furnace and the properties of the iron modified by the addition of other materials. The resulting iron is then cast. Cast irons have between about 2 and 4% carbon and often significant amounts of silicon, as well as smaller amounts of other elements. This range of carbon content gives the materials a high fluidity. Also the materials when solidifying show no significant volume contraction. There is a whole range of cast irons with their different properties and applications, but all can be cast into a wide variety of simple or complex shapes.

Grey and white cast irons

The iron–carbon phase diagram (Figure 3.5) shows that at a temperature of 1147°C a eutectic occurs for alloys with 4.3% carbon. Between 2.0 and 4.3% carbon, increasing the carbon percentage reduces the temperature at which solidification occurs. Figure 3.20 shows the relevant part of the phase diagram.

Consider the cooling of an iron with between 2.0 and 4.3% carbon from the liquid state. When solidification starts to occur, the result is the formation of austenite in the liquid. At 1147°C the solidification is complete. With further cooling, austenite plus cementite can form. However, if there is very slow cooling, the cementite is not stable and graphite flakes can form. At 723°C the remaining austenite changes to alpha ferrite and the result at room temperature is a structure of pearlite plus graphite flakes (Figure 3.21). This type of cast iron is known as *grey iron* because of the grey appearance of its freshly fractured surface.

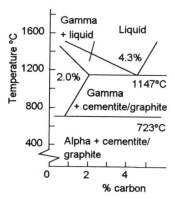

Figure 3.20 *Part of iron–carbon phase diagram*

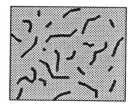

Figure 3.21 *Grey cast iron: graphite flakes in a matrix of pearlite*

Graphite has negligible tensile strength. The strength of grey iron increases as the amount of free graphite is reduced and as the fineness of the graphite flakes is increased. An increase in the carbon content up to 4.3% carbon, the eutectic value, lowers the melting point of the iron and, in doing so, favours the production of iron with graphite rather than cementite because the cooling from the liquid state is from a lower temperature and hence the cooling rate is slower. Some elements when included with the carbon in the iron affect the formation of graphite. Silicon and phosphorus affect the composition of the eutectic point and are considered to have a carbon equivalence in the formation of graphite. The combined effect of carbon, silicon and phosphorus is represented by the value of the *carbon equivalent* where:

$$\text{carbon equivalent } \% = \% \text{ C} + \text{one-third}(\% \text{ Si} + \% \text{ P}) \qquad [1]$$

The carbon equivalent determines how close the composition of an iron is to the eutectic value of 4.3% carbon and hence how likely grey iron is to be produced. For example, if there is 1.5% silicon and 0.3% phosphorus with an iron containing 3.5% carbon, then the carbon equivalent is 4.1% and thus the iron is very close to the eutectic point.

A fast cooling rate gives a structure which solidifies at 1147°C to give austenite and cementite which then, at 723°C, gives a structure of cementite and pearlite. This structure is known as *white iron* because of the white appearance of its freshly fractured surface. White iron, because of its high cementite content, is hard and brittle. This makes it difficult to machine and hence of limited use. The main use is where a wear-resistant surface is required. The tendency of cast iron to solidify with the carbon present in the form of cementite rather then graphite increases as the carbon percentage, or carbon equivalent percentage, is reduced and as the cooling rate increases. The iron is said to *chill* if such changes occur. Chilled structures are hard and brittle.

Sulphur often exists in cast iron and has the effect of stabilising cementite, thus favouring the production of white iron rather than grey iron and hence increasing hardness and brittleness. The amount of sulphur in an iron has, therefore, to be controlled. The addition of small amounts of manganese to an iron containing sulphur enables the sulphur to form manganese sulphide. The removal of the free sulphur has the effect of increasing the chance of grey iron being produced. A typical composition of a grey iron is 3.2 to 3.5% carbon, 1.3 to 2.3% silicon, 0.15 to 1.0% phosphorus, 0.1% sulphur and 0.5 to 0.7% manganese. This gives a carbon equivalent approaching 4.3%, the eutectic value.

A casting will often have sections of varying thickness. The rate of cooling of a part of a casting will depend on the thickness of the section concerned, the thinner the section the more rapidly it will cool. This means that there is likely to be a variation in the properties of different parts of a casting. Thus, for example, a very thin section might be a white iron and so very hard and virtually unmachinable. A different section might be grey with white edges. A thicker section might be a fine-grained grey iron while an even thicker section might be, at least internally, a coarse-grained grey iron. Because the properties of grey iron depend on the rate of cooling and this is influenced by section thickness, the British Standard specifications for grey iron are in terms of the properties of a test piece machined from a specific size casting, namely a 30 mm diameter cast bar.

Grey cast irons are specified in British Standards by a grade number which gives the minimum tensile strength of the iron in the standard test piece. Table 3.11 gives some examples.

Table 3.11 *Properties of grey cast irons*

Grade	% C equivalent	Tensile strength MPa	% elongation	Brinell hardness	Impact strength* J
150	4.5	150	0.6	100–170	8–13
200	4.2	200	0.4	120–190	8–16
250	3.85	250	0.5	145–220	13–23
300	3.65	300	0.5	165–240	16–31
350	3.5	350	0.5	185–260	24–47

* Unnotched 20 mm diameter Izod test pieces.

Figure 9.22 shows the effect on the tensile strength of the different grades of section thickness. This has to be taken into account in considering the likely strengths at different thicknesses within a casting produced to a particular grade. Thus an iron, which in the standard test piece bar of 30 mm diameter, gives a tensile strength of 150 MPa, i.e. grade 150, would give this same strength in the centre of a casting section which is 15 mm thick (the cooling of effectively a flat plate is lower than a bar). However, if the thickness were 50 mm the tensile strength would be only about 100 MPa. This lower value occurs because the centre of this thickness section cools slower than the test piece.

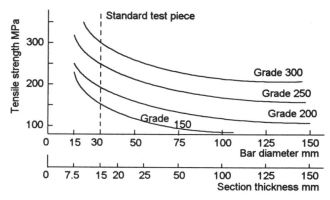

Figure 3.22 *Effect of section thickness on tensile strength of grey cast iron*

Grey iron has excellent machinability since the graphite flakes act as chip breakers. With coarse graphite flakes, grey iron is very good at damping out vibrations. The network of graphite flakes also gives good thermal conductivity. Typical uses are motor cylinders and pistons, machine castings, crankcases and machine tool beds.

Annealing is used with grey cast irons to provide optimum machinability and remove stresses. Annealing involves heating the cast iron to about 760°C, soaking at that temperature and then cooling very slowly. Grey cast iron can be hardened by quenching, such a treatment generally being followed by tempering. A tempering temperature of 475°C is common. A stress relieving treatment is often used before a significant amount of machining takes place, this involving heating to about 550°C.

Malleable cast irons

Malleable cast irons are produced by the heat treatment of white cast irons. Three forms of malleable iron occur: *whiteheart*, *blackheart* and *pearlitic*. Malleable irons have better ductility than grey cast irons and this, combined with their higher tensile strengths, makes them useful materials.

In the blackheart process, white iron castings are heated in a non-oxidising atmosphere to 900°C (Figure 3.23) and soaked at that temperature for two days or more. This causes the cementite to break down. The result is spherical aggregates of graphite in a

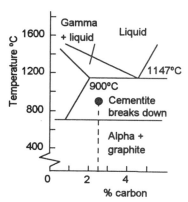

Figure 3.23 *Production of blackheart malleable iron*

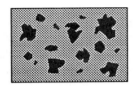

Figure 3.24 *Blackheart malleable iron*

matrix of austenite. The casting is then cooled very slowly, resulting in the austenite changing into ferrite and more graphite. The result at room temperature is a structure consisting of 'rosettes' of graphite in a matrix of ferrite (Figure 3.24).

The whiteheart process also involves heating white iron castings to about 900°C and soaking at that temperature. But in this process the castings are packed in canisters with haematite iron ore. This gives an oxidising atmosphere. Where the casting is thin, the carbon is oxidised and forms a gas which leaves the casting. In the thicker sections of the casting, only the carbon in the surface layers leave. The result, after very slow cooling, is a ferrite structure in the thin sections of the casting and a ferritic outer layer with a ferrite plus pearlitic inner core for the thick sections.

Pearlitic malleable iron is produced by heating white iron castings in a non-oxidising atmosphere to 900°C and soaking at that temperature. This causes the cementite to break down and give, as with the blackheart iron, spherical aggregates of graphite in an austenite matrix. If more rapid cooling is used than with the blackheart iron the austenite gives a pearlitic structure and thus graphite 'rosettes' occur in a pearlitic structure. This pearlitic malleable iron has a higher tensile strength than blackheart iron. Alternatively, pearlitic malleable irons can be produced by adding about 1% manganese. The manganese inhibits the production of graphite. Higher strength pearlitic malleable irons are produced by quenching the iron from 900°C and then tempering to give graphite 'rosettes' in a tempered martensite structure; a better term for this form is thus *tempered martensite malleable iron*.

Table 3.12 gives the properties of blackheart, whiteheart and pearlitic malleable cast irons.

Table 3.12 *Properties of blackheart, whiteheart and pearlitic malleable cast irons*

BS grade	Min. tensile strength MPa	Min. 0.2% proof stress	Min. percent. elongation
Blackheart			
B30-06	300		6
B32-10	320	190	10
B35-12	350	200	12
Whiteheart			
W35-04	360		3
W38-12	400	210	8
W40-05	420	230	4
W45-07	480	280	4
Pearlitic			
P45-06	450	270	6
P50-06	500	300	5
P55-04	550	340	4
P60-03	600	390	3
P65-02	650	430	2
P70-02	700	530*	2

* If air quenched and tempered, the 0.2% proof stress is 430 MPa min.

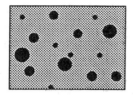

Figure 3.25   *Spheroidal-graphite iron*

Activity

Make a flow chart showing the forms of cast iron and how they are produced.

Spheroidal-graphite cast irons

Spheroidal-graphite (SG) iron, or sometimes termed *nodular iron* or *ductile iron*, has the graphite in the iron in the form of spheres or nodules. Magnesium or cerium is added to the iron before casting. The effect of this is to prevent the formation of graphite flakes during the cooling, the graphite forming spheres instead. At room temperature the structure of the cast iron is mainly pearlitic with spheres of graphite (Figure 3.25). This structure reduces the ductility and strength much less than the graphite flakes in grey iron. The resulting material is thus more ductile and stronger than a grey iron.

An annealing heat treatment process of heating to 900°C, soaking and then slowly cooling can be applied to give a structure of graphite spheres in ferrite. The result is an increase in ductility but a reduction in tensile strength. Quenching from 800–900°C in oil, with tempering, can be used to give a martensitic structure. Table 3.13 gives the properties of spheroidal- graphite cast irons.

Table 3.13 *Properties of spheroidal-graphite cast irons*

BS grade	Matrix	Heat treatment	Min. tensile strength MPa	Min. 0.2% proof stress MPa	Percent. elong.
350/22	Ferrite	Annealed	350	220	22
420/12	Mainly ferrite	Annealed	420	270	12
500/7	Ferrite + pearlite	Annealed	500	320	7
800/2	Tempered martensite	Quenched + tempered	800	480	2

3.1.9 Case studies

The following case studies then illustrate the use of ferrous alloys in structures.

Bridges

The Romans built bridges of stone in the form of arches (Figure 3.26). This was because stone is strong in compression but weak in tension and the arch form put the stone in compression. The term *architecture of compression* is often used for such types of structures since they have always to be designed to put the materials into compression. Cast iron is strong in a compression and weak in tension. Thus, at the end of the eighteenth century when such material was first used in building bridges, the iron bridge followed virtually the same form of design as a stone bridge and was an arch. The first iron bridge built in 1779 over the River Severn was thus an arch; it is about 8 m wide and 100 m long and still standing. A common material used nowadays for bridges is reinforced and prestressed concrete. Such bridges tend to be in the form of an arch in order to keep the material predominantly in compression. The reinforcement is to enable the concrete, which is weak in tension but strong in compression, to withstand tensile forces.

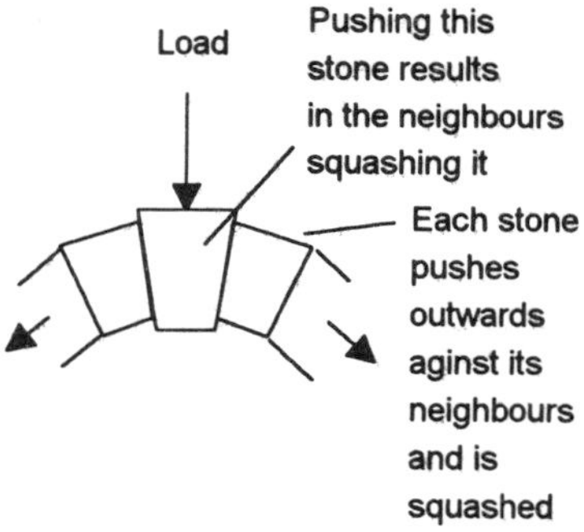

Load

Pushing this stone results in the neighbours squashing it

Each stone pushes outwards aginst its neighbours and is squashed

Figure 3.26   *The arch bridges gaps by putting the stone in compression*

The introduction of steel, which was strong in tension, enabled the basic design to be changed for bridges and enabled the *architecture of tension*. It was no longer necessary to have arches to keep the material in compression. The result was the emergence of truss structures; Figure 3.27 shows one form of truss bridge. Essentially the truss is a hollow beam and the loading results in the upper part of this structure being in compression and the lower part in tension; some of the diagonal struts are in compression and some in tension. The steel used is likely to be a high-strength low-alloy steel. So that welding can be used in the bridge construction, the carbon content is kept low and manganese added to enable solution hardening of the ferrite, e.g. 0.2% carbon with 1.5% manganese.

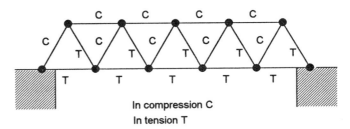

Figure 3.27 *The basic form of a truss bridge*

Suspension bridges depend on the use of materials that are strong in tension (Figure 3.28) and use steel cables, the cables being in tension. Since the forces acting on the cable have components which pull inwards on the supporting towers, firm anchorage points are required for the cables. High-tensile steel wire is basically a medium to high-carbon (0.8 to 0.9% carbon) steel with about 0.8% manganese. The wire is produced from hot-rolled rod and the cooling of the rod results in a coarse pearlitic structure. Heat treatment is used to obtain a fine pearlitic structure, the treatment being termed *patenting*. The wire work hardens during the wire drawing process from rod and tensile strengths in excess of 2000 MPa are possible. The steel has also some ductility, typically about 1 to 2%.

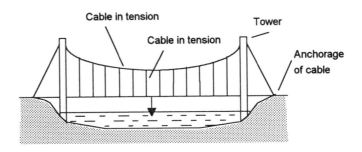

Figure 3.28 *The basic form of a suspension bridge*

Steel for car bodies

The requirement of the body structure of a car are that it must be rigid and strong enough to support the weight of the passengers and the car itself, protect against impact, remain durable at temperatures in the range –40°C to +50°C and all weather conditions, be suitable for mass production and as light as possible to give good fuel economy. The early cars were based on the use of a chassis to take the loading with body panels attached to it; however, in 1934 this type of structure was replaced by the chassisless body in which the body shell took the load. The material that enabled this change to occur was sheet steel. The steel was weldable, could be pressed into the required shapes, had a ductile–brittle transition temperature below the minimum service temperature and was cheap. The cost factor suggests a plain carbon steel. The formability and weldability criteria indicate that a carbon steel with less than about 0.25% carbon could be used; however, the requirement for a low ductile–brittle transition temperature indicates a low-carbon steel with about 0.1% carbon.

Steel for train rails

The material used for rails needs high resistance to wear and plastic flow and is met by a steel having a fully pearlitic structure. The normal grade steel used has thus 0.4 to 0.5% carbon, 0.05 to 0.35% silicon and 0.8 to 1.25% manganese; this gives a steel with a tensile strength of about 700 MPa and a percentage elongation of about 10%. The hardness and wear resistance is increased by refining the pearlitic structure and so higher wear-resistance grades have about 0.65 to 0.75% carbon, 0.1 to 0.5% silicon, 0.8 to 1.3% manganese and give a tensile strength of about 880 MPa and a percentage elongation of 8%. For very arduous conditions, e.g. where high axle loads are involved or tightly curved track, the pearlitic structure is even more refined by heat treatment.

Problems 3.1

1 How do the structures of hypoeutectoid and hypereutectoid steels differ at room temperature as a result of being slowly cooled from the austenitic state?
2 Determine the percentages of alpha ferrite and cementite present just below 723°C in slowly cooled (a) 0.8% carbon steel, (b) 2.0% carbon steel.
3 A carbon steel contains 0.1% carbon. What will be the percentage of pearlite present at just below 723°C if it is slowly cooled from the austenitic state?
4 Describe the form of the microstructure of a slowly cooled steel having the eutectoid structure.
5 Explain how the percentage of carbon present in a carbon steel affects the mechanical properties of the steel.

6 Explain the significance of the martensite start temperature and martensite finish temperature for the hardness possible with a carbon steel.

7 What is the effect of tempering on martensite and hence the hardness of a steel?

8 What factors limit the engineering applications of plain carbon steels?

9 Which type of plain carbon steel would be most suitable for (a) railway wagon wheels, (b) nails, (c) hammers, (d) reinforcement bars for concrete work, (e) knives?

10 In what ways do alloying elements, other than carbon, affect the properties of steel?

11 In what ways do the elements (a) manganese, (b) chromium, (c) molybdenum, (d) sulphur, (e) silicon affect the properties of steel when they are alloyed with it?

12 What elements are usually added to a steel to improve its machinability?

13 State two elements that, when added to steel, restrict grain growth.

14 A commonly used steel has the following composition: 0.40% carbon, 0.55% manganese, 1.50% nickel, 1.20% chromium, 0.30% molybdenum. What are the effects of nickel, chromium and molybdenum on the properties of the steel?

15 What are stainless steels?

16 Explain how the ferritic, martensitic and austenitic forms of stainless steel are produced and how their properties differ.

17 Explain the significance of sigma phase on the properties of stainless steels.

18 Explain what is meant by weld decay and how it can be prevented.

19 Explain the effects on the microstructure and properties of cast iron of (a) cooling rate, (b) carbon content, (c) the addition of silicon, manganese, sulphur and phosphorus.

20 Describe the conditions under which (a) grey iron, (b) white iron are produced.

21 In what way does the section thickness of a casting affect the structure of the cast iron?

22 How do the mechanical properties of malleable irons compare with those of grey irons?

23 Describe the way in which the structures of nodular irons and malleable irons differ from those of grey irons.

24 A grey cast iron is to be produced for casting and is required to solidify with no primary austenite or graphite. If the carbon content of the iron is 3.5%, what percentage of silicon can be added to give the required structure?

25 How does the presence of graphite as flakes or nodules affect the properties of cast iron?

26 Which types of cast irons would you suggest for (a) a sewage pipe, (b) the crankshaft in an internal combustion engine, (c) grinding mill parts, (d) a manhole cover, (e) a machine tool bed?

3.2 Heat treatment of steels

Key point

Heat treatment is the process of controlled heating and cooling of a material in order to change its micro-structure and properties

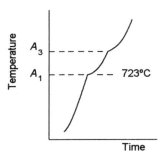

Figure 3.29 *Heating curve with a phase change occurring*

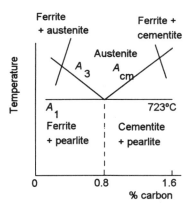

Figure 3.30 *Critical points*

The term *heat treatment* is used to describe a process involving controlled heating and cooling of a material in order to change its microstructure and properties. Vital factors in determining the outcome of heat treatment are the temperature to which the material is heated and the rate at which it is cooled back to room temperature. In this section the processes of hardening, tempering, austempering, martempering, surface hardening and the production of wear-resistant surfaces, full annealing, normalising, subcritical annealing and spheroidising annealing are considered.

3.2.2 Microstructural transformations

The iron–carbon phase diagram of Figure 3.5 shows the phases that will be present in iron–carbon alloys for a range of compositions at different temperatures if sufficient time is allowed for all the movements of atoms by diffusion at any temperature to proceed to completion. When an alloy is cooled, or heated, and a phase change occurs, a graph of temperature against time shows a discontinuity (see Figure 3.1). The temperatures at which such discontinuities occur are known as *arrest points* or *critical points*.

For example, for a hypoeutectoid steel a critical temperature is 723°C which marks the transformation of the steel from a ferrite plus pearlitic structure to ferrite plus austenite (Figure 3.29). This lower critical point, i.e. 723°C, is denoted by A_1 and is always used to denote the transformation of the steel from a ferrite plus pearlitic structure to ferrite plus austenite. The upper critical point on the figure is denoted by A_3 and marks the transformation from a ferrite plus austenite structure to just one of austenite. Figure 3.30 shows these critical points on a simplified iron–carbon phase diagram. A hypereutectoid steel has two critical points, the lower critical point A_1 marking the transformation from a steel with a structure of ferrite plus cementite to cementite plus austenite and the upper critical point A_{cm} the transformation from the cementite plus austenite structure to austenite.

Note that the values of the critical points obtained from a heating curve differ slightly from those given by a cooling curve, the cooling values being slightly greater than the heating values.

The time scale for the transformations given by the critical points in Figure 3.30 is long enough for all diffusion processes to be completed and equilibrium attained. Faster temperature changes can, however, lead to the diffusion processes not being completed and so equilibrium not being attained; these thus lead to the microstructure being in a non-equilibrium state and so not that forecast by the iron–carbon phase diagram.

3.2.2 TTT and CCT diagrams

The effect of time and temperature on the microstructure of a steel can be expressed in terms of a *time–temperature transformation*

Key point

The combined effect of time and temperature on the microstructure of a steel can be expressed by a time–temperature transformation (TTT) diagram.

Sample heated to just above 723°C and completely transformed to austenite

723°C

Sample quenched to, say, 500°C and kept at this temperature for, say, 2 s. Some of the austenite changes during this time

500°C

Sample quenched in cold water so that remaining austenite is transformed to martensite

Cold water

Microscopic examination to determine the amount of martensite

Figure 3.31 *Obtaining one point on a TTT diagram*

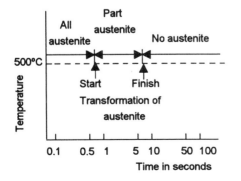

Figure 3.32 *The transformation at a particular temperature*

(TTT) diagram. Such a diagram is sometimes referred to as an *isothermal transformation (IT) diagram* or *Bain S curve.*

To obtain the TTT diagram for a particular steel, a small sample is heated to a temperature at which it is completely austenite. The sample is then quickly transferred to a bath of liquid at some predetermined temperature, i.e. quenched to this temperature. The time for which the sample is at this temperature is measured. This is the time during which the sample is transforming from being completely austenitic and the change is taking place at the constant temperature of the bath of liquid, hence the term isothermal since this term means constant temperature. After the required time, the sample is quickly quenched in cold water. This halts the transformation that was occurring and has the effect of converting the remaining austenite into martensite. The amount of martensite produced can be determined by microscopic examination of the sample. Hence the amount of the sample that has not been transformed from austenite, at the temperature of the bath, can be determined. The investigation is carried out for a number of different times at the particular bath temperature (Figure 3.31). The way in which the microstructure of the steel changes with time at that temperature can then be represented by the line drawn in Figure 3.32.

By repeating the above sequence of measurements for different temperatures, a composite diagram can be built up which shows the time taken for the transformation from austenite to take place at different temperatures. This is the TTT diagram.

Figure 3.33 shows an example for a 0.8% plain carbon steel. As the phase diagram indicates, this is the eutectoid composition. For comparison with the conditions that would occur with very slow cooling, the relevant part of the phase diagram is also included with the graph of the martensitic start and finish temperatures superimposed.

1 ***Above 723°C***
    Austenite is the only stable phase. For temperatures below this the austenite is unstable and given sufficient time suffers a transformation.

2 ***From 723°C down to about 566°C***
    The lower the temperature the shorter the time taken for the austenite to begin to transform and complete the transformation. This transformation is to pearlite, the higher temperature giving a coarser pearlite than the lower temperature. At the 'nose' of the diagram the conditions for the transformation of austenite are at their optimum and it takes the shortest time to start transforming.

3 ***From 566°C down to about 215°C***
    The time taken for the austenite to begin to transform and complete the transformation increases. Diffusion is now too slow at these temperatures for pearlite to form and the result is known as *bainite.*

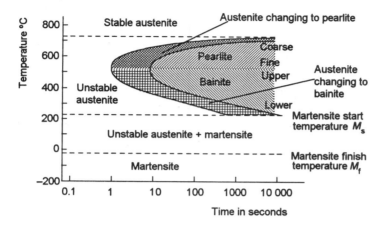
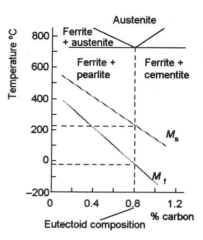

Figure 3.33 *TTT and phase diagrams for a 0.8% plain carbon steel*

Bainite, like pearlite, is a mixture of ferrite and cementite. However, while pearlite consists of alternate layers of ferrite and cementite, bainite consists of a dispersion of fine cementite particles in ferrite. There are two major forms of bainite; upper bainite which forms in the temperature range just below that of pearlite formation and lower bainite which forms at temperatures closer to the martensitic start temperature.

4   *From 215°C down to –20°C*
The unstable austenite transforms to martensite. The martensitic start temperature is about 215°C. The amount of martensite formed is practically independent of time, being determined by the temperature of the steel.

5   *Below –20°C*
The structure is now entirely martensite, this being about the martensitic finish temperature at this composition.

Different TTT diagrams are produced for other carbon steels with different carbon content. For example, Figure 3.34 shows the TTT diagram for a 0.45% carbon steel. This differs from that for the 0.8% carbon steel, i.e. the eutectoid composition steel, in that there is an additional region due to ferrite transformation. Also with the lower percentage carbon, the TTT diagram has effectively been shifted to the left, i.e. the transformations begin more quickly and are completed more quickly at any given temperature. For example, compare the transformations at 500°C for the 0.45% carbon steel (Figure 3.35) with those given in Figure 3.32 for the 0.8% carbon steel.

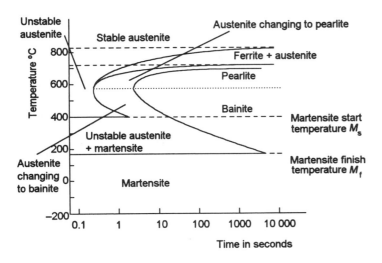

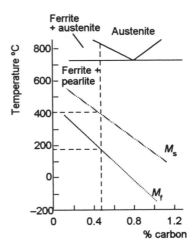

Figure 3.34 *TTT and phase diagrams for a 0.45% plain carbon steel*

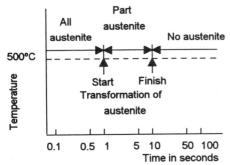

Figure 3.35 *The transformations at 500°C for a 0.45% plain carbon steel*

Figure 3.36 shows the TTT and phase diagrams for a plain carbon steel with 1.2% carbon. The increasing amount of carbon has pushed the 'nose' of the TTT diagram to the right.

The TTT diagram is also changed by the addition of other elements. Figure 3.37 shows the TTT diagram for an alloy steel having 0.4% carbon and 0.8% chromium. The effect of the chromium is to shift the 'nose' of the TTT diagram to the right, i.e. increase the time needed at any particular temperature for the transformation to pearlite or bainite, compare Figure 3.38 with Figure 3.35. Nickel, molybdenum and manganese all have similar effects to that shown by the chromium (see later, Figure 3.40).

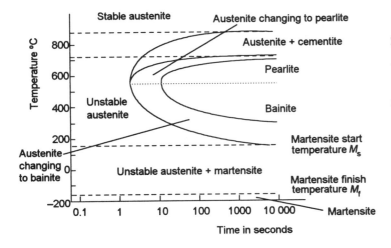

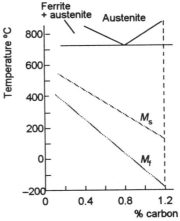

Figure 3.36 *TTT and phase diagrams for a 1.2% plain carbon steel*

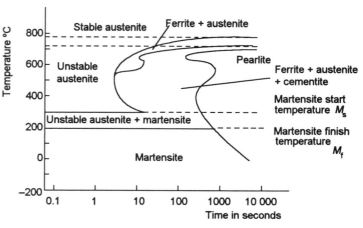

Figure 3.37   TTT diagram for a 0.4% carbon, 0.8% chromium steel

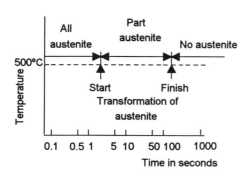

Figure 3.38   *The transformations at 500°C for a 0.4% carbon 0.8% chromium steel*

Key point

A continuous cooling transformation (CCT) diagram can be produced by determining the microstructures produced in a steel at various rates of cooling. The CCT diagram differs from the TTT diagram in that for CCT diagrams the transformations are determined for continuous cooling whereas TTT diagrams are obtained for transformations at constant temperatures.

CCT diagrams

A hard, strong, steel can be produced by heating a plain carbon steel to a temperature above its A_3 critical temperature so that the microstructure is entirely changed to austenite, then cooling quickly so that the austenite is transformed into martensite. The treatment thus involves continuous cooling. The cooling rate depends on the medium used for the quenching, water giving a faster cooling rate than oil, oil a faster rate than air cooling.

A *continuous cooling transformation (CCT) diagram* can be produced by determining the microstructures produced in a steel at various rates of cooling. The CCT diagram differs from the TTT diagram in that for CCT diagrams the transformations are determined for continuous cooling whereas TTT diagrams are obtained for transformations at constant temperatures.

The procedure for determining a CCT diagram is to heat a sample to above the austenising temperature, hold it there until all the sample is austenite and then place it in the appropriate quenching medium to give the required rate of cooling. It is cooled down to a particular temperature and then cooled very rapidly by quenching in water. This transforms the untransformed austenite into martensite. Examination of the sample then allows the amount of martensite to be estimated and hence the amount of untransformed austenite at the temperature. This is then repeated with the same rate of cooling for a number of temperatures. The entire procedure is then repeated for different cooling rates. Hence the CCT diagram can be obtained. Figure 10.11 shows an example.

Figure 3.39 is the CCT diagram for a 0.8% plain carbon steel. It is very similar to the TTT diagram (see Figure 3.33), the differences being that the TTT diagram is effectively displaced slightly to the right and downwards, also the CCT diagram does not show all the TTT lines. Superimposed on the CCT diagram are lines showing different rates of cooling.

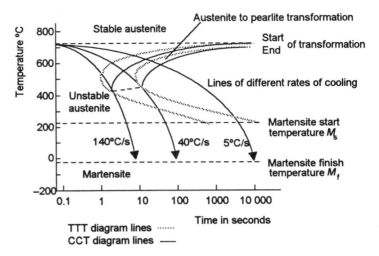

Figure 3.39 *CCT diagram for a 0.8% plain carbon steel*

For the fast rate of cooling of 140°C/s from the stable austenite state, the only transformation that occurs is from austenite to martensite, the cooling line not passing through the area indicating the austenite to pearlite transformation. For a cooling rate of between 140°C/s and 40°C/s, the austenite starts to transform to pearlite but the transformation is not completed. The remaining austenite then transforming to martensite. For the slower rate of cooling of 40°C/s, the austenite starts to transform to pearlite and the transformation is completed. We thus end up with a completely pearlite structure. Thus for the structure to be entirely martensitic, and hence show the maximum hardness, the cooling rate must be such that the only transformation is from austenite to martensite. This means that the cooling rate must be faster than 140°C/s. This is thus the *critical cooling rate*. For the structure to be entirely pearlite, we must have a cooling rate slower than 40°C/s.

The critical cooling rate thus depends on the form of the CCT diagram, and consequently the TTT diagram. The result of decreasing the amount of carbon in a plain carbon steel is to shift the 'nose' of the diagram to the left. This means that the critical cooling rate for a 0.45% plain carbon steel is higher than that for the 0.8% plain carbon steel. The more the CCT/TTT diagram is shifted to the left the faster the cooling rate needed to obtain a completely martensitic structure. In fact, by about 0.3% carbon the TTT diagram has been shifted so far to the left that the cooling rate is too fast to be achieved and so such steels cannot be hardened by quenching.

However, if elements such as chromium, nickel, molybdenum and manganese are added to a steel, the TTT 'nose' is shifted to the right. This means that the critical cooling rate is reduced. Figure 3.40 illustrates this, showing how the total percentage, up to about 7%, of these alloying elements changes the TTT diagram.

Thus it might be possible to use oil or salt quenching rather than water quenching to achieve a fully martensitic structure. The *hardenability* is said to have been improved.

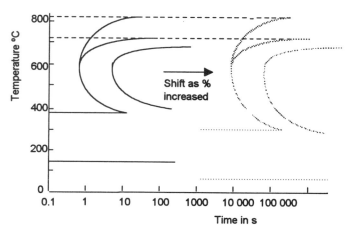

Figure 3.40 *Effect of alloying elements on TTT diagram*

CCTs as a function of bar diameter

The rate of cooling that occurs for a piece of steel when quenched in, say, water depends on the size of the piece of steel. The bigger the piece of steel the slower its rate of cooling. The quenching medium has thus to be chosen with a particular size of steel in mind if the critical cooling rate is to be achieved. Figure 10.13 shows the continuous cooling transformations that occur for specific quenching media as a function of the diameters of the bars being quenched. Thus for a 10 mm diameter bar, air cooling will result in a microstructure involving ferrite, pearlite and a small amount of bainite. With oil quenching, the result is bainite plus martensite. With water quenching, the entire microstructure is martensite. Thus for this particular steel and diameter bar, water quenching must be used if a completely martensitic structure is required. The microstructures given by the diagram are those at the centre of the quenched bar and those at the surface may differ due to the surface cooling at a different rate.

Figure 3.41 shows the continuous cooling transformations for a plain carbon steel. Figure 3.42, however, is for a steel with a similar percentage of carbon but with nickel, chromium and molybdenum alloying elements; about 1.5% nickel, 1.2% chromium and 0.3% molybdenum. As will be apparent from a comparison of the two figures, a much less severe quenching can be used with the alloy steel but a martensitic structure can still be obtained. For a 10 mm diameter bar, air cooling will suffice.

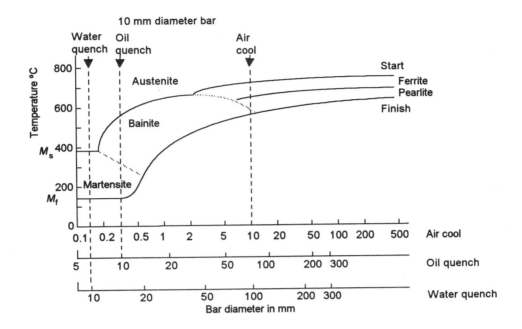

Figure 3.41 *Continuous cooling transformations as a function of bar diameter of a 0.38% plain carbon steel*

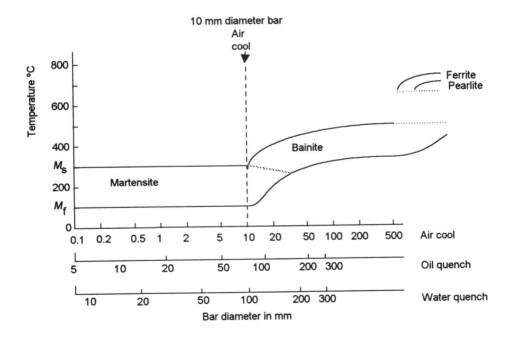

Figure 3.42 *Continuous cooling transformations as a function of bar diameter of a 0.4% alloy steel*

> **Example**
>
> Use Figure 3.41 to determine the structure that will occur with (a) a 5 mm diameter bar of 0.38% plain carbon steel when air cooled, (b) a 10 mm diameter of the bar when oil quenched.
>
> (a) Drawing a vertical line on the graph from the axis for air cooling at 5 mm gives austenite changing to ferrite at about 780°C, not all of it being transformed before the austenite to bainite transformation starts. The transformation is complete at about 540°C and the final structure is one of ferrite plus bainite.
> (b) Drawing a vertical line on the graph from the axis for oil-quenching at 10 mm gives bainite starting to form at about 650°C, not all of it being transformed before the austenite to martensite transformation starts to occur at about 550°C. The transformation is complete at about 450°C and the final structure is bainite plus martensite.

3.2.3 Hardenability

When a block of steel is quenched, the surface can show a different rate of cooling to that of the inner core of the block. Figure 3.43 illustrates one possibility with a CCT diagram for a steel. The surface has a faster rate of cooling than the core and the result is that the surface layers are martensite and the inner core pearlite plus martensite.

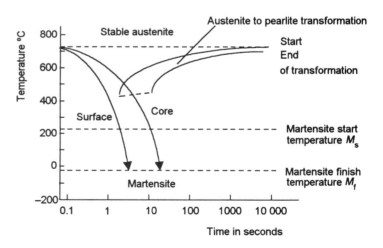

Figure 3.43 *Quenching a bar*

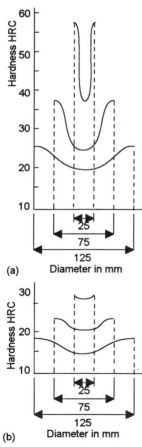

(a)

(b)

Figure 3.44 *Variation of hardness with depth: (a) water, (b) oil quench*

Figure 3.44 shows how the hardness varies with depths for a number of different bars of a 0.48% plain carbon steel when quenched in water and oil. With the water quenching, the hardness in the inner core is significantly different from the

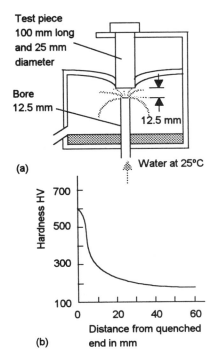

(a)

(b)

Figure 3.45 *(a) Jominey test,*
(b) results for a 0.4% plain carbon steel

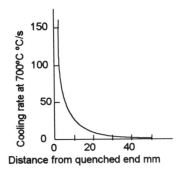

Figure 3.46 *Cooling rate*

surface hardness. The larger diameters also show lower surface and lower core hardness as their increased mass has resulted in a lower overall rate of cooling. With the oil quenching, the cooling rates are lower than with water quenching and thus the hardness values are lower. This is because the quenched bars are not entirely martensitic, even at the surface with the smaller diameter bar.

The term *hardenability* is used as a measure of the depth of martensitic hardening introduced into a steel section by quenching and is measured by the response of the steel to a standard test. The *Jominy test* involves heating a standard test piece of the steel to the austenitic state, fixing it in a vertical position and then quenching the lower end by means of a jet of water (Figure 3.45(a)). This method of quenching results in different rate of cooling along the length of the test piece. After the quenching, a flat portion is ground along one side of the test piece, 0.38 mm deep, and hardness measurements made along the length of the test piece. Figure 3.45(b) shows the types of results that are produced.

The significant point about the Jominy test results is not that they give the hardness at different distances along the test piece but that they give the hardness at different cooling rates. Each distance along the test piece corresponds to a different rate of cooling (Figure 3.46). The hardness at these cooling rates applies to both points on the surface and points inside a steel sample, provided we know the cooling rates at those points. This also applies regardless of the quenching medium used. For example, Figure 3.47 shows how the cooling rate at the centre of circular cross-section bars at 700°C when quenched in different quench to distances along the Jominy test pieces these distances are also given in the figure, thus allowing the Jominy hardness results to be used to predict the hardness at the centre of the bars. Thus, using the test result quoted in Figure 3.45, the hardness at 20 mm from the quenched end of the test piece is about 230 HV. This means that, using Figure 3.47, a circular cross-section bar of diameter about 75 mm would have this hardness at its centre when quenched in still water from 700°C. If still oil had been the quenching medium, this would be the hardness at the centre of a 60 mm diameter bar.

Figure 3.48 shows in a similar manner how the cooling rate is related to the diameter of a circular cross-section bar for points on the surface. The Jominy test result of 230 HV at 20 mm from the quenched end of the test piece means, using Figure 3.48, a circular cross-section bar of diameter 50 mm would have this hardness at its surface when quenched from 700°C in mildly agitated molten salt.

To illustrate the use of Jominy test results in the selections of steels, consider Figure 3.49 which shows Jominy test results for two different steels. In order to enable the significance of the results to be seen in terms of the hardness at the centre of different diameter bars, the cooling rates have been transposed into diameter values by means of Figure 3.47. The alloy steel can be said to have better hardenability than the plain carbon steel. The plain carbon steel cannot be used in a diameter greater than about

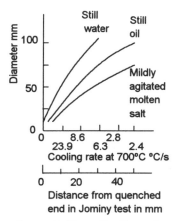

Figure 3.47 *Cooling rates at the centres of different diameter bars*

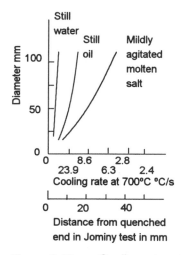

Figure 3.48 *Cooling rates at the surfaces of different diameter bars*

25 mm with a still water quench if the bar is to be fully hardened. However, the alloy steel is fully hardened for bars with diameters in excess of 100 mm.

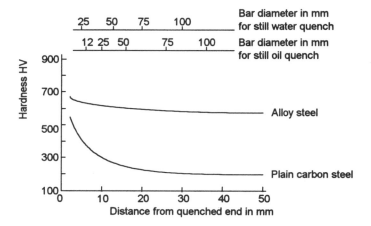

Figure 3.49 *Results of Jominy tests for an alloy steel and a plain carbon steel*

Ruling section

If you look up the mechanical properties of a steel in the data supplied by manufacturers or standard tables, you will find that different values are quoted for different limiting ruling sections. The *limiting ruling section* is the maximum diameter of round bar at the centre of which the specified properties may be obtained. Table 3.14 shows an example. The reason for the difference of mechanical properties for different size bars of the same steel is that, during the heat treatment, different rates of cooling occur at the centres of such bars and this results in differences in microstructure and hence mechanical properties.

Table 3.14 *Effect on properties of limiting ruling section*

Steel	Condition	Limiting ruling section mm	Tensile strength MPa	Minimum % elongation
070M55	Hardened and tempered	19	850 to 1000	12
		63	770 to 930	14
		100	700 to 850	14

Factors affecting hardenability

Factors other than quench rate affecting the hardenability are the grain size and composition of a steel.

In a homogeneous austenitic structure, pearlite starts to grow almost exclusively at the grain boundaries and thus the greater the amount of grain boundary the greater the sites for pearlite to start growing. Thus a coarse grain structure will have greater

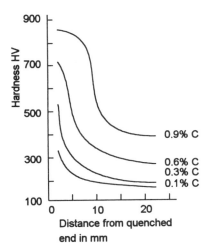

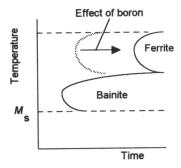

Figure 3.50 *Effect of carbon content on hardenability*

Figure 3.51 *Effect of boron on TTT diagram*

hardenability as a consequence of having a smaller amount of grain boundaries and so less pearlite.

Alloying elements affect hardenability. Carbon, because it is the element that controls the hardness of martensite, determines the maximum hardness that can be obtained with a particular steel. It also has an effect on the hardenability (Figure 3.50). Achieving an increased hardenability by increasing the amount of carbon in a steel is generally not used since the increased amount of carbon has other effects, e.g. a decrease in ductility, and hardenability is better controlled by the addition of other alloying elements. Elements such as molybdenum, chromium, manganese, silicon, copper and nickel all are used to improve the hardenability. Boron in very small percentages, 0.002 to 0.003%, has a marked effect on hardenability, considerably improving it. The boron inhibits ferrite formation and provides a structure with a higher percentage of bainite. Figure 3.51 shows the effect on the TTT diagram for a low carbon steel of including boron, the 'nose' of the ferrite part of the curve being pushed to the right and so making it more likely that the cooling curve would only pass through the bainite loop. The resulting steels are termed *boron steels*.

3.2.4 Tempering

Martensite is a hard and brittle substance. Thus a steel which has been quenched so that its entire structure is martensite will not only be hard but probably very brittle. It will not be tough. The toughness of such a steel can be improved by tempering. The grain in ductility is, however, balanced by a reduction in strength and hardness.

Tempering is the name given to the process in which steel, hardened as a result of quenching, is reheated to a temperature below the A_1 temperature in order to modify the structure of the steel. Martensite is a highly stressed solid solution of carbon in a distorted ferrite lattice. Heating such a structure enables carbon atoms that are trapped in this distorted ferrite lattice to diffuse out and form fine cementite. The amount of carbon that diffuses out of the martensite lattice, and hence the amount of softer ferrite structure left behind, depends on the temperature to which the steel has been heated and the time for which it is held at that temperature. Thus the mechanical properties of the steel can be controlled by the tempering process.

Figure 3.52 shows how, for an oil-quenched steel (0.4% carbon, 0.7% manganese, 1.8% nickel, 0.8% chromium, 0.25% manganese), the tempering temperature affects the hardness, tensile strength, yield stress and percentage elongation. The higher the tempering temperature, the lower the tensile stress, yield stress and hardness but the higher the percentage elongation.

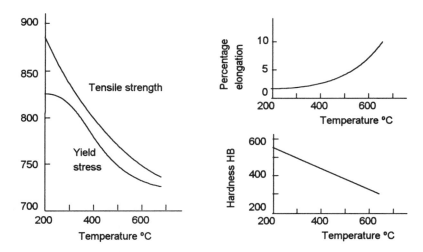

Figure 3.52 *The effect of tempering on the properties of a steel*

3.2.4 Austempering and martempering

The *austempering process* involves heating a steel to the austenite state, i.e. above A_3, and then quenching it in a bath held at a temperature above the martensite start temperature M_s. The steel is then held at that temperature until the austenite is completely transformed into bainite, after which it is allowed to cool. The initial quenching must be at a cooling rate greater than or equal to the critical cooling rate so that no pearlite is formed. The final cooling rate for the bainite structure can be at any rate as no further changes take place. Figure 3.53 shows the above sequence of events on a TTT diagram.

Bainite structures have lower strengths than martensitic structures but have better ductility and impact toughness. Because the steel does not suffer a severe quench treatment it is less likely to crack and distort. Table 3.15 shows how the properties of an austempered steel compare with those of the same steel (0.95% carbon) when quenched to martensite and then tempered.

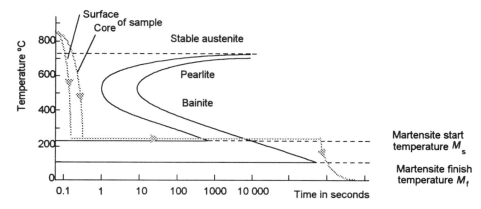

Figure 3.53 *Austempering*

Table 3.15 *The effect on properties of austempering*

Properties	Austempered	Quenched and tempered
Hardness HV	545	560
% elongation	11	1
Impact test J	58	16

Martempering

Martempering is a hardening treatment that consists of heating a steel to its austenitic state, i.e. above A_3, and then quenching it to a temperature just above M_s. The quenching must give a cooling rate faster than the critical cooling rate. The steel is then held at that temperature sufficiently long for the entire piece of steel to come to the same temperature, without any transformation to bainite occurring. It is then cooled in air to change the austenite to martensite. No transformation product other than martensite should result from the process. Figure 3.54 shows the above sequence of events on a TTT diagram. After such a treatment the steel is tempered in the usual way.

The effect of such a treatment is to minimise cracking and distortion, the thermal shock of the quenching having been reduced. The hardness and ductility are generally similar to those obtained by direct quenching to the martensitic state followed by tempering. The impact toughness may, however, be better.

3.2.5 Surface hardening

There are many situations where the need for good wear or fatigue resistance may indicate the need for a hard surface to a component while the requirement for a tough material with good resistance to impact failure would indicate the need for a soft, ductile core. Various methods are used for surface hardening and are grouped into two categories:

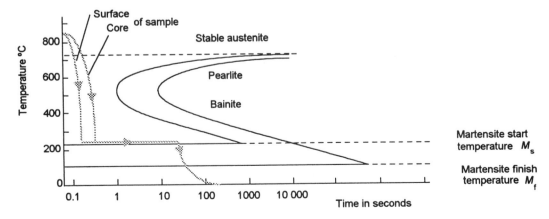

Figure 3.54 *Martempering*

- *Selective heating of the surface layers, i.e. surface heat treatment*

  The composition of the steel to be surface hardened by selective heating has to be chosen with certain criteria in mind. Before the surface treatment, the steel must have its inner core with the right mechanical properties. These properties will be unaffected by the surface treatment. The selective heating has to be able to change the microstructure of the steel to austenite in a very short amount of time so that only the surface is brought up to the austenising temperature. After the steel is quenched, provided the steel has less than about 0.4% carbon, the surface austenite changes to martensite and so the surface becomes harder.

- *Changing the composition of the surface layers*

  The composition of the surface layers may be changed by increasing the carbon content, this being termed *carburising*; diffusing nitrogen into the surface to produce hard nitrides, this being termed *nitriding*; or introducing both carbon and nitrogen into the surface layers, *carbonitriding* or *ferritic nitrocarburising*.

Local surface heat treatment

One selective heating method is called *flame hardening*. This involves heating the surface of a steel with an oxyacetylene flame and then immediately quenching the surface with cold water (Figure 3.55). The heating transforms the structure of the surface layers to austenite and the quenching changes this austenite to martensite. The depth of hardening depends on the heat supplied per unit surface area per unit time. Thus the faster the burner is moved over the surface, the less the depth of hardening. The temperatures used in this method are typically of the order of 850°C or more, i.e. above the A_3 temperature.

Another method of selective heating is *induction hardening*. This method involves placing the steel component within a coil through which a high frequency current is passed (Figure 3.56). The form of the induction coil depends on the shape of the component being hardened, also the size of area to be hardened. The alternating current induces another alternating current to flow within the surface layers of the steel component, the induced electrical currents heating the surface layers. The temperatures so produced cause the surface layers to change to austenite. When the surface has reached the austenising temperature, the surface is sprayed with cold water to transform the austenite to martensite. The depth of heating produced by this method, and hence the depth of hardening, is related to the frequency of the alternating current used. The higher the frequency the less the hardened depth (Table 3.16).

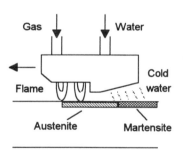

Figure 3.55   *Flame hardening*

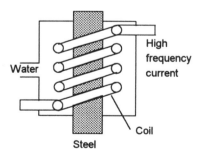

Figure 3.56   *Principle of induction heating*

Table 3.16 *The effect of frequency on depth of hardening*

Frequency kHz	Depth of hardening mm
3	4.0 to 5.0
10	3.9 to 4.0
450	0.5 to 1.1

Carburising

Carburising involves changing the carbon content of the surface, followed by a quenching process to convert the surface layers to martensite. This process is normally carried out on a steel containing less than about 0.2% carbon, the carburising treatment being used to give about 0.7 to 0.8% carbon in the surface layers. This wide difference in carbon content is needed because the quenching process following the carburising will affect both the inner core and the surface layers and the surface layers are to be converted to martensite but the inner core remain soft and ductile.

There are a number of carburising methods. With *pack carburising* the steel component is heated to above the A_3 temperature while in a sealed metal box which contains the carburising medium, e.g. a carbon-rich material such as charcoal and an energiser such as barium carbonate. The oxygen present in the box reacts with the carbon to produce carbon monoxide. This carbon-rich atmosphere in contact with the hot steel results in carbon diffusing into the surface austenite layers.

In *gas carburising* the component is heated to above the A_3 temperature in a furnace in an atmosphere of carbon-rich gas. The result is that carbon diffuses into the surface austenitic layers. Gas carburising is the most widely used method of carburising.

Salt bath carburising, or *cyaniding*, involves heating the component in a bath of suitable carbon-rich salts. Sodium cyanide is mainly used. The carbon from the molten salt diffuses into the component. In addition there is also diffusion of some nitrogen into the component. Both carbon and nitrogen can result in a microstructure which can be hardened. This method tends to produce relatively thin, hardened layers with high carbon content. This occurs because carburising takes place very quickly. One of the problems with this method is the health and safety hazard posed by the poisonous cyanide. Another problem is the removal of salt from the hardened component after treatment. This can be particularly difficult with threaded parts and blind holes.

Carburising can result in the production of a large grain structure due to the time for which the material is held at temperature in the austenitic state. Though the final product may have a hard surface, there may be poor impact properties due to the large grain size. A heat treatment might thus be used to refine the grains. Because the core and the surface have different compositions a two stage process is often used. The first stage involves a heat treatment to refine the grains in the core. The component is heated to just above the A_3 temperature for the carbon content of the core (Figure 3.57). For a core with 0.2%

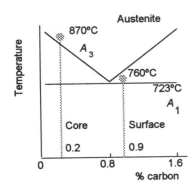

Figure 3.57 *Heat treatment for a carburised steel*

carbon, this is about 870°C. The component is then quenched in oil. The result is a fine grain core, but the surface layers are rather coarse martensite. The martensite is refined by heating to above the A_1 temperature for the carbon content of the surface layers. If the surface layers have 0.9% carbon, this is about 760°C. The component is then water quenched. The second stage treatment has little effect on the core but refines the martensite in the outer layers. The treatment may then be followed by a low temperature tempering, e.g. about 150°C, to relieve internal stresses produced by the treatment.

10.7.3 Nitriding

Nitriding involves changing the surface composition of a steel by diffusing nitrogen into it to produce hard nitride compounds. The process is used with those alloy steels that contain elements that form stable nitrides, e.g. steels containing aluminium, chromium, molybdenum, tungsten and vanadium. Prior to the nitriding treatment the steel is hardened and tempered to the properties required of the core. The tempering temperature does, however, need to be in the region 560 to 750°C. This is because the nitriding process requires a temperature up to about 530°C and this must not be greater than the tempering temperature as the nitriding process would temper the steel and so change the properties of the core.

Unlike carburising, nitriding is carried out at temperatures below the stable austenitic state. The process consists of heating a component in an atmosphere of ammonia gas and hydrogen, the temperatures being of the order of 500 to 530°C. The time taken for the nitrogen to react with the elements in the surface of the steel is often as much as 100 hours. The depth to which the nitrides are formed in the steel depends on the temperature and the time allowed for the reaction. Even with such long times, the depth of hardening is unlikely to exceed about 0.7 mm. After the treatment, the component is allowed to cool slowly in the ammonia– hydrogen atmosphere. With most nitriding conditions, a thin white layer of iron nitrides is formed on the surface of the component. This layer adversely affects the mechanical properties of the steel, being brittle and generally containing cracks. It is therefore removed by mechanical means or chemical solutions.

Because with nitriding no quenching treatments are involved, cracking and distortion are less likely than with other surface hardening treatments. Very high surface hardnesses can be obtained with special alloys. The hardness is retained at temperatures up to about 500°C, whereas that produced by carburising tends to decrease and the surface becomes softer at temperatures of the order of 200°C. The capital cost of the plant is, however, higher than that associated with pack carburising.

Carbonitriding

Carbonitriding is the name given to the surface hardening process in which both carbon and nitrogen are allowed to diffuse into a steel when it is in the austenitic–ferritic condition. The component is heated in an atmosphere containing carbon and ammonia and the temperatures used are about 800 to 850°C. The nitrogen inhibits the diffusion of carbon into steel and with the temperatures and times used being smaller than with carburising, the result is relatively shallow hardening. Though this process can be used with any steel that is suitable for carburising, it tends to be used only for mild steels and low-alloy steels.

Ferritic nitrocarburising

This treatment involves the diffusion of both carbon and nitrogen into the surface of a steel. The treatment involves a temperature below the A_1 temperature when the steel is in a ferritic condition. A very thin layer of a compound of iron, nitrogen and carbon is produced at the surface of the steel. This gives excellent wear and anti-scuffing properties. The process is mainly used on mild steel in the rolled or normalised condition.

Comparison of processes

In comparing surface hardening processes, the term *case depth* is used. This is the depth below the surface of a steel to which hardening occurs by a surface hardening process. It is that portion of the steel whose composition has been measurably altered from the original composition. Table 3.17 shows the comparison.

Table 3.17 *Surface hardening treatments*

Process	Temperature °C	Case depth mm	Case hardness HRC	Main uses
Pack carburising	810–1100	0.25–3	45–65	Low-carbon and carburising alloy steels. Large case depths, large components. E.g. gear teeth, bearing surfaces
Gas carburising	810–980	0.07–3	45–85	Low-carbon and carburising alloy steels. Large number of components. E.g. gear teeth, bearing surfaces
Cyaniding	760–870	0.02–0.07	50–60	Low-carbon and low-alloy steels.
Nitriding	500–530	0.07–0.7	50–70	Alloy steels. Lowest distortion. E.g. pistons, valves, gauges, dies, moulds
Carbo-nitriding	700–900	0.02–0.7	50–60	Low-carbon and low-alloy steels. E.g. gears, pulleys, screws, washers
Flame hardening	850–1000	Up to 0.8	55–65	0.4 to 0.7% carbon steels, selective heating and hence hardening. E.g. gears, bearing surfaces
Induction hardening	850–1000	0.5–5	55–65	0.4 to 0.7% carbon steels, selective heating and hence hardening. E.g. long drive shafts, crankshafts

Wear resistant treatments

In addition to the surface hardening treatments, there are a number of other treatments which can be applied to surfaces to reduce wear:

- *Siliconising*

  This process is used with low-carbon and medium-carbon steels. Silicon dissolves in iron at a temperature of about 1000°C to form a solid solution. Surfaces exposed to the silicon acquire a low coefficient of friction and can retain some lubricant. Adhesive wear is thus reduced. The process, however, involves high temperatures and the surface is not suitable where high pressures are involved, e.g. ball bearings.

- *Sulfinuz*

  This process can be used with all ferrous metals and titanium alloys. The material is heated in a salt bath to about 540 to 600°C. The treatment introduces carbon, nitrogen and sulphur into the surfaces. The result is good scuffing resistance with some reduction in the coefficient of friction.

- *Sulf BT*

  This process can be used with all ferrous metals, except 13% chromium stainless steels. The material is heated in a salt bath to about 180 to 200°C. The treatment introduces sulphur into the surfaces. The low temperature process minimises distortion and the surfaces have good scuffing resistance.

- *Noskuf*

  This process is used with case hardening and direct hardening steels. It is applied after carburising and involves heating the material in a salt bath at about 700 to 760°C. The process adds nitrogen and carbon to the surface layers and gives good scuffing resistance.

- *Phosphating*

  This is used with all ferrous metals. A film is produced on the surface of the material by either chemical or electrochemical treatments at about 40 to 100°C. The porous nature of the film helps to retain lubricants and resists scuffing. The treatment is less effective than nitriding or Sulf BT in improving wear resistance.

- *Boriding*

  This process is used with low-carbon, medium-carbon and low-alloy steels. Boron is allowed to diffuse into the surfaces at temperatures of the order of 900 to 1000°C. An iron boride compound is produced which gives the surface a good resistance to abrasive wear. The surface is very hard and is difficult to grind or polish; diamond tipped tools and silicon carbide or alumina wheels have to be used.

3.2.6 Annealing

The term *full annealing* is used for the treatment that involves heating a steel to its austenitic state before subjecting it to very slow cooling, i.e. allowing to cool in the furnace. For carbon steels with less than about 0.8% carbon, the temperature to which the

Key point

Full annealing is the heat treatment that involves heating a steel to its austenitic state before subjecting it to very slow cooling in order to give a reduction in hardness and increase in ductility.

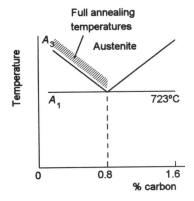

Figure 3.58 *Full annealing temperatures*

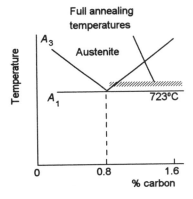

Figure 3.60 *Full annealing temperatures*

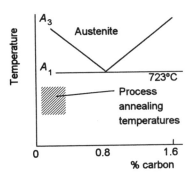

Figure 3.60 *Process annealing temperatures*

steel is heated is about 40°C above the A_3 temperature. The result of such a treatment is a microstructure of ferrite and pearlite and hence a very soft steel. Figure 3.58 shows the relevant austenising temperature region on the iron–carbon phase diagram and Figure 3.59 the temperature–time relationship for the process on a CCT diagram.

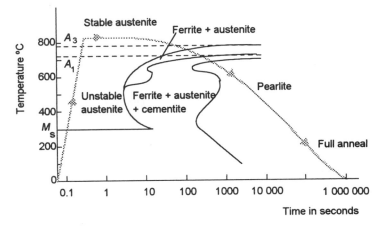

Figure 3.59 *Full annealing and the CCT diagram*

For a carbon steel with more than 0.8% carbon, the steel is heated to about 40°C above the A_1 temperature (Figure 3.60) before being slowly cooled. The result of such a treatment is to produce a microstructure of austenite plus excess cementite in dispersed spheroidal form. The overall result is a soft steel. The reason for not heating this steel to above the A_3 temperature is that slow cooling of such a steel results in a network of cementite surrounding pearlite and so a relatively brittle steel.

Sub-critical annealing

Sub-critical annealing, generally referred to as *process annealing*, is often used during cold-working processes with low-carbon steels (less than about 0.3% carbon), where the material has to be made more ductile and stresses have to be relieved for the process to continue. The process involves heating the steel to a temperature some 80 to 170°C below the A_1 temperature (Figure 3.60), holding it at that temperature for a while and then allowing the material to cool in air. This is a faster rate of cooling than that employed with full annealing where the material is cooled in the furnace. This process leads to recrystallisation. Prior to this treatment the grains may have been deformed by the cold-working process, afterwards having a new grain structure with no deformation. The effect of this treatment is to give a reduction in hardness and an increase in percentage elongation (Table 3.18).

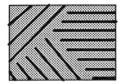

Laminar structure of
pearlite breaks down
to give spheroidal
cementite in ferrite matrix

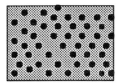

Figure 3.62   *Spheroidising*

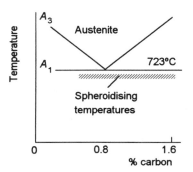

Figure 3.63   *Spheroidising
annealing temperatures*

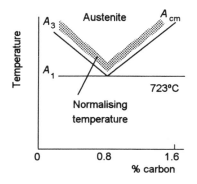

Figure 3.65   *Normalising
temperatures*

Table 3.18 *The effect of process annealing on the properties of a
0.15% plain carbon steel*

Condition	Hardness HV	% elongation
Cold worked	187	22
Process annealed	162	25

10.8.2 Spheroidising annealing

When sub-critical annealing is applied to steels that have
percentages of carbon greater than 0.3%, the effect of the heating
is to cause the cementite to assume spherical shapes (Figure 3.62).
Because of this the process is referred to as *spheroidising
annealing*. The steel is heated to about 30°C below the A_1
temperature (Figure 3.63), held there for several hours to allow the
cementite to change into spherical forms and then allowed to cool,
generally in the furnace. The result is spheroidal cementite in a
matrix of ferrite and a material which is more machinable.

Normalising

The term *normalising* is used to describe an annealing process
which involves heating a steel to about 50°C above the A_3
temperature and then allowing the steel to cool in air. The process
is thus similar to full annealing but has a faster rate of cooling
(Figure 3.64). It also differs from full annealing in that for
hypereutectoid steels the steel is cooled from 50°C above the A_{cm}
temperature (Figure 3.65).

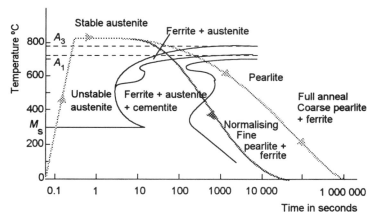

Figure 3.64 *Normalising compared with full annealing*

The result of such a treatment is a ferrite and pearlitic
microstructure, the hardness and strength being slightly greater
than that which would have occurred with full anealing. Figure
3.66 illustrates the effect on the properties of plain carbon steels of
full annealing and normalising, Table 3.17 giving data on the
properties of some plain carbon steels.

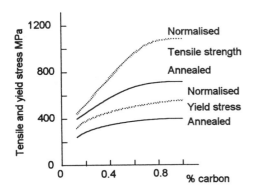 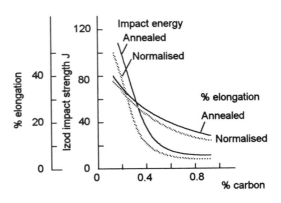

Figure 3.66 *Effect of annealing and normalising on plain carbon steels*

Table 3.17 *Effect on properties of plain carbon steels of normalising and full annealing*

% carbon	Treatment	Austenising temp. °C	Tensile strength MPa	Yield strength MPa	Percentage elongation	Hardness HB	Izod impact strength J
0.2	As rolled		450	330	36	143	87
	Normalised	870	440	346	36	130	118
	Annealed	870	395	295	37	111	123
0.4	As rolled		620	413	25	200	49
	Normalised	900	590	374	28	170	65
	Annealed	790	520	353	30	150	44
0.6	As rolled		814	483	17	240	18
	Normalised	900	775	420	18	230	13
	Annealed	790	625	373	23	180	11

3.2.7 Case study

Steels used for tools required to operate at high speeds have to maintain a high level of hardness at the high temperatures realised in operation. Such high-speed steels typically have a high carbon content, up to about 1.5%, in order to give a hard martensitic structure. Chromium, typically about 4%, is used to promote the formation of martensite and as an hardenability agent, up to 10% molybdenum and 20% tungsten may be present to promote the production of carbides. Also there may be about 1 to 5% vanadium; this is a strong carbide forming agent. Cobalt also may be present.

The heat treatment of such a steel involves annealing to produce a softened condition to facilitate machining, relieve internal stresses and provide a suitable structure for subsequent hardening treatments. The annealing involves heating to a temperature above A_1, about 850 to 900°C, soaking at that temperature for two to four hours, and then furnace cooling to a temperature below 600°C. This results in the formation of a ferrite matrix with finely dispersed carbide particles. The hardening treatment to produce the martensite structure involves heating to near to the solidus temperature, i.e. about 1200 to 1300°C, in order that a significant amount of the carbides formed by the alloying elements can be taken into solid solution. Following a short soaking at this temperature, only a few minutes in order to minimise decarburisation and grain growth, the material is cooled in air or by quenching in an oil or salt bath. The material is then tempered by heating to between 530 and 570°C. This gives a tempered martensite structure which is hard and stable at elevated temperatures and destabilises the retained austenite so that it forms martensite on cooling. A second tempering treatment is then used to temper the newly formed martensite.

Problems 3.2

1 Explain what is meant by the term critical points and, in this context, the symbols A_1, A_3 and A_{cm}.
2 What is shown by a TTT diagram and how is it obtained?
3 Sketch typical TTT diagrams for plain carbon steels having (a) the eutectoid composition, (b) less carbon than the eutectoid composition and (c) more carbon than the eutectoid composition.
4 Using Figure 10.5, estimate how long it will take the transformation from austenite to pearlite to take at 650°C for a 0.8% plain carbon steel.
5 What is the effect on the TTT diagram of a carbon steel of the addition of chromium to the alloy?
6 What are continuous cooling transformation diagrams and how is the critical cooling rate determined by such graphs?
7 Explain, with the aid of a CCT diagram, how the structures produced by cooling a plain carbon steel are determined by the rate of cooling. Consider cooling rates both greater and less than the critical cooling rate.
8 Figure 3.67 shows a CCT diagram for a steel. Estimate (a) the critical cooling rate, (b) the cooling rate needed to give a pearlitic structure with no bainite or martensite.
9 Describe the Jominy test and explain how test results can be used to determine the hardness distribution that would occur across a section of a steel component when it is quenched.
10 Estimate, for the steel giving the Jominy test result shown in Figure 3.68, the hardness at the surface and in the core of a circular cross-section bar with a diameter of 40 mm when oil quenched. Also use Figures 3.47 and 3.48.
11 Using Figure 3.49 for the plain carbon steel given there, what would be the hardness at the centre of a 25 mm diameter bar of the plain carbon steel when it is quenched in (a) still water, (b) oil?
12 Explain the term 'limiting ruling section' and why it is necessary to quote it when giving mechanical properties of steels.

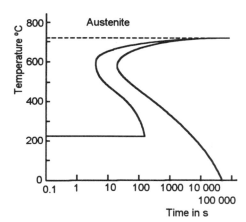

Figure 3.67   *Problem 8*

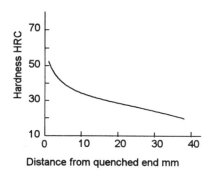

Figure 3.68   *Problem 10*

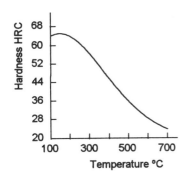

Figure 3.69   *Problem 13*

13 Figure 3.69 shows the tempering curve for a 1.4% plain carbon steel. What would be the hardness of such a steel after tempering at 300°C?

14 Describe the following heat treatment processes: (a) full annealing, (b) process annealing, (c) spheroidising annealing, (d) normalising.

15 Explain, using a TTT diagram, how full annealing and normalising can lead to different microstructures.

16 Describe the microstructure produced in the plain carbon steel giving the CCT diagram of Figure 3.70 when subject to the following heat treatments:
    (a) heated to 820°C and water quenched to 25°C,
    (b) heated to 820°C, quenched to 650°C, held at that temperature for 90 s and then quenched to 25°C,
    (c) heated to 820°C, water quenched to 25°C, heated to 500°C, held for 1000 s and then air cooled to 25°C,
    (d) heated to 820°C, water quenched to 350°C, held for 1000 s and then cooled to 25°C,
    (e) heated to 820°C, furnace cooled to 25°C,
    (f) heated to 820°C, water quenched to 350°C, held for 6 s and then air cooled to 25°C.

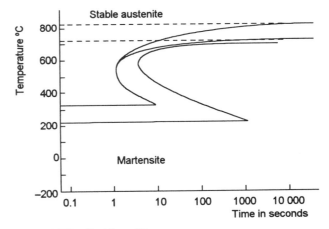

Figure 3.70   *Problem 16*

17 Using the TTT diagram of Figure 3.33 for a 0.8% plain carbon steel, describe the microstructures produced by the following heat treatments:
    (a) heated to 820°C and then water quenched to 25°C,
    (b) heated to 820°C, water quenched to 25°C, heated to 350°C, held for one hour, cooled to 25°C,
    (c) heated to 820°C, quenched to 350°C, held for one hour, air cooled to 25°C,
    (d) heated to 820°C, quenched to 700°C, held for two hours, water quenched to 25°C,
    (e) heated to 820°C, furnace cooled to 25°C.

18 Compare the surface hardening processes of flame hardening, carburising and nitriding.

19 A steel with 0.6% carbon is to be surface hardened. Which type of process, selective heating or carburising, would be suitable?

3.3 Non-ferrous alloys

The term *non-ferrous alloy* is used for those alloys which do not have iron as their base element. The following are some of the non-ferrous alloys in common use in engineering and which are discussed in more detail in this section:

Light alloys

These alloys are characterised by low density, in general from about 1.7 to 4.5 Mg/m^3 with zinc alloys about 6.6 Mg/m^3; this is much lower than the density of steel.

- *Aluminium alloys*
  These alloys have low density, good electrical and thermal conductivity, and high corrosion resistance, tensile strengths of the order of 150 to 400 MPa and tensile modulus of about 70 GPa. Typical uses are metal boxes, cooking utensils, aircraft bodywork and parts.

- *Magnesium alloys*
  These alloys have a low density, good electrical and thermal conductivity, tensile strengths of the order of 250 MPa and tensile modulus about 40 GPa. Typical uses are castings and forgings in the aircraft industry.

- *Titanium alloys*
  These alloys have low density, high strength, high corrosion resistance, can be used at high temperatures, tensile strengths of the order of 1000 MPa and tensile modulus about 110 GPa. Typical uses are in aircraft for compressor discs, blades and castings, and in chemical plant where high resistance to corrosive atmospheres is required.

- *Zinc alloys*
  These alloys have good electrical and thermal conductivity, high corrosion resistance, low melting points, tensile strengths of about 300 MPa and tensile modulus about 100 GPa. Typical uses are as car door handles, toys, car carburettor bodies and components that, in general, are produced by die casting.

Heavy alloys

These have higher densities, typically about 8.8 Mg/m^3, than the light metal alloys

- *Copper alloys*
  These include such alloys as brass and bronze and have good electrical and thermal conductivity, high corrosion resistance, tensile strengths of the order of 180 to 300 MPa and tensile modulus about 20 to 28 GPa. Typical uses are electrical components, pump and valve parts, coins, instrument parts, springs, screws.

• *Nickel alloys*
These alloys have good electrical and thermal conductivity, high corrosion resistance, high melting points, high strengths, can maintain their properties at high temperatures, tensile strengths between about 350 and 1400 MPa and tensile modulus about 220 GPa. Typical uses are pipes and containers in the chemical industry where high resistance to corrosive atmospheres is required, food processing equipment, gas turbine blades. The names Monel, Inconel and Nimonic are given to some forms of nickel alloys.

• *Cobalt alloys*
The cobalt alloys have properties and uses similar to those of nickel alloys.

3.3.1 Aluminium and its alloys

Aluminium has a density of 2.7 Mg/m^3, compared with 7.9 Mg/m^3 for iron. Thus for the same size component, the aluminium will be about one-third of the mass of the iron version. Aluminium alloys have a modulus of elasticity of about 70 GPa whereas iron alloys have a modulus of about 210 GPa. Thus the same size strip of an aluminium alloy will be much less stiff than that of an iron alloy, though the stiffness per unit mass is about the same. Although aluminium alloys have a low tensile strength compared with iron alloys, their strength to weight ratios are comparable.

Aluminium has a great affinity for oxygen and any fresh metal in air rapidly oxidises to give a thin layer of the oxide on the surface. This surface layer is not penetrated by oxygen and so protects the metal from further attack. The good corrosion resistance of aluminium is due to this oxide layer on its surface.

Pure aluminium is a weak, very ductile, material. It has an electrical conductivity about two-thirds that of copper but weight for weight is a better conductor. Aluminium of high purity (99.5% or greater) is too weak a material to be used in any other capacity than a lining for vessels. It is used in this way to give a high corrosion resistant surface. Aluminium of commercial purity (99.0 to 99.5%) is widely used as a foil for sealing milk bottles, thermal insulation and kitchen foil for cooking. The presence of relatively small percentages of impurities in aluminium considerably increases the tensile strength and hardness of the material.

The mechanical properties of aluminium, and its alloys, depend not on the purity of the aluminium but on the amount of work to which it has been subject. The effect of working is to fragment the grains. This results in an increase in tensile strength and hardness and a decrease in ductility. By controlling the amount of working, different degrees of strength and hardness can be produced. These are said to be different *tempers*. In materials specifications, the temper of aluminium and its non-heat treatable alloys is indicated as shown in Table 3.18. Table 3.19 shows typical properties of 'pure' aluminium.

Table 3.18 *Temper codes for work hardening*

British code	American code	Condition
M	F	As manufactured, e.g. rolled
O	O	In the annealed or soft condition
H		Work hardened, the degree of work hardening being indicated by numbers
H2		Quarter work hardened
H4		Half work hardened
H6		Three-quarters work hardened
H8		Fully work hardened
	H	Work hardened, the letter being followed by two numbers, the first indicating the processes used and the second the degree of hardening
	H1x	Cold worked only
	H2x	Cold worked and partly annealed
	H3x	Cold worked and stabilised at a low temperature to prevent age hardening
	Hx2	Quarter work hardened
	Hx4	Half work hardened
	Hx8	Fully work hardened
	Hx9	Extra hard temper

Table 3.19 *Typical properties of aluminium*

% purity	Condition	Tensile strength MPa	Percentage elongation	Hardness HB
99.99	Annealed (O)	45	60	15
	Half hard (H4)	82	24	22
	Fully hard (H8)	105	12	30
99.8	Annealed (O)	66	50	19
	Half hard (H4)	99	17	31
	Fully hard (H8)	134	11	38
99.5	Annealed (O)	78	47	21
	Half hard (H4)	110	13	33
	Fully hard (H8)	140	10	40
99	Annealed (O)	87	43	22
	Half hard (H4)	120	12	35
	Fully hard (H8)	150	10	42

Aluminium alloys are designated by their chemical composition, Section 1.2.2 giving details. They can be divided into two groups: wrought alloys and cast alloys. Each of these can be further divided into two groups: alloys which are not heat treated and those which are.

The term *wrought alloy* is used for an alloy that is suitable for shaping by a working process, e.g. forging, extrusion, rolling. The term *cast alloy* is used for an alloy that is suitable for shaping by a casting process. In the specifications of non-heat treatable alloys, their temper is indicated in the same way as described in Table 3.18. Such alloys have properties which are affected by work hardening. Heat treatable alloys have their tempers specified as shown in Table 3.20.

Table 3.20 *Tempers for heat treated aluminium alloys*

British code	American code	Condition
T	T	Heat treated letter followed by digit or letter to indicate form of heat treatment
	T1	Cooled from fabrication temperature and naturally aged
	T2	Cooled from fabrication temperature, cold worked and naturally aged
TD	T3	Solution treated, cold worked and naturally aged
TB	T4	Solution treated and naturally aged
	T5	Cooled from fabrication temperature and artificially aged
TF	T6	Solution treated and artificially aged
	T7	Solution treated and stabilised by over-ageing
TH	T8	Solution treated, cold worked and artificially aged
	T9	Solution treated, artificially aged and cold worked
TE	T10	Cooled from fabrication temperature, cold worked and artificially aged
	W	Solution treated. This temper is unstable and ageing occurs at room temperature. Added digits 51, 52, 54 are used to indicate a stress relieving process

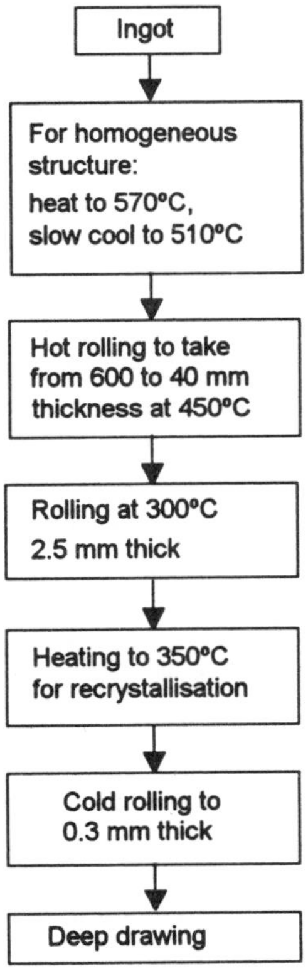
Ingot
For homogeneous structure:
heat to 570°C, slow cool to 510°C
Hot rolling to take from 600 to 40 mm thickness at 450°C
Rolling at 300°C
2.5 mm thick
Heating to 350°C for recrystallisation
Cold rolling to 0.3 mm thick
Deep drawing

Figure 3.71 *Stages in the production of beverage cans*

Wrought alloys

Common alloys in this category are aluminium with manganese and/or magnesium. The non-heat treatable wrought alloys of aluminium do not significantly respond to heat treatment but have their strength developed by the extent of the working to which they are subject.

A common aluminium–manganese alloy (3003) has 1.25% manganese. The effect of the manganese is to increase the tensile strength by predominantly dispersion hardening from intermetallic compounds. The alloy has high ductility and good corrosion properties, finding uses as kitchen utensils, tubing and corrugated sheet for building. The addition of some magnesium provides solid solution strengthening. The alloy (3004) with 1.2% manganese and 1.0% magnesium is used for beverage cans, combining strength with ductility so that cans can be deep drawn. Figure 3.71 shows the sequence of operations. After the cold rolling the temper is H19, a tensile strength of about 310 MPa and percentage elongation 10%.

Aluminium and magnesium form solid solutions (Figure 3.72) and wrought alloys containing from about 0.8% to just over 5% magnesium are widely used. Aluminium–magnesium alloys, e.g. 5454, are used for welded applications such as tanks for liquids and pressure vessels. The alloy 5182, 4.5% magnesium, 0.35% manganese, is used for the lids of aluminium beverage cans, this alloy being harder than the 3004 used for the cans and so more able to withstand the forces involved with ring pulls. These alloys have excellent corrosion resistance and thus find considerable use in marine environments, e.g. constructional materials for boats and ships.

Table 3.21 shows properties of commonly used forms of these alloys.

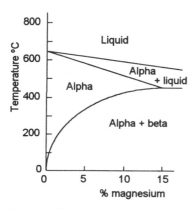

Figure 3.72 *Phase diagram for aluminium–magnesium alloys*

Table 3.21 *Properties of wrought non-heat treatable aluminium alloys*

Alloy	Composition %	Temper	0.2% proof stress MPa	Tensile strength MPa	Percent. elong.
3003	1–1.5 Mn	O	25	110	30
		H18	185	200	4
3004	1–1.5 Mn,	O	70	180	20
	0.8–1.3 Mg	H38	250	280	5
5050	0.2 Mn,	O	55	145	24
	0.5–1.1 Mg	H38	200	220	6
5182	0.35 Mn,	O	140	275	25
	4.5 Mg	H34	285	340	10
5454	0.5–1 Mn,	O	120	250	22
	2.4–3.0 Mg	H34	240	305	10

Note: the composition only gives the main alloying elements.

The heat treatable wrought alloys can have their properties changed by heat treatment. Copper, magnesium, zinc and silicon are common additions to aluminium to give such alloys, the three series being 2xxx (aluminium–copper and aluminium–copper–magnesium), 6xxx (aluminium–magnesium–silicon) and 7xxx (aluminium–zinc–magnesium and aluminium–zinc–magnesium–copper). All depend on age hardening to develop their full strength properties.

Figure 3.73 shows the phase diagram for aluminium–copper alloys. When such an alloy, say 3% copper, is slowly cooled, the structure at about 540°C is a solid solution of the alpha phase. When the temperature falls below the solvus temperature a copper–aluminium compound is precipitated. The result at room temperature is an alpha solid solution with this copper–aluminium precipitate. The precipitate is rather coarse, but this can be changed by heating to about 500°C, soaking at that temperature, and then quenching to give a supersaturated solid solution of just the alpha phase with no precipitate. This treatment is known as *solution treatment* and gives an unstable situation. With time a fine precipitate will be produced. Heating to, say, 165°C for about ten hours hastens the production of this fine precipitate (Figure 3.74). The alloy with this fine precipitate is both stronger and harder, the treatment being termed *precipitation hardening*.

Aluminium with 5.5% copper (2011) also contains small amounts of lead and bismuth to assist in chip formation and is used where good machining characteristics are required. Aluminium with 6.3% copper (2219) is weldable, has relatively good properties at elevated temperatures and good toughness at low temperatures. It is used for tanks storing liquefied gases and aircraft parts.

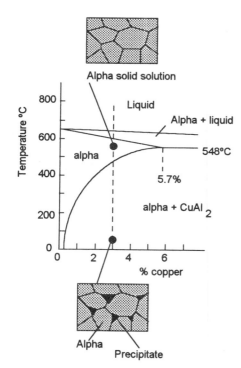

Figure 3.73 *Phase diagram for aluminium–copper alloys*

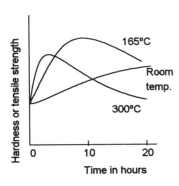

Figure 3.74 *The effect of time and temperature*

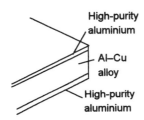

Figure 3.75 *A clad sheet*

High-purity aluminium

Al–Cu alloy

High-purity aluminium

Aluminium–copper–magnesium alloys (2xxx) are more widely used. The addition of magnesium to aluminium–copper alloys accelerates and intensifies precipitation hardening. These were the first precipitation hardening alloys to be discovered (Alfred Wilm in 1911) and led to the production of an alloy known as *Duralumin*. A modified form of this alloy (2017) with 4.0% copper, 0.4–0.6% magnesium and 0.7% manganese is still widely used. Higher strength alloys have, however, been developed, e.g. 2014 with 4.4% copper, 0.5% magnesium, 0.9% silicon and 0.8% manganese. The high silicon content increases the response to hardening during artificial ageing. Such an alloy is widely used in aircraft bodywork. The heat treatment process used with such alloys is solution treatment at 480°C, quenching and then precipitation hardening at either room temperature for about four days or at 165°C for ten hours.

The presence of the copper in these alloys does, however, reduce the corrosion resistance and thus the alloy is often clad (Figure 3.75) with a thin layer of high-purity aluminium to improve the corrosion resistance.

Aluminium–magnesium–silicon alloys (6xxx) are widely used as structural alloys with medium strength and good weldability. Typical applications are truck and marine structures, pipes, railings and furniture. Such alloys are normally aged at about 170°C. Another group of alloys are the aluminium–zinc–magnesium alloys (7xxx), these having medium to high strength and find a use in welded structures such as aircraft.

Table 3.22 shows the properties of representative examples of the more widely used wrought, heat treatable, alloys.

Table 3.22 *Properties of wrought, heat treatable, aluminium alloys*

Alloy	Composition %	Temper	0.2% proof stress MPa	Tensile strength MPa	Percentage elongation
2011	5–6 Cu, 0.4 Si, 0.3 Zn	T6	295	390	17
2219	5.6–6.8 Cu, 0.2–0.3 Mn, 0.1 Zn, 0.2 Si	T62	290	415	10
		T87	395	475	10
2017	3.5–4.5 Cu, 0.4–0.8 Mg, 0.4–1.0 Mn, 0.2–0.8 Si, 0.25 Zn	T4	275	425	22
2014	3.9–5.0 Cu, 0.2–0.8 Mg, 0.4–1.2 Mn, 0.5–1.2 Si, 0.25 Zn	T6	410	480	13
6061	0.15–0.4 Cu, 0.8–1.2 Mg, 0.15 Mn, 0.4–0.8 Si, 0.25 Zn	T6	275	310	12
6063	0.1 Cu, 0.45–0.9 Mg, 0.1 Mn, 0.2–0.6 Si, 0.1 Zn	T6	215	240	12
7001	1.6–2.6 Cu, 2.6–3.4 Mg, 6.8–8.0 Zn, 0.2 Mn, 0.35 Si	T6	625	675	9

Note: where a second digit is used with T tempers, it indicates a specific treatment.

Cast alloys

An alloy for use in the casting process must flow readily to all parts of the mould and on solidifying should not shrink too much, with any shrinking not resulting in fracture. Special advantages of aluminium alloys for castings are light weight, relatively low melting temperatures, good surface finish and, in general, good fluidity. However, there is a relatively high shrinkage, between about 3.5 and 8.5%.

In choosing an alloy for casting, the type of casting process being used needs to be taken into account. In sand casting the mould is made of sand bonded with clay or a resin and the cooling rate is relatively slow. An alloy for casting by this method must give suitable strength after a slow cooling process. Permanent mould, gravity die, casting with its metal moulds gives a faster cooling rate. With pressure die casting the mould is made of metal and the hot metal being injected into the die under pressure. This results in even faster cooling and thus alloys for use with this casting method must develop suitable strength after fast cooling.

Alloys based on aluminium–silicon with between 9 and 13% silicon are widely used as casting alloys with both sand and die casting and can be used in the 'as-cast' condition, i.e. no heat treatment is required. Figure 3.76 shows the phase diagram for aluminium–silicon alloys. The addition of silicon to aluminium increases its fluidity, between about 9 to 13% giving suitable fluidity for casting. The eutectic for aluminium–silicon alloys has a composition of 11.6% silicon. An alloy of this composition changes from the liquid to solid state without any change in temperature and alloys close to this composition solidify over a small temperature range, making them particularly suitable for die casting where a quick change from liquid to solid is required for rapid ejection from the die to give high output rates.

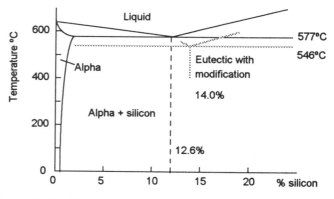

Figure 3.76 *The aluminium–silicon phase diagram*

For the eutectic composition, aluminium–silicon alloys show a rather coarse structure of alpha phase and silicon. For an alloy having more silicon than the 12.6% eutectic value, the microstructure consists of silicon crystals in eutectic structure. The coarse eutectic structure, together with the presence of the embrittling silicon crystals, results in rather poor mechanical

Activity

Draw a flow chart in which the wrought and cast, heat treatable and non-heat-treatable alloys of aluminium are listed.

properties for the casting. The structure can, however, be made finer and the silicon crystal formation prevented by a process known as *modification*. This involves adding about 0.005 to 0.015% metallic sodium or sodium salts to the melt before casting. This produces a considerable refinement of the eutectic structure and also causes the eutectic composition to change to about 14.0% silicon, this displacement being indicated in Figure 3.76. Thus for a silicon content below 14%, the structure, as modified, has alpha phase crystals in a finer eutectic structure. The result is an increase in both tensile strength and ductility and so a much better casting material. Table 3.23 gives the mechanical properties of 13% silicon alloys when unmodified and modified. Aluminium–silicon alloys are widely used for both sand and die casting with such applications as car sumps, gear boxes and radiators and a wide variety of thin-walled and complex castings.

Table 3.23 *Properties of 13% silicon alloys*

Condition	Form of casting	Tensile strength MPa	Percentage elongation	Hardness HRC
Unmodified	Sand	125	2	50
	Die	195	3.5	63
Modified	Sand	195	13	58
	Die	220	8	72

The addition of copper to aluminium–silicon alloys results in increased strength and better machinability though reducing castability, ductility and corrosion resistance. Such alloys have between about 3 to 10.5% silicon and 1.5 to 4.5% copper, the higher silicon alloys being used for die casting and the lower silicon alloys for sand casting. Other casting alloys are aluminium–magnesium alloys (typically about 4% magnesium), these being widely used for sand casting. They have excellent corrosion resistance and are often used for marine applications. The addition of copper, magnesium and other elements to aluminium alloys, either singly or in some suitable combinations, can enable the alloy to be heat treated. Thus an alloy having 5.5% silicon and 0.6% magnesium can be subject to solution treatment followed by precipitation hardening to give a high-strength casting material. Another heat treatable casting alloy has 4.0% copper, 2.0% nickel and 1.5% magnesium. Table 3.24 shows the properties of commonly used casting alloys.

Anodising

Aluminium develops a hard oxide surface layer in air, hard because the oxide (alumina) is a ceramic material. This layer can be thickened by an electrolytic process, the treatment being known as *anodising*. The freshly formed anodic layer is porous and has to be sealed, by immersion in boiling water or a special solution, to give maximum protection against atmospheric corrosion. The film can be coloured by pigments for decorative purposes.

Table 3.24 *Properties of commonly used aluminium casting alloys*

Al. Ass. number	BS LM number	Composition %	Temper	Casting process	0.2% proof stress MPa	Tensile strength MPa	Percent. elong.
	LM6	10–13 Si	M/F	S	65	185	8
				PM	90	205	9
				D	130	250	2.5
208.0		2.5–3.5 Si, 3.5–4.5 Cu	T533	S	105	185	1.5
	LM4	4–6 Si, 2–4 Cu, 0.2–0.6 Mn	T21	S	95	175	3
			T6	PM	230	295	2
	LM2	9–11.5 Si, 0.7–1.5 Cu	M/F	D	150	250	3
413.0	LM20	10–13 Si, 0.7–2.5 Cu	M/F	D	140	265	2
319.0	LM21	5.5–6.5 Si, 3–4 Cu	T21	S	125	185	1
			T6	PM	125	200	2
380.0	LM24	7.5–9.5 Si, 3–5 Cu	M/F	D	160	315	3
356.0	LM29	6.5–7.5 Si, 0.2–0.4 Mg	T6	S	205	230	4
			T6	PM	225	240	4
514.0	LM5	3–6 Mg, 0.3–0.7 Mn	M/F	S	80	170	5
			M/F	PM	80	230	10

Note: S = sand cast, PM = permanent mould cast (gravity die cast), D = pressure die cast.

3.3.2 Magnesium and its alloys

Key points

Magnesium has a very low density compared with other metals. The strength to weight ratio is high, being greater than that of aluminium. The alloys have good machinability and weld readily.

Magnesium has a density of 1.7 Mg/m³ and thus a very low density compared with other metals. It has an electrical conductivity of about 60% of that of copper, as well as a high thermal conductivity. It has a low tensile strength and needs to be alloyed with other metals to improve its strength. Under ordinary atmospheric conditions, magnesium has good corrosion resistance. This is provided by an oxide layer that develops on the surface. However, this oxide layer is not completely impervious, particularly in air containing salts, and thus the corrosion resistance can be low under adverse conditions.

Pure magnesium has a hexagonal close-packed structure and this leads to less ductility than the face-centred cubic structure of aluminium. Its atomic radius is about 0.160 nm, which is close to a range of other elements, e.g. aluminium 0.143 nm, zinc 0.133 nm, zirconium 0.161 nm, and so readily forms alloys. Alloying increases the number of slip planes and so increases the ductility. However, the basic hexagonal close-packed structure of magnesium does limit the amount of deformation that can be produced and so limits the use of magnesium alloys to produce wrought products.

Because of the low density of magnesium, the magnesium-base alloys have low densities. Thus magnesium alloys are used in applications where lightness is a primary consideration, e.g. in aircraft. Aluminium alloys have higher densities than magnesium alloys but can have greater strength. The strength to weight ratio for magnesium alloys is, however, greater than that of aluminium alloys. Magnesium alloys also have the advantage of good machinability and weld readily.

Magnesium alloys for casting predominate over wrought alloys, most magnesium products being formed by high pressure die casting. Most of the alloys used show high fluidity and so can be used for castings with intricate and thin-walled parts. Magnesium also has, when compared with other metals, a low specific heat per unit volume and so magnesium castings cool more quickly and so faster cycle times are possible with die casting. Magnesium–aluminium–zinc and magnesium–zinc–zirconium are the main two groups of alloys in general use.

The magnesium–aluminium system forms the basis of a group of magnesium alloys. Most contain about 8 to 9% aluminium with small amounts of zinc to improve the strength and manganese to improve corrosion resistance. AZ91 is a very widely used alloy, having 9% aluminium, 0.7% zinc and 0.2% manganese. Note that the code for designating the alloy is that of the American Society of Testing Materials and this is becoming increasingly used throughout the world. The two letters indicate the principal alloying elements with the first letter indicating the element present in the greatest quantity: A = aluminium, B = bismuth, C = copper, D = cadmium, E = rare earths, F = iron, G = magnesium, H = thorium, K = zirconium, L = lithium, M = manganese, N = nickel, P = lead, Q = silver, R = chromium, S = silicon, T = tin, W = yttrium, Y = antimony and Z = zinc. The two following numbers indicate the nominal, rounded, percentages of these elements.

Magnesium–zinc alloys can be age hardened but are susceptible to micro porosity and consequently are not usable for castings. However, the introduction of zirconium enables grains to be refined, so increasing strength, and the addition of rare earths tends to suppress micro porosity.

Because of ductility problems, wrought products are produced from magnesium alloys by mainly hot working processes, typically extrusion, rolling and press forging, at temperatures in the range 300 to 500°C. The most widely used sheet alloy is AZ31 (3% aluminium–1% zinc–0.3% manganese) and this is strengthened by work hardening and is weldable. For extrusion AZ61 (8% aluminium–1% zinc–0.7% manganese) is widely used and can be age hardened.

Table 3.25 shows the properties of commonly used magnesium alloys.

3.3.3 Titanium and its alloys

Titanium has a relatively low density of 4.5 Mg/m^3, just over half that of steel. It has a relatively low strength when pure but alloying gives a considerable increase in strength. Because of the low density of titanium, its alloys have a high strength to weight ratio. It has a high melting point (1660°C) and excellent corrosion resistance. However, titanium is an expensive metal, its high cost reflecting the difficulties experienced in the extraction and formation of the material; the ores are quite plentiful.

Key points

Titanium alloys have high melting points, excellent corrosion resistance, high strength and low density.

Table 3.25 *Properties of commonly used magnesium alloys*

ASTM code	BS code	Composition %	Condition	0.2% proof stress MPa	Tensile strength MPa	Percentage elongation
Casting alloys						
AZ81	A8	8 Al, 0.5 Zn, 0.3 Mn	As sand cast	80	140	3
			T6	80	220	5
AZ91	AZ91	9.5 Al, 0.5 Zn, 0.3 Mn	As sand cast	95	135	2
			T4	80	230	4
			As chill cast	100	170	2
Wrought alloys						
AZ31	AZ31	3 Al, 1 Zn, 0.3 Mn	Sheet O	120	240	16
			H24	160	250	4
			Extrusion F/M	130	230	4
			Forging F/M	105	200	7
AZ61	AZM	6.5 Al, 1 Zn, 0.3 Mn	Extrusion F/M	180	260	7
			Forging F/M	160	275	7

Note: the temper designations used for magnesium are the same as those used for aluminium.

Titanium can exist in two crystal forms, alpha which is a hexagonal close-packed structure and beta which is a body-centred cubic. In pure titanium, the alpha structure is the stable phase up to 883°C and is transformed into the beta form above this temperature. This beta form then remains stable up to the melting point.

Commercially pure titanium ranges in purity from 99.0 to 99.5%, the main impurities being iron, carbon, oxygen, nitrogen and hydrogen. Such material is lower in strength than titanium alloys but more corrosion resistant. The properties of the commercially pure titanium are largely determined by the oxygen content. Table 3.26 shows the composition and properties of such materials. Because of its excellent corrosion resistance, commercially pure titanium is used for chemical plant components, surgical implants, marine and aircraft engine parts.

Table 3.26 *Composition and properties of commercially pure annealed titanium*

Composition %	Temperature °C	Tensile strength MPa	Yield stress MPa	Percent. elong.
99.5 Ti	20	330	240	30
	300	150	95	32
99.2 Ti	20	440	350	28
	300	220	120	35
99.1 Ti	20	520	450	25
	300	240	140	34
99.0 Ti	20	670	590	20
	300	310	170	37

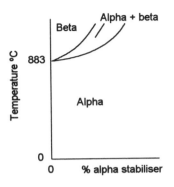

Figure 3.77 *Effect of alpha stabiliser*

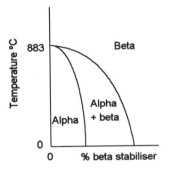

Figure 3.78 *Effect of beta stabliser*

Classification of titanium alloys

Titanium alloys can be grouped into categories according to the phases present in their structure. The addition of elements such as aluminium, tin, oxygen or nitrogen results in the enlargement of the alpha phase region on the phase diagram, such elements being referred to as *alpha-stabilising elements*. Figure 3.77 shows the effect of increasing amounts of alpha-stabilising elements on the phases present. The alpha phase exists to much higher temperatures. Other elements, such as vanadium, molybdenum, silicon and copper, enlarge the beta phase region and are termed *beta-stabilising elements*. Figure 3.78 illustrates their effect on the phases present. Increasing the amounts of beta stabiliser means that beta phase can exist at room temperature. Other elements added to titanium alloys, e.g. zirconium, can contribute solid solution strengthening.

The four categories into which titanium alloys are grouped are:

- **Alpha-titanium alloys**

   These are composed entirely of alpha phase. An example of such an alloy is 92.5% titanium–5% aluminium–2.5% tin. Both the aluminium and tin are alpha stabilisers. Such alloys have the hexagonal close-packed structure and, as a consequence, are strong, maintain their strength at high temperatures but are difficult to work. The alloys have good weldability and are used where high temperature strength is required, e.g. steam turbine blades.

- **Near alpha-titanium alloys**

   These are composed of almost all alpha phase with a small amount of beta phase dispersed throughout the alpha. Such alloys are achieved by adding small amounts, about 1 to 2%, or beta-stabilising elements such as molybdenum and vanadium to what is otherwise an alpha-stabilised alloy. An example of such an alloy is 90% titanium, 8% aluminium, 1% molybdenum and 1% vanadium. This alloy is normally used in the annealed condition. There are two forms of annealing; mill annealing and duplex annealing. *Mill annealing* involves heating the alloy to 790°C, soaking for eight hours and then furnace cooling. *Duplex annealing* involves mill annealing followed by reheating to 790°C, soaking for quarter of an hour and then air cooling. The result of such annealing is beta particles dispersed throughout an alpha matrix. The alloy in the annealed state is used for airframe and jet engine parts which require high strengths, good creep resistance and toughness up to temperatures of about 850°C. The alloy has good weldability.

- **Alpha–beta-titanium alloys**

   These contain sufficient quantities of beta-stabilising elements for there to be appreciable amounts of beta phase at room temperature. An example of such an alloy is 90% titanium–6% aluminium–4% vanadium. The aluminium stabilises the alpha phase while the vanadium stabilises the beta phase. These alloys can be solution treated, quenched and aged for increased strength. The microstructure of the alloys depends on their composition and heat treatment. Thus, a fast cooling rate from a temperature where the material was all beta, e.g. quenching in cold water, produces a martensitic structure with some increase in hardness. Ageing can then produce some further increase in strength as a result of beta precipitates. Figure 3.79 shows the effects on the properties of quenching from different temperatures. The maximum strength is obtained by solution treatment to a temperature just below the beta transformation temperature.

The 90% titanium–6% aluminium–4% vanadium alloy can be readily welded, forged and machined. At 482°C the alloy shows a metallurgical change in structure and there is a change in its mechanical properties, the strength decreasing and the ductility increasing. The alloy is used for both high- and low-temperature applications, e.g. rocket motor cases, turbine blades and cryogenic vessels.

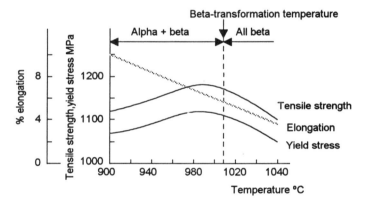

Figure 3.79 *Effect of the solution heat treatment temperature on the properties of 90% titanium–6% aluminium–4% vanadium alloy after ageing for eight hours at 593°C*

• *Beta-titanium alloys*

When sufficiently high amounts of beta-stabilising elements are added to titanium, the resulting structure can be made entirely beta at room temperature after quenching, in some cases by air cooling. Unlike alpha-titanium alloys, beta-titanium alloys are readily cold worked in the solution treated and quenched condition and can be subsequently aged to give very high strengths. In the high-strength condition the alloys have low ductilities. They can also suffer from poor fatigue performance. The alloys are thus not so widely used as the alpha–beta alloys.

A typical beta-titanium alloy has 77% titanium–13% vanadium–11% chromium–3% aluminium. The alloy is usually used in the solution treated, quenched and aged condition in order to obtain the very high tensile strength. It is used for aerospace components, honeycomb panels and high strength fasteners.

Table 3.27 shows the properties of a range of commonly used titanium alloys. In considering such alloys, account needs to be taken of the possibility of stress corrosion cracking in salt water environments for near alpha alloys and the poor fatigue performance of beta alloys.

Table 3.27 *Properties of titanium alloys*

Common designations	Composition	Condition	Temperature °C	Tensile strength MPa	Yield stress MPa	Percentage elongation
Alpha alloy						
IMI317	5 Al, 2.5 Sn	Annealed	20	860	800	16
			315	565	450	18
Near alpha alloy						
8-1-1	8 Al, 1 Mo,	Duplex	20	1000	950	15
	1 V	annealed	315	795	620	20
			540	620	515	25
6-2-4-2S	6 Al, 2 Sn,	Duplex	20	980	895	15
	4 Zr, 2 Mo	annealed	315	770	585	16
			540	650	490	26
Alpha–beta alloy						
IMI318	6 Al, 4 V	Annealed	20	990	925	14
6-4			315	725	625	14
			540	530	425	35
		Sol. + aged	20	1170	1100	10
			315	860	705	10
			540	655	485	22
6-6-2	6 Al, 2 Sn,	Annealed	20	1070	1000	14
	6 V		315	930	805	18
		Sol. + aged	20	1275	1170	10
			315	980	895	12
Beta alloy						
13-11-3	3 Al, 13 V,	Sol. + aged	20	1220	1170	8
	11 Cr		315	885	795	19
8-8-2-3	3 Al, 8 Mo,	Sol. + aged	20	1310	1240	8
	8 V, 2 Fe		315	1130	980	15

3.3.4 Zinc and its alloys

Zinc has a density of 7.1 Mg/m^3 and a melting point of only 419°C. It is a relatively weak metal with good corrosion resistance due to the formation of an impervious oxide layer on the surface. Zinc is frequently used as a coating on steel in order to protect that material against corrosion, the product being known as *galvanised steel*.

Key points

Zinc is relatively weak, with low melting point and good corrosion resistance.

The main use of zinc alloys is for die casting. They are excellent for this by virtue of their low melting points and the lack of corrosion of dies used with them. The two alloys that have been widely used for this purpose are known as MAZAK 3 and MAZAK 5. MAZAK 3 has the composition of 3.8 to 4.3% aluminium, 0.01% copper, 0.03 to 0.06% magnesium and 1.0% iron. MAZAK 5 has the composition 3.8 to 4.3% aluminium, 0.75 to 1.25% copper, 0.03 to 0.06% magnesium and 1.0% iron. Table 3.28 shows the properties.

Table 3.28 *Properties of zinc pressure die casting alloys*

Alloy	Condition	Tensile strength MPa	Percent. elong.	Impact strength J
MAZAK 3	As cast	286	15	57
	Stabilised	273	17	61
MAZAK 5	As cast	335	9	58
	Stabilised	312	10	60

The zinc used in the alloys has to be extremely pure so that little, if any, impurities are introduced into the alloys; typically the required purity is 99.99%. This is because the presence of very small amounts of cadmium, lead or tin renders the alloy susceptible to inter crystalline corrosion. The products of this corrosion cause a casting to swell and may lead to failure in service. After casting the alloys undergo a shrinkage which takes about a month to complete; after that there is a slight expansion. A casting can be *stabilised* by heating to 100°C for about six hours.

Zinc alloys can be machined and, to a limited extent, worked. Soldering and welding are not generally feasible. Zinc alloy die castings are widely used in domestic appliances, for toys, car parts such as door handles and fuel pump bodies, optical instrument cases.

3.3.5 Copper and its alloys

Copper has a density of 8.93 Mg/m³. It has a very high electrical and thermal conductivity and can be manipulated by either hot or cold working. Pure copper is very ductile and relatively weak. The tensile strength and hardness can be increased by working; this does, however, decrease the ductility. Copper has good corrosion resistance. This is because there is a surface reaction between copper and the oxygen in the air which results in the formation of a thin protective oxide layer.

Copper and low-alloyed coppers

Very pure copper can be produced by an electrolytic refining process. An impure slab of copper is used as the anode while a pure thin sheet of copper is used as the cathode and the two electrodes suspended in a warm solution of dilute sulphuric acid (Figure 3.80). The passage of an electric current through the cell

Key points

Copper has very high electrical and thermal conductivity, good corrosion resistance, and when pure is ductile and fairly weak.

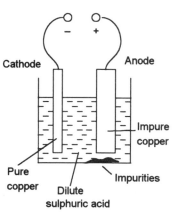

Figure 3.80 *Basic arrangement for refining copper*

causes copper to leave the anode and become deposited on the cathode. The result is a thicker, pure copper cathode while the anode effectively disappears; the impurities having fallen to the bottom of the container. The copper produced by this process is often called *cathode copper* and has a purity greater than 99.99%. It is used mainly as the raw material for the production of alloys, though there is some use as a casting material.

Electrolytic tough pitch high-conductivity copper is produced from cathode copper which has been melted and cast into billets, and other suitable shapes, for working. It contains a small amount of oxygen, present in the form of cuprous oxide, which has little effect on the electrical conductivity of the copper. The copper has a minimum of 99.9% copper and nominally 0.05% oxygen; copper can form an alloy system with oxygen. This type of copper should not be heated in an atmosphere where it can combine with hydrogen because the hydrogen can diffuse into the metal and combine with the cuprous oxide to generate steam, this then being able to exert sufficient pressure to cause cracking of the copper.

Fire refined tough pitch high-conductivity copper is produced from impure copper. In the fire refining process, the impure copper is melted in an oxidising atmosphere. The impurities react with the oxygen to give a slag which is removed. The remaining oxygen is partially removed by poles of green hardwood being thrust into the liquid metal, the resulting combustion removing oxygen from the metal. The resulting copper has an electrical conductivity almost as good as that of the electrolytic tough pitch high-conductivity copper. The copper has a minimum of 99.85% copper and nominally 0.05% oxygen.

Oxygen-free high-conductivity copper can be produced if, when cathode copper is melted and cast into billets, there is no oxygen present in the atmosphere, the operation taking place under a reducing atmosphere of carbon monoxide and nitrogen. It has a minimum of 99.95% copper. Another method of producing oxygen-free copper is to add phosphorus during the refining. This gives a decrease of about 20% in the electrical conductivity. Such copper is known as *phosphorus deoxidised copper* and, unlike other forms of copper, it can give good welds.

The addition of about 0.5% arsenic to copper increases its tensile strength, especially at temperatures of about 400°C. It also improves its corrosion resistance but greatly reduces the electrical and thermal conductivities. This type of copper is known as *arsenical copper*. The addition of small amounts of cadmium improves the strength of copper for overhead electrical conductors without significant loss of electrical conductivity, the copper being termed *cadmium copper*.

The mechanical properties of tough pitch, oxygen-free and deoxidised coppers are similar but their electrical and thermal conductivity properties differ significantly. Table 3.29 shows the properties of these coppers. Note that the electrical conductivities have been expressed on the IACS scale; on this scale the value of 100% corresponds to the electrical conductivity of annealed copper at 20°C with its conductivity of 5.800×10^7 S/m.

Table 3.29 *Properties of coppers*

Ref.	Copper	Composition %	Condition	Tensile strength MPa	Percent. elong.	Hardness HB	Electrical conductivity IACS %
C101	Electrolytic tough pitch hc copper	Cu 99.90 min., 0.05 O	Annealed	220	50	45	101.5–100
			Hard	400	4	115	
C102	Fire-refined tough pitch hc copper	Cu 99.90 min., 0.05 O	Annealed	220	50	45	101.5–100
			Hard	400	4	115	
C103	Oxygen-free hc copper	Cu 99.95 min.	Annealed	220	60	45	101.5–100
			Hard	400	6	115	
C104	Fire refined tough-pitch hc copper	Cu 99.85 min., 0.05 O	Annealed	220	50	45	95–89
			Hard	400	4	115	
C105	Tough pitch arsenical copper	Cu 99.20 min., 0.05 O, 0.3–0.5 As	Annealed	220	50	45	95–89
			Hard	440	4	115	
C106	Phosphorus deoxidised copper	Cu 99.85 min., 0.013–0.05 P	Annealed	220	60	45	90–70
			Hard	400	4	115	
C107	Phosphorus deoxidised arsenical copper	Cu 99.20 min., 0.3–0.5 As, 0.013–0.5 P	Annealed	220	60	45	50–35
			Hard	400	4	115	
C108	Cadmium copper	Cu 99.0, 1 Cd	Annealed	280	45	95	75–92
			Hard	700	4	145	
C109	Tellurium copper	Cu 99.5, 0.3–0.8 Te	Soft bar	230	60	50	
			Hard bar	330	270	105	92–98
C110	Oxygen-free hc copper	Cu 99.99 min.	Annealed	220	60	45	101.5–100
			Hard	400	6	115	

Key point

Brasses are copper–zinc alloys with up to about 43% zinc.

Electrolytic tough pitch high-conductivity copper finds use in high-grade electrical applications, e.g. wiring and bus bars. Fire-refined tough pitch high-conductivity copper is used for standard electrical applications. Tough pitch copper is also used for heat exchangers and chemical plant. Oxygen-free high-conductivity copper is used for high-conductivity applications where hydrogen may be present and electronic components. Phosphorus deoxidised copper is used for chemical plant where good weldability is necessary and for plumbing and general pipe work. Arsenical copper is used for general engineering work.

Brasses

The *brasses* are copper–zinc alloys containing up to about 43% zinc. Figure 3.81 shows the relevant part of the phase diagram.

Brasses with between 0 and 35% zinc solidify as alpha solid solutions, usually cored. These brasses have high ductility and can readily be cold worked. *Gilding brass*, 10% zinc, is used for jewellery because it has a colour resembling that of gold and can easily be worked. *Cartridge brass*, 30% zinc and frequently referred to as *70/30 brass*, is used where high ductility is required with relatively high strength. It is called cartridge brass because of its use in the production of cartridge and shell cases. The brasses in 0 to 30% zinc range all have their tensile strengths and hardness increased by working, their ductilities decreasing.

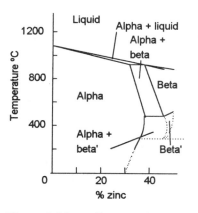

Figure 3.81 *Copper–zinc alloys*

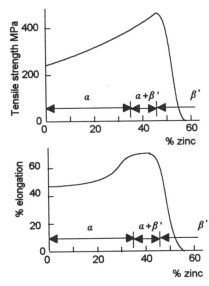

Figure 3.82 *Strength and ductility for copper–zinc alloys*

Brasses with between 35 and 46% zinc solidify as a mixture of two phases. Between about 900°C and 453°C the two phases are alpha and beta, At 453°C the beta phase transforms to a low temperature modification which is referred to as beta' phase. Thus at room temperature the two phases present are alpha and beta'. The presence of the beta' phase produces a drop in ductility but an increase in tensile strength. These brasses are known as *alpha–beta* or *duplex brasses*. They are not generally cold worked but have good properties for hot forming processes, e.g. extrusion. This is because when hot we have the beta phase and this is more ductile than the beta' phase and hence the combination gives, when hot, a very ductile material. Thus the hot working needs to take place at a temperature in excess of 453°C. The name *Muntz metal* is given to a brass with 60% copper–40% zinc. The addition of lead to Muntz metal improves considerably the machining properties, without significantly changing the strength and ductility. *Leaded Muntz metal* has 60% copper, 0.3 to 0.8% lead and the remainder zinc.

Figure 3.82 shows how the tensile strength and ductility of brasses depends on the phases present and hence the percentage of zinc. Table 3.30 gives the properties of commonly used brasses.

Example

On the basis of its microstructure, what properties might be expected from a brass containing 20% zinc?

Such a brass will be entirely alpha phase with a strength about 320 MPa and a percentage elongation about 50%.

Table 3.30 *Properties of brasses*

Ref.	Name	Composition %	Condition	Tensile strength MPa	0.2% pr. stress MPa	Percent. elong.	Electrical conduct. IACS %
CZ101	Gilding metal	90 Cu, 10 Zn	Soft	280	100	48	44
CZ102	Red brass	85 Cu, 15 Zn	Soft	310	130	40	37
CZ106	Cartridge brass	70 Cu, 30 Zn	Soft	330	120	70	27
CZ108	Basis brass	63 Cu, 37 Zn	Soft	340	130	56	23
			Hard	530	180	5	26
CZ109	Muntz metal	60 Cu, 40 Zn	Soft	380	160	40	28
CZ112	Naval brass	62 Cu, 37 Zn, 1 Sn	Soft	370	140	45	26

Key points

Tin bronzes are copper–tin alloys. When phosphorus has been added they are called phosphor bronzes. Casting bronzes containing zinc are called gunmetals.

Tin bronzes

Copper–tin alloys are known as *tin bronzes*. Figure 3.83 shows the phase diagram for such alloys. The structure that normally occurs with up to about 10% tin is predominantly alpha solid solution, this phase being ductile. Higher percentage tins alloys will invariably include a significant amount of the delta phase. This is a brittle intermetallic compound.

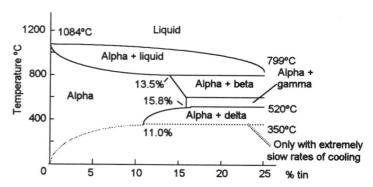

Figure 3.83 *Phase diagram for tin bronzes*

Tin bronzes that contain up to about 9% tin are alpha bronzes at room temperature and can be cold worked. In making such a wrought bronze alloy, oxygen can react with the metal and lead to a weak alloy. Phosphorus is normally added to the liquid metals to act as a deoxidiser. Some of the phosphorus remains in the final alloy; such alloys are termed *phosphor bronzes*. A typical wrought phosphor bronze alloy might have about 95% copper, 5% tin and 0.02 to 0.4% phosphorus. These alloys are used for springs, bellows, electrical contacts, clips and instrument components. Wrought tin bronzes are stronger than the brasses, in particular in the cold-worked condition, and have better corrosion resistance.

Cast phosphor bronze alloys contain between 5 and 13% tin with 0.5% phosphorus. High tin contents over about 10% make the tin unworkable because of the presence of both delta phase and a copper–phosphorus compound. However, such compositions can be cast and are particularly useful for bearing surfaces, having a low coefficient of friction and being able to withstand heavy loads. A typical cast phosphor bronze has about 90% copper, 10% tin and a maximum of 0.5% phosphorus.

Casting bronzes that contain zinc are called *gunmetals*. The zinc reduces the cost of the alloy and also makes unnecessary the use of phosphorus for deoxidisation as this function is performed by the zinc. *Admiralty gunmetal* contains 88% copper, 10% tin and 2% zinc. This alloy finds general use for marine components, hence the term 'Admiralty'. Lead is often added to improve machinability.

Table 3.31 gives the properties of commonly used tin bronzes.

Aluminium bronzes

Copper–aluminium alloys are known as *aluminium bronzes*. Figure 3.84 shows the phase diagram for such alloys. Up to about 9% aluminium gives *alpha bronzes*, such alloys containing just the alpha phase. Alloys with about 7% aluminium can be cold worked easily. When the aluminium content is above 9% and the temperature above 900°C, the beta phase is introduced into the structure and results in alloys termed *duplex alloys*. As the amount of beta phase increases, the tensile strength increases but the ductility decreases. Such alloys are mainly used for casting.

Key point

Copper–aluminium alloys are called aluminium bronzes and have high strength and good resistance to corrosion and wear.

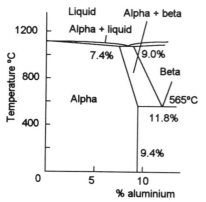

Figure 3.84 *Phase diagram for aluminium bronzes*

Table 3.31 *Properties of tin bronzes*

Ref.	Name	Composition %	Condition	Tensile strength MPa	0.2% pr. stress MPa	Percent. elong.	Electr. conduct. IACS %
Wrought alloys							
PB101	3% phosphor bronze	97 Cu, 3 Sn, 0.02–0.40 P	Soft	320	110	55	15–25
			Hard	580	450	8	
PB102	5% phosphor bronze	95 Cu, 5 Sn, 0.02–0.40 P	Soft	340	160	60	13–18
			Hard	630	500	8	
PB103	7% phosphor bronze	95 Cu, 7 Sn, 0.01–0.4 P	Soft	370	130	65	11–15
			Hard	650	570	14	
Cast gunmetals							
G1	Admiralty gunmetal	88 Cu, 10 Sn, 2 Zn	Sand cast	270–340	130–160	13–25	10–11
			Chill cast	250–310	130–170	3–8	
LG1	Leaded gunmetal	83 Cu, 5 Sn, 5 Zn, 5 Pb	Sand cast	180–220	80–130	11–15	12–16
			Chill cast	180–270	80–130	2–8	
LG2	Leaded gunmetal, eighty-five-three fives	85 Cu, 7 Sn, 3 Zn, 3 Pb	Sand cast	200–270	100–130	13–25	10–15
			Chill cast	200–280	110–140	6–15	

Aluminium bronzes have high strength and good resistance to corrosion and wear because of the thin films of aluminium oxide formed on the surfaces. Typical applications of such alloys are high strength and high corrosion-resistant items in marine and chemical environments, e.g. pump casings, gears and valve parts.

Table 3.32 gives the properties of commonly used aluminium bronzes.

Table 3.32 *Properties of aluminium bronzes*

Ref.	Name	Composition %	Condition	Tensile strength MPa	0.2% pr. stress MPa	Percent. elong.	Electr. conduct. IACS %
Wrought alloys							
CA101	5% aluminium bronze	95 Cu, 5 Al	Soft	370	140	65	15–18
			Hard	650	540	15	
CA102	8% aluminium bronze	92 Cu, 8 Al	Soft	420	90	50	13–15
			Hard	540	230	10	
CA103	9% aluminium bronze	88 Cu, 9 Al, 3 Fe	Soft	570	260	30	7
			Hard	650	340	22	
Cast alloys							
AB1	Aluminium bronze	88 Cu, 9.5 Al, 2.5 Fe	Sand cast	500–590	170–200	40–18	8–12
			Die cast	540–620	200–270		
AB2	Aluminium bronze	80.5 Cu, 9.5 Al, 5 Fe, 5 Ni	Sand cast	640–700	250–300	20–13	6–8
			Die cast	650–700	250–310		

Table 3.33 *Properties of a beryllium bronze*

Ref.	Name	Composition %	Condition	Tensile strength MPa	0.2% pr. stress MPa	Percent. elong.	Electr. conduct. IACS %
CB101	Beryllium bronze, beryllium copper	98 Cu, 1.7 Be, 0.2–0.6 Co + Ni	Sol. treated	480–500	185–190	45–50	16–78
			Sol. treated + cold worked	730–750	580–620	5	
			Sol. treated + precipitation treated	1150–1160	930–940	5	22–32
			Cold worked + precipitation treated	1300–1340	1100–1140	2	

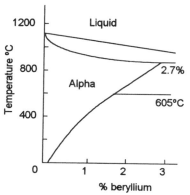

Figure 3.85 *Phase diagram for beryllium bronzes*

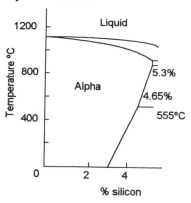

Figure 3.86 *Phase diagram for silicon bronzes*

Beryllium bronzes

Copper alloyed with percentages of beryllium between 0.6 and 2% can be precipitation hardened to give alloys with very high tensile strengths, such alloys being known as *beryllium bronzes* or *beryllium copper*. Figure 3.85 shows the phase diagram. The alloys are relatively expensive and used for high-conductivity, high-strength electrical components, springs, clips and fastenings. Table 3.33 gives the properties of a commonly used beryllium bronze.

Silicon bronzes

Copper–silicon alloys are called *silicon bronzes* and typically contain 1 to 3% silicon, in some cases with small amounts of iron or manganese. Figure 3.86 shows the phase diagram. The alloys can be cast or cold or hot worked and have a high corrosion resistance, good strength and good weldability. Table 3.34 shows the properties of a commonly used silicon bronze.

Cupronickels

Alloys of copper and nickel are known as *cupronickels*, though if zinc is also present they are referred to as *nickel silvers*. Copper and nickel are soluble in each other in both the liquid and solid states and thus form a solid solution whatever the proportions of the two elements. They are thus alpha phase over the entire range and suitable for both cold and hot working over the entire range. The 'silver' coinage in use in Britain is a 75% copper–25% nickel alloy.

Table 3.34 *Properties of a silicon bronze*

Ref.	Name	Composition %	Condition	Tensile strength MPa	0.2% pr. stress MPa	Percent. elong.	Electr. conduct. IACS %
CS101	Silicon bronze, copper silicon	2.7–3.5 Si, 0.7–1.5 Mn, rem. Cu	Soft sheet	370–390	120–140	60	7
			Hard sheet	600–650	450–490	15–12	6

Table 3.35 *Properties of cupronickels and nickel silvers*

Ref.	Name	Composition %	Condition	Tensile strength MPa	0.2% pr. stress MPa	Percent. elong.	Electr. conduct. IACS %
Cupronickels							
CN101	95/5 cupronickel	94 Cu, 5 Ni, 1 Fe, 0.6 Mn	Soft	280	90	40	
			Cold worked	320–380	300–350	14–10	14
CN102	90/10 cupronickel	87.5 Cu, 10 Ni, 1.5 Fe, 1 Mn	Soft	320	120	42	
			Hard	360–420	340–380	30–12	10
CN104	80/20 cupronickel	80 Cu, 20 Ni, 0.5 Mn	Soft	340	120	40	
			Hard	400–550	310–490	32–20	6
CN105	75/25 cupronickel	75 Cu, 25 Ni, 0.5 Mn	Soft	360	140	40	
			Cold rolled	450–590	390–530	15–3	5
CN107	70/30 cupronickel	68 Cu, 30 Ni, 1.5 Mn, 0.5 Fe	Soft	390	150	45	5
Nickel silvers							
NS103	10% nickel silver	63 Cu, 27 Zn, 10 Ni	Soft	350–370	100–140	45–65	
			Hard	470–540	430–460	10–12	8
NS105	15% nickel silver	64 Cu, 21 Zn, 15 Ni	Soft	390–420	130–180	45–55	7
			Hard	490–710	370–600	2–15	

Key points

Copper–nickel alloys are called cupronickels, and nickel silvers if zinc is present.

Activity

Compare the properties of copper alloys by making a grid with rows labelled with the groups of alloying elements used with copper alloys and columns labelled tensile strength, ductility, corrosion resistance and electrical conductivity. Using the tables given in this section, suggest a particular grade of alloy which typifies the alloy performances indicated in the table.

The addition to copper–nickel alloys of 1 to 2% iron increases their corrosion resistance. Nickel silvers have a silvery appearance and find use for such items as knives, forks and spoons. The alloys can be cold worked.

Table 3.35 gives the properties of commonly used cupronickels and nickel silvers.

Corrosion and copper alloys

Pure copper exposed to the atmosphere, e.g. as roofing sheet, acquires a green coloration due mainly to a reaction of the copper with sulphur in the atmosphere. The removal of copper by this means is very slow and indeed the coloration is generally one of the architectural aims.

Copper alloys are susceptible to a number of forms of corrosion. Brasses, particularly with more than about 15% zinc, are susceptible to dezincification and stress corrosion (see Chapter 23 for more details). High zinc brasses, however, are more resistant to erosion than low zinc ones, e.g. Muntz metal is used for tubing if there are high water velocities and the chance of erosion. Bronzes have better resistance to stress corrosion and erosion than brasses. Cupronickels are very corrosion resistant.

Which copper alloy?

There are many copper alloys. The following indicates the type of selection that could be made for specific applications:

- *Electrical power cables, bus bars, domestic wiring*
  High-conductivity copper (C101, C102)

- *Deep drawn items*
  Brass (CZ105, CZ106), cupronickel (CN104), nickel silver (NS104, NS105)

- *Cold-headed rivets, screws, pins, etc.*
  Brass (CZ105, CZ106)

- *Extruded sections*
  Copper (C104, C106), brass (CZ119, CZ123), aluminium bronze (CA103, CA104, CA105), silicon bronze (CS101)

- *Bellows and diaphragms*
  Phosphor bronze (PB101, PB102)

- *Welded pressure vessels*
  Deoxidised copper (C106, C107), silicon bronze (CS101)

- *Hot forgings and stampings*
  Brass (CZ109, CVZ120, CZ122, CZ123)

- *Bearings*
  For hard shafts: phosphor bronze (PB104, PB1, PB2, PB4), gunmetal (G1), leaded brass (CZ124); for soft shafts: phosphor bronze (PB103, PB104), leaded bronze (CZ124), leaded gunmetal (LG2, LG3); for average conditions: phosphor bronze (PB104), leaded bronze (CZ124), leaded gunmetal (LG2, LG3); marine bearings: gunmetal (G1)

- *Gears*
  For light duty: gunmetals (G1, LG2, LG3, LG4), brasses (DCB1, DCB2, DCB3), aluminium bronze (AB1); for moderate duty: phosphor bronze (PB103, PB104), aluminium bronze (CA103, CA104, CA105); for very heavy duty: phosphor bronze (PB2), aluminium bronze (AB2)

- *Heat exchangers and tubes*
  For clean fresh water: copper (C107); general use: copper (C101, C102, C104)

- *Pipes*
  For gases: copper (C106); for sea-water: phosphor bronze (PB1, PB4), gunmetal (G1, LG2, LG4), brass (HTB1, CZ112), cupronickel (CN107), silicon bronze (CS101), aluminium bronze (CA104, CA105); for refrigerants: copper (C106), brass (CZ102); for domestic water services: deoxidised copper (C106)

- *Springs*
  Phosphor bronze (PB102, PB103), nickel silver (NS104, NS106, NS107), brass (CZ106, CZ107, CZ108), beryllium bronze (CB101)

3.3.6 Nickel and its alloys

Key points

Nickel and its alloys have good corrosion resistance with good strength which is maintained at high temperatures.

Nickel has a density of 8.88 Mg/m³ and a melting point of 1455°C. It possesses excellent corrosion resistance, hence it is often used as a cladding on a steel base. This combination allows the corrosion resistance of the nickel to be realised without the high cost involved in using entirely nickel. Nickel has good tensile strength and maintains it at quite elevated temperatures (Figure 3.87). It can be both cold and hot worked, has good machining properties and can be joined by welding, brazing and soldering. Table 3.36 shows the properties of commercially pure nickels.

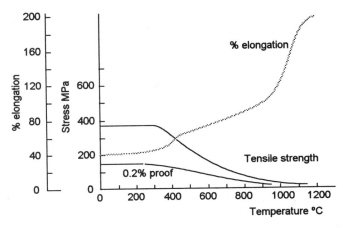

Figure 3.87 *Effect of temperature on the properties of nickel 200 (hot rolled and annealed)*

Because of its corrosion resistance and strength, commercially pure nickel (nickel 200) is used in the food processing industry in chemical plant and in the petroleum industry. Nickel 201 is similar to nickel 200 but has a lower carbon content, this lowering the work hardening rate and increasing the ductility so that it is more readily used for cold-working operations. Nickel is also used in the production of chromium-plated mild steel, the nickel forming an intermediate layer between the steel and the chromium. The nickel is electroplated onto the steel.

Table 3.36 *Properties of commercially pure nickel*

Trade name	Specif- ications	Composition %	Condition	Tensile strength MPa	0.2% pr.stress MPa	Percent. elong.
Nickel 200	NA 11, B160, SB160	Min. 99.0 Ni	Cold rolled and annealed	380	105	30–40
			Hot rolled and annealed	380	130	40
Nickel 201	NA 12, B127, SB127	Min. 99.0 Ni, less than 0.02 C	Cold rolled and annealed	350	85	30–40
			Hot rolled and annealed	350	85	30

Nickel–copper alloys

Nickel and copper are completely soluble in each other. Nickel–copper alloys containing about two-thirds nickel and one-third copper are called *Monels*. Monel 400 has 66.5% nickel and 31.5% copper. It has high strength, toughness and weldability. It is highly resistant to sea water, alkalis, many acids and superheated steam, hence its use for marine fixtures and fasteners, food processing equipment and chemical engineering plant components. Monel K-500 has a basic Monel composition but includes 3.0% aluminium and 0.6% titanium. These enable age hardening precipitates to be formed and so give higher strength, while still maintaining the excellent corrosion properties. All the Monels retain their strength and toughness to temperatures of 500°C. Table 3.37 lists the properties at room temperature. Figure 3.88 shows how the properties vary with temperature for Monel 400 and Monel K-500.

Table 3.37 *Typical properties of Monels at 20°C*

Alloy	Composition %	Condition	Tensile strength MPa	0.2% pr. stress MPa	Percent. elong.
Monel 400, NA 13	Ni 66.5, Cu 31, Fe 1.5, Mn 1.0	Cold worked + annealed	480	170	35
		Cold worked + stress relief	600	400	20
Monel K-500, NA 18	Ni 66.5, Cu 30, Al 2.8, Ti 0.5	Cold worked + solution tr. + precipitation treated	900	620	20
		Hot worked + solution tr. + precipitation treated	900	600	20

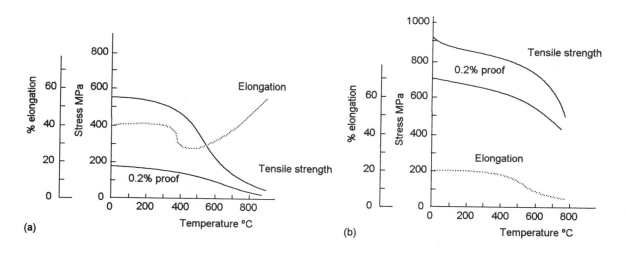

(a)   (b)

Figure 3.88 *Properties of (a) Monel 400 after hot rolling and annealing and (b) Monel K-500 after hot rolling and age hardening*

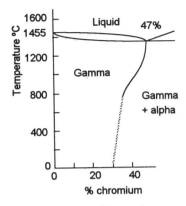

Figure 3.89 *Phase diagram for nickel–chromium alloys*

Nickel–chromium alloys

Another group of nickel alloys are based on nickel combined with chromium. Figure 3.89 shows the phase diagram. Chromium has a solid solubility in nickel of up to about 30% at room temperature. The dissolved chromium forms a layer of chromium oxide on the surface which then protects the alloy from oxidation.

Inconel 600 is a basic engineering alloy with 76.5% nickel, 15.5% chromium and 8% iron, having high strength and excellent resistance to corrosion at both normal and high temperatures. The alloy is not heat treatable but can be work hardened. Figure 3.90 shows how its properties vary with temperature. Table 3.38 gives the properties of the alloy at a number of temperatures.

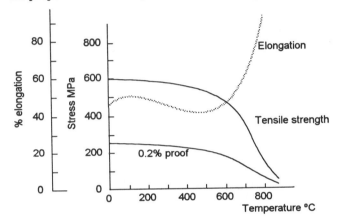

Figure 3.90 *Properties of hot rolled, annealed Inconel 600*

Inconel 601 contains 61.6% nickel, 23% chromium, 14% iron and 1.4% aluminium. The high chromium content gives it good corrosion resistance and high-temperature resistance to oxidisation.

The *Nimonic* series of alloys are based on nickel alloys with about 20% chromium and other elements which give precipitates for precipitation hardening. They have high strengths and good creep resistance at high temperatures and are used in gas turbines for discs and blades. *Nimonic 80* is essentially a 76.5% nickel–20% chromium solid solution with 2.25% titanium and 1% aluminium forming precipitates. The result is an alloy that can be precipitation hardened. Nimonic 80 was developed in 1941 and has since been followed by many other Nimonic and related alloys, both wrought and cast, with yet better properties at high temperatures. Table 3.38 gives the properties at various temperatures for examples of Nimonic alloys.

Another group of heat resisting alloys is based on nickel and iron. They generally contain from 25 to 45% nickel and 15 to 60% iron with chromium added for oxidisation resistance, molybdenum for solid solution strengthening and other metals such as aluminium and titanium for precipitates. Table 3.38 gives the properties for two such alloys, *Incoloy 800* and *Incoloy 801*. Such alloys find use as heat exchanger and furnace parts.

Table 3.38 *Properties of nickel-base heat-resisting alloys*

Alloy	Composition %	Condition	Temp. °C	Tensile strength MPa	0.2% proof stress MPa	Percent. elong.
Wrought alloys						
Inconel 600, NA 14	Ni 76, Cr 15.5, Fe 8	Hot rolled	20	590	250	50
			400	560	185	50
			600	530	150	40
Nimonic 80A, NA 20, HR 1	Ni 73, Cr 19.5, Co 1, Ti 2.25, Al 1.4, Fe 1.5	Hot rolled, solution tr. and precipitation	20	1240	740	24
			400	1150	680	26
			600	1080	620	20
			800	620	490	24
Nimonic 115, HR 4	Ni 55, Cr 15, Co 15, Mo 4, Ti 4, Al 5, Fe 1	Hot rolled, solution tr. and precipitation	20	1230	860	27
			600	1100	790	20
			800	1020	760	19
			1000	420	200	26
Incoloy 800, NA 15	32.5 Ni, 45.7 Fe, 21 Cr, 0.38 Ti, 0.38 Al, 0.05 C	Hot worked	20	450	170	30
			700	300	180	70
Incoloy 825, NA 16	42 Ni, 30 Fe, 21 Cr, 2.3 Cu, 0.09 Ti, 3 Mo	Hot worked and solution treated	20	590	220	30
			700	335	210	49
			1000	65	35	75
Cast alloys						
Nimocast 80, ANC 9	70 Ni, 20 Cr, 0.6 Si, 2.0 Co, 1.2 Al, 2.6 Ti, 0.6 Mn, 0.08 C	As cast	700	970	680	12
			1000	75	40	82
Nimocast 90, ANC 10	57.5 Ni, 19.5 Cr, 0.6 Si, 16.5 Co, 1.3 Al, 2.4 Ti, 0.09 C, 0.6 Mn, 2.0 Fe	As cast	700	540	400	18
			1000	154	77	40
Nimocast PK24, HC204, VMA 12	61 Ni, 9.5 Cr, 15 Co, 3 Mo, 5.5 Al, 4.7 Ti, 1 V	As cast	700	965	825	6
			1000	500	380	5

Note: the above data represents short-term properties, see Chapter 6 for a discussion of creep.

The name *superalloys* is used for alloys that show far greater strength at high temperatures than conventional alloys. Thus the nickel alloys discussed above and the cobalt alloys in the next section are examples of superalloys.

3.3.7 Cobalt alloys

Cobalt has a density of 8.85 Mg/m^3 and a melting point of 1495°C. By alloying cobalt with chromium, tungsten, nickel, carbon and other elements, high temperature alloys have been produced. Strengthening in such alloys is primarily as a result of solid solution strengthening and carbide precipitation. The original development of these alloys arose from the need to develop an alloy which could be used for turbine blades for aircraft during

World War II and that could be cast, so enabling such blades to be produced more easily in quantity. Later developments have led to the production of wrought alloys. Table 3.39 shows the properties of a small selection of these alloys. The data represents short-term properties, see Chapter 6 for a discussion of creep.

Table 3.39 *Properties of cobalt-base heat-resisting alloys*

Alloy	Composition %	Condition	Temp. °C	Tensile strength MPa	0.2% proof stress MPa	Percent. elong.
Wrought alloys						
Haynes 25	50 Co, 20 Cr, 10 Ni, 15 W, 3 Fe, 0.1 C	Solution treated*	20	1010	460	64
			650	710	240	35
			870	325	240	30
Haynes 188	37 Co, 22 Cr, 22 Ni, 14.5 W, 3 Fe, 0.1 C	Solution treated (precipitation occurs in use at temp.)	20	960	485	56
			650	710	305	61
			870	420	260	73
S-816	42 Co, 20 Cr, 20 Ni, 4 W, 4 Mo, 4 Fe, 0.38 C, 4 Nb	Solution treated + precipitation treated	20	965	385	30
			650	765	305	25
			870	360	240	16
Cast alloys						
MAR-M509	54.5 Co, 23.5 Cr, 7 W, 10 Ni, 3.5 Ta, 0.6 C, 0.5 Zr, 0.2 Ti	As cast	20		570	
			760		365	
			980		180	

3.3.8 Case studies

The following are some case studies illustrating the uses made of non-ferrous alloys.

Scaffold tubing

Scaffold tubing needs to be made from a material which is relatively strong, capable of being extruded into tube form, corrosion resistance and not heavy. Aluminium alloys fit this requirement and the alloy 6082 is used. This alloy is widely used for structural purposes. It is an aluminium–magnesium–silicon alloy with between 0.45 and 0.9% magnesium and between 0.7 and 1.3 % silicon and is hot extruded to give tubes. It is heat treated by a solution treatment involving heating and cold water quenching and then aged to give a precipitation hardened material with a 0.2% proof stress of about 250 MPa and a tensile strength of about 300 MPa. It has good weldability, corrosion resistance and immunity to stress-corrosion cracking.

Electrical switch contacts

Electrical switch contacts, e.g. as in Figure 3.91, and sockets into which electronic device pins are push-fits need:

- To give a low electrical contact resistance and maintain it after repeated use and exposure to air and hence oxidation.

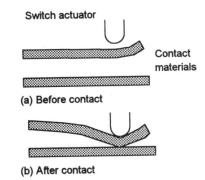

Switch actuator

Contact materials

(a) Before contact

(b) After contact

Figure 3.91 *Electrical contact*

- A material which has reasonably good electrical conductivity and which is elastic with no plastic deformation over the range of stresses likely to be experienced; this is a necessary condition if the contacts are to be pressed together as a result of elasticity and if the contacts are to spring back when the contact force is removed.

The obvious material for good conductivity is copper. However, a relatively pure copper is ductile with a low proof stress so would deform very easily. Table 3.40 shows data for coppers and copper-based alloys. All the metals in the table have an acceptably conductivity for their use as strips. Brasses and the nickel silver have higher proof stresses than coppers but even better are the aluminium and phosphor bronzes. They have sufficiently ductility when soft for forming but work hardening gives a material with good elasticity. All the metals listed in the table are suitable for forming to the required shapes.

The requirement for low contact resistance, with this being maintained over time and after repeated contacts having being made, requires testing of the materials in the conditions they would be used in. A problem with coppers and many copper-based alloys is that they have excellent corrosion resistance but that this results from thin surface layers of oxide being formed on exposure to the atmosphere. It is these layers which prevent further corrosion. However, they do have a relatively high resistance and so will affect contact resistance. The brasses have good corrosion resistance but surface layers result in a significant contact resistance. The nickel silver has reasonable resistance to corrosion but significant contact resistance. The aluminium bronze and phosphor bronze have very good corrosion resistance and maintains a particularly low contact resistance and are thus widely used for switch contacts.

Table 3.40 *Data for coppers and copper-based alloys*

Material	0.2% proof stress MPa	Tensile strength MPa	Percent. elong.	Electrical cond. % IACS
Electrolytic tough-pitch h.c. copper C101	50–320	220–400	50–4	101.5–100
Oxygen-free h.c. copper C103	50–320	220–400	60–6	101.5–100
Red brass CZ102	120–140	310–330	40	32
Common brass CZ108, soft sheet	130	340	56	27
hard sheet	180	530	5	26
Aluminium brass CZ110, soft sheet	125	340	60	23
15% nickel silver NS105, soft sheet	130–180	390–420	55–45	7
hard sheet	370–600	490–710	15–2	7
95/5 Aluminium bronze CA101, soft sheet	370	140	65	15–18
hard sheet	650	540	15	15–18
3% Phosphor bronze PB101, soft strip	110	320	55	15–25
hard strip	450–460	580–590	8	15–25

Coins

The properties required of the materials used for coins are:

- Must be durable with good resistance to brittle fracture, good wear resistance, i.e. hard, and good chemical stability in high humidity, coastal atmospheres, to perspiration, perfumes, etc.

- Have good electrical conductivity but be non-magnetic; this is because of the use of coins in automatic vending machines.

- Be capable of being economically formed into coins and the faces stamped with the required designs.

- Be made of a material which is worth less than the value of the coin so no advantage is gained by melting the coins.

The materials that are used are 'coinage bronze', a copper alloy with 2.5% zinc and 0.5% tin, for copper-coloured coins and a cupronickel with 25% nickel for silver-coloured coins. Coinage bronze is an alpha alloy which is ductile and can be cold worked; it has a tensile strength of about 500 MPa, a hardness of about 170 HV and an electrical conductivity of 15 to 25% IACS. The cupronickel can be cold worked and has a tensile strength of about 700 MPa, a hardness of about 200 HV and an electrical conductivity of 5% IACS. Gold and silver are not used because the materials are too expensive. The coins are made by cold rolling to form sheets the thickness of a coin, punching out coin-side discs, annealing and pickling to soften and clean the discs, forming a raised rim to the discs in order to protect the faces from damage, and then stamping the faces of the coins between two dies to give the required face designs.

Problems 3.3

1 What is the effect on the properties of aluminium of (a) its purity, (b) its temper?

2 For aluminium–magnesium alloys, describe the effect on the strength of the alloy of increasing the percentage of magnesium.

3 Describe the changes in microstructure that occur with the solution treatment and precipitation hardening processes for aluminium–copper alloys.

4 Aluminium beverage cans are recycled. The body of the can is made of the 3004 alloy and the lids of 5182 alloy. What problems might this pose for making use of the recycled material?

5 Describe the features of aluminium–silicon alloys which make them suitable for use with die casting.

6 Explain how the addition of a small amount of sodium to the melt of an aluminium–silicon alloy, e.g. a 12% silicon alloy, changes its properties.

7 Use the phase diagram in Figure 3.76 to determine (a) the temperatures at which solidification of an 85%

aluminium–15% silicon alloy will start to solidify and solidification will be completed, and (b) the temperatures when the alloy is modified by the addition of a small amount of sodium.

8 Magnesium alloys have a high strength-to-weight ratio; of what significance is this in the uses to which alloys of magnesium are put?

9 Though titanium alloys are expensive compared with other non-ferrous alloys, they are used in aircraft such as Concorde. What advantages do such alloys possess which outweighs their cost?

10 Explain the differences in microstructure and properties of alpha, near alpha, alpha–beta and beta titanium alloys.

11 What problems can arise when impurities are present in zinc die casting alloys?

12 What problems can occur if oxygen is present in high-conductivity copper?

13 What is the effect of the percentage of zinc in a copper–zinc alloy on its microstructure and hence strength and ductility?

14 Which phases would you expect to be present in the following brasses (see phase diagram in Figure 11.11) and how would they affect the properties: (a) 10% zinc–90% copper, (b) 20% zinc–80% copper, (c) 40% zinc–60% copper?

15 What brass composition is likely to be most suitable for applications requiring (a) maximum tensile strength, (b) maximum ductility, (c) the best combined tensile strength and ductility?

16 In general, what are the differences in (a) composition, (b) properties of alpha phase and duplex alloys of copper alloys?

17 What are the general properties of phosphor bronze?

18 For copper–nickel alloys: (a) how does the phase diagram differ from that of copper–zinc alloys and (b) over what range of compositions will copper–nickel alloys be alpha solid solutions at room temperature?

19 What are the general characteristics of the nickel–copper alloys known as Monels?

20 Justify the choice of particular alloy given for the following applications:
(a) Zinc alloys for die casting small items.
(b) Magnesium alloys in aircraft structures.
(c) Aluminium for milk bottle caps.
(d) Titanium alloys for high-speed aircraft.
(e) Copper for domestic water pipes.
(f) Cartridge brass for cartridge cases.
(g) Aluminium alloy for ribs of hang gliders.
(h) Nickel alloys for gas-turbine blades.
(i) Brass for cylinder lock keys.
(j) Aluminium or copper alloys for kitchen pans.
(k) Copper for electrical cables.
(l) Cupronickels for tubes in desalination plants.

4 Polymers, ceramics and composites

Summary

This chapter is a consideration of the properties of non-metallic materials, namely polymers, ceramics and composites, that are commonly used by engineers and how they might be used in engineering applications.

Objectives

By the end of this chapter, the reader should be able to:

- explain the behaviour of commonly encountered thermoplastics, thermosets and elastomers;
- explain the properties of clay-based ceramics, engineering ceramics, glasses and cement;
- describe the form of fibre-reinforced materials, particle reinforced materials, dispersion strengthened materials and laminates.

4.1 Polymers

Key point

Compared with metals, polymers have lower densities, a lower stiffness, stretch more, are not so hard, have lower thermal conductivities, have higher specific heat capacities and are electrical insulators.

This section is a discussion of the properties and uses of commonly used thermoplastics, thermosets and elastomers. Compared with metals, polymers have lower densities, a lower stiffness, stretch more and are not so hard. Solid polymers have very low thermal conductivities when compared with metals, and foamed polymers even lower, e.g. polyurethane foam has a thermal conductivity of about 0.4 W m^{-1} K^{-1} compared with values of hundreds for metals. They have higher thermal expansivity than metals, in general about 2 to 10 times greater, e.g. polypropylene has a linear thermal expansivity of about 9×10^{-5} K^{-1} while copper is 1.7×10^{-5} K^{-1}. Solid polymers have specific heat capacities two to five times those of metals, e.g. low density polyethylene has a specific heat capacity of 1.9 kJ kg^{-1} K^{-1} while copper is 0.38 kJ kg^{-1} K^{-1}. Polymers are electrical insulators, having resistivities which are about 10^{20} times greater than those of metals. In general, polymers are resistant to weak acids, weak alkalis and salt solutions but not necessarily resistant to organic solvents, oils and fuels. Polymers are generally affected by exposure to the atmosphere and to sunlight. The effect can show as a slow ageing process with the material becoming more brittle. This is due to bonds being formed between neighbouring molecular chains. Such effects can be reduced by the inclusion of suitable additives with the polymer. Some polymers when stressed and in contact with certain environments can develop brittle cracking, e.g. polyethylene in the presence of detergents. Elastomers can be particularly affected, when stressed, by atmospheric ozone. Stabilisers are usually included to inhibit such cracking.

4.1.1 Thermoplastics

In general, the properties of thermoplastic polymers can be changed by changing the length of individual chains, changing the form of the individual chains, e.g. putting branches on the chain of 'lumpy molecules', changing the strength of bonds within chains and changing the strength of bonds between chains. Linear polymers are quite flexible but molecular rings in the chain and side groups or 'lumpy molecules' on the chain have a stiffening effect. Side groups on chains, molecular rings in the backbone and strong van der Waals forces between chains all increase the melting temperature.

Crystallinity is influenced by the nature of the molecular chains and the ease with which they can be packed together, linear chains with no side groups or irregular 'lumpy molecules' on the chain crystallise most easily. Crystallinity also increases the melt temperature.

Non-crystalline polymers have excellent transparency but, since the crystals in a polymer scatter light, crystallinity reduces transparency (Figure 4.1). Polymer transparency thus ranges from highly transparent to completely opaque to visible light.

Moisture absorption by polymers is largely dependent on the atoms making up the polymer. The presence of oxygen and chlorine atoms gives rise to some absorption while nitrogen considerably increases the absorption. Polyamides, i.e. nylons, contain nitrogen and so have significant water absorbing properties. Moisture absorption increases the volume of a polymer and generally reduces the strength and stiffness. It also causes an increase in the electrical conductivity and dielectric constant. Solvents attack thermoplastics by separating the molecular chains.

Tables 4.1 and 4.2 give a general overview of the main properties of thermoplastics. It needs, however, to be recognised that additives and processing can significantly change the properties. With reference to the tables: State: SC is semi-crystalline, G is glass. Max. temperature is the approximate maximum temperature at which the polymer can be in continuous use; higher temperatures are possible for intermittent use. Impact property: 1 is brittle, 2 is tough when unnotched but brittle when bluntly notched, 3 is tough under all conditions, all referring to about 20°C. For water absorption: L is low, less than 0.1% by weight in 24 h immersion; M is medium, between about 0.1 and 0.4%; H is high, greater than 0.4% and often about 1%. R is resistant, A is attacked with AO being attacked by oxidising acids.

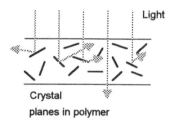

Figure 4.1 *Effect of crystallinity on transparency, the crystal scattering light and so reducing transparency*

Table 4.1 *Properties of thermoplastics commonly used in engineering*

Polymer	State	T_g °C	Max. temp. °C	Tensile strength MPa	Tensile modulus GPa	Percentage elongation	Impact property
Polyethylene							
High density	SC	−120	125	22–38	0.4–1.3	50–800	2
Low density	SC	−90	85	8–16	0.1–0.3	100–600	3
Polypropylene	SC	−10	150	30–40	1.1–1.6	50–600	2
Polyvinyl chloride							
With no plasticiser	G	87	70	52–58	2.4–4.1	2–40	2
With low plasticiser	G		100	28–42		200–250	2
Polystyrene							
No additives	G	100	70	35–60	2.5–4.1	2–40	1
Toughened	G		70	17–24	1.8–3.1	8–50	2
ABS	G	100	70	17–58	1.4–3.1	10–140	2–3
Polycarbonate	G	150	120	55–65	2.1–2.4	60–100	2–3
Acrylic	G	100	100	50–70	2.7–3.5	5–8	1
Polyamides							
Nylon 6	SC	50	110	75	1.1–3.1	60–320	2–3
Nylon 6.6	SC	55	110	80	2.8–3.3	60–300	2–3
Nylon 6.10	SC	50	110	60	1.9–2.1	85–230	2–3
Nylon 11	SC	46	110	50	0.6–1.5	70–300	2–3
Polyethylene terephthalate	SC	69	120	50–70	2.1–4.4	60–100	2
Polyphenylene oxide, modified	G	210	130	55	2.5	50	2–3
Polyacetals							
Copolymer	SC		100	60	2.9	60–75	2
Homopolymer	SC	−76	100	60	3.6	15–75	2
PTFE	SC	−120	260	14–35	0.4	200–600	3
Cellulose acetate	G	120	70	24–65	1.0–2.0	5–55	2
Cellulose acetate butyrate	G	120	70	17–20	1.0–2.0	8–80	2

Note: impact properties 1 = brittle, 2 = tough but brittle when notched, 3 = tough

Table 4.2 *Chemical stability of thermoplastics at 20°C*

Polymer	Water absorption	Acids		Alkalis		Organic solvents
		Weak	Strong	Weak	Strong	
Polythene: high density	L	R	AO	R	R	R
low density	L	R	AO	R	R	R
Polypropylene	L	R	AO	R	R	R
PVC, unplasticised	M	R	R	R	R	A
Polystyrene	L	R	AO	R	R	A
ABS	M	R	AO	R	R	A
Polycarbonate	L	R	A	A	A	A
Acrylic	M	R	AO	R	R	A
Polyamides	H	A	A	R	R	R
Polyester	L	R	A	R	A	A
Polyacetal: copolymer	M	A	A	R	R	R
homopolymer	M	R	A	R	A	R
PTFE	L	R	R	R	R	R
Cellulosics	H	R	A	R	A	A

The following are some of the more commonly used thermoplastics, their properties and uses.

Polyethylene

Polyethylene (PE) is referred to as high density (HDPE) when the molecular chain is linear and as low density (LDPE) when the molecular chain has branches and the chains cannot be so tightly packed and a less crystalline structure is feasible. LDPE has a density of about 918 to 935 kg/m³. It is flexible, tough and has good chemical resistance. Figure 4.2 shows the stress–strain graph and how it is affected by temperature. HDPE is much stronger and stiffer, though not as tough. It also has good chemical resistance. The two forms of polyethylene can be blended to give properties intermediate between the two. Both forms can be extruded, blow moulded, rotationally moulded and injection moulded.

Low-density polyethylene is used mainly in the form of films and sheeting, e.g. polyethylene bags, 'squeeze' bottles, ball-point pen tubing, wire and cable insulation. A major use is for heavy duty sacks for fertilisers and other items. Shrink wrapping is another use for LDPE. High-density polyethylene is used for piping, toys, household ware. It is also used as ultra thin film for the wrapping of meat, fish, etc. in supermarkets and as the carrier bag. Additives commonly used with polyethylene are carbon black as a stabiliser, pigments to give coloured forms, glass fibres to give increased strength and butyl rubber to prevent in-service cracking.

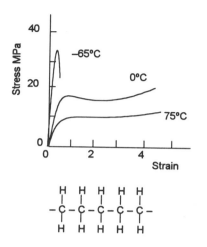

Figure 4.2 *Low-density polyethylene, also see Figures 2.117 and 118*

Polypropylene

Polypropylene (PP) (Figure 4.3) is used mainly in its crystalline form, this being a linear polymer with side-groups regularly arranged along the chain. The presence of the side-groups gives a more rigid and stronger polymer than polyethylene in its linear form. The degree of crystallinity affects the physical properties. In addition, the properties are affected by the lengths of the molecular chain, i.e. the molecular mass. Low molecular mass tends to be associated with a high degree of crystallinity. The greater the crystallinity, the stiffer the polypropylene. An increase in molecular mass increases the impact strength. Polypropylene has good fatigue resistance, chemical resistance and electrical insulation properties. It can be extruded, blow moulded, injection moulded and sheet can be thermoformed. It is used for crates, containers, fans, car fascia panels, cabinets for radios and TV sets, toys and chair shells. Polypropylene film can be made by blow extrusion or by slit-die extrusion with subsequent rapid cooling to avoid the formation of crystalline regions and so give good transparency.

A copolymer of polypropylene with ethylene, *polypropylene-ethylene copolymer (EPM)*, has properties determined by the proportion of ethylene present and how it is introduced into the chain structure. The effect of quite small percentages of ethylene, of the order of a few percent, giving block copolymers and considerably increasing the impact strength. Random copolymers

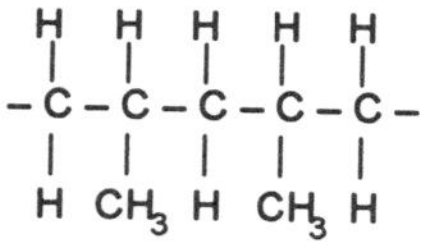

Figure 4.3 *Basic polypropylene, also see Figure 2.124*

of ethylene and propylene with higher percentages of ethylene produce an elastomer. Such an elastomer has good resistance to atmospheric degradation.

Polyvinyl chloride

Polyvinyl chloride (PVC) is a linear-chain polymer with bulky chlorine side-groups (Figure 4.4) which prevent crystalline regions occurring. The polymer is mixed with a variety of additives to give a range of plastics. PVC is a hard and rigid material but plasticiser can be added to give flexible forms. It is extruded to give sheet, film, pipe and cable covering, and calendered into sheet. It can be injection moulded, blow moulded, rotationally moulded and thermoformed.

The rigid form of PVC, i.e. the unplasticised form, is used for piping for waste and soil drainage systems, rainwater pipes, lighting fittings and curtain rails. Plasticised PVC is used for the fabric of 'plastic' raincoats, bottles, shoe soles, garden hose piping, gaskets, sachets for shampoos, and inflatable toys.

A copolymer of vinyl chloride with vinyl acetate in a mass ratio of about 85 to 15 gives a rigid material, while a ratio of 95 to 5 gives a flexible material. The acetate grouping is bulkier than the chlorine atom and so serves to prevent close contact between the resulting polymer chains. The addition of plasticiser can also be used to modify the properties. It has properties similar to those of PVC but is easier to mould and thermoform. Like PVC it is non-crystalline. The non-plasticised copolymer is used for gramophone records, while heavily plasticised copolymer is calendered to produce floor tiles. The material in this form has good abrasion and impact resistance.

Ethylene-vinyl acetate (EVA)

Ethylene-vinyl acetate copolymer (EVA) is a linear chain polymer with short side branches and can show crystallinity. The properties depend on the relative proportions of the two constituents. Increasing the vinyl acetate component increases the flexibility, large amounts producing a polymer with properties more like those of an elastomer than a thermoplastic. EVA copolymers are flexible, resilient, tough and have good resistance to atmospheric degradation. They can be extruded, injection moulded, blow moulded and rotationally moulded.

EVA is used for road-marker cones, ice-cube trays, medical and surgical ware and as a major constituent in hot-melt adhesives.

Polystyrene

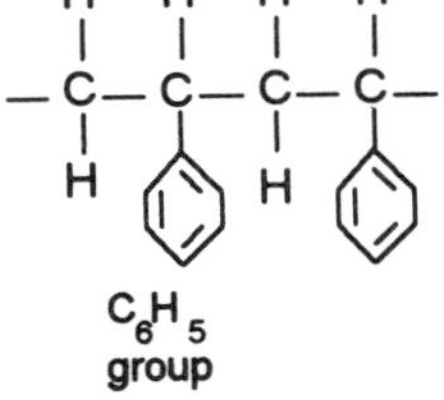

Polystyrene (PS) is a linear-chain polymer with bulky side-groups (Figure 4.5) which prevent crystalline regions occurring. Polystyrene is available in many forms. General purpose polystyrene is a rather brittle, transparent material with a smooth surface finish that can be printed on. It is used for injection moulding containers for cosmetics, light fittings, boxes and ball-point pen barrels.

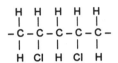

Figure 4.4 *Polyvinyl chloride, also see Figure 2.125*

Copolymer

Figure 4.5 *Polystyrene*

Toughened or high-impact polystyrene is a blend of polystyrene with rubber particles. This blending improves the impact resistance but results in a decrease in tensile modulus, tensile strength and transparency. Injection moulding or thermoforming from extruded sheet is used to produce cups for vending machines and casings for cameras, projectors, radios, television sets and vacuum cleaners.

A widely used form of polystyrene is a foamed polymer known as expanded polystyrene. This is a rigid form which is used for insulation and packaging.

Polystyrenes are attacked by many solvents, e.g. petrol, dry cleaning agents, greases, oxidising acids and some oils. Exposure to detergents can lead to stress-cracking.

Acrylonitrile-butadiene-styrene terpolymer

Styrene-acrylonitrile copolymer (SAN) is a brittle, glass, copolymer of styrene and acrylonitrile. It can, however, be toughened with polybutadiene. Some of the polybutadiene forms a graft terpolymer with the styrene acrylonitrile while some produces small rubber spheres which become dispersed through the terpolymer and SAN matrix. Figure 4.6 shows the stress–strain graphs for SAN and the acrylonitrile-butadiene-styrene terpolymer (ABS). ABS is an amorphous material which is tough, stiff and abrasion resistant. It can be injection moulded, extruded, rotationally moulded and thermoformed.

ABS is widely used as the casing for telephones, vacuum cleaners, hair dryers, radios, television sets, typewriters, luggage, boat shells and food containers.

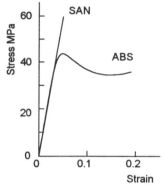

Figure 4.6 *SAN and ABS*

Polycarbonate

Polycarbonate (PC) is a linear chain polymer (Figure 4.7) which at room temperature is well below its glass transition temperature and hence amorphous. The form of the chain makes for a very stiff structure. Polycarbonate is tough, stiff, strong and transparent and retains its properties well with increasing temperature, e.g. at 125°C polycarbonate film still gives a tensile strength of about 30 MPa. It also has reasonable outdoor weathering resistance and good electrical insulation properties. It may be injection moulded, blow moulded, extruded and vacuum formed.

Polycarbonate is used for applications where resistance to impact abuse and relatively high temperatures can be encountered. Typical applications are transparent street lamp covers, infant-feeding bottles, machine housings, safety helmets, housing for car lights and tableware such as cups and saucers.

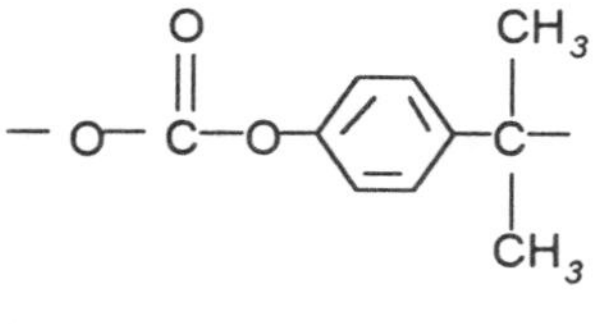

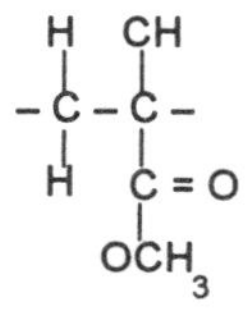

Figure 4.7 *Polycarbonate*

Acrylics

Acrylics are completely transparent thermoplastics, mostly based on *polymethyl methacrylate (PMMA)* which has linear chains with bulky side-groups (Figure 4.8) and so gives an amorphous structure. They give a stiff, strong material with outstanding

Figure 4.8 *Polymethyl methacrylate*

weather resistance. They can be cast, extruded, injection moulded and thermoformed.

Because of its transparency and weather resistance, acrylics are used for light fittings, canopies, lenses for car lights, signs and nameplates. Opaque acrylic sheet is used for domestic baths, shower cabinets, basins and lavatory cisterns.

Polyamides

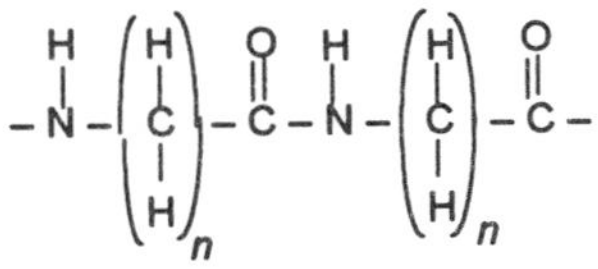

Polyamides (PA) are commonly known as nylons and are linear polymers (Figure 4.9) which give crystalline structures. There are a number of common polyamides: nylon 6, nylon 6.6, nylon 6.10 and nylon 11. Where there are two numbers, the full stops separating them are sometimes omitted, e.g. nylon 66 is nylon 6.6. The numbers refer to the number of carbon atoms in each of the reacting substances used to give the polymer. The first number, in fact, indicates the number of carbon atoms in the resulting polymer chain before the chain repeats itself. The two most used nylons are nylon 6 and nylon 6.6. Nylon 6.6 has a higher melting point than nylon 6 and is also stronger and stiffer. Nylon 11 has a lower melting point and is more flexible.

In general, nylons are strong, tough materials with relatively high melting points. But they do tend to absorb moisture. The effect of this is to reduce tensile strength and stiffness. It, however, increases toughness. Nylon 6.6 can absorb quite large amounts of moisture; nylon 11, however, absorbs considerably less. Nylons often contain additives, e.g. a stabiliser or flame retardent. Glass spheres or fibres are added to give improved strength and stiffness. Molybdenum disulphide is an additive to nylon 6 to give a material with very low frictional properties. Nylons are usually injection moulded, though they can be extruded and extrusion blow moulded.

Nylons are used for the manufacture of fibres for clothing, gears, bearings, bushes, housings for domestic and power tools, electric plugs and sockets.

Polyesters

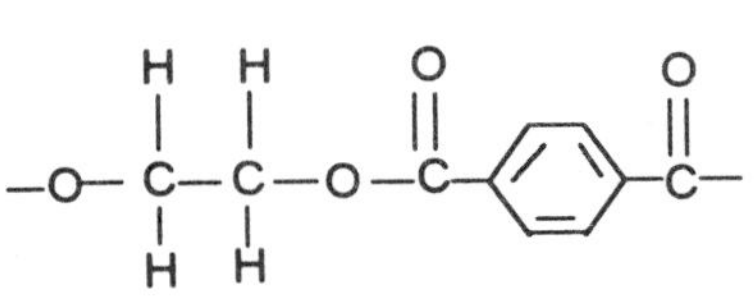

Polyesters are available in both thermoset and thermoplastic form. The main thermoplastic form is *polyethylene terephthalate (PET)*. This is a linear chain polymer with a C_6H_5 group in the backbone and side groups (Figure 4.10). It gives crystalline structures and is below its glass transition temperature at room temperature. If it is rapidly quenched from a melt to below its glass transition temperature, an amorphous structure is produced, the molecular chains not having sufficient time to become orderly packed. The polyester has properties similar to nylon and is usually injection moulded.

It is widely used in fibre form for the production of clothes. Other uses are for fizzy drink bottles, electrical plugs and sockets, push-button switches, wire insulation, recording tapes, insulating tapes and gaskets.

Figure 4.9 *Nylon, also see Figure 2.126*

Figure 4.10 *PET*

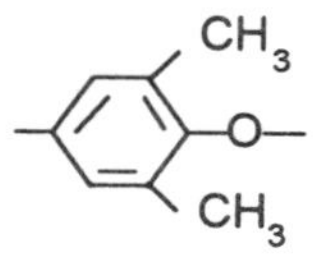

Figure 4.11 *PPO*

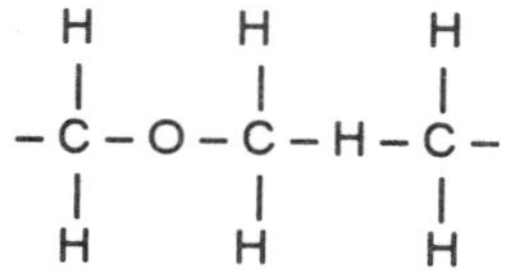

(a)

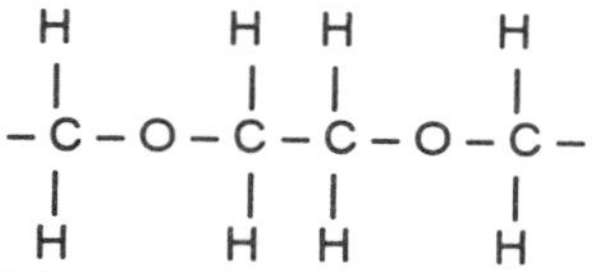

(b)

Figure 4.12 *Acetal: (a) homopolymer, (b) copolymer*

Polyphenylene oxide

Polyphenylene oxide (PPO) has a linear structure with a C_6H_5 ring and oxygen atoms in the backbone (Figure 4.7). The rings in the backbone make for a stiff molecular chain and so a material which is highly rigid, strong and has good dimensional stability. It is completely miscible with polystyrene and such modified PPO is widely used for electrical fittings, dishwasher and washing machine components, car fascia panels and VDU cabinets.

Polyacetals

Polyacetals are linear-chain polymers with a backbone which consists of alternate carbon and oxygen atoms. One of the main forms is *polyoxymethylene (POM)*, which is sometimes just referred to as *acetal homopolymer* (Figure 4.12(a)). A copolymer has a modified backbone structure, the chain consisting of occasional ethylene units (Figure 4.12(b)). The homopolymer is slightly stronger and stiffer than the copolymer but the copolymer has the advantage of better long-term strength at high temperatures. In general, acetals are strong, stiff and have good impact resistance. They have low coefficients of friction and good abrasion resistance. Glass-filled acetal is used where even higher stiffness is required. Processing is by injection moulding, extrusion or extrusion blow moulding.

Typical applications are pipe fittings, parts for water pumps and washing machines, car instrument housings, bearings, gears, hinges and window catches, and seat belt buckles.

PTFE

Polytetrafluoroethylene (PTFE) is a linear polymer like polyethylene, the only difference being that instead of hydrogen atoms there are fluorine atoms. It has a very high crystallinity as manufactured, about 90%, though this degree of crystallinity can be reduced during processing to about 50% if quench cooled, or 75% if just slowly cooled. PTFE is tough and flexible and can be used over a wide range of temperatures, 250°C down to almost absolute zero, and still retain the very important property of not being attacked by any reagent or solvent. It also has a very low coefficient of friction.

PTFE is a relatively expensive material and is not processed as easily as other thermoplastics. It tends to be used where its special properties of resistance to chemical attack and low coefficient of friction are needed. Journal bearings with a PTFE surface can be used without lubrication because of the low coefficient of friction; PTFE can even be used at temperatures up to about 250°C. Piping carrying corrosive chemicals up to 250°C are made of PTFE. Other applications are gaskets, diaphragms, valves, O-rings, bellows, couplings, dry and self-lubricating bearings, coatings for frying pans and other cooking utensils (known as 'non-stick'), coverings for rollers handling sticky materials, linings for hoppers and chutes, and electrical insulating tape.

Activity

Select a thermoplastic, using the data given in Tables 4.1 and 4.2, for each of the following sets of requirements:
(a) transparent with high rigidity and good water resistance,
(b) high strength, high toughness and good water resistance.

Key point

Thermosetting polymers are stronger and stiffer than thermoplastics and generally can be used at higher temperatures.

Thermosetting polymers are three-dimensional covalently bonded networks whose rigidity increases as the level of cross-linking increases.

Cellulosics

The most common cellulosic materials are *cellulose acetate (CA)*, *cellulose acetate butyrate (CAB)* and *cellulose acetate propionate (CAP)*. Cellulose acetate is hard, stiff and tough, but has poor dimensional stability due to a high absorption of water. CAB is tougher and more resistant to water uptake and hence more dimensionally stable. CAP is slightly harder, stiffer and stronger. All three can be extruded and injection moulded.

CA is widely used for spectacle frames, having the advantage that it can be softened by a slight increase in temperature and hence adapted to facilitate adjustment to fit the wearer. Other applications are tool handles, keyboard keys and toys. CAB is used for internally illuminated roadside signs, extruded piping, pens and containers. CAP is used for toothbrush handles, pens, knobs, steering wheels, toys and film for blister packing. In its simplest form a blister pack consists of a plastic sheet which is contoured to fit the shape of an object and fixed to a card backing. Blisters are made by thermoforming techniques.

4.1.2 Thermosets

Thermosetting polymers are stronger and stiffer than thermoplastics and generally can be used at higher temperatures. As they cannot be shaped after the initial reaction in which the polymer chains are formed and linked, the processes by which thermosetting polymers can be shaped are limited to those where the product is directly formed to the required shape from the raw polymer materials. No further process is possible other than possibly some machining, and this limits the processes available to just moulding. A number of different moulding methods are used but essentially all involve the combination together of the chemicals in a mould so that the cross-linked chains are produced while the material is in the mould.

In general, thermosets have high thermal stability, high dimensional stability, high stiffness, good resistance to creep, low densities and high electrical and thermal insulating properties. The properties and uses of commonly used thermosets are discussed below; an overview of the properties is given in Table 4.3.

Phenolics

Phenolics give highly cross-linked polymers. *Phenol formaldehyde* was the first synthetic polymer and is known as *Bakelite*. The polymer is opaque and initially light in colour. It does, however, darken with time and so is always mixed with dark pigments to give dark-coloured materials. It is supplied in the form of a moulding powder which includes the resin, fillers and other additives such as pigments. When this powder is heated in a mould the cross-linked polymer chain is produced.

Table 4.3 *Properties of thermosets*

Polymer	Density Mg/m³	Tensile strength MPa	Tensile modulus GPa	Percentage elongation	Max. service temp. °C
Phenol formaldehyde					
Unfilled	1.25–1.30	35–55	5.2–7.0	1.0–1.5	120
Wood flour filler	1.32–1.45	40–55	5.5–8.0	0.5–1	150
Asbestos filler	1.60–1.85	30–55	9.0–11.5	0.1–0.2	180
Urea formaldehyde					
Cellulose filler	1.5–1.6	50–80	7.0–13.5	0.5–1	80
Wood flour filler	1.5–1.6	40–55	7.0–10	0.5–1	
Melamine formaldehyde					
Cellulose filler	1.5–1.6	55–85	7.0–10.5	0.5–1	95
Glass filled	2.0	35–138	12–17	0.5	
Epoxy resin					
Cast	1.15	60–100	3.2		
60% glass fabric	1.8	200–240	21–25		200
Polyester					
Unfilled	1.3	55	2.4		200
30% glass fibre	1.5	120	7.7	3	
Polyurethane					
Foam	0.016–0.1	55			150

The fillers account for some 50 to 80% of the total weight of the moulding powder. Wood flour, a very fine soft wood sawdust, when used as a filler increases the impact strength of the plastic; asbestos fibres improve the heat properties and mica the electrical resistance.

Phenol formaldehyde mouldings are used for electrical plugs and sockets, switches, door knobs and handles, camera bodies and ashtrays. Composite materials involving the phenolic resin being used with paper or open-weave fabric, e.g. a glass fibre fabric, are used for gears, bearings and electrical insulator parts.

Amino-formaldehydes

Amino-formaldehyde materials, generally *urea formaldehyde* and *melamine formaldehyde*, give highly cross-linked polymers. Both are widely used as moulding powders, like the phenolics. Cellulose and wood flour are widely used as fillers. Hard, rigid, high-strength materials are produced, with the melamine being harder and having better heat and stain resistance than the urea.

Both materials are used for tableware, e.g. cups and saucers, knobs, handles, light fittings and toys. Composites with open-weave fabrics are used as building panels and for electrical components.

Epoxides

Epoxy resins form thermosetting materials by being combined with a hardener which enables the cross-links to be established between the epoxy molecules and so give a thermoset material.

They are generally used in conjunction with glass, or other, fibres to give hard and strong composite materials. Epoxy resins are also excellent adhesives, giving very high adhesive strengths.

Epoxy resins, because of their hardness and chemical resistance, are used as protective coatings for cans, drums and wires. Because of their high electrical resistivity, dielectric strength, and ability to retain these properties in wet conditions, the resins are used for encapsulation of transistors and other electrical components. Composites with glass fibre fabrics are used for boat hulls and table tops.

Polyesters

Polyesters can be used to produce either thermosets of thermoplastics. If an unsaturated polyester, i.e. one which has molecules containing double carbon–carbon covalent bonds, is combined with suitable compounds, e.g. a styrene, the double bond breaks to allow cross-linkages to be established by means of the styrene to other polyester molecules and so give a cross-linked polymer. The thermoset form is mainly used with glass, or other, fibres to form hard, strong composites. Such composites are used for boat hulls, architectural panels, car bodies, panels in aircraft, and stackable chairs. They have a maximum service temperature of the order of 200°C.

Polyurethanes

Polyurethanes can have properties ranging from elastomers to thermosets, depending on the degree of cross-linking produced between the molecular chains. As thermosets they are widely used to produce rigid foams. The foams have advantages over expanded polystyrene in having lower density, lower thermal conductivity and better oil, grease and heat resistance. The rigid foam can be formed in situ, e.g. in wall cavities for thermal insulation. Main uses are in refrigerators, structural sandwich panels in buildings and marine buoyancy applications.

4.1.3 Elastomers

Elastomers are polymers which show very large strains when subject to stress and which will return to their original dimensions when the stress is removed. They are essentially amorphous polymers, having a glass transition temperature well below their service temperature. The polymer structure is that of linear-chain molecules with some cross-linking between chains. One way of classifying elastomers is in terms of the form of the polymer chains:

- ***Only carbon in the backbone of the polymer chain***
  E.g. natural rubber, butadiene-styrene, butadiene-acrylonitrile, butyl rubbers, polychloroprene, ethylene-propylene

Key point

Elastomers are polymers which show very large strains when subject to stress and which will return to their original dimensions when the stress is removed.

Key point

Thermoplastic elastomers, as the name indicates, can be repeatedly heated and reformed whereas other elastomers, once cured, cannot be so reformed.

- ***Polymer chains with non-carbon atoms in the backbone***
  (a) Oxygen, e.g. polypropylene oxide, (b) Silicon, e.g. fluorosilicone, (c) Sulphur, e.g. polysulphide
- ***Thermoplastic elastomers***
  These are block copolymers with alternating hard and soft blocks, e.g. polyurethane, styrene-butadiene-styrene

The following sections discuss the properties and uses of a range of elastomers, Table 4.4 giving an overview of their properties.

Table 4.4 *Properties of elastomers*

Elastomer	Tensile strength MPa	Percentage elongation	T_g °C	Service temps. °C	Resistance to oil and greases	Resilience
Natural rubber	20	800	–73	–50 to +80	Poor	Good
Butadiene-styrene	24	600	–58	–50 to +80	Poor	Good
Butadiene-acrilonitrile	28	700	–55	–50 to +100	Excellent	Fair
Butyl	20	900	–79	–50 to +100	Poor	Fair
Polychloroprene	25	1000	–48	–50 to +100	Good	Good
Ethyl-propylene	20	300	–75	–50 to +100	Poor	Good
Polypropylene oxide	14	300		–20 to +170	Poor	
Fluorosilicone	8	300	–123	–100 to +200	Good	Fair
Polysulphide	9	500	–50	–50 to + 80	Good	Fair
Polyurethane	40	650	–60	–55 to + 80	Poor	Poor
Styrene-butadiene-styrene	14	700		–60 to +80	Poor	

Note: resilience is the capacity of a material to absorb energy in the elastic range, being measured from the area under the elastic portion of the stress–strain graph. To illustrate this property; a high resilience rubber ball will bounce to a much higher height than a low resilience one.

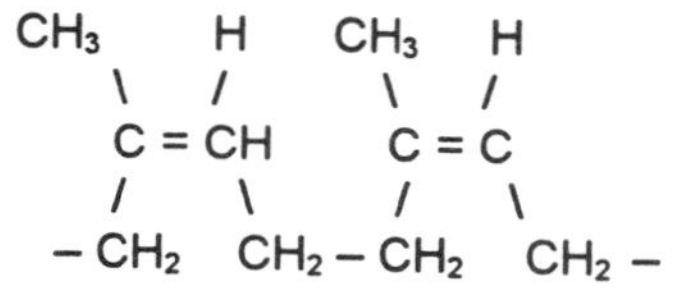

Figure 4.13 *Part of the natural rubber chain*

Natural rubber

Natural rubber (NR) is, in its crude form, just the sap from a particular tree, the Hevea braziliensis. It consists of very long chain molecules, some 100 000 carbon atoms long, and is mainly polyisoprene (Figure 4.13). This is processed with sulphur to produce cross-links between chains, the amount of linkage being determined by the amount of sulphur added. This is called *vulcanisation*. A soft rubber is likely to have about 3% sulphur. In addition, antioxidants, plasticisers and reinforcing fillers are added. Temperature has a marked effect on the properties of natural rubber; at low temperatures the rubber can lose its elasticity and become brittle and at high temperatures it can lose all its stiffness. Natural rubber is inferior to synthetic rubbers in oil and solvent resistance, oils causing it to swell and deteriorate.

Natural rubber is used for balloons and rubber bands and as a high percentage of the materials used for tyre treads, conveyor belts and seats. Rubber foams are manufactured from natural rubber and are used as carpet backing and car seats.

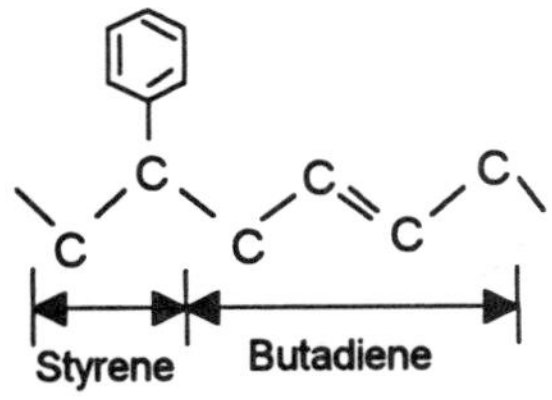

12.4.2 Butadiene-styrene

Butadiene-styrene rubbers (SBR) are copolymers produced from butadiene and styrene (Figure 4.14). Like natural rubber, the molecular chains are cross-linked by vulcanisation with sulphur and reinforced with carbon black. SBR is cheaper than natural rubber and is widely used as a synthetic substitute for it. SBR has good low-temperature properties, good wear and weather resistance and good tensile properties. It has, however, poor resistance to fuels and oils and poor fatigue resistance.

SBR is used in the manufacture of tyres, hose pipes, conveyor belts, footwear and cable insulation. It has poorer resilience than natural rubber and so has been unable to replace it for larger tyres where such properties are essential.

Butadiene acrylonitrile

Butadiene acrylonitrile rubbers (NBR) are commonly referred to as *nitrile rubbers*. Like SBR they are copolymers (Figure 4.15) produced from butadiene with, this time, acrylonitrile. The acrylonitrile component is varied between about 20 to 50% by weight of copolymer. Such a rubber has excellent resistance to fuels and oils, even at comparatively high temperatures. The greater the acrylonitrile content the greater the resistance to fuels and oils. NBR is vulcanised by either sulphur or peroxides and reinforced by carbon black.

NBR is a high-cost rubber and thus tends to be used where its excellent resistance to fuels and oils is a vital requirement. It thus finds uses as hoses, gaskets, seals, tank linings and rollers.

Butyl rubbers

Butyl rubbers (BUTYL) are copolymers of isobutylene and isoprene (Figure 4.16) with vulcanisation by sulphur and reinforcement by carbon black. The rubber has the important property of extreme impermeability to gases, with low resilience. Hence it is used for inner linings of tubeless tyres, steam hoses and diaphragms.

Polychloroprene

Polychloroprene (CR) is commonly known as *neoprene*. It has a structure (Figure 4.17) similar to isoprene, i.e. natural rubber. Zinc and magnesium oxides are used to give vulcanisation, the cross-links probably being with oxygen atoms. As a result of the presence of the chlorine atoms, it has good resistance to oils and a variety of other chemicals, and also good weathering characteristics. It is widely used in the chemical industry, because of these properties, for oil and petrol hoses, gaskets, seals, diaphragms and chemical tank linings.

Ethylene-propylene

Ethylene-propylene rubber (EPM) is a random copolymer formed from ethylene and propylene. When cross-linked it gives a rubber

Figure 4.14   *SBR chain*

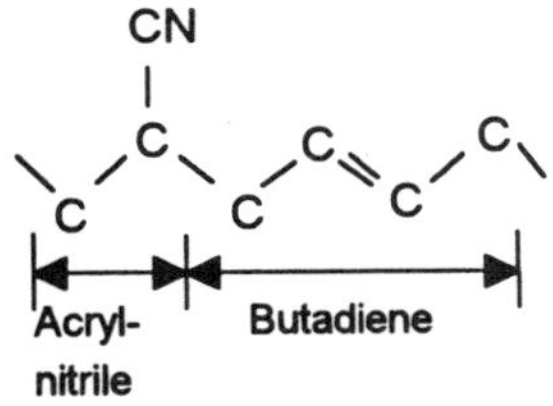

Figure 4.15   *NBR chain*

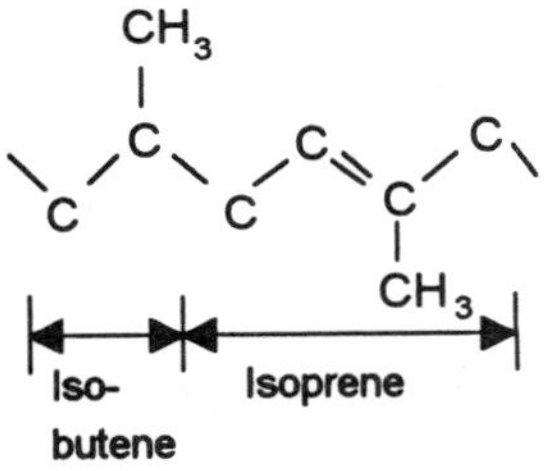

Figure 4.16   *Part of BUTYL chain*

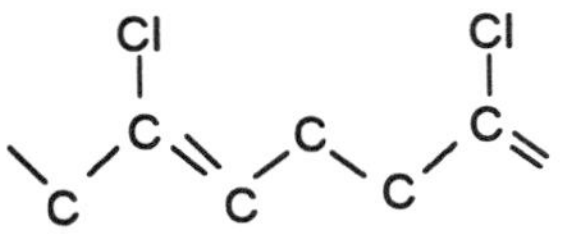

Figure 4.17   *CR chain*

which is very highly resistant to heat and attack by oxygen and ozone. This high resistance and its good electrical insulation properties, gives it an application as electrical cable covering.

A terpolymer of propylene, ethylene and a diene, after vulcanisation by sulphur, gives a rubber (EDPM) with very high resistance to heat and attack by oxygen and ozone. It is widely used in the car industry.

Polypropylene oxide

Polypropylene oxide rubber (PPO) has polymer chains containing oxygen. Such rubbers have good low-temperature properties and good weathering properties but are not resistant to acids and hydrocarbons. The mechanical properties and resilience are good. The main uses are for electrical insulation and moulded mechanical parts.

Fluorosilicones

Fluorosilicones (FVMQ) have polymer chains containing silicon instead of carbon. Because of this they have, for elastomers, exceptional high-temperature properties, being usable up to temperatures of the order of 200°C. They have good resistance to oils, fuels and organic solvents, good electrical insulation properties and good weathering properties. Because such elastomers are expensive compared with others, uses are restricted to parts where resistance to oils and solvents is required at high temperatures, e.g. O-rings, seals, gaskets and hoses.

Polysulphides

Polysulphides have polymer chains with backbones containing both oxygen and sulphur atoms. Without fillers the strength of the rubber is low. It has poor abrasion resistance, swells in hot water, has low resistance to high temperatures and will burn rapidly. The rubber also has a strong odour. It has good resistance to oils and solvents and low permeability to gases; it can, however, be attacked by micro-organisms.

It has a limited number of uses, e.g. lining of oil and paint hoses, printing rolls and a coating for fabrics. In liquid form it is used as a sealant in building work and for paints and adhesives.

Polyurethanes

Polyurethane rubbers (PUR) are thermoplastic elastomers. Such elastomers have the characteristic form of block polymers with hard and soft blocks alternating down the polymer chain (Figure 4.18). With polyurethane rubbers, the hard blocks are polyurethane and the soft blocks polyether or polyester. The hard blocks are glass polymers and act as physical links in forming the network links between chains (Figure 4.19). The effectiveness of the links diminishes rapidly above the glass transition temperature of the polyurethane. The hard blocks also act as a fine particle reinforcing filler for the material.

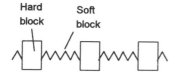

Hard Soft
block block

Figure 4.18 *Thermoplastic rubber chain*

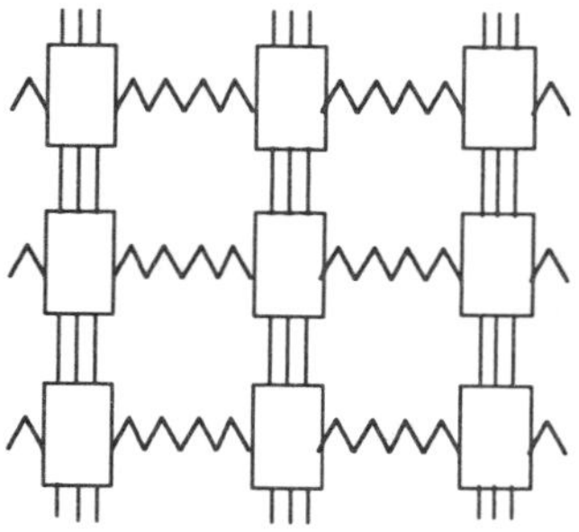

Figure 4.19 *Thermoplastic rubber*

Polyurethane elastomers are widely used for flexible foams, having properties similar to those of natural rubbers but with better resistance to ageing. Hence they are widely used as cushioning or for packaging. Polyurethane rubbers have higher tensile strengths than other rubbers.

Styrene-butadiene-styrene

Styrene-butadiene-styrene (SBS) is another example of a thermoplastic elastomer. It is a block copolymer with a soft block, polybutadiene, terminated at each end by a hard block, polystyrene. The polystyrene blocks form the links between polymer chains. The properties of the material are controlled by the ratio of styrene to butadiene, the percentage of styrene being in the range 15 to 30%. The properties are similar to those of natural rubber, with reasonable abrasion resistance and weathering properties, but poor resilience and flame resistance. The main uses are in adhesives, carpet backing and as an additive to other polymers in the production of plastics.

4.1.4 Case studies

The following case studies illustrate how the uses that can be made of polymers depend on their properties.

Underground pipes

Consider the requirements for pipes to carry water, gas or sewage and be buried in the ground. The pipes must resist degradation from within the pipe and external attack, must be capable of being joined, must be flexible enough to go round corners and be fed easily into trenches in the ground. Polymers are widely used for such pipes. Medium density polyethylene (MDPE) and unplasticised PVC are the main thermoplastic materials used. Thermoset pipes manufactured from glass fibre-reinforced polyester are also used. MDPE pipes may be thermofusion welded or coupled with compression joints; PVC and thermoset pipes are generally coupled by compression joints.

Plastic kettles

Many modern kettles are made from a plastic. The properties that have to be met by the plastic are that they must not absorb water, must not contaminate water, must not soften below 100°C, must remain stiff when the inner wall of the kettle is at 100°C (the mean temperature of the kettle wall is likely to be about 80°C), must be capable of being formed into the required shape and must be cheap. The requirement that water should not be absorbed eliminates nylons. Most thermosetting materials can contaminate water and so are eliminated. The requirement that the material should not soften below 100°C eliminates all the amorphous polymers which have a glass transition temperature below 100°C

Activity

Using the data in Table 4.4, suggest an elastomer for each of the following requirements:
(a) an oil seal with good resilience up to 80°C,
(b) a dry seal working at −30°C.

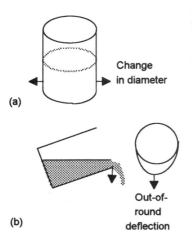

(a)

(b)

Figure 4.20   *Loading of a kettle*

and also crystalline polymers with a melting temperature below 100°C.

The stiffness of a kettle body will depend on the elastic modulus of the material used and the shape of the kettle. Consider the problem of the kettle bellying out, i.e. changing its diameter, as a result of the weight of the water (Figure 4.20(a)). By considering the kettle as a thin-walled cylinder we can calculate that if the required stiffness is to be where the diameter does not change by more than 1 mm with a wall thickness of 5 mm, a diameter of 120 mm, and a water height of 200 mm then a tensile modulus which does not drop below about 0.002 GPa is required and so is not likely to present a problem. A greater problem occurs, however, when water is being poured from the kettle and we have its weight being taken by one side of a diameter (Figure 4.20(b)). If the out-of-round deflection is to be kept to below 1 mm with a wall thickness of 5 mm, then the tensile modulus has not to fall below about 0.4 GPa. The polymer widely used for kettles is polypropylene containing an additive of about 10% talc. This gives a tensile modulus at room temperature of about 2 GPa which drops to about 0.5 GPa by 80°C.

Problems 4.1

1 How do the properties of high- and low-density polyethylene differ?
2 How do the properties of polystyrene with no additives and toughened polystyrene differ?
3 How does the structure of polypropylene chains differ from those of polyethylene and how does this affect the properties?
4 Which is stiffer at room temperature, PVC with no plasticiser or polypropylene?
5 The casing of a telephone is made of ABS. How would you expect the casing to behave if someone left a burning match or cigarette against it?
6 What are the special properties of PTFE which render it useful despite its high price and processing problems?
7 Figure 4.21 shows the effect on the Charpy impact strength for nylon 6 of the percentage of water absorbed. As the percentage of water absorbed increases, is the material becoming more or less tough?
8 How do the mechanical properties of thermosets differ, in general, from those of thermoplastics?
9 What is Bakelite and what are its mechanical properties?
10 Describe how cups made of melamine formaldehyde with a cellulose filler might be expected to behave in service.
11 Describe the basic structure of a thermoplastic elastomer.
12 Suggest polymers for the following applications: (a) an ashtray, (b) a garden hose pipe, (c) a steam hose, (d) insulation for electric wires, (e) the transparent top of an electrical meter, (f) a plastic raincoat, (g) a toothbrush

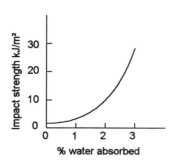

Figure 4.21   *Problem 7*

handle, (h) a camera body, (i) a road marker cone, (j) picnic cups and saucers, (k) a door knob, (l) a conveyor belt, (m) a fuel hose, (n) cushioning.

13 For each of the following applications, suggestions are given for polymers. Select the most appropriate one in each case: (a) the casing for a computer: polyethylene, polypropylene, ABS, PVC, (b) a self-lubricating gear that is not exposed to moisture: polyethylene, ABS, nylon 6.6, polycarbonate, (c) an electrical insulator at room temperature: polyethylene, PVC, ABS, PET, (d) a gasket exposed to fuel and oils: natural rubber, SBR, NBR.

4.2 Ceramics

The term *ceramics* covers a wide range of materials, e.g. brick, concrete, clay, industrial ceramics such as oxides, carbides and nitrides, refractory materials and glasses. This section considers the properties and uses of such materials in engineering.

Ceramics are usually hard and brittle, good electrical and thermal insulators, have good resistance to chemical attack and tend to have low thermal shock resistance, because of their low thermal conductivity and low thermal expansivity. They can be grouped as:

- *Domestic ceramics*
  These include porcelain, china, earthenware, stoneware and cement.

- *Natural ceramics*
  Stone is a natural ceramic.

- *Engineering ceramics*
  These include oxides, nitrides, carbides, borides and silicates. Such materials are widely used in engineering as furnace components, combustion tubes, tool tips and grinding tools.

- *Glasses and glass ceramics*
  Ceramics are generally crystalline, though amorphous states are possible. If silica in the molten state is cooled very slowly it crystallises at the freezing point. However, if the molten silica is cooled more rapidly it is unable to get all its atoms into the orderly arrangements required of a crystal and the resulting solid is a disorderly arrangement which is called a *glass*. *Glass ceramics* are fine-grained polycrystalline materials produced by a controlled crystallisation of glasses

- *Electronic materials*
  These include ferrites, ferroelectrics and semiconductors.

The properties of a ceramic are determined by its composition and its microstructure produced as a result of its fabrication. Unlike metallic materials, the microstructure cannot in general be changed by working or further treatment. Most ceramics start as powders or mixtures of powders which are shaped and then subject to the required temperature in order to consolidate them.

Layers and bonding

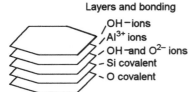

Figure 4.22 *Kaolinite*

Initially separate particles

Particles pack closer together

Bridges develop between
the particles

Coherent solid formed

Figure 4.23 *Stages in the sintering process*

4.2.1 Clay-based products

The term *clay-based products* covers a very wide-ranging array of items, including porcelains, earthenware and stoneware. Clay consists mainly of the mineral kaolinite, this being hydrated aluminium silicate $Al_2O_3.2SiO_2.2H_2O$, with small amounts of other oxides. The clays are the element that enables the material to be formed to the required shape. The plastic property of clay occurs because of the crystal structure of kaolinite, the unit cell being a hexagonal assembly of thin ionic and covalently bonded layers (Figure 4.22). The surfaces of the crystal can hydrogen bond to water. Thus with water added to the clay, the plates of the crystal can be made to slide easily over each other. When the clay is fired to a temperature of about 800°C, the crystal structure is destroyed and sintering occurs. This process results in particles coalescing as a result of diffusion occurring between particles (Figure 4.23). The result is what we term earthenware or low-quality bricks.

Porcelains are well-vitrified ceramics based on a mixture of clays, feldspar and a filler. The term *vitrification* is used to describe the tendency of some of the components to melt to a viscous liquid, this cooling to a glassy phase. The term well vitrified is used when the process of melting has resulted in a dense impervious material being obtained. Feldspar has the basic composition $K_2O.Al_2O_3.6SiO_2$ and is termed a flux since it makes a glass when the materials are fired, flows between the particles and binds the materials together. The filler is often a siliceous material such as ground sand or flint.

Most porcelains used in engineering for electrical insulator components, e.g. the insulators that support electric power cables, are made from typically about 40% clay, 30% feldspar and 20% quartz. This composition is chosen so that the resulting material does not shrink excessively during drying, distort excessively as it shrinks after firing and does not require too high a firing temperature. The resulting microstructure after vitrification is, in general terms, quartz grains in a matrix composed of the glassy phase produced by the feldspar and interpenetrated by a mass of mullite ($3Al_2O_3.2SiO_2$) crystals.

Another form of clay-based product called *steatites* is made from clay, feldspar and talc, this being magnesium silicates in a plate-like structure. Such a product mainly finds a use in electrical components. A low-voltage steatite might have 80 to 90% talc, 5 to 10% clay and 5 to 10% feldspar. A steatite for use at higher voltages might have the feldspar replaced by barium or calcium carbonate. An important property is their low dielectric losses.

Materials for use for refractory purposes where low-voltage electrical insulation is also required are made by combining clay with quartz and alumina, with in some cases added talc. The result is a relatively open, porous structure, the porosity assisting in improving resistance to thermal shock. Such materials find use as heating elements for electric fires, car exhaust catalyst supports, thermocouple sheaths.

Table 4.5 gives data on the properties of a number of ceramics.

Table 4.5 *Properties of clay-based ceramics*

Material	Density Mg/m^3	Coeff. of expansion 10^{-6} K^{-1} at 25°C	Elastic modulus GPa	Short-term strength MPa	Max. use temp. °C	Resist-ivity Ω m	Dielectric constant at 50 Hz, 25°C	Loss tangent 10^{-4} at 25°C	Dielectric strength kV/mm
Porcelain, siliceous	2.2–2.3	3–6	70	50–100	1100	10^{11}	6–7	250	20
Steatite, low voltage	2.3–2.5	6–9	60	<50	1000	10^{10}	6	250	
Steatite, normal	2.6–2.65	6–9	80	50–100	1000	10^{11}	6–6.2	50	15
Refractory, alumina	1.9–2.4	1.5–3.5	Depends on density	<50	1500	10^7			

4.2.2 Engineering ceramics

Unlike the clay-based products which are based on processed minerals, engineering ceramics are essentially pure compounds, being generally oxides, nitrides, carbides, borides or silicides.

Oxide ceramics

Materials based on alumina, Al_2O_3, are very widely used. Moderate amounts of other compounds, e.g. magnesium oxide, silicon oxide and calcium oxide, are added to form a glassy phase. The aluminium oxide with, for example, magnesium oxide is compacted and then sintered to produce the alumina ceramic in a glassy matrix. The amount of this glassy phase has a pronounced effect on the properties.

A range of ceramics are produced with alumina content varying between almost 100% and about 80%. Ceramics with more than 99.7% alumina are used for refractory purposes. The addition of 0.1% magnesium oxide results in a ceramic with a finer grain size and hence stronger product. Ceramics with 96.5% to 99.0% alumina and added magnesium oxide are widely used for electrical and engineering applications and will often have other compounds added to control the electrical properties. Table 13.2 shows the properties of this type of ceramic.

Table 4.6 *Properties of alumina-based ceramics*

Material	Density Mg/m^3	Coeff. of expansion 10^{-6} K^{-1} at 25°C	Elastic modulus GPa	Short-term strength MPa	Max. use temp. °C	Resist-ivity Ω m	Dielectric constant at 50 Hz at 25°C	Loss tangent 10^{-4} at 25°C	Dielectric strength kV/mm
Alumina 99%	3.7	5.9	380	*	1600	10^{12}	9	5	17
Alumina 95%	3.5	5.6–5.9	320	*	1400	10^{12}	9	5	15
Alumina 90%	3.4	5.0–6.0	260	*	1200	10^{12}	9	5	15

Note: * depends on grain size, 200–400 MPa for fine grains, 100–200 MPa medium to coarse grains

For high-temperature applications, there needs to be a minimum amount of the glassy phase and a large grain size, i.e. alumina typically more than 99%. For high strength and wear resistance, fine grain size is required. Such ceramics are used for spark plug insulators, fuse bodies, electronic packaging substrates, chute linings for ore and abrasive materials handling, shot blast nozzles, for processing equipment and insulators for lamp elements. Spark plug insulators use 95% alumina and combine good insulation properties with the ability to withstand the rough treatment they encounter, thermal shocks experienced each time the engine fires with temperatures rising to as high as 900°C and pressures to 100 atmospheres.

Silicon carbide ceramics

Silicon carbide is a compound of silicon with carbon and is a hard material and because of a protective oxide layer is stable in air up to about 1650°C. It is widely used as an abrasive, as a refractory, for furnace heating elements and for voltage-sensitive resistors (termed *varistors*).

Silicon carbide is covalently bonded and exists in a number of forms with various stacking arrangements of the planes of atoms. Beta silicon carbide is a cubic structure and is produced by low-temperature vapour phase reactions. This form changes irreversibly to alpha silicon carbide at temperatures of about 2000°C. Pure cubic beta silicon carbide is a semiconductor with a band gap of approximately 2.2 eV.

There are a number of methods used to manufacture silicon carbide products:

- **Reaction bonding**
  This is sometimes termed *self-bonding*. A mixture of silicon carbide powder and powdered carbon is formed to the required shape. It is then brought into contact with molten silicon which then permeates the pore space in the compacted powders, reacts with the carbon to produce further silicon carbide which then bonds the original silicon carbide grains together.

- **Sintering**
  High-temperature firing, without the application of external pressure, of silicon carbide powder with a suitable additive can be used, the additive being to improve the otherwise very slow sintering that occurs with just silicon carbide. Boron and aluminium compounds are commonly used additives.

- **Hot pressing**
  Hot-pressed products are produced by adding a small amount of an oxide, usually about 2% alumina, to the silicon carbide. The alumina is more readily deformed and helps to bind the carbide particles together.

Table 4.7 *Properties of silicon carbide*

Density Mg/m³	Coeff. of expansion 10^{-6} K^{-1}	Elastic modulus GPa	Fracture toughness MN m$^{-3/2}$	Hardness HV0.1	Short-term strength MPa	Max. use temp. °C
3.0–3.15	2.8	440	4	750*	200–400	2000

Note: * the other values given in the table are virtually independent of the method of producing the silicon carbide; however, the hardness value given this is for reaction-bonded silicon carbide, hot-pressed silicon carbide has a higher value. The hardness scale is the Vickers test with a 100 g load.

Silicon carbide is a very hard material and very wear resistant. These properties enable it to be used for such applications as an abrasive powder and for precision bearings. Table 4.7 shows the mechanical and thermal properties of silicon carbide.

Nitride ceramics

Silicon nitride Si_3N_4 is a widely used engineering ceramic, having reasonable strength and fracture toughness and can be used to high temperatures. It has the advantage of a low coefficient of expansion and hence good resistance to thermal shock. The methods used for the production of silicon nitride products are primarily:

- **Reaction bonding**
  Reaction-bonded silicon nitride (RBSN) products are produced by heating the compacted silicon powder in a nitrogen atmosphere. Since the gas has to penetrate to the centre of the compressed silicon, a relatively porous structure has to be maintained.

- **Hot pressing**
  Hot-pressed silicon nitride (HPSN) does not have the porous structure of the reaction-bonded product and so is much more dense. It is produced by hot pressing, at about 1700 to 1800°C, silicon nitride with sintering additives such as magnesium oxide. This oxide is liquid at this temperature and aids in the bringing together of the silicon nitride particles. The presence of the oxide imposes a limit on the temperature to which the silicon nitride can be used.

- **Sintering**
  Sintered silicon nitride (SSN) is produced by heating silicon nitride with a suitable additive, the additive being to improve the otherwise very slow sintering that occurs with just silicon nitride. Magnesium oxide, yttrium oxide and aluminium oxide are commonly used additives. High-density products are formed in this way.

The name *sialon* is used for ceramics formed when silicon nitride, alumina, silica and aluminium nitride are combined, the name deriving from the components Si, Al, O and N. The product can be hot pressed or sintered. In hot pressing, the components react together in the presence of a liquid phase produced by the silica that is present.

The hard-wearing properties and ability to be used at relatively high temperatures give applications for silicon nitride and sialon ceramics as precision bearings, stamping dies, and for cutting tool tips. Table 4.8 shows the properties of silicon nitride products produced by the above methods.

Boride ceramics

Boron carbide is a very hard ceramic material and is used in highly abrasive conditions such as shot-blast nozzles. Boron carbide products are generally made by hot pressing. Table 13.5 shows the properties of such a material. It oxidises in air above about 1000°C.

Boron nitride has a crystal structure rather like that of graphite, i.e. plate like (see Figure 2.43). As a consequence it is soft and machinable, unlike the other ceramics. Boron nitride is usually made by hot pressing boron nitride powder. The resulting product has a structure with all the plates of the boron nitride lined up in planes perpendicular to the direction of pressing and thus the properties of the product differ in different directions. Table 4.9 shows the properties of boron nitride products produced by hot pressing in directions parallel to and at right angles to the direction of pressing.

Some products, e.g. thin walled shapes, can be made by chemical vapour deposition, the vapour being deposited on a suitably shaped carbon substrate which is subsequently removed.

> **Activity**
>
> Select a ceramic that will give minimal dimensional change (a) as the temperature changes between 0°C and 100°C and (b) under load at fixed temperature.

Table 4.8 *Properties of silicon nitride*

Form of silicon nitride	% porosity	Density Mg/m^3	Coeff. of expansion 10^{-6} K^{-1}	Elastic modulus GPa	Fracture toughness MN m$^{-3/2}$	Hardness HV0.1	Short-term strength MPa	Max. use temp. °C
Reaction bonded	15–40	1.9–2.8	1.5	120–250	2	750	300	1200
Hot pressed	0	3.1–3.2	1.5	310	6	1700	900	1600
Sialon	0	3.3	1.6	300	6.5	1840*	800	1200

Note: HV0.1 is the Vickers test with 100 g load; * is the Knoop test HK0.05 with 50 g load.

Table 4.9 *Properties of boron ceramics*

Boron ceramic	Density Mg/m^3	Coeff. of expansion 10^{-6} K^{-1}	Elastic modulus GPa	Fracture toughness MN m$^{-3/2}$	Hardness HV0.1	Short-term strength MPa	Max. use temp. °C
Carbide, hot pressed	2.3–2.5	3.3	450	2.5–5.0	3200	>400	1000
Nitride, hot pressed, //	1.9–2.1	0–2.5	100		Very soft		1500
⊥		0–1	20				
Nitride, vapour deposition, //	2.1		150		Very soft		1500
⊥		−2					

Note: HV0.1 is the Vickers test with a 100 g load; // is parallel to axis, ⊥ is perpendicular to it.

4.2.3 Glasses

The basic ingredient of most glasses is sand, i.e. silica. Ordinary window glass, termed *soda lime glass*, is made from a mixture of sand, limestone (calcium carbonate) and soda ash (sodium carbonate). Heat resistant glasses, such as Pyrex, are made by replacing the soda ash by boric oxide, such glasses being termed *borosilicate glasses*.

Many of the mechanical properties of glasses are almost independent of their chemical composition. They tend to have a tensile modulus of about 70 GPa. The tensile strength in practice is markedly affected by microscopic defects and surface scratches and for design purposes a value of about 50 MPa is generally used. Glasses have low ductility, being brittle. They have low thermal expansivity and low thermal conductivity and thus have low resistance to thermal shock; think of what happens when hot water is poured into a cold glass, there is a strong chance that the glass will crack. Glasses, in general, are electrical insulators with resistivities of the order of 10^{14} Ω m or higher. They are resistant to many acids, solvents and other chemicals. The maximum service temperature tends to about 500 to 1300°C, depending on the composition of the glass. Table 4.10 gives the properties of a selection of glasses.

Key points

Glasses tend to have a tensile modulus of about 70 GPa, a tensile strength of about 50 MPa, low ductility, low thermal expansivity, low thermal conductivity, low resistance to thermal shock, are generally electrical insulators, and are resistant to many acids, solvents and other chemicals. The maximum service temperature tends to be about 500 to 1300°C, depending on the composition of the glass.

Table 4.10 *Properties of glasses*

Glass	Composition %	Density Mg/m³	Coeff. of expansion 10^{-6} K^{-1}	Elastic modulus GPa	Fracture toughness MN m$^{-3/2}$	Hardness HK0.1	Max. use temp. °C
Soda lime glass	70–75 SiO$_2$, 0.5–3 Al$_2$O$_3$, 12–17 Na$_2$O, 4.5–12 CaO, 0–3 MgO	2.5	9.2	70	<1	460	460
Borosilicate glass	60–80 SiO$_2$, 1–4 Al$_2$O$_3$, 10–12 B$_2$O$_3$	2.2	3.2	67	<1	420	490

Note: the hardness scale HK0.1 is the Knoop test with a 100 g load.

Improving the crack-resistance of glass

Glass is brittle and as a consequence tends to shatter if subject to a sudden impact load. Car windscreens have this problem when stones thrown up by the wheels of cars strike the windscreen with a high velocity. The need is to stop the glass completely shattering and all the jagged glass hitting the driver and passengers in the car.

This can be achieved by using *laminated glass*, two panes of glass being used with a layer of a polymer, polyvinyl-buterate (PVB), sandwiched between them. For the bullet-proof glass, multiple layers of glass and polymer, generally polycarbonate, are used. When the stone travelling at more than about 15 m/s hits a sheet of glass, a cone-shaped cracked area is likely to be produced. This will extend from the hit surface through the thickness of the glass. The polymer interlayer stops it from extending through to the second sheet of glass and so prevents the windscreen completely shattering on impact. However, the blow can cause the screen to flex and if severe enough result in the rear glass of the laminate breaking.

Toughened glass might be used for the side windows of cars. With surface cracks, tensile stresses act in such a way as to open them while compressive stresses close them. Thus compressive stresses result in cracks not propagating while tensile stresses cause cracks to propagate. To reduce the chance of glass fracturing, treatments are applied that build in compressive stresses in the surfaces of the glass. Thus when a tensile stress is applied to the glass, it has to exceed the built-in compressive stress before the surface is in tension and a crack can propagate. As a result, glass with built-in surface compressive stresses can withstand the application of a larger tensile stress without fracturing and is termed *toughened glass*.

There are two types of method used for producing surface internal compressive stresses. *Thermal toughening* involves heating a sheet of glass to a uniform temperature close to its softening point. The surfaces are then cooled uniformly and rapidly by cold air jets. The thermal contraction of the surfaces occurs freely because the interior of the glass is still hot and soft. The result is cooler, hardened, surface layers with a hotter, softer core. When the core begins to cool, it can only contract by pulling against the already hardened surface layers. This results in the surface layers being put in compression and the core in tension (Figure 4.24). *Chemical toughening* involves some of the ions in the surface layers of the glass being replaced by larger diameter ions, e.g. sodium ions with a diameter of 0.196 nm might be replaced by potassium ions with a diameter of 0.266 nm. The larger ions cause the surface layers to expand. The expansion is, however, resisted by the untreated core with the result that the surface is in compression and the core in tension.

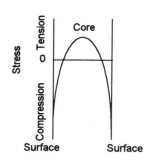

Figure 4.24 *Internal stresses with toughened glass*

Glass ceramics

Where glass is to be used for its transparency it is necessary for the glass not to crystallise during its processing, the surfaces of the crystal grains scattering light and so reducing the light transmitted. *Glass ceramics*, however, are so designed that crystallisation occurs to give a fine-grained polycrystalline material. Such a material has considerably higher strength than most glasses and retains its strength to much higher temperatures. It also has excellent resistance to thermal shock.

Glass ceramics are produced by using a raw material containing a large number of nuclei on which crystal growth can start, e.g. small amounts of oxides such as those of titanium, phosphorus or zirconium. The material is heated to form a glass, e.g. 1650°C, to the required product shape and then cooled. The product might be in the form of sheets which are then edge finished and surface decoration applied before being heated to a high enough temperature, e.g. 900°C, to give controlled grain growth until the required grain size is obtained. Figure 4.25 illustrates this

Key point

Glass ceramics are glasses where crystallisation has been allowed to occur and give a fine-grained polycrystalline material.

sequence for the production of cooker tops. Most forms of glass ceramic are based on $Li_2O–Al_2O_3–SiO_2$ and $MgO–Al_2O_3–SiO_2$. Such a ceramic has a very low coefficient of expansion. Table 4.11 gives the properties. Glass ceramics are used for cooker tops, cooking ware and telescope mirrors.

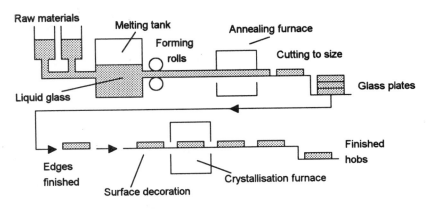

Figure 4.25 *The stages in the production of a ceramic glass cooker top*

Table 4.11 *Properties of glass ceramics*

Glass ceramic	Density Mg/m^3	Coeff. of expansion $10^{-6} K^{-1}$	Elastic modulus GPa	Fracture toughness $MN\ m^{-3/2}$	Hardness HK0.1	Max. use temp. °C
Pyroceram 9606 (Corning)	2.6	6.5	120	1–2.5	660	900
Pyroceram 9608 (Corning)	2.6	0.7–2.0	88	1–2.5	590	700
EE1087 (GEC Ceramics)	2.4	10–11.5	92	1–2.5		

Note: HK0.1 is the Knoop test with a 100 g load. Trade names are given for the glass ceramics.

4.2.4 Cement

Portland cement is a mixture of about 75% limestone, $CaCO_3$, and 25% clay, mainly aluminium silicate with iron and alkali oxides. These are ground together and heated in a kiln at a temperature of between 1400 and 1500°C. A series of reactions occur until the mixture coalesces together into lumps of clinker about 5 to 10 mm in size. After cooling this clinker is mixed with 3 to 5% by weight of gypsum, $CaSO_4$, and ground to give the cement powder. The function of the gypsum is to control the initial setting rate. Apart from the gypsum, the cement contains tricalcium silicate ($CaO.Ca_2SiO_4 = 3CaO.SiO_2$, termed alite), dicalcium silicate ($Ca_2SiO_4 = 2CaO.SiO_2$, termed belite), tricalcium aluminate ($2CaO.Ca(AlO_2)_2 = 3CaO. Al_2O$, termed aluminate), tetracalcium aluminferrite ($4CaO.Al_2O_3.Fe_2O$, termed ferrite) and alkali oxides (K_2O, Na_2O, CaO).

When water is added to cement it turns into a paste which is workable and pourable. A series of reactions occur. In the first stage a considerable amount of heat is evolved by the reactions as the cement sets to a rigid material. This takes a few hours. After

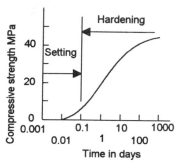

Figure 4.26   *Hydration of cement*

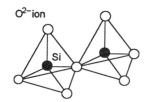

Figure 4.27   *Polymerisation of silicate ions*

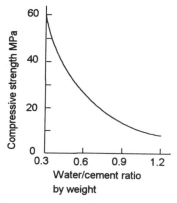

Figure 4.28   *Effect of water/cement ratio on strength*

Key point

Concrete is a composite structure with the cement being used to bind together particles of sand and aggregate.

setting the next stage has the cement hardening and its compressive strength increasing. This takes many days. Figure 4.26 shows how the compressive strength varies with time.

In general terms the reaction that occurs when the water is added is that initially there is a high rate of heat evolution when the alite and aluminate hydrate. This is accompanied by silicate polymerisation in the grains of cement, i.e. the joining together of silica tetrahedra to form chains (Figure 4.27). The grains then become coated with gels, a gel being a swollen polymer network of high viscosity. It is this gel coating which gives hydrated cement its paste-like consistency. Fine needle-like crystals then begin to form and interlock. The cement is then set. During hardening, growth and interlocking of these needle-like crystals continue, along with the formation of plate-like hexagonal crystals known as portlandite. The resulting structure of hydrated and hardened cement is thus an interlocked mass of crystals and amorphous fibres consisting of partially dried-out silicate gel in the fibres and portlandite in the crystals.

The cement contains pores between fibres, between fibres and crystals and larger spaces where the gel did not grow as a result of air pockets or trapped water. A totally hydrated cement requires a water to cement ratio of about 40%. Below that value the cement will contain unreacted grains while above it there will be excess water. The excess water will lead to greater porosity as a result of becoming trapped within the hardened cement. Figure 4.28 shows how the compressive strength of the hardened cement depends on the water/cement ratio after a hardening time of about 28 days.

Concrete is a composite structure with the cement being used to bind together particles of sand and aggregate. Composites are discussed in more detail in Section 4.3. The properties of the concrete depend on the relative proportions of water, cement, sand and aggregate in the material, the average size of the aggregate particles, the type of aggregate used and its surface texture. Typically, concrete with a Portland cement/water ratio of 0.5 and using crushed aggregate will have a compressive strength after 28 days of about 50 MPa, its tensile strength being about one-tenth of the compressive strength. The modulus of elasticity is typically about 30 GPa.

4.2.5 Thermal shock

If you pour hot water into a cold glass there is every chance that the glass will shatter. This is because glass is a poor conductor of heat and one side of the glass is at a much higher temperature than the other and its desire to expand is restrained by the colder glass. The term *thermal shock* is used and we are thus concerned with the thermal shock resistance of such materials.

Maths in action

When the temperature changes, an unconstrained material will expand or contract. If, however, it is constrained so that its natural expansion or contraction is prevented, then stresses are set up in the material. Suppose we have an unconstrained rod of material of length L, then when the temperature rises from T_0 to T_1 its length will increase by $aL(T_1 - T_0)$, where a is the coefficient of linear expansion. If the rod is constrained (Figure 4.29) then this expansion cannot occur and it is as if the rod has been squashed by this amount. Thus we have a compressive strain of $a(T_1 - T_0)$ produced. Assuming Hooke's law, the compressive stress σ is:

$$\sigma = aE(T_1 - T_0) \qquad [1]$$

In considering thermal shock we are often concerned with material which has a hot object put in contact with one surface, e.g. hot water poured into a cold glass. Suppose the surface layer immediately in contact with the hot object instantly attains the higher temperature while there is not sufficient time for any heat to flow to the underlying layers and increase their temperature (Figure 4.30). The expansion of the surface layer in a direction parallel to the surface and also in a direction at right angles to the surface is thus constrained by the colder underlying layers. The result is a biaxial compressive stress σ:

$$\sigma = \frac{aE(T_1 - T_0)}{1 - v} \qquad [2]$$

where v is Poisson's ratio. The *resistance to thermal shock R* can be defined as the temperature change required to give fracture. Thus:

$$R = \frac{\sigma_f(1 - v)}{aE} \qquad [3]$$

where σ_f is the stress required to give failure.

Equation [3] only applies when we consider the underlying layers to have an unchanged temperature because no heat has flowed through from the hotter surface. Because of thermal conduction some heat will flow and so the underlying layers will not all be at T_0, the situation being like that shown in Figure 4.31. The rate at which heat flows through to the underlying layers is proportional to the thermal conductivity λ of the material. Thus we modify the definition of thermal shock resistance in equation [3] to take account of this and use a resistance R_1 as a better measure of shock resistance:

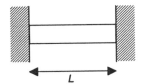

Figure 4.29 *Constrained rod*

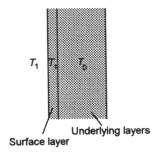

Figure 4.30 *Underlying layers not heated*

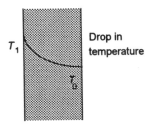

Figure 4.31 *Some heat flows to underlying layers*

$$R_1 = \frac{\sigma_f(1-v)\lambda}{\alpha E} \qquad [4]$$

Under conditions where the surface temperature changes at a constant rate, the surface stress is found to depend on the thermal diffusivity D. Thermal diffusivity is a measure of how fast heat is transmitted through a solid and is given by:

$$D = \frac{\lambda}{\rho c} \qquad [5]$$

where ρ is the density of the material and c the specific heat. Under such a situation a resistance R_2 is used as a measure of thermal resistance:

$$R_2 = \frac{\sigma_f(1-v)D}{\alpha E} \qquad [6]$$

Table 4.12 shows the properties of a number of ceramic materials and the calculated thermal shock resistance values, the properties are for a temperature of about 200°C. It can be seen that high fracture strength, high thermal conductivity and low coefficient of expansion lead to the materials with the best shock resistance.

Table 4.12 *Thermal shock resistance*

Property	Soda glass	Pyrex glass	Glass ceramic	Alumina 99.5%	Silicon carbide, RBSC	Silicon nitride, RBSN
Density Mg/m^3	2.5	2.4	2.6	3.9	3.1	2.4
Specific heat J g^{-1} K^{-1}	0.98	0.94	0.83	0.99	0.88	0.86
Fracture strength MPa	70	70	110	380	300	160
Elastic modulus GPa	72	70	90	400	400	200
Coeff. of expansion 10^{-6}K^{-1}	7.8	3.2	1.0	7.7	4.5	2.6
Therm. conductivity W m^{-1} K^{-1}	1.8	1.5	2.4	25	100	9
Therm. diffusivity 10^{-6} m^2 s^{-1}	0.73	0.66	1.1	6.5	37	4.4
Poisson's ratio	0.2	0.2	0.2	0.2	0.2	0.2
Resistance R K	100	250	978	99	133	246
Resistance R_1 W/m	180	375	2347	2475	13 300	2214
Resistance R_2 10^{-6} m^2 K s^{-1}	73	165	1076	644	4921	1082

4.2.6 Case studies

In addition to the cases discussed earlier in this chapter, the following case studies illustrate further applications of ceramics.

Transmission line insulator

The insulators used to support the cables used to distribute electricity around the country have to be able to maintain electrical insulation at voltages of 400 kV or higher. Porcelains are used,

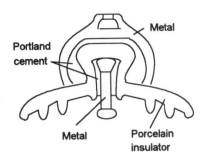

Figure 4.32   *Insulator*

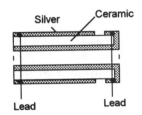

Figure 4.33   *Ceramic tube capacitor*

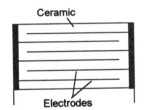

Figure 4.34   *Multi-layer capacitor*

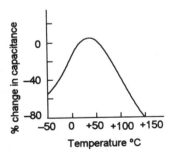

Figure 4.35   *Effect of temperature for a Z5U capacitor*

being based on clay, i.e. kaolinite, feldspar and a filler such as sand or flint. The raw materials are blended with water to form a slip. Water is removed from the slip by pressing against a filter. Air is also removed by a vacuum process. The material is then formed into the required shape, generally by pressing. It is then dried and coated with a glaze slip comprising a water slurry of quartz and feldspar with a stain material to give the required colour. The component is then fired at about 1200°C. Figure 4.32 shows the typical form of such an insulator.

Ceramic capacitors

Ceramic capacitors are formed either as discs, tubes or multilayers. Discs can be formed by dry pressing a slip formed of the powdered ceramic with a 5 to 10% organic binder. The pieces are fired, then coated on the flat surfaces with silver paint before a further brief firing. Solder and leads are then applied before the assembly is coated in a polymer. Tubes are formed by extrusion. After sintering the tubes are then completely coated, on both the inside and outside, with silver. Machines are then used to grind the silver from one end and remove a ring of silver from the outer surface near the other end. Leads are then looped round each end and soldered in place. Figure 4.33 shows the basic form of such a capacitor. Multilayer capacitors (Figure 4.34) typically have a dielectric ceramic layer about 20 mm thick, with the overall dimensions ranging from about 1.25 mm by 1 mm by 1 mm to 6 mm by 6 mm by 2.25 mm and capacitance values from 1 pf to a few microfarads. One method of producing such a capacitor involves forming the dielectric powder into a slip with an organic binder and firing it to form a continuous strip which is then cut into 150 mm square sheets. The electrode pattern is then screen printed onto the surface, the electrode sheets then stacked and consolidated under pressure. The stack is then cut into the capacitor-sized pieces so that a structure of the form shown in Figure 4.34 is produced. Heating is then used to remove the organic binder and the structure then fired.

In considering the materials used for capacitor dielectrics, since the generally requirement is for capacitors which are as compact as possible, a useful measure is the capacitance that can be produced by unit volume. Polymer capacitors typically have about 5 to 1000 pF/mm³, mica capacitors about 5 pF/mm³, single-layer ceramic capacitors, e.g. Z5U, about 5 pF/mm³ and multi-layer ceramic capacitors, e.g. Z5U, have about 30 000 pF/mm³. The designations used for commercial ceramic dielectrics are codes for performance rather than particular materials, referring particular to temperature sensitivity (Figure 4.35). Z5U is based on barium titanate and alloying it with $CaZrO_3$ so that Ca ions are substituted for some of the Ba ions. Z5U has a dielectric constant of about 6000, a dielectric strength of 10 kV/mm and a loss factor, i.e. tan δ, at a frequency of 1 kHz of 20×10^{-3}.

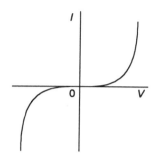

Figure 4.36 *Current–voltage relationship for a varistor*

Heating element

For such items as heating elements, silicon carbide is used with a range of manufacturing methods employed. Commonly used methods are reaction bonding and sintering. Another method that is also used is *in situ formation from carbon and silicon oxide*. A carbon tube is heated to about 1900°C in a bed of sand, i.e. SiO_2 and coke, i.e. carbon. Silicon monoxide is generated from the reaction between the sand and coke and transforms the carbon tube to beta-silicon carbide. The electrical resistance of the resulting heating element can be adjusted by cutting a spiral through the tube wall and adjusting the pitch of the spiral.

Varistor

Voltage dependent resistors, varistors, offer a high resistance at low voltages and a low resistance at high voltages, having a current–voltage relationship of the form shown in Figure 4.36. Silicon carbide can be used for such a resistor. Silicon carbide powder is mixed with clay and other siliceous compounds and then pressed to the required shape before being fired at about 1300°C. The result is silicon carbide particles in a siliceous glass matrix.

Problems 4.2

1 How does the addition of water to clay-based products give rise to the plasticity which enables them to be easily moulded to the required shapes?
2 What are the materials used to make porcelain?
3 Explain what is meant by vitrification.
4 What are the properties of silicon nitride which result in it being used for cutting tool tips?
5 Boron nitride can be easily machined, silicon nitride cannot. What structural differences account for this?
6 Why is chemically toughened glass stronger in tension than the untoughened glass?
7 Toughened soda glass has a density of 2.5 Mg/m³, specific heat capacity of 0.98 J g⁻¹ K⁻¹, fracture strength of 250 MPa, modulus of elasticity of 72 GPa, coefficient of expansion of 7.8 × 10⁻⁶ K⁻¹, thermal conductivity of 1.8 W m⁻¹ K⁻¹ and Poisson's ratio of 0.25. Determine the thermal shock resistance, (a) assuming no heat is conducted through to lower layers of the material from the surface, (b) taking into account such conduction, (c) assuming a steady rate of surface temperature change.
8 Glass ceramic is used for cooker tops. Why not use (a) a glass, (b) stainless steel?
9 A glass ceramic cooker top is being manufactured and has a surface 500 mm by 500 mm. It is heated to 620°C and then quenched along one edge in cold water at 20°C. Determine the largest thermal stress that occurs if heat conduction through the material is neglected. The coefficient of expansion of the material is 1 × 10⁻⁶ K⁻¹, the tensile modulus 92 GPa and Poisson's ratio 0.24.

4.3 Composites

The term *composite* is used for a material composed of two different materials bonded together with one serving as the matrix surrounding fibres or particles of the other. This section is a consideration of the various forms composites can take and the resulting properties.

There are many examples of composite materials encountered in everyday products. Composites can be classified into four categories:

- *Fibre reinforced*

  The fibres may be continuous throughout the matrix or short fibres, and aligned in all the same direction or randomly arranged. For example, there are composites involving glass or carbon fibres in polymers, ceramic fibres in metals and metal fibres in ceramics. Many plastics are glass fibre reinforced, the result being a much stronger and stiffer material than given by the plastic alone. Polymer composites are used for such applications as car instrument panels, domestic shower units and crash helmets. Vehicle tyres are rubber reinforced with woven cords. A common example of a metal-reinforced ceramic composite is reinforced concrete. The composite material enables loads to be carried that otherwise could not have been carried by the concrete alone. Ceramic-reinforced metal composites are used for rocket nozzles, wire-drawing dies, cutting tools and other applications where hardness and performance at high temperatures might be required.

- *Particle reinforced*

  Cermets, composites involving ceramic particles in a metal matrix, are widely used for the tips of cutting tools. Ceramic particles are hard but brittle and lack toughness; the metal is soft and ductile. Embedding the ceramic particles in the metal gives a material that is strong, hard and tough. Glass spheres are widely used with polymers to give a composite which is stronger and stiffer than the polymer alone.

- *Dispersion strengthened*

  The strength of a metal can be increased by small particles dispersed throughout it. Thus solution treatment followed by precipitation hardening for an aluminium–copper alloy can lead to a fine dispersion of an aluminium–copper compound throughout the alloy. The result is a higher tensile strength because the movement of dislocations through the alloy is hindered. Another way of introducing a dispersion of small particles throughout a metal is sintering.

- *Laminates*

  Laminates are composites in which materials are sandwiched together. Plywood is an example where thin sheets of wood are bonded together to give a stronger laminated structure.

4.3.1 Fibre-reinforced materials

The main functions of the fibres in a composite are to carry most of the load applied to the composite and provide stiffness. For this reason, fibre materials have high tensile strength and a high elastic modulus. Ceramics are frequently used for the fibres in composites. Ceramics have high values of tensile strength and tensile modulus, the useful asset of low density, but are brittle and the presence of quite small surface flaws can markedly reduce the tensile strength. By incorporating such fibres in a ductile matrix it is possible to form a composite which makes use of the high strength–high elastic modulus properties of the fibres and the protective properties of the matrix material to give a composite with properties considerably better than with just the matrix material alone or the properties of damaged fibre material.

The properties required of a suitable matrix material are that:

- It adheres to the fibre surfaces so that forces applied to the composite are transmitted to the fibres so that they can assume the primary responsibility for the strength of the composite.

- It protects the fibre surfaces from damage.

- It keeps the fibres apart to hinder crack propagation.

The fibres used may be continuous, i.e. in lengths running the full length of the composite, or discontinuous, i.e. in short lengths. They may be aligned so that they are all lying in the same direction or randomly orientated (Figure 4.37). Aligning them all in the same direction gives a directionality to the properties of the composite. Table 4.14 gives the properties of some commonly used fibre reinforcing materials.

(a) Continuous, aligned

(b) Discontinuous, aligned

(c) Discontinuous, random

Figure 4.37 *Forms of fibre reinforcements*

Table 4.14 *Properties of fibres*

Fibre	Density Mg/m^3	Tensile modulus GPa	Tensile strength MPa
Alumina	3.2	170	2100
Silicon carbide (Nicalon)	2.6	250	2200
Boron	2.65	420	3500
Carbon	1.8	250	2700
E-Glass	2.5	70	2200*
Polyethylene (Spectra 1000)	0.97	172	2964
Polyamide (Aramid) (Kevlar 49)	1.45	125	3000

Note: *3500 MPa freshly drawn.

Glass fibres are widely used for reinforcing polymers. The fibres may be long lengths running through the length of the composite, or discontinuous short lengths randomly orientated within the composite. Another form of composite uses glass fibre mats or cloth in the plastic. These may be in the form of randomly

Chopped strand mat

Continuous filament mat

Bi-directional fabric

Figure 4.38 *Examples of mats and fabrics*

orientated fibre bundles which may be chopped or continuous and are loosely held together with a binder or a woven fabric (Figure 4.38). The effect of the fibres is to increase both the tensile strength and tensile modulus, the amount of change depending on both the form the fibres take and the amount. The continuous fibres give the highest tensile strength and tensile modulus composite but with a high directionality of properties. The strength along the direction of the fibres could be as high as 800 MPa while that at right angles to the fibre direction may be as low as 30 MPa, i.e. just about the strength of the plastic alone. Randomly orientated short fibres do not lead to this directionality of properties but do not give such high strengths and tensile modulus. Table 4.15 gives examples of the strength and modulus values obtained with glass fibre-reinforced polymer.

Table 4.15 *Properties of reinforced polyester*

Composite	% by weight of glass fibre	Tensile modulus GPa	Tensile strength MPa
Polyester alone		2 to 4	20 to 70
With short random fibres	10 to 45	5 to 14	40 to 180
With plain weave cloth	45 to 65	10 to 20	250 to 350
With long fibres	50 to 80	20 to 50	400 to 1200

Fibre-reinforced metals are a later development than fibre-reinforced polymers, having only been developed since the 1970s when the production of fibres of boron and silicon carbide became feasible. Because the coefficient of thermal expansion of ceramics is less than that of metals, e.g. that for silicon carbide being about one-fifth that of aluminium, such composites lead to a reduction in the coefficient. Likewise the thermal and electrical conductivities of ceramics is less than that of metals and so the composites have lower thermal and electrical conductivities. For example, the thermal conductivity of aluminium is 201 W m^{-1} K^{-1} and an aluminium–silicon carbide composite with 15% of silicon carbide has a thermal conductivity of 140 W m^{-1} K^{-1}. Significant increases in the tensile modulus of elasticity can be obtained by including ceramic fibres in metals, the improvement being particularly significant for those metals such as aluminium and magnesium that have low modulus values. For example, the modulus of elasticity for aluminium is about 70 GPa, while that of an aluminium–silicon carbide composite with 50% continuous silicon carbide fibres is about 200 GPa. Significant increases in the tensile strength of metals can be obtained by the inclusion of ceramic fibres. Table 4.16 gives some examples of fibre-reinforced metals and the tensile strengths achieved.

Key points

In designing fibre-based composites, the designer has to consider where the fibres should be placed and their directions. Does the product require high stiffness and high strength in some particular direction or directions? If so, continuous aligned fibres in such directions needs to be considered.

Table 4.16 *Properties of reinforced metals*

Composite	Tensile strength MPa
Aluminium + 50% silica fibres	900
Aluminium + 50% boron fibres	1100
Nickel + 8% boron fibres	2700
Nickel + 40% tungsten fibres	1100
Copper + 50% tungsten fibres	1200
Copper + 80% tungsten fibres	1800

Ceramic matrix composites, such as silicon carbide fibre-reinforced glass ceramics and silicon carbide whisker-reinforced alumina, enable the elastic modulus, strength and toughness of the ceramic matrix material to be improved. For example, a $Li_2O–Al_2O_3–MgO–SiO_2$ glass ceramic with a tensile modulus of 86 GPa has as a composite with 50% silicon carbide fibre reinforcement a modulus of about 130 GPa, its toughness improved from 1.5 MN $m^{-3/2}$ to 17 MN $m^{-3/2}$ and its strength improved from 180 MPa to 620 MPa. Silicon carbide-reinforced alumina uses whiskers of silicon carbide, these being single crystals rather than the polycrystalline form of fibres. With 25% silicon carbide whiskers, the tensile modulus is improved from 340 GPa to 390 GPa, the strength from 300 MPa to 900 MPa and the fracture toughness from 4.5 MN $m^{-3/2}$ to 8.0 MN $m^{-3/2}$.

Continuous fibre composites

Composites made with continuous aligned fibres all aligned in the same direction have properties which depend on the direction in which the material is stretched, the modulus of elasticity and strength in the direction which the fibres are aligned being markedly different from its modulus and strength in the direction at right angles to the fibres. The term *isotropic material* is used for one which has the same properties in all directions, the term *anisotropic material* for one which has properties which depend on the direction. A composite with the continuous aligned fibres is an anisotropic material.

Maths in action

The law of mixtures

The mass m_c of a composite is made up of the masses of the matrix m_m and the fibres m_f, i.e.

$$m_c = m_m + m_f \qquad [7]$$

Since mass is volume v times density ρ, then equation [7] can be written as:

$$v_c\rho_c = v_m\rho_m + v_f\rho_f$$

and so:

$$\rho_c = \frac{v_m}{v_c}\rho_m + \frac{v_f}{v_c}\rho_f$$

v_m/v_c is the volume fraction V_m that is matrix and v_f/v_c is the volume fraction V_f that is fibre. Thus:

$$\rho_c = V_m\rho_m + V_f\rho_f \qquad [8]$$

Note that since $v_m = v_c - v_f$ we must have $V_m = 1 - V_f$. Equation [8] can be termed a *law of mixtures*.

Elastic modulus

Now consider the law of mixtures for the elastic modulus of a composite having aligned continuous fibres with the direction of the fibres parallel to the loading direction (Figure 4.39). The load F_c on the composite is the sum of the loads on the matrix F_m and on the fibres F_f:

$$F_c = F_m + F_f \qquad [9]$$

But stress σ is force/area and so for the composite $\sigma_c = F_c/a_c$, for the matrix $\sigma_m = F_m/a_m$ and for the fibres $\sigma_f = F_f/a_f$ where a_c is the cross-sectional area of the composite, a_m the cross-sectional area of that part of the composite that is matrix and A_f that part that is fibre. But $\sigma = E\varepsilon$, where E is the modulus of elasticity and ε the strain, thus $F_c = \sigma_c a_c = E_c\varepsilon a_c$, $F_m = \sigma_m a_m = E_m\varepsilon a_m$ and $F_f = \sigma_f a_f = E_f\varepsilon a_f$. The strain is the same for the matrix and the fibres will be the same as that for the composite if we assume that the fibres are firmly bonded to the matrix. Equation [9] can thus be written as:

$$E_c\varepsilon a_c = E_m\varepsilon a_m + E_f\varepsilon a_f$$

and thus, if we write A_m as the area fraction a_m/a_c and A_f as a_f/a_c:

$$E_c = A_m E_m + A_f E_f \qquad [10]$$

Since $a_m = a_c - a_f$, then $A_m = 1 - A_c$. Equation [10] is the law of mixtures for the tensile modulus when we have continuous fibres aligned parallel to the loading direction. The area fractions are equivalent to the volume fractions and thus equation [10] becomes:

$$E_c = V_m E_m + V_f E_f \qquad [11]$$

Now consider the elastic modulus when the fibres are at right angles to the direction of loading (Figure 4.40).

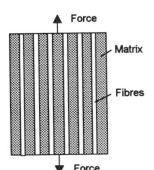

Figure 4.39 *Continuous aligned fibres parallel to load direction*

Force

Matrix

Fibres

Force

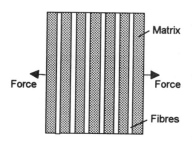

Figure 4.40 *Continuous aligned fibes at right angles to load direction*

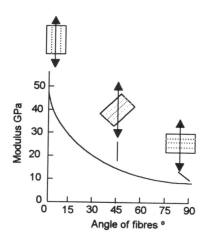

Figure 4.41 *Modulus variation with angle*

The total extension δ_c of the composite in the direction of the load will be the sum of the extensions of the matrix δ_m and the fibres δ_f:

$$\delta_c = \delta_m + \delta_f \tag{12}$$

The stress on the matrix will be the same as on the fibres and on the composite as a whole. The strain $\varepsilon = \delta/l$, where l is the original length and thus $\delta = \varepsilon l = (\sigma/E)l$, where σ is the stress. Thus equation [12] becomes:

$$\frac{\sigma}{E_c} l_c = \frac{\sigma}{E_m} l_m + \frac{\sigma}{E_f} l_f$$

and so:

$$\frac{1}{E_c} = \frac{L_m}{E_m} + \frac{L_f}{E_f} \tag{13}$$

where L_m is the matrix length fraction l_m/l_c and L_f is the fibre length fraction l_f/l_c. The length fractions are equivalent to the volume fractions and thus equation [13] becomes:

$$\frac{1}{E_c} = \frac{V_m}{E_m} + \frac{V_f}{E_f} \tag{14}$$

Equation [11] will give a different elastic modulus value to that given by equation [14]. For an epoxy-glass composite with the volume fraction of fibres as 65%, equation [11] gives for the modulus of elasticity when the load is at 0° to the fibres (modulus for epoxy 3.5 GPa, modulus for the glass fibres 70 GPa):

$$E_c = V_m E_m + V_f E_f = 0.35 \times 3.5 + 0.65 \times 70 = 47 \text{ GPa}$$

and equation [14] gives

$$\frac{1}{E_c} = \frac{V_m}{E_m} + \frac{V_f}{E_f} = \frac{0.35}{3.5} + \frac{0.65}{70} = 0.11$$

and hence $E_c = 9$ GPa when the load is at 90° to the fibres. Figure 4.41 shows, for a glass fibre epoxy composite, how the modulus of elasticity varies with the angle between the load direction and the direction of the fibres.

Strength

Now consider the strength of fibre-reinforced composites. This is more difficult to predict than the stiffness because there are a number of ways the composite can fail: fibre failure, matrix failure or fibre–matrix interface failure.

The matrix has a lower strength than the fibres. Thus if we have a very small volume fraction then when the matrix fails, the load has to be carried by the fibres but as there are not many fibres to carry the load complete failure follows. Thus when we have the load applied in a direction parallel to the fibres, equation [9], $F_c = F_m + F_f$ gives:

$$\sigma_c a_c = \sigma_m a_m + \sigma_f a_f$$

where σ_c is the composite failure stress, σ_m the matrix failure stress and σ_f is the fibre stress when the matrix fails. Thus:

$$\sigma_c = \sigma_m A_m + \sigma_f A_f = \sigma_m V_m + \sigma_f V_f \qquad [15]$$

and because there are so few fibres we can generally neglect their contribution and write:

$$\sigma_c = \sigma_m V_m \qquad [16]$$

At large fibre volume fractions, the fibres will take most of the load and after matrix failure will still be able to carry a load and so determine the strength of the composite. Then we must have:

$$\sigma_c = \sigma_f V_f \qquad [17]$$

where σ_f is the fibre failure stress.

As with the modulus of elasticity, the strength depends on the orientation of the fibres relative to the load direction. Figure 4.42 shows the effect of fibre orientation on the strength of a typical continuous fibre composite, glass fibre-reinforced epoxy.

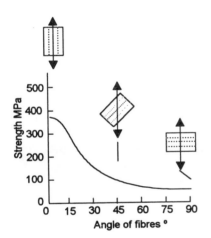

Figure 4.42 *Strength variatiom with angle*

Example

A composite has continuous aligned glass fibres with a tensile modulus of 76 GPa and strength of 1500 MPa in a matrix of polyester with a tensile modulus of 3 GPa and strength of 50 MPa. Determine the tensile modulus and strength of the composite when the load is applied in a direction parallel to the fibres and the volume fraction is 60%.

Using equation [11]: $E_c = V_m E_m + V_f E_f = 0.4 \times 3 + 0.6 \times 76 = 46.8$ GPa. If we assume that the fibres will continue to take the load after matrix failure, then equation [17] gives for the composite strength: $\sigma_c = \sigma_f V_f = 1500 \times 0.6 = 900$ MPa.

Example

A column of reinforced concrete has steel reinforcing rods running through its entire length and parallel to the column axis. If the concrete has a modulus of elasticity of 15 GPa and a tensile strength of 2.8 MPa and the steel rods a modulus of elasticity of 210 GPa and a tensile strength of 400 MPa, determine the modulus of elasticity and tensile strength of the composite if the steel rods occupy 10% of the cross-sectional area of the column.

Using equation [11]: $E_c = V_m E_m + V_f E_f = 0.9 \times 15 + 0.1 \times 210 = 34.5$ GPa. Using equation [15]: $\sigma_c = \sigma_m V_m + \sigma_f V_f = 0.9 \times 2.8 + 0.1 \times 400 = 42.5$ MPa.

Discontinuous fibres

The properties of composites containing short discontinuous fibres depends on the lengths of the fibres, their orientation and volume fraction. The tensile modulus of composites with short length fibres all aligned is, in a direction parallel to the fibres, less than that of the comparable aligned continuous fibre composite. The tensile modulus of composites with non-aligned short length fibres is less than that for a composite with the same fibres when aligned.

Maths in action

Stress transfer

To consider the effects of the length of short length fibres we need to consider the stress transfer that takes place between the matrix and fibres. When a load is applied to a composite it is applied to the matrix and transferred to the fibres by some combination of shear and tensile or compressive stresses acting across the interface. The nature of the bond between the matrix material and the fibres thus plays a critical role in determining the properties of the composite

For a composite under tension, we can think of a discontinuous fibre in the matrix as being stretched as a result of interfacial shear stresses acting on the surfaces of the fibre (Figure 4.43). These shear stresses will be a maximum at the ends of the fibre where it stretches the most and fall as we consider points on the fibre nearer to the middle. The tensile stress in the fibre will be zero at the ends and build up as the shear stress decreases. Provided the fibre is long enough we can consider the shear stresses decreasing to zero and the tensile stress rising to a maximum value in the central region of the fibre.

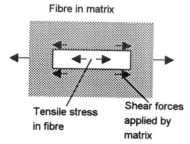

Fibre in matrix

Tensile stress in fibre

Shear forces applied by matrix

Figure 4.43 *Single fibre in a matrix*

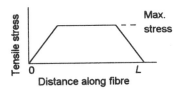

(a) Length > critical length

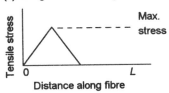

(b) Length = critical length

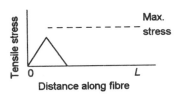

(c) Length < critical length

Figure 4.44 *Stress transfer for different length fibres*

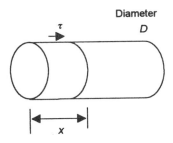

Figure 4.45 *Shear force at interface*

Table 4.17 *Typical values of L_c/D ratios*

Fibre	Matrix	L_c/D ratio
Glass	Polyester	40
Glass	Polypropylene	140
Carbon	Epoxy	35
Boron	Epoxy	35
Silicon carbide	Alumina	10
Boron	Aluminium	20

If the fibre is not long enough, the shear stress will not fall to zero and the tensile stress will not reach the maximum possible value. In order to achieve this maximum level of tensile stress, the fibre length must be at least equal to a *critical length* L_c. Figure 4.44 shows the tensile stress patterns we might expect for different fibre lengths.

Consider the interfacial shear stress acting on a single fibre in a matrix (Figure 4.45). If τ is the average interfacial shear stress then the shear force acting on a section of the fibre, length x and of uniform cross-sectional diameter D, is:

$$\text{shear force} = \text{shear stress} \times \text{area} = \tau \pi D x$$

This shear force results in a longitudinal stress in the fibre of σ_f. Thus:

$$\sigma_f \tfrac{1}{4}\pi D^2 = \tau \pi D x$$

and so:

$$\sigma_f = \frac{4\tau x}{D}$$

The stress increases from zero at the end of a fibre, i.e. when $x = 0$, to its maximum possible value when $x = \tfrac{1}{2}L_c$. Hence, the maximum value of the tensile stress is:

$$\text{maximum } \sigma_f = \frac{2\tau L_c}{D} \qquad [18]$$

The critical length to diameter ratio L_c/D must not be less than $\sigma_f/2\tau$ if the composite is to realise the potential of the fibre. If, for a glass fibre–polyester composite the maximum value of stress the fibres can withstand is 1500 MPa and the shear strength is 25 MPa, then L_c/D is 30. For fibres of diameter 5 μm then L_c is 0.15 mm. If the fibres used are of greater diameter then the critical length is increased. Table 4.17 shows some typical values of the L_c/D ratio.

If the fibre length L is longer than the critical length and we assume a linear variation of stress with distance, as in Figure 4.44(a), then the average tensile stress in the fibre is given the area under the stress–fibre length graph divided by the fibre length:

$$\text{average stress} = \left[(L - L_c) + \tfrac{1}{2}L_c\right]\frac{\text{max. }\sigma_f}{L}$$

$$= \left[1 - \frac{L_c}{2L}\right]\text{max. }\sigma_f \qquad [19]$$

$$= \left[1 - \frac{L_c}{2L}\right]\frac{2\tau L_c}{D} \qquad [20]$$

When the fibre length is equal to the critical length, as in Figure 4.44(b), then the average tensile stress in the fibre is half the maximum stress:

$$\text{average stress} = \frac{\tau L_c}{D} \qquad [21]$$

When the fibre length is less than the critical length, as in Figure 4.44(c), the tensile stress in the fibre will never reach the maximum value. The area under the tensile stress–fibre length graph is half the maximum stress value attained, this being $2\tau L/D$ and so:

$$\text{average stress} = \frac{\tau L}{D} \qquad [22]$$

As the above equations indicate, the average stress in a short fibre will be less than the maximum stress the fibre can withstand. With a continuous fibre the fibre is assumed to be entirely at the maximum stress value (Figure 4.46) when it fails, this not being the case with a short fibre because of the drop off of stress near the fibre ends. Compare Figure 4.46(a) with Figure 4.44. Figure 4.46(b) shows how the average stress in a short fibre compares with the maximum stress value for a fibre, i.e. the value at which continuous fibres would fail, as the length of fibre, in terms of multiples of the critical length, is increased. The average stress value can be used with the rule of mixtures, equation [15], to give a value for the failure stress of the composite.

Tensile modulus

A modified rule of mixtures can be used, incorporating a length efficiency parameter η_L to take account of the fibres not being continuous. Thus, equation [11] becomes:

$$E_c = V_m E_m + \eta_L V_f E_f \qquad [23]$$

For carbon fibres of length 0.1 mm and diameter 8 μm in epoxy the efficiency parameter is 0.20. With these fibres of length 1.0 mm, the efficiency parameter is 0.89 and as the length increases so the parameter approaches 1.

For non-aligned short length fibres an orientation efficiency parameter η_O can be used to take account of this, equation [18] then becoming:

$$E_c = V_m E_m + \eta_O \eta_L V_f E_f \qquad [24]$$

For completely three-dimensionally random fibres, the orientation efficiency parameter has the value 0.2.

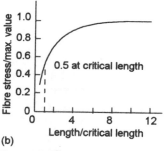

(a)

(b)

Figure 4.46 *(a) Stress transfer for continuous fibres, (b) average stress in short fibres*

Example

A glass fibre–polyester composite contains 60% by volume of fibres, the fibres being of length 3 mm. If the failure stress for the fibres is 1500 MPa, the shear strength 25 MPa and the matrix has a tensile strength of 50 MPa, determine the strength of the composite.

The critical length is, as derived above, 0.15 mm and so the fibres have a length which is greater than the critical length. Using equation [19], the maximum value of the average stress acting on the fibres is:

$$\text{max. average stress} = \left[1 - \frac{L_c}{2L}\right]\text{max.}\,\sigma_f$$

$$= \left[1 - \frac{0.15}{6}\right]1500 = 1462.5 \text{ MPa}$$

Hence, using equation [15], the strength of the composite is:

$$\text{strength} = \sigma_m V_m + \sigma_f V_f = 50 \times 0.4 + 1462.5 \times 0.6$$

$$= 897.5 \text{ MPa}$$

This compares with a strength of 920 MPa that occurs with the continuous fibres and 475 MPa if the fibres had been equal to the critical length.

4.3.2 Particle-reinforced materials

Particle-reinforced materials have particles with diameters of the order of 1 μm or more dispersed throughout the matrix, the particles often accounting for a quarter to half, or more, of the volume of the composite.

Cermets

Cermets, or cemented carbides, are examples of particle-reinforced composites in which hard ceramic particles are in a metal matrix. The ceramics used have high strengths, high values of tensile modulus and high hardness, but are brittle. By comparison, the metals are weaker and less stiff, but ductile. By incorporating ceramic particles, often about 80% by volume, in a metal matrix, a composite can be produced which is strong, hard and tough and can be used as a tool material.

For example, tungsten carbide is a very hard (about 2000 HV) ceramic with a high tensile modulus but also very brittle. Tools made from this material would thus be extremely brittle. A cermet involving tungsten carbide in a metal matrix, cobalt, can be made by mixing tungsten carbide powder with cobalt powder and heating the compacted powders to a temperature above the melting

Key point

Cermets are an example of a particle-reinforced composite, having hard ceramic particles in a metal matrix.

point of the cobalt. The liquid cobalt then melts and flows round each tungsten carbide particle. After solidification, the cobalt acts as a binder for the tungsten carbide. The composite has a better toughness than the tungsten carbide alone, since crack propagation through the material is hindered. When in use, the tungsten carbide particles in the surface of the material provide the tool with its cutting ability. As the tungsten carbide particles at the cutting surface become blunted, they either fracture or pull out of the cobalt matrix and expose fresh tungsten carbide particles which can continue to provide cutting ability. For a fine cutting tool, the amount of cobalt in the composite is low and the tungsten carbide particles fine so that the tungsten carbide particles pull out easily and the tool remains sharp. For a rough cutting tool, the amount of cobalt is increased to improve toughness and coarser tungsten carbide particles are used. Table 4.18 shows the composition and application of some cobalt–tungsten carbide composites used as tool materials.

Particle-reinforced polymers

Many polymeric materials incorporate fillers, these being particulate. Examples of such fillers are glass beads, silica flour and rubber particles. Such fillers can be regarded as discontinuous fibres which have lengths comparable with their diameter. Their effect on the tensile strength and modulus of elasticity thus tends to be smaller.

The toughness of some polymers is increased by incorporating tiny rubber particles in the polymer matrix. Polystyrene is toughened in this way by polybutadiene particles to give a product referred to as *high impact polystyrene* (HIPS). The rubber particles increase the toughness from about 1 to 1.7 MN m$^{-3/2}$. They block the transmission of cracks and, since they deform readily, absorb energy. The rubber has a lower tensile modulus and tensile strength than that of the matrix material and the net result is a lowering of the tensile modulus and tensile strength, but much greater elongations before breaking and a tougher material.

> **Key point**
>
> Most plastics are polymers with either particulate fillers or discontinuous fibres. Many of the additives are just there to 'add bulk', and have little effect on the strength or modulus of elasticity. However, in some cases, the additives are used to increase strength and modulus of elasticity.

Table 4.18 *Cobalt–tungsten carbide materials*

Cobalt %	Tungsten carbide %	Tungsten carbide grain size	Typical applications
3	97	Medium	Machining of cast iron, non-ferrous metals and non-metallic materials
6	94	Fine	Machining of non-ferrous and high-temperature alloys
6	94	Medium	General purpose machining for metals other than steels, small and medium size compacting dies and nozzles
6	94	Coarse	Machining of cast iron, non-ferrous materials, compacting dies
10	90	Fine	Machining steel, milling, form tools
10	90	Coarse	Percussive drilling bits
16	84	Fine	Mining and metal forming tools
16	84	Coarse	Mining and metal forming tools, medium and large size dies
25	75	Medium	Heavy impact metal forming tools, e.g. heading dies, cold extrusion dies

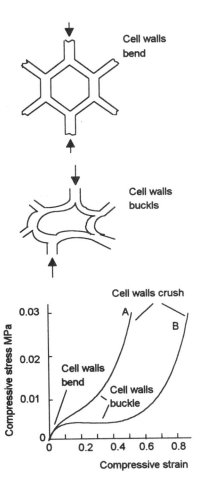

Figure 4.47 *Stress–strain graphs for foamed polymers*

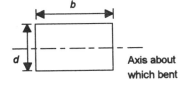

Figure 4.48 *Cross-section of bent beam*

Carbon black, which consists of very fine particles of carbon, is widely used as a filler with vulcanised rubber. The carbon black enhances the strength, stiffness, hardness, wear resistance and heat resistance of the rubber.

Foams

Foams are a particulate composite in which the component bound by the matrix is not a solid but bubbles of a gas. Such foams are used as cushioning in furniture, energy-absorbent packaging and padding, for thermal insulation, for buoyancy, for structural panels and as the filling in sandwich panels (see later in this chapter). The parameters determining the characteristics of foams are the ratio of bulk density of the foam to that of the unfoamed matrix material, and the cellular structure of the foam. The foam can be open cell, closed cell or a mixture of the two. With a closed-cell structure, the gas bubbles in the foam are discrete and not interconnected, whereas in an open-cell structure the bubbles have coalesced and are interconnected. Structural foams and sandwich foams have solid skins covering the foamed core.

Figure 4.47 shows typical forms of compressive stress–strain graphs for polymeric foams. Over the initial straight-line part of the graph, the cell walls just bend under the action of the applied stress. The next stage is when the walls elastically buckle, often giving a plateau of deformation at almost constant stress. The deformation is still elastic and so recoverable. Finally the cell walls suffer irrecoverable buckling collapse.

With a material used for, say, a cushion, the foam is required to give continually increasing resistance to increasing load and so a plateau on the stress–strain graph is not required. Thus a stress–strain graph of the form shown as A is required. Foams used for packaging need to absorb the energy involved when packages are dropped and so a plateau is highly desirable, like that indicated for B in the figure, since it indicates a high energy absorption. A low-density packaging form with a density ratio, bulk foam to unformed plastic, of 0.01 might be used for packaging small delicate instruments; heavier density foams being used for packaging heavier components.

Foamed plastics have a lower modulus of elasticity, and so are more flexible, than when unfoamed. However, in bending this reduction in modulus can be more than offset by the ability to increase the second moment of area I. The second moment of area of a rectangular cross-section (Figure 4.48) of breadth b and depth d is $bd^3/12$. Thus because of the lower density of the foamed plastic, we can have a much larger cross-section for the same mass and thus a much larger second moment of area. The important parameter in determining the stiffness of a beam subject to bending is the product EI. For example, for a cantilever of length L subject to a force F at its free end, the deflection y of that free end is $FL^3/3EI$. Thus we can minimise the deflection per unit force, i.e. y/F, and so have a stiffer cantilever, if there is a large value of EI. With the foamed plastic, the reduction in E can be more than offset by the increase in I.

Maths in action

For structural and sandwich foams, to a reasonable approximation, the elastic modulus E_f of the foam is related to the modulus E_s of the solid matrix material by:

$$E_f = V_s E_s \qquad [25]$$

where V_s is the fraction of the bulk volume of the foam that is solid matrix and is given by:

$$V_s = \frac{v_s}{v_s + v_g} \qquad [26]$$

where v_s is the volume of the foam that is solid and v_g the volume that is gas. The bulk density ρ_f of the foam is

$$\rho_f = \frac{m_s + m_g}{v_s + v_g}$$

with m_s being the mass of the foam that is solid and m_g the mass of the gas in the foam. Thus equation [26] can be written as:

$$V_s = \frac{v_s \rho_f}{m_s + m_g}$$

Since m_g is very small in comparison with m_s, we can neglect the m_g term. Since $\rho_s = m_s/v_s$ then:

$$V_s \approx \frac{\rho_f}{\rho_s}$$

and so equation [25] can be written as:

$$E_f \approx \frac{\rho_f}{\rho_s} E_s \qquad [27]$$

For foams with uniform densities but without the structural skin, a better relationship has been found experimentally to be:

$$E_f = \left(\frac{\rho_f}{\rho_s}\right)^n E_s \qquad [28]$$

where n has the approximate value of 1.5 for such foams in tension and 2 in compression.

Thus, for example, foamed polystyrene with a volume fraction of polymer of 0.5 would have a tensile modulus of about $0.5^{1.5} = 0.35$ times that of the unfoamed polystyrene. The compression modulus would be about $0.5^2 = 0.25$ times that of the unfoamed polystyrene.

Expanded polystyrene used for thermal insulation and packaging has a volume fraction of about 0.05 and thus a tensile modulus of about $0.05^{1.5} = 0.011$ of that of the unfoamed polystyrene and a compression modulus of $0.05^2 = 0.0025$ that of the unfoamed polystyrene.

4.3.3 Dispersion strengthened metals

One way of introducing a dispersion of small particles throughout a metal uses sintering. This process involves compacting a powdered metal powder in a die and then heating it to a temperature high enough to knit together the particles in the powder. If this is done with aluminium the result is a fine dispersion of aluminium oxide, about 10%, throughout an aluminium matrix. The aluminium oxide occurs because aluminium in the presence of oxygen is coated with aluminium oxide and when the aluminium powder is compacted, much of the surface oxide film becomes separated from the aluminium and becomes dispersed through the metal. The aluminium oxide powder, a ceramic, dispersed throughout the aluminium matrix gives a stronger material than that which would have been given by the aluminium alone. At room temperature, the tensile strength of the sintered aluminium powder (SAP) is about 400 MPa, compared with that of about 90 MPa for the aluminium. The sintered aluminium has an advantage over precipitation-hardened aluminium alloys in that it retains its strength better at high temperatures (Figure 4.49). This is because at the higher temperatures, the precipitate particles in precipitation hardened alloys tend to coalesce or go into solution in the metal.

4.3.4 Laminates

Plywood is an example of a laminated material. It is made by gluing together thin sheets of wood with their grain directions at right angles to each other (Figure 4.50). The grain directions are the directions of the cellulose fibres in wood, a natural composite, and thus the resulting structure, the plywood, has fibres in mutually perpendicular directions. Thus, whereas the thin sheet had properties that were directional, the resulting laminate has no such directionality.

It is not only wood that is laminated, metals are too. The *cladding* (Figure 4.51) of an aluminium–copper alloy with aluminium to give a material with a better corrosion resistance is another example. Galvanised steel can be considered a further example, the layer of zinc on the steel giving better corrosion resistance. Steel for use in food containers is often plated with tin to improve its corrosion resistance.

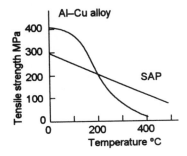

Figure 4.49   *The effect of temperature on strength*

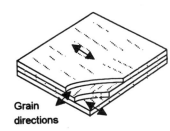

Figure 4.50   *Plywood, 3-ply*

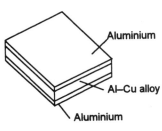

Figure 4.51   *Clad alloy*

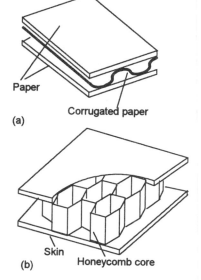

Paper

Corrugated paper

(a)

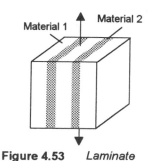

Skin

(b) Honeycomb core

Figure 4.52 *(a) Corrugated cardboard, (b) honeycomb structure*

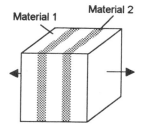

Material 1 · Material 2

Figure 4.53 *Laminate*

Material 1 · Material 2

Figure 4.54 *Laminate*

Corrugated cardboard is another form of laminated structure (Figure 4.52(a)), consisting of paper corrugations sandwiched between layers of paper. The resulting structure is much stiffer, in the direction parallel to the corrugations, than the paper alone. A structure with a different sandwiched core has an aluminium, polymer or paper *honeycomb structure* (Figure 4.52(b)) sandwiched between thin sheets of metal or polymer. Such a structure has good stiffness and is very light and is often used for structural panels. Another form of sandwich panel has a core of low density, e.g. a foamed polymer, sandwiched between two stronger and stiffer panels.

Maths in action

Consider the elastic modulus for a laminate made up of alternate layers of materials with different stiffnesses and loaded in the way shown in Figure 4.53. If we assume that the bonding between the layers is such that each of the layers is equally strained then the problem of determining the modulus of elasticity of the laminate is just the same as that carried out earlier in this chapter for continuous fibres in a matrix. We can apply the law of mixtures [11] to give:

$$E_c = V_1 E_1 + V_2 E_2 \qquad [29]$$

where E_c is the elastic modulus of the composite, E_1 that of material 1 and E_2 that of material 2. V_1 is the volume ratio of material 1 and V_2 the volume ratio of material 2.

If the material is loaded as shown in Figure 4.54 then we assume that each of the layers is equally stressed and thus, as in equation [14]:

$$\frac{1}{E_c} = \frac{V_1}{E_1} + \frac{V_2}{E_2} \qquad [20]$$

Example

A sheet of plywood consists of three equally thick sheets, the upper and lower sheets having their grain in the same direction and the middle sheet with its grain at right angles. The wood has a tensile modulus for forces in the direction parallel to the grain of 10 GPa and in the transverse direction 0.4 GPa. Determine the tensile modulus of the laminate when loaded in a direction parallel to the grain direction of the outer sheets.

The situation is similar to that shown in Figure 4.53 and thus, using equation [29]: $E_c = V_1 E_1 + V_2 E_2 = (2/3) \times 10 + (1/3) \times 0.4 = 6.8$ GPa.

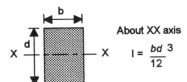

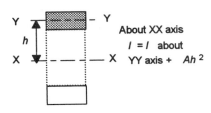

Figure 4.55 *Second moment of area*

Example

How thick would a rectangular strip of steel, modulus of elasticity 210 GPa, have to be to have the same bending stiffness as a laminate consisting of two 1.5 mm thick surfaces with an elastic modulus of elasticity 7 GPa and a core of foamed polymer of thickness 6 mm and a density which is 50% of the unfoamed polymer, it having a modulus of elasticity of 7 GPa.

For equal bending stiffness we must have the product EI the same for the laminate as for the steel. For the laminate, the core will have a tensile modulus given by equation [28] as:

$$E_f = \left(\frac{\rho_f}{\rho_s}\right)^{1.5} E_s = 0.5^{1.5} \times 7 = 2.47 \text{ GPa}$$

and a compression modulus of:

$$E_f = \left(\frac{\rho_f}{\rho_s}\right)^{2} E_s = 0.5^2 \times 7 = 1.75 \text{ GPa}$$

One surface of a bent beam is in tension and one in compression, thus we will use the average modulus of 2.11 GPa for our comparison.

The bending stiffness of the laminate is the sum of the bending stiffness of the core and the surface sheets about the central axis. A rectangular section has a second moment of area about a central axis of $bd^3/12$, where b is the breadth and d the depth (Figure 4.55). The second moment of area for a rectangular section about a parallel axis a distance y from the axis through the centre of the rectangular section is $I + Ay^2$, where A is the area of the section (theorem of parallel axes). Thus for the laminate we have:

$$(EI)_{laminate} = 2.11 \times \frac{b \times 6^3}{12} + 7 \times 2 \times b \left[\frac{1.5^3}{12} + 1.5 \times 3.75^2 \right]$$

$$= 360b$$

For the steel to have the same value bending stiffness we must have $(EI)_{steel} = 360b$. This means $I = 360b/210 = 1.7b$. Hence:

$$\frac{bd^3}{12} = 1.7b$$

and so $d = 2.7$ mm.

4.3.6 Case studies

In additions to the applications already discussed in this chapter, the following case studies illustrate the use of reinforced materials.

Reinforced concrete

Consider the design of a reinforced concrete beam, e.g. for use as a lintel over a window. Because concrete is weak in tension it is its underside which is most likely to crack and so steel reinforcement bars are put in the lower half of the beam (Figure 4.56). There must be a strong bond between the concrete and the steel as it is the bonded combination which resists the tensile stresses. The bond arises from the ability of concrete to stick to steel, the frictional resistance between steel and concrete and mechanical locking arising from surface irregularities on the steel becoming locked in the concrete. Plain steel bars are usually hooked at the ends as anchorage within the ends of the beam.

Another type of stress that can occur with beams is shear stress and so reinforcement can be included within a beam to strengthen it against shear. This takes the form of stirrups inclined which are either vertical or at 45° (Figure 4.57). If you think of a beam as being like a truss then the 45° stirrups are along the 45° tension members.

Figure 4.56   *Reinforced concrete beam*

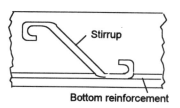

Stirrup

Bottom reinforcement

Figure 4.57   *Reinforcement against shear*

Polypropylene structural foam

Consider the problem of the material for use as the drum on a front-loading domestic washing machine. It has to withstand hot soapy water which, with a load of wet washing, is agitated first one way and then the other, water being pumped in and out, and spinning for drying at perhaps 800 rev/min. The drum has also to support the weight of water and washing, and carry counterbalance weights, hoses and mounting elements. Materials that are used are stainless steel and vitreous-enamelled mild steel; there are, however, heavy costs involved in setting up the tooling to manufacture the drums with these materials. Another possibility is to use a plastic such as polypropylene. However, this material is not stiff enough for use as the end wall of the drum and withstand the dynamic, intermittent, loading that occurs. Glass-fibre filled polypropylene gives an improved stiffness, by a factor of about 2.2, and particularly if the glass fibres have been pre-coated with a chemical which improves the adhesion with the polypropylene (the factor is now about 3.5). The stiffness of the drum is also improved if foamed glass-filled polypropylene is used.

Problems 4.3

1 List some of the reasons that might be invoked for using composite materials.
2 Calculate the tensile modulus in the aligned direction and right angles direction for a composite consisting of 45% by volume of continuous aligned glass fibres,

tensile modulus 76 GPa, in a polyester matrix with a tensile modulus of 4 GPa.

3 Calculate the tensile modulus in the aligned direction and right angles direction for a composite consisting of 45% by volume of continuous aligned carbon fibres, tensile modulus 400 GPa, in a polyester matrix with a tensile modulus of 4 GPa.

4 Calculate the tensile modulus in the aligned direction and right angles direction for a composite consisting of 40% by volume of continuous aligned boron fibres, tensile modulus 380 GPa, in an aluminium matrix with a tensile modulus of 70 GPa.

5 Estimate the tensile strength in the aligned direction of a composite consisting of 50% by volume of continuous aligned carbon fibres, tensile strength 3000 MPa, in a polyester matrix of tensile strength 30 MPa.

6 Estimate the tensile strength in the aligned direction of a composite consisting of 60% by volume of continuous aligned glass fibres, tensile strength 2500 MPa, in a polyester matrix of tensile strength 35 MPa.

7 Estimate the fraction of the load carried by glass fibres in a composite if they are continuous and aligned in the direction of the load and constitute 45% of the volume. The glass fibres have a tensile modulus of 70 GPa and the matrix 2.8 MPa.

8 What is the ratio of the longitudinal modulus of elasticity to the transverse modulus for a composite with continuous aligned fibres constituting 50% of the volume if the tensile modulus of the fibres is 50 times that of the matrix?

9 Explain the significance of the 'critical length' when discontinuous fibres are used for a composite.

10 Calculate the fibre critical length and tensile strength of a composite made with discontinuous fibres of diameter 6 mm, length 2 mm and tensile strength 2000 MPa in a matrix which gives a shear strength of 22 MPa.

11 Calculate the average density of a composite which is made of 80% by volume of silicon carbide whiskers, density 3.10 Mg/m^3, and an aluminium matrix of density 2.77 Mg/m^3.

12 Estimate the longitudinal modulus of elasticity for an aligned discontinuous fibre composite if the fibres constitute 40% of the volume, the fibres have a modulus of elasticity of 400 GPa and the matrix a modulus of 5 GPa and the length efficiency parameter is 0.8.

13 How is the stiffness of a foam related to its density?

14 Estimate the density and compressive modulus of a polypropylene foam which has a tensile modulus of 0.2 GPa. The unfoamed polypropylene has a density of 910 kg/m^3.

15 A laminate consists of one sheet of thickness 3 mm of woven glass fibre-reinforced resin with an elastic modulus of 9.2 GPa, then a layer of thickness 6 mm of random discontinuous fibre reinforced resin with an elastic modulus of 6.2 GPa and finally a layer of

thickness 1.5 mm of glass cloth-reinforced resin with an elastic modulus of 9.7 GPa. Determine the elastic modulus of the laminate when the forces are applied in a direction parallel to the faces of the sheets.

16 A laminate consists of a sheet of polyurethane foam of thickness 10 mm between two sheets of random discontinuous fibre-reinforced polyester, each of thickness 2 mm. Determine the elastic modulus of the laminate when the forces are applied in a direction parallel to the faces of the laminate. The elastic modulus of the foam can be taken as 0.3 GPa and that of the polyester composite sheets 7 GPa.

17 A 5-ply plywood has outer plies of walnut, each of thickness 1 mm and having a modulus of elasticity parallel to the grain direction of 11.2 GPa, and three inner plies of Douglas fir, each of thickness 2 mm. The Douglas fir layers are arranged with the two outer ones with grain directions at right angles to that of the walnut and the inner layer with grain direction the same as that of the walnut. The Douglas fir has a modulus of elasticity of 16.4 GPa parallel to the grain and 0.9 GPa at right angles to it. Determine the elastic modulus of the plywood for forces applied in a direction parallel to the grain direction of the walnut.

18 Show that for a laminate the electrical conductivity of the laminate r_c in a direction perpendicular to the constituent layers is given by:

$$\frac{1}{\sigma_c} = \frac{V_1}{\sigma_1} + \frac{V_2}{\sigma_2} + \frac{V_3}{\sigma_3} + \dots$$

where σ_1, σ_2, σ_3 are the electrical conductivities of the layers and V_1, V_2, V_3 their volume fractions and that in a direction parallel to the plates the conductivity is:

$$\sigma_c = V_1 r_1 + V_2 r_2 + V_3 r_3 + \dots$$

19 Show that for a laminate the thermal conductivity of the laminate λ_c in a direction perpendicular to the constituent layers is given by:

$$\frac{1}{\lambda_c} = \frac{V_1}{\lambda_1} + \frac{V_2}{\lambda_2} + \frac{V_3}{\lambda_3} + \dots$$

where λ_1, λ_2, λ_3 are the thermal conductivities of the layers and V_1, V_2, V_3 their volume fractions and that in a direction parallel to the plates the conductivity is:

$$\lambda_c = V_1 \lambda_1 + V_2 \lambda_2 + V_3 \lambda_3 + \dots$$

5 Forming and joining

Summary

In selecting materials for particular products, due regard has to be paid to the methods that can be used to form the products. For example, with the production of a domestic kettle we might wish to consider a metal and whether such a shape can be produced by deep drawing or casting, or we might consider a polymer and whether it can be produced by injection moulding. We also have to consider whether the spout can be formed with the kettle body or has to be joined to it. This chapter is thus a consideration of the various processes that are commonly used to form metals, polymers, ceramics and composites and methods that are used for joining materials.

Objectives

By the end of this part, the reader should be able to:

- appreciate the range of processes and their characteristics used to form metals, polymers, ceramics and composites;
- appreciate the range of methods that can be used to join materials.

5.1 Forming processes with metals

This section presents an overview of the main forming processes used with metals, the limitations they put on the form of the products and the metals that can be used. The processes can be divided into three main categories:

1 *Casting*
Shapes are produced by pouring liquid metal into a mould.

2 *Forming*
Metals are shaped by plastic deformation, e.g. sheet metal forming, forging, rolling, extrusion.

3 *Material removal*
Shapes are produced by metal removal, e.g. mechanical machining and electromachining.

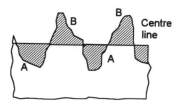

Figure 5.1 *Terms*

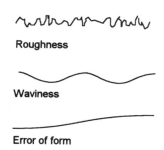

Figure 5.2 *Area between surface above centre line equals area below*

Table 5.1 Relation of roughness to surface texture

Surface texture	Roughness R_a μm
Very rough	50
Rough	25
Semi-rough	12.5
Medium	6.3
Semi-fine	3.2
Fine	1.6
Coarse-ground	0.8
Medium-ground	0.4
Fine-ground	0.2
Super-fine	0.1

Key points

Casting is the shaping of an object by pouring the liquid metal into a mould and then allowing it to solidify. The resulting shape may be that of the final manufactured object, or one that requires some machining, or even an ingot which is then further processed by manipulative processes.

Roughness

In discussing processes we need to consider the roughness of the surfaces produced. *Roughness* is defined as the irregularities in the surface texture which are inherent in the production process but excluding waviness and errors of form. Roughness takes the form of a series of peaks and valleys which may vary in both height and spacing and is a characteristic of the process used. Waviness may arise from such factors as machine or work deflections, vibrations, heat treatment or warping strains. Roughness and waviness may be superimposed on departures of the surface from the true geometrical form. Figure 5.1 illustrates the terms roughness, waviness and error of form.

One measure of roughness is the *arithmetical mean deviation*, denoted by the symbol R_a. This is the arithmetical average of the variation of the profile above and below a reference line throughout the prescribed sampling length. The reference line may be the centre line, this being a line chosen so that the sums of the areas contained between it and those parts of the surface profile which lie on either side of it are equal (Figure 5.2). Thus:

$$R_a = \frac{\text{sum of areas A} + \text{sum of areas B}}{\text{sample length}} \times 1000 \qquad [1]$$

where the sample length is in millimetres and the areas in square millimetres.

Table 5.1 indicates the significance of R_a values in terms of the surface texture. The degree of roughness that can be tolerated for a component depends on its use. Thus, for example, precision sliding surfaces will require R_a values of the order of 0.2 to 0.8 μm with more general sliding surfaces 0.8 to 3 μm. Gear teeth are likely to require R_a values of 0.4 to 1.6 μm, friction surfaces such as clutch plates 0.4 to 1.5 μm, mating surfaces 1.5 to 3 μm.

5.1.1 Casting

Most metal products have at some stage in their manufacture been cast. *Casting* is the shaping of an object by pouring the liquid metal into a mould and then allowing it to solidify. The resulting shape may be that of the final manufactured object, or one that requires some machining, or even an ingot which is then further processed by manipulative processes.

The mould used to form the shape into which the liquid metal is poured has to be designed in such a way that, however complicated the shape, the liquid metals flows easily and quickly to all parts. This has implications for the finished casting in that sharp corners and re-entrant sections have to be avoided and gradually tapered changes in section used. The fluidity of a casting alloy tends to be higher the higher its temperature above the liquidus and for alloy compositions with a narrow freezing range, i.e. pure metals and eutectic composition, than those having a wide temperature region over which they are a mixture of solid and liquid.

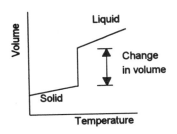

Figure 5.3   *Volume change on
solidification*

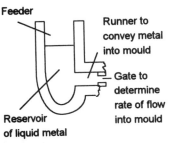

Figure 5.4   *Feeder*

When the liquid metal solidifies there is a change in volume
(Figure 5.3). The volume change depends on the alloy concerned
and is of the order of 1 to 6%. A consequence of this shrinkage is
that a reservoir of liquid metal must be provided to feed into the
mould as the metal in the mould solidifies and shrinks. Such a
reservoir is termed a feeder (Figure 5.4). The ideal feeder is one
that can deliver its full charge of metal into the casting before
solidification is complete.

The grain structure in the cast product is determined by the rate
of cooling (Figure 5.5). The metal in contact with the mould
surfaces cools faster than that in the centre of the casting. As a
consequence small crystals, termed *chill crystals*, start to form at
the surfaces. These are small because the metal has cooled too
rapidly for them to grow to any size. The cooling rate nearer the
centre is, however, much slower and so some of the chill crystals
have time to grow into large elongated crystals perpendicular to
the mould wall, these being called *columnar crystals*. In the centre
of the mould the cooling rate is the slowest. While growth of the
columnar crystals is taking place, small crystals can start to grow
in the central region. These grow in the liquid metal which is
constantly on the move due to convection currents. The final result
is a central region of medium-sized, almost spherical, crystals
called *equiaxed crystals*.

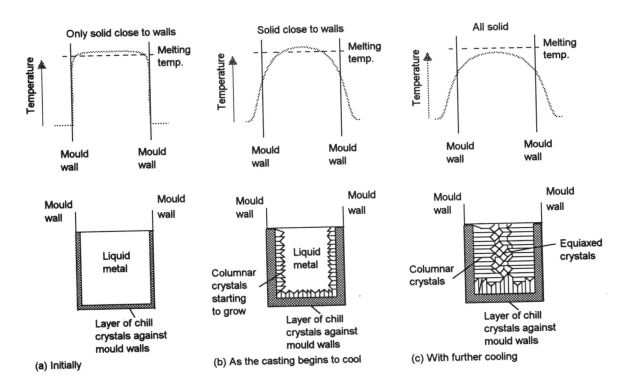

Figure 5.5   *Stages in the solidification of a casting*

In general, a casting structure having entirely small equiaxed crystals is preferred. This type of structure can be promoted by a more rapid rate of cooling for the casting. Castings in which the mould is made of sand tend to have a slower rate of cooling as sand has a low thermal conductivity. Thus sand castings tend to have large columnar grains and hence relatively low strength. Die casting involving metal moulds has a much faster rate of cooling and so gives castings having a bigger zone of equiaxed crystals. As these are smaller than columnar crystals, the casting has better properties. Table 5.2 shows differences that can occur with aluminium casting alloys.

Table 5.2 *Effect of casting process on properties of aluminium alloy castings*

Aluminium alloy	Tensile strength MPa		Percentage elongation	
	Sand cast	Die cast	Sand cast	Die cast
5% Si, 3% Cu	140	150	2	2
12% Si	160	185	5	7

Castings do not show directionality of properties, the properties being the same in all directions. They do, however, have the problems produced by blowholes and other voids occurring during the solidification, e.g. from trapped air, the evolution of gas from the liquid metal as it cools and from local solidification shrinkage.

In making a mould, account has also to be taken of the fact that the dimensions of the finished casting will be less than those of the mould due to the contraction occurring when the metal cools to room temperature. Moulds are generally made in two or more parts, which are clamped together while the liquid metal is poured into them, then separated when the metal has solidified to enable the finished casting to be extracted. Complex castings can be achieved by the use of moulds having a number of parts. Hollow castings or holes or cavities can be achieved by incorporating separate loose pieces inside the mould, these being termed cores.

There are a number of casting methods possible and the factors determining the choice of a particular method are:

- Size, complexity and dimensional accuracy required.

- The number of castings required.

- The cycle time, i.e. the time taken to complete a casting and then be ready to repeat the process.

- The flexibility of the casting process to be adaptable to different forms of product.

- The operating cost per casting.

- The mechanical properties required of the casting.

- The quality of the casting, i.e. surface finish, porosity, non-metallic inclusions.

Sand casting

Sand casting involves the making of a mould using a mixture of sand with clay. This is packed around a pattern of the casting, generally of a hard wood and larger than the required casting to allow for shrinkage. The mould is made in two or more parts so that the pattern can be extracted after the sand has been packed round it. Each mould is destroyed after making just one casting, though the pattern which was used to make the mould can be used repeatedly to make further moulds. Figure 5.6 shows a sectional view of an example of a mould.

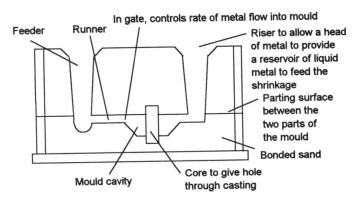

Figure 5.6 *Sectional view of a mould*

Sand casting can be used for a wide range of casting sizes and for small- or large-number production. Patterns are relatively cheap to make and mould making is also relatively easy. It is thus the cheapest process for small-number production and a reasonably priced process for large-number production. It does, however, have a long cycle time since the rate at which heat transfers out of the casting is slow and thus quite some time can elapse after pouring the metal into the mould before the casting is ready. The use of multiple moulds can, however, increase the production rate. Because of the low thermal conductivity of the mould material, large columnar crystals occur and thus the mechanical properties tend to be poor. Surface finish is also poor and porosity and non-metallic inclusions in the casting tend to be common. A wide range of alloys can be cast by this process.

Die casting

In this method the same mould is used for a large number of castings. Such moulds have to be made of a material which will withstand the temperature changes and wear associated with repeated castings and thus are made of metal. This tends to restrict the metals that can be cast to relatively low melting points, e.g. light alloys and some steels and cast irons.

Gravity die casting has the liquid metal poured into the mould and the head of liquid metal is responsible for forcing the metal to flow into the various parts of the mould (Figure 5.7). This method

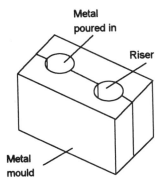

Figure 5.7 *Gravity die casting*

is mainly used for small, simple shapes with only the use of simple cores for items such as holes. The cycle time is limited by the rate of heat transfer out of the casting and, since the mould is metal, this is faster than occurs with sand casting. The surface texture is good. The making of the metal mould is relatively difficult and so expensive.

With *pressure die casting* the liquid metal is injected into a water-cooled mould under pressure (Figure 5.8). This has the advantage that the metal can be forced into all parts of the mould cavity and thus very complex shapes with high dimensional accuracy can be produced. The metal solidifies quickly, since it is in contact with metal walls which have high thermal conductivity. There is a need for high fluidity and thus the method is restricted to low melting point alloys, usually eutectic composition, such as those of aluminium, copper, magnesium and zinc.

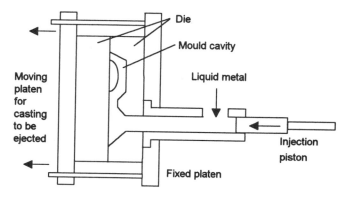

Figure 5.8 *Pressure die casting*

Squeeze casting involves an accurately metered quantity of molten metal being poured into a preheated mould. No runners or feeding system are used. The mould is then closed, pressure applied, and the metal squeezed. This facilitates the flow of the metal into all parts of the mould and also stops entrapped gases bubbling out. As a result, there is a low level of porosity, a good surface texture and, because of the rapid rate of solidification, a fine microstructure. Costs are high because of the tooling required and the accuracy of control required.

There are limitations to the size of the casting that can be produced by die casting, that for pressure die casting being smaller than that for gravity die casting. The cost of the mould is high and thus the process is relatively uneconomic for small-number production. Large-number production is necessary to spread the cost of the mould. These initial high costs may, however, be more than compensated for with large-number production by the reduction or elimination of machining or finishing costs.

Centrifugal casting

Another method which is used to force the liquid metal into the various parts of the mould is known as *centrifugal casting*. The mould is rotated (Figure 5.9) and the forces resulting from this rotation force the metal to coat the walls of the mould. This method is used for relatively long, hollow objects without the need for cores, e.g. large diameter pipes. The method is not suitable for complex castings.

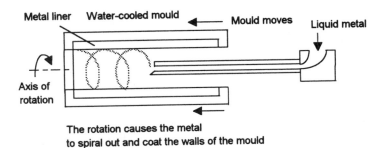

The rotation causes the metal
to spiral out and coat the walls of the mould

Figure 5.9 *Centrifugal casting*

Investment casting

Investment or *lost wax casting* is a process that can be used for metals that have to withstand very high temperatures, and so have high melting points, and for which high dimensional accuracy is required, e.g. aeroplane engine blades. The process is not restricted to high melting point metals but can be used with most metals.

Investment casting (Figure 5.10) involves the following steps:

1 A master pattern is made.

2 A master die is made from the master pattern.

3 Wax patterns are made from the master die by pouring, or injecting under pressure, molten wax into the master die.

4 When cold, the wax patterns are removed from the die and assembled in a tree-like manner on a common feeding and gating system.

5 The cluster of wax patterns is then coated with a refractory slurry. This is repeated until a shell about 5 to 10 mm thick is produced. The result is a refractory mould with a very smooth layer of investment material adjacent to the wax pattern and following every detail.

6 The investment is allowed to harden.

7 The investment is heated to melt the wax pattern and permit it to run out of the mould.

8 The ceramic mould is fired and preheated to the temperature at which the liquid metal will be poured into it.

9 The liquid metal is poured into the mould.

10 When the metal is solidified, the mould is broken away from the casting.

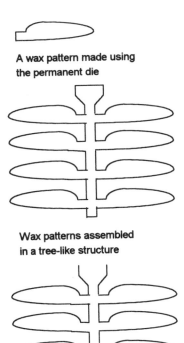

A wax pattern made using
the permanent die

Wax patterns assembled
in a tree-like structure

Tree-like structure mould after
pattern coated with ceramic slurry,
hardened and then wax removed

Figure 5.10 *Investment casting*

The size of castings that can be produced by this method is limited and it is an expensive process for large-number production. Labour costs are high due to the many stages in the process and production rates are fairly low because of the process complexity.

Full mould casting

Full mould casting has many similarities with investment casting but produces a sand casting as the final outcome. The steps involved are:

1   Make a permanent metal die.

2   Inject polystyrene beads into the metal die. Then pass steam through the mould to fuse and expand the polystyrene beads to form a solid polystyrene pattern.

3   After cooling, open the die and extract the polystyrene pattern.

4   Coat the pattern with refractory slurry and dry to form a refractory coating.

5   Place the pattern in loose dry unbonded sand and consolidate the sand by vibration. Pour metal into the polystyrene. The polystyrene melts and the metal fills the resulting mould in the sand. This method avoids the necessity for withdrawal of the pattern from the mould that is necessary with conventional sand casting.

This method can be used for very complex shapes, is very cheap and is suitable for small-number production. The same quality of casting is produced as with conventional sand casting.

Characteristics of casting processes

Each casting method has important characteristics which determine its appropriateness in a particular situation. Table 5.3 illustrates some of the key differences between the casting methods. The following factors largely determine the type of casting process used:

1   *Large heavy casting*
Sand casting can be used for very large castings.

Table 5.3 *Casting processes*

Casting process	Usual materials	Section thickness mm	Size kg	Roughness R_a μm	Production rate, items per hour
Sand	Most	> 4	0.1–200 000	25 to 12.5	1–60
Gravity die	Non-ferrous	3 to 50	0.1–200	3.2 to 1.6	5–100
Investment	All	1 to 75	0.005–700	3.2 to 1.6	Up to 1000
Centrifugal	Most	3 to 100	(25 mm–1.8 m dia.)	25 to 12.5	Up to 50
Pressure die: high pressure	Non-ferrous	1 to 8	0.0001–5	1.6 to 0.8	Up to 200
Pressure die: low pressure	Non-ferrous	2 to 10	0.1–200	1.6 to 0.8	Up to 200

2 *Complex design*

Sand casting is the most flexible method and can be used for very complex castings.

3 *Thin walls*

Investment casting or pressure die casting can cope with walls as thin as 1 mm. Sand casting cannot cope with such thin walls.

4 *Good reproduction of detail*

Pressure die casting or investment casting gives good reproduction of detail, sand casting being the worst.

5 *Good surface finish*

Pressure die casting or investment casting gives the best finish, sand casting being the worst.

6 *High melting point alloys*

Sand casting or investment casting can be used.

7 *Tooling cost*

This is highest with pressure die casting. Sand casting is cheapest. However, with large number production, the tooling costs for metal moulds can be defrayed over a large number of castings, whereas the cost of the mould for sand casting is the same no matter how many castings are made since a new mould is required for each casting.

5.1.2 Manipulative methods

Manipulative processes involve the shaping of a material by plastic deformation processes. Figure 5.11 shows a stress–strain graph for a ductile material. Plastic deformation starts to occur when the stress exceeds the yield stress so the forces required for manipulative processes are those that give rise to stresses in excess of the yield stress, but below that of the tensile strength. Where the deformation is carried out at a temperature in excess of the recrystallisation temperature of the metal, the process is said to involve *hot working*. Figure 5.12 shows the general effect on the stress–strain graph of increasing the temperature. The yield stress is reduced and the amount of plastic deformation possible. It also reduces the work hardening rate. Work hardening occurs as a result of the working of the material increasing the dislocation density. Hot-working processes are rolling, forging and extrusion. Plastic deformation at temperatures below the recrystallisation temperature is called *cold working*. Cold-working processes are cold rolling, drawing, pressing, spinning and impact extrusion.

Alloys listed in tables as *wrought alloys* are ones that have, following casting, been through a hot-working process and had their grains refined to give these improved properties. Most of the materials used for manipulative processes, whether hot or cold working, and cutting processes have been through such processes. However, the forming processes used to give such wrought alloys, and further manipulative processing, can give rise to a

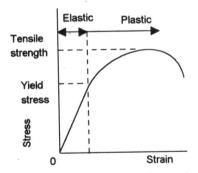

Figure 5.11 *Stress–strain graph*

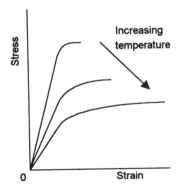

Figure 5.12 *Stress–strain graph*

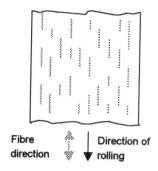

Fibre direction | Direction of rolling

Figure 5.13   *Fibres*

directionality of properties. In addition to the alloying elements present in a metal there are impurities from the fluxes and stages used in the melting process. With the cast ingot these impurities are reasonably randomly distributed but with working they tend to 'flow' with the structure and become oriented as *fibres* in the direction of the working. Thus with rolled products, e.g. sheet, the fibre lines tend to be in a direction parallel to the direction of rolling (Figure 5.13).

Hot and cold working

An increase in the temperature at which a metal is worked means less energy is required to work the metal since it is more malleable at higher temperatures, the work per unit volume required to produce a particular strain being the area under the stress–strain graph up to that strain. High temperatures can mean, however, surface scaling or damage occurring. The initial cast material has coarse grains; hot working breaks the grains down to give a finer structure and thus better mechanical properties. Hot-working processes are rolling, forging and extrusion.

During cold working, the crystal structure becomes broken up and distorted, leading to an increase in mechanical strength and hardness and a decrease in ductility, this being termed *work hardening*. The more the material is worked the harder and more brittle the material becomes. Table 5.4 shows the effect on the mechanical properties of work hardening when a sheet of annealed aluminium is cold rolled and its thickness reduced. A stage in the working can be reached when the material becomes too hard and too brittle to be further worked. With the rolled aluminium sheet referred to in Table 5.4, this condition has been reached with about a 60% reduction in sheet thickness. The material is then said to be *fully work hardened*.

Table 5.4 *Effect of work hardening on mechanical properties*

% reduction in sheet thickness	Tensile strength MPa	Percentage elongation	Hardness HV
0	92	40	20
15	107	15	28
30	125	8	33
40	140	5	38
60	155	3	43

When a cold-worked metal is heated the events that occur depend on the temperature to which it is heated. The events can be broken down into three phases:

1   *Recovery*
    When a cold-worked metal is heated to temperatures up to about $0.3T_m$, where T_m is the melting point on the Kelvin temperature scale of the metal concerned, then the internal stresses resulting from the working start to become relieved.

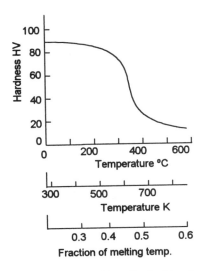

Figure 5.14 *The effect of heat treatment on cold-worked copper*

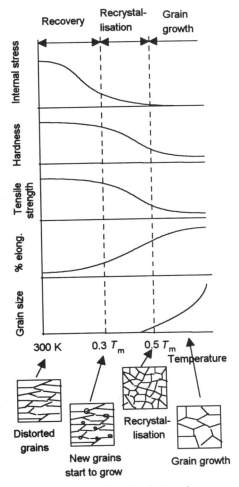

Figure 5.15 *The effect of an increase in temperature on cold-worked materials*

There are no changes in grain structure during this but just some slight rearrangement of atoms in order that the stresses become relieved. This process is known as *recovery*. Copper has a melting point of 1083°C, or 1356 K. Hence stress relief with copper requires heating to about 407 K, i.e. 134°C.

2 ***Recrystallisation***
If the heating is continued to a temperature of about 0.3 to $0.5T_m$ there is a very large decrease in hardness, decrease in strength and increase in elongation. Figure 5.14 shows the effect on the hardness. The grain structure of the metal changes, the metal recrystallising. With *recrystallisation*, crystals begin to grow from nuclei in the most heavily deformed parts of the metal. The onset of recrystallisation is about 150°C for aluminium, 200°C for copper, 450°C for iron and 620°C for nickel.

3 ***Grain growth***
As the temperature is further increased from the recrystallisation temperature, so the crystals grow until they have completely replaced the original distorted cold-worked structure. The hardness, tensile strength and percentage elongation change little during this phase, the only change being that the grains grow.

Activity

Using information from the previous chapter, determine the temperature at which a plain carbon steel with 0.3% carbon will melt. Also estimate the temperatures at which a cold-rolled bar of 0.3% carbon steel will experience recovery, recrystallisation and grain growth.

Figure 5.15 summarises the above effects. The term *annealing* is used for the heat treatment of changing the properties of an alloy by recrystallisation. The factors affecting recrystallisation are:

1 A minimum amount of deformation is necessary before recrystallisation can occur. The amount of permanent deformation necessary depends on the metal concerned.

2 The greater the amount of cold work the lower the crystallisation temperature for a particular metal.

3 Alloying increases the recrystallisation temperature.

4 No recrystallisation takes place below the recrystallisation temperature. The higher the temperature above the recrystallisation temperature the shorter the time needed at that temperature for a given crystal condition to be attained.

5 The resulting grain size depends on the temperature, the higher the temperature the larger the grain size.

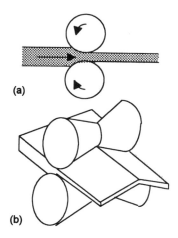

(a)

(b)

Figure 5.16 *Rolling: (a) flat sheet between cylindrical rollers, (b) shaped rolling between contoured rollers*

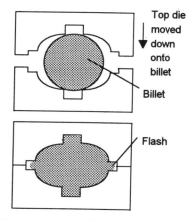

Top die moved down onto billet

Billet

Flash

Figure 5.17 *Closed die forging*

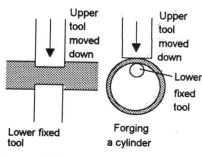

Upper tool moved down

Lower fixed tool

Upper tool moved down

Lower fixed tool

Forging a cylinder

Figure 5.18 *Open die forging*

6 The amount of cold work prior to the heat treatment affects the size of the grains. The greater the amount of cold work the smaller the resulting grain size. The greater the amount of cold work the more centres are produced for crystal growth to be initiated.

Manipulative methods commonly used include rolling, forging, extrusion, drawing and sheet forming.

Rolling

Rolling is the continuous shaping of a metal by passing the metal between the gap between a pair of rotating rollers (Figure 5.16) and can be a hot- or cold-working process depending on the temperature of the material being rolled. When cylindrical rollers are used, the product is in the form of a bar or sheet, but profiled rollers can be used to produce contoured surfaces, e.g. structural sections used for window frames with the rolled product only having to be trimmed to size and joined with other rolled shapes to make the frame. Any metal can be rolled, provided it is ductile at the rolling temperature. Hot rolling is usually the first step in converting ingots and billets to the required shape. Sheet and strip are often cold rolled as a cleaner, smoother finish to the metal surfaces is produced than hot rolling. It also gives, as a result of work hardening, a harder product.

Forging

Forging is a hot-working process and involves the metal being squeezed between a pair of dies, the metal having to be ductile at the forging temperature. Forging can be either closed die forging or open die forging. Closed die forging is used for small components, open die forging normally being used for larger components.

With *closed die forging* the hot metal is squeezed between two shaped dies which effectively form a complete mould (Figure 5.17). The metal flows under the pressure into the die cavity. In order to completely fill the die cavity, a small excess of metal is allowed and this is squeezed outwards to form a flash which is later trimmed away. Closed die forging can be automated and can thus have a very short cycle time. It can be used to produce large numbers of components with high dimensional accuracy and better mechanical properties than would be produced by casting or machining. This is because the fibre direction resulting from the flow of the metal can give useful directionality of properties. The term *drop forging* is used for closed die forging when the impact of a hammer or falling weight is used to force the upper die against the lower die.

With *open die forging* the hot metal is hammered by a vertically moving tool trapping the metal against a stationary tool and squeezing it (Figure 5.18). This type of forging is like that once carried out by the village blacksmith. By repeated hammering a section can be thinned.

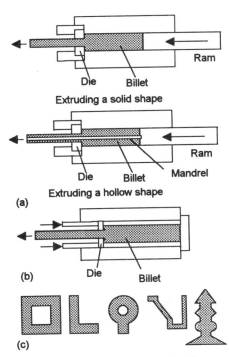

(a)

(b)

(c)

Figure 5.19 *(a) Direct extrusion, (b) indirect extrustion, (c) examples of sections*

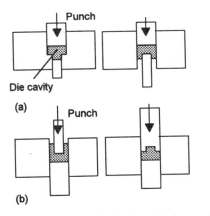

(a)

(b)

Figure 5.20 *Extrusion: (a) forward (b) backward*

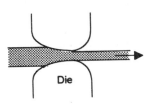

Figure 5.21 *Wire drawing*

Forgings can be produced from any metal that is ductile at the forging temperature concerned. Forging temperatures range from 920 to 1370°C for steels, 590 to 920°C for copper alloys and from 340 to 480°C for aluminium and magnesium alloys.

Extrusion

Extrusion involves metal being forced, under pressure, to flow through a die, i.e. a shaped orifice, to give a product of a small cross-sectional area. The materials used must be ductile and this can mean hot extrusion, the higher temperature often being needed to reduce to manageable levels the pressure required. There are a number of forms the extrusion process can take.

For hot extrusion we have direct extrusion and indirect extrusion. With *direct extrusion* (Figure 5.19(a)) a heated billet is forced through a die. It is rather like squeezing toothpaste out of its tube. Long lengths of quite complex section, including hollow sections, can be extruded. With *indirect extrusion* (Figure 5.19(b)) a heated billet of metal is extruded through a die by the die being pushed into the billet. A wide variety of sections can be produced by hot extrusion, many of which would be uneconomic to produce by any other method. Figure 5.19(c) shows some examples.

Figures 5.20(a) and (b) shows some other variations of the extrusion process, these often being termed *impact extrusion*, in which in (a) a punch is forced into a billet of the material while it is in the die and causes it to flow in the same direction as the punch motion and so produce the required component. This is termed *forward extrusion*. In (b) the metal is made to flow in a direction opposite to that of the punch movement. This is termed *backward extrusion*. Impact extrusion is a cold-working process and so is restricted to the softer more ductile materials. Forward and backward extrusion are used to produce small components such as can-shaped products.

Wire drawing

Wire drawing involves the pulling in of metal through a die (Figure 5.21) in order to reduce the diameter to that required. A number of stages are often used as cold working hardens the metal and so there may be annealing operations between the various drawing stages in order to soften the material so that further drawing can take place.

Sheet formation

The term *deep drawing* is used for the forming of sheets in which a sheet metal blank is pushed into a die aperture by a punch (Figure 5.22), the action causing the metal to flow into the die. The blank is not clamped round its rim and so the sheet is pulled into the die by the action. The term deep drawing tends to be just used for when the depth of the product is one or more times its diameter and the term *stamping* is often used if it is less, though in practice there is no real difference in the operation. The more ductile materials such as aluminium, brass and mild steel are used

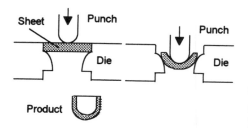

Figure 5.22 *Deep drawing*

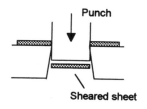

Figure 5.23 *Blanking*

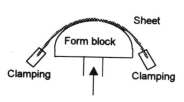

Figure 5.25 *Stretch forming*

and the products are typically deep-recessed parts with vertical walls, e.g. cup- and box-shaped articles, without flanges. Cartridge cases are an example of such a product.

Often two distinct operations such as blanking and deep drawing may be carried out with the same die. *Blanking* is when a shaped blank is sheared from a continuous sheet by the action of a punch (Figure 5.23). Figure 5.24 shows a combination press tool for blanking and then deep drawing a cup-shaped product.

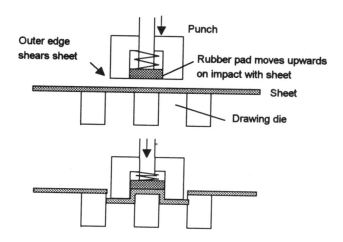

Figure 5.24 *Combination tool for blanking and deep drawing*

Stretch forming involves sheet being gripped round its edges and then a form block pushed into the sheet to deform it (Figure 15.25). Figure 15.26 shows such stretching combined with drawing. Because the blank is clamped round its edges, the action of the form block causes the sheet to be stretched in order to be pushed into the required shape. Ductile materials have to be used. The term *pressing* is often used for this process, it being used to form such products as car body panels, kitchen pans and other cooking utensils.

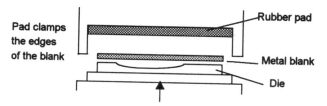

Figure 5.26 *Stretch drawing*

Characteristics of manipulative processes

Table 5.5 shows some of the characteristics of the processes. Compared with casting, wrought products tend to have a greater degree of uniformity and reliability of mechanical properties. The manipulative processes do, however, tend to give a directionality of properties which is not the case with casting.

Activity

Look at products made by rolling, forging, deep drawing and stretch forming. Suggest reasons why the particular manufacturing process has been chosen in each case.

Table 5.5 *Manipulative processes*

Process	Usual materials	Section thickness mm	Minimum size	Maximum size	Roughness R_a μm	Production rate per hour
Closed-die forging	Steels, Al, Cu, Mg alloys	3 upwards	10 cm²	7000 cm²	3.2 to 12.5	Up to 300 items
Roll forming	Any ductile material	0.2 to 6			0.8 to 3.2	
Drawing	Any ductile material	0.1 to 25	3 mm dia.	6 m dia.	0.8 to 3.2	Up to 3000 items
Impact extrusion	Any ductile material	0.1 to 20	6 mm dia.	0.15 m dia.	0.8 to 3.2	Up to 2000 items
Hot extrusion	Most ductile materials	1 to 100	8 mm dia.	500 mm dia.	0.8 to 3.2	Up to 720 m
Cold extrusion	Most ductile materials	0.1 to 100	8 mm dia	4 m long	0.8 to 3.2	Up to 720 m

Note: ductile materials are commonly aluminium, copper and magnesium alloys and to a lesser extent carbon steels and titanium alloys.

5.1.3 Powder processing

Powder processing involves (Figure 5.27):

1 ***Blending the metal powder***
Iron, copper and graphite particles might be blended so that when sintered the required alloy is produced. A lubricant might also be added. This is to enable the particles to flow more easily over each other and so pack better together and give a higher density product with less voids.

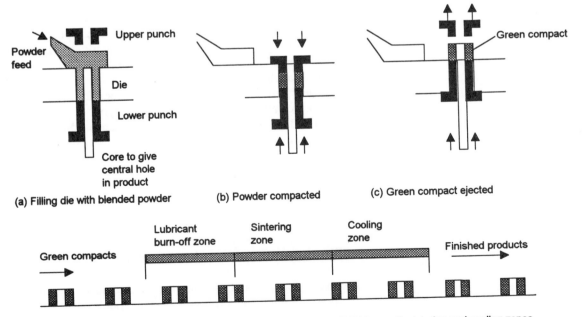

Powder feed
Upper punch
Die
Lower punch
Core to give central hole in product

(a) Filling die with blended powder

(b) Powder compacted

(c) Green compact ejected
Green compact

Green compacts
Lubricant burn-off zone
Sintering zone
Cooling zone
Finished products

(d) Green compacts loaded onto a belt which moves them through lubricant burn-off, sintering and cooling zones

Figure 5.27 *Compacting and sintering*

2 *Compacting the powder*

The powder is placed in a die and compacted, at room temperature, to form the required shape component. Some plastic flow of the particles occurs as a result of this pressure and the component retains its shape when removed from the die. This is termed a *green compact*.

3 *Sintering to produce the solid component*

The green compact is then heated to a high temperature so that the particles coalesce to form the solid component. If the compacted powder had been heated in the die there would have been the problem of having a die material which could withstand the high temperature. If a lubricant is used, then this would be burnt off.

In order to achieve higher densities and to overcome the variations in density that arise from compacting in just two directions, *isostatic pressing* can be used. This involves placing the powder in a deformable container and subjecting it to hydrostatic pressure. With *cold isostatic pressing*, a plastic or rubber mould or 'bag' is filled with the powder and tightly sealed. The bag is then placed in a vessel and subject to hydrostatic pressure to consolidate the powder and produce a green compact.

Powder processing eliminates the need for any machining since the dimensional accuracy and surface finish obtained is good. There is no scrap since there is 100% material utilisation. However, because of the time taken for the sintering, the time to produce a product can be fairly long. The machinery and dies are relatively expensive, particularly if the process is automated. Density variations can occur in the product due to the difficulty of obtaining uniform compression in the compacting of the powder. It is a useful method for the production of components which would be difficult to cast because of high melting point alloys or difficult to manipulate because of being brittle.

Typical applications include iron–8% carbon for moderately loaded gears, iron–5–20% copper–0.8% carbon for medium loaded gears, nickel alloy steel 4% nickel–1% copper–0.7% carbon for wear resisting, high stress components such as differential and transmission gears, copper–nickel–zinc for corrosion resisting conditions such as gears for use in marine environments.

5.1.4 Material removal methods

Material removal methods are the shaping of products by the selective removal of material. Material removal methods are generally a secondary process, following a primary process such as casting or forging, and are used to produce the final shape to the required accuracy and surface finish. Invariably waste material is produced and thus any costing of a material removal process has to allow for this. To keep the waste to a minimum, the primary process should give a product as near the final required dimensions as possible, bearing in mind any need to remove material to give a good surface finish.

Material removal methods can be grouped as:

- **Cutting**

  In this category we have methods which create chips, this including single-point cutting methods, e.g. planing and turning, and multi-point cutting methods, e.g. drilling, sawing and filing, and those methods which do not create chips, e.g. blanking and shearing.

- **Abrasive**

  This includes grinding and honing.

- **Non-mechanical**

  This includes electrical discharge machining and electrochemical machining.

Cutting

Cutting involves using a tool that is of a harder material than that of the workpiece. Tools may be made of high-speed steels, metal carbides or ceramics. In cutting, the tool causes the workpiece material at the cutting edge to become highly stressed and subject to plastic deformation. The more ductile the material the greater the amount of plastic deformation and the more the material of the workpiece spreads along the face of the tool. The deformed chip flows over the tool surface, generating heat as a result of friction (Figure 5.28). It is difficult to lubricate between the chip and the tool face and high frictional forces and secondary shear occur. The more the plastic deformation prior to fracture of a chip the greater the force needed to machine the material and so the greater the expenditure of energy in the machining process. Thus a ductile material on machining gives rise to a continuous chip while a more brittle material leads to small discontinuous chips being produced. Less energy is needed for cutting when this happens.

The term *machinability* is used to describe the ease of machining. A material with good machinability will produce small chips, need low cutting forces and energy expenditure, be capable of being machined quickly and give a long tool life. Ductile and soft materials have poor machinability. A relative measure of the machinability is given by the *machinability index*. With British standards, for steels the plain carbon steel 070M20 is rated as 100% and others compared with it. In the AISI system, an index of 100% is specified for the 1212 plain carbon steel. It is only a rough guide to machinability, but the higher the index the better the machinability. Table 5.6 gives some typical values for steels. The free cutting steels have additives, e.g. sulphur or lead, to aid the formation of chips. Sulphur combines with manganese to give manganese sulphide inclusions which shear more easily. The manganese sulphide also acts as a lubricant.

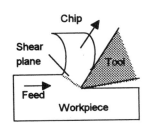

Figure 5.28   *Chip formation*

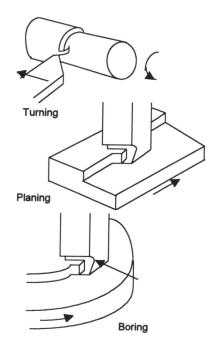

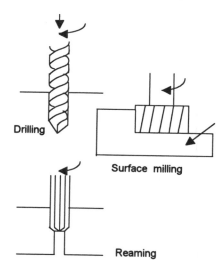

Figure 5.29 *Single point cutting*

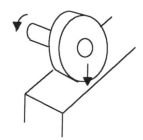

Figure 5.30 *Multiple point cutting*

Figure 5.31 *Grinding*

Table 5.6 *Machinability*

Material	Machinability index
Plain carbon steel	
070M20, 0.20% carbon	100
080M30, 0.30% carbon	70
080M40, 0.40% carbon	70
080M50, 0.50% carbon	50
070M55, 0.55% carbon	50
Alloy steels	
120M36, 1.2% manganese steel	65
150M19, 1.5% manganese steel	70
150M36, 1.5% manganese steel	65
530M40, chromium steel	40
605M36, manganese–molybdenum steel	50
708M40, chromium–molybdenum steel	40
722M24, chromium–molybdenum steel	35
817M40, nickel–chromium–molybdenum steel	35
Free-cutting steels	
210M15	200
212M36	70
214M15	140
220M07	200

Figure 5.29 shows examples of *single point cutting methods*. The tools used have one major cutting edge. Turning involves the workpiece rotating against the tool, the outcome having a circular cross-section. Planing is used to produce flat surfaces or slots. Boring involves the workpiece moving in a circular path against the tool and can be used to reduce the external diameter of a circular form or increase the internal diameter.

Figure 5.30 shows examples of *multiple point cutting methods*. The tools have more than one major cutting edge. Drilling has a tool with multiple cutting edges and, because the cutting action takes place within the workpiece, the chips have to come out of the hole past the drill which itself largely fills the hole. Milling tools have multiple edges with each edge taking its share of the cutting as the workpiece is fed past them. A wide variety of different forms of tools exist with tools designed to produce plane surfaces parallel to or at right angles to the base face, plane surfaces at an angle relative to the base face, key ways or slots, helical flutes and grooves, holes, etc. Reaming involves a multiple cutting edge tool and is designed to increase the diameter of holes.

Abrasive methods

Grinding is a multi-edge cutting operation (Figure 5.31) employing what could be considered as a self-sharpening tool. The grinding wheel has abrasive particles, such as carborundum, bonded in a matrix. As the wheel rotates, these small, hard, brittle particles cut small chips from the workpiece. As the edges of these particles become blunted, the forces acting on them due to friction increase and eventually the force may become large enough to fracture them or tear them free from the matrix. The result, either way, is to expose new cutting surfaces.

Because each cutting edge is very small and the edges are numerous, close dimensional control of the cutting action is possible and very fine surfaces can be produced. Also, because the abrasive particles are very hard materials, the grinding wheel can be used to machine very hard material workpieces. Grinding is used to remove surplus material from a workpiece, improve the dimensional accuracy, obtain the required surface finish or to machine very hard materials which are not so readily turned or milled.

No chip methods

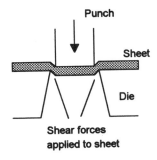

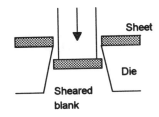

Figure 5.32   *Shearing*

Shearing can be used to cut shapes from metal sheet with no chips being formed. Figure 5.32 shows the basic shearing action. The punch descends onto the metal sheet and deforms it plastically into the die. When the applied stress exceeds the shear strength of the metal, the metal shears through and the 'cutting' occurs. *Piercing* and *blanking* are shearing operations. With piercing, the piece removed from the sheet is the waste item and the perforated sheet the required product. With blanking, the piece removed from the sheet is the required product and the perforated sheet the waste item.

Comparison of machining methods

Machining operations vary quite significantly in cost, particularly if the operation is considered in terms of the cost necessary to achieve particular tolerances, e.g. for a tolerance of 0.10 mm:

Shaping	Most expensive
Planing	
Horizontal boring	
Milling	
Turret (capstan)	Least expensive

The cost of all processes increases as the required tolerance is increased. At high tolerances, grinding is one of the cheapest processes. Machining, in general, is a relatively expensive process when compared with many other methods of forming materials. The machining process is, however, a very flexible process which allows the generation of a wide variety of forms. A significant part of the total machining cost of a product is due to setting-up times when there is a change from one machining step to another. By reducing the number of machining steps and hence the number of setting-up times, a significant saving becomes possible. Thus the careful sequencing of machining operations and the choice of machine to be used is important.

As Table 5.7 indicates, the different machining operations also produce different surface finishes. The choice of process will also depend on the geometric form required for the product. Table 5.8 indicates the processes that can be used for different geometric forms.

Table 5.7 *Surface finishes*

Machining process	R_a μm	
Planing and shaping	25 to 0.8	Least smooth
Drilling	8 to 1.6	
Milling	6.3 to 0.8	
Turning	6.3 to 0.4	
Grinding	1.6 to 0.1	Most smooth

Table 5.8 *Machining processes for particular geometric forms*

Type of surface	Suitable process
Plane surface	Shaping, planing, face milling, surface grinding
Externally cylindrical surface	Turning, grinding
Internally cylindrical surface	Drilling, boring, grinding
Flat and contoured surfaces and slots	Milling, grinding

Non-mechanical methods

Electrical discharge machining (EDM) (Figure 5.33) involves the removal of material by the action of electrical discharges, i.e. sparks. The tool and the workpiece are submerged in a fluid, such as paraffin or light oil. This fluid normally acts as an insulator. A voltage is then applied between the workpiece and the tool and increased until the insulating properties of the fluid break down and a massive pulse of current flows between the tool and the workpiece. This causes part of the workpiece to be vaporised and hence metal to be removed. The fluid cools the vaporised material into 'chips' which are then flushed away and filtered out from the fluid. EDM has the advantages over traditional machining of the hardness of the material being machined not being a factor, very complicated shapes can be produced and there are no mechanical stresses in the workpiece during the metal removal. It has the disadvantages of the workpiece having a surface layer which is different from the parent metal, the process is relatively slow and tool wear is significant.

Electrochemical machining (ECM) involves the removal of material by electrolysis. The tool and the workpiece are immersed in an electrolyte and connected to a power supply (Figure 5.34). The tool is a shaped electrode which becomes the cathode of the electrolytic cell, the workpiece being the anode. Electrically conductive material is removed by electrolysis from the anode and enters the stream of electrolyte and is swept out for filtration. The workpiece becomes machined to a mirror image of the tool.

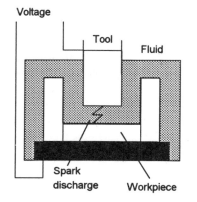

Figure 5.33 *Electrical discharge machining*

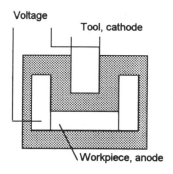

Figure 5.34 *Electrochemical machining*

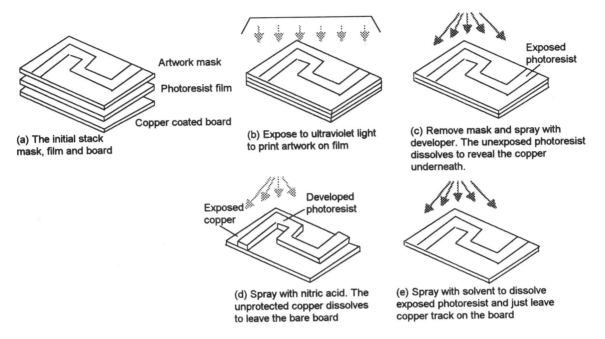

Figure 5.35 *Stages in the production of metal tracks on a printed circuit board*

The term *chemical machining* is used for the removal of material from a workpiece by exposing it to a suitable chemical reagent, the workpiece remaining exposed to the reagent until the required amount of material has been removed. In order to just selectively remove material from some parts of the workpiece, parts are masked to prevent the reagent coming into contact with those parts. This method is used in the production of electrical circuit boards (Figure 5.35) and integrated circuits.

5.1.5 Tool materials

The term 'tool' is taken to include cutting tools, press tools, moulds and dies, drawing and extrusion dies, electrodes for electrical discharge machining, etc. The choice of material for a tool is dictated by the properties required of the tool and the cost. In considering cost, account has to be taken not only of the material and production costs for the tool but also tool life.

The main properties required of a cutting tool are likely to be:

- Room temperature hardness.

- Resistance to thermal softening, i.e. the ability to remain hard at operating temperatures.

- Toughness so that the sudden loading of the tool does not chip or fracture it.

- Wear resistance to maximise the life of the tool.

- Chemical inertness with respect to the workpiece.

The range of tool materials available includes steels; non-ferrous alloys containing chromium, cobalt and tungsten; carbides; diamond and ceramics. Tool steels are grouped into a number of categories (the British code being that of the AISI code prefixed by the letter B). The following are the general categories of tool materials and their properties:

1 *Water-hardening tool steels (code BW)*

These are generally plain carbon steels with a carbon content of about 0.6 to 1.2%. The steels are hardened by quenching in water. Unfortunately they are fairly brittle when very hard, lacking toughness. Where medium hardness with reasonable toughness is required, the steels will have about 0.7% carbon. Such a steel might be used for a cold chisel where the requirement to withstand shocks is essential. Where hardness is the primary consideration and toughness not so important, e.g. taps and dies, a steel with about 1.2% carbon might be used.

2 *Shock-resistant tool steels (code BS)*

The most important property required of this category of steel is toughness since the materials have to withstand shock loading, e.g. as hand and pneumatic chisels, shear blades, punches, blanking dies, etc. Such steels thus have a relatively low carbon content of about 0.5%. A general purpose shock-resistant tool steel is BS5 with a composition of 0.55% carbon, 0.40% molybdenum, 0.8% manganese and 2.0% silicon.

3 *Cold-work tool steels (codes BO, BA, BD)*

Cold-work tool steels are used where the tool is operating cold and where toughness and resistance to wear are important. This category is subdivided into oil-hardening (code BO or O), air-hardening (code BA or A) and high-carbon–high-chromium types (code BD or D).

Oil-hardening cold-work steels are widely used tool steels. They have a high cold hardness and a high hardenability from relatively low quenching temperatures, but cannot be used for hot working or high-speed cutting (where high temperatures result) since they do not remain hard at high temperatures. An example of such a steel is BO1 which has 0.90% carbon, 0.50% tungsten, 0.50% chromium and 1.0% manganese. Such a material is used for blanking dies, bending and forming dies, taps, etc. Because oil quenching is used there is less distortion, cracking and dimensional changes than with water quenching. However, these changes are even further reduced with air-hardening tool steels. They are used where toughness is the main requirement, e.g. for blanking dies, forming rolls, stamping dies, drawing dies, etc., and, since they give smaller dimensional changes after hardening and tempering than other cold-work tool steels, are better for intricate die detail. An example of such a steel is BA2 which has 1.0% carbon, 1.0% molybdenum and 5% chromium.

Cold-work, high-carbon, high-chromium, tool steels have excellent wear resistance. They do, however, thermally soften and so cannot be used for high-speed cutting. Also they are brittle. The high wear resistance is a consequence of the high chromium and high carbon content. For example, BD2 has 1.5% carbon, 1.0% molybdenum, 12.0% chromium and 1.0% vanadium. The molybdenum increases the hardenability of the steel; the vanadium refines the grain size.

4 *Hot-work tool steels (code BH)*

These are tool steels for use in applications, such as hot extrusion and hot forging, where resistance to thermal softening is required and the material is to remain hard and strong at the hot-working temperature. The material must also have good resistance to wear at temperature and be able to withstand thermal and often mechanical shock. There are three main types of hot-work steels, the distinction being in terms of the main alloying element used. The types are those based mainly on chromium (codes BH1 to BH19), on tungsten (BH20 to BH39) and on molybdenum (BH40 to BH59). Chromium, tungsten and molybdenum form carbides which are both stable and hard.

An example of a hot-work chromium tool steel is BH10 which has 0.40% carbon, 2.5% molybdenum, 3.25% chromium and 0.40% vanadium. This steel is used for extrusion and forging dies, hot shears, aluminium die casting dies, etc. An example of a hot-worked tungsten tool steel is BH21 which has 0.35% carbon, 9.0% tungsten and 3.5% chromium. It is used for hot blanking dies, extrusion dies for brass, etc. An example of a hot-work molybdenum tool steel is BH42 which has 0.60% carbon, 6.0% tungsten, 5.0% molybdenum, 4.0% chromium and 2.0% vanadium. It is used for hot extrusion dies, hot upsetting dies, etc.

5 *High-speed tool steels (codes BT, BM)*

Steels that are able to be used for tools which work at high speeds are called high-speed steels. The high speed results in the tool becoming hot and such steels have to be able to maintain their properties up to temperatures as high as 500°C. This means they must be resistant to tempering at such temperatures. The ability of a steel to resist thermal softening in the red-hot temperature range is called *red hardness*. In addition to red hardness, high-speed tool steels must also have good wear resistance and high hardness.

There are two basic types of high-speed tool steels; one where the main alloying element is tungsten (code BT or T) and the other where molybdenum is the main element (code BM or M). Group M steels have slightly greater toughness than group T steels at the same hardness, otherwise the properties are very similar. Group M steels have, however, the advantage of lower cost and so tend to be used in preference to group T steels.

An example of a tungsten high-speed steel is BT4 with 0.75% carbon, 18.0% tungsten, 4.0% chromium, 1.0% vanadium and 5.0% chromium. The cobalt increases the red hardness of the alloy but lowers toughness. Increasing the carbon and vanadium increases wear resistance. Such a steel is used for lathe tools, drills, boring tools, milling cutters, etc. An example of a molybdenum high-speed steel is BM1 with 0.85% carbon, 1.5% tungsten, 8.5% molybdenum, 4.0% chromium and 1.0% vanadium. Such a steel is used for drills, reamers, milling cutters, lathe tools, woodworking tools, saws, etc.

6 *Special-purpose tool steels (code BL)*
In addition to the steels mentioned above, there are some special-purpose low-alloy steels intended for specific tasks. An example of such a steel is BL2 which has 0.45–1.00% carbon, 0.7–1.20% chromium, 0.1–0.3% vanadium and 0.10–0.90% manganese. It is an oil-hardening steel of fine grain size.

7 *Stellites*
Stellites are cobalt alloys and have very high resistance to oxidation and chemicals with very high hardness at both room and high temperatures. Stellite grade 4 has 53% cobalt, 31% chromium, 1.0% carbon and 14% tungsten and is used for extrusion dies for copper and brass.

8 *Carbides*
Tungsten, titanium and tantalum carbides are very hard with high melting points. By themselves they are brittle but when fine particles of the carbide are bonded together with cobalt, a useful tool material, a *cermet* or *cemented carbide*, is produced. The properties can be varied by varying the proportions of carbide and cobalt, the grain size of the carbide and the carbides concerned. The materials produced with tungsten carbide are used for tools for cutting cast iron and non-ferrous alloys. By including titanium and tantalum carbides the wear resistance is improved and tools using these materials are used for machining carbon and alloy steels at high speeds. The high carbide content makes them harder than either the high-speed tool steels or stellites. Many carbide tools are made in the form of 'throwaway' inserts so that when the cutting edge becomes blunt the insert is thrown away.

9 *Diamond*
Diamond is the hardest known material. Its use as a tool is largely limited to the production of highly accurate, smooth, bearing surfaces. Very high cutting speeds are possible.

10 *Ceramics*
These are generally alumina with fine particles sintered to give cutting tips. They are usually used as 'throwaway' tips. High cutting speeds are possible.

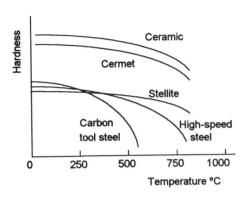

Figure 5.36 *Effect of temperature on hardness*

Figure 5.36 shows in general the effect of temperature on the hardness of the various tool materials. Table 15.4 shows the comparative properties of commonly used tool materials.

Table 15.4 *Properties of tool materials*

Material	Working hardness HRC	Resistance to thermal softening	Resistance to abrasive wear	Relative toughness	Relative cost
BW1	68	Low	Fair	High	1
BS1	45	Medium	Fair	V. good	3
BO1	60	Low	Medium	Medium	3
BA2	62	High	Good	Medium	4
BD2	62	High	V. good	Low	5
BH13	50	High	Fair	V. good	4
BT1	52	V. high	V. good	Low	8
BM2	60	V. high	V. good	Low	7
Stellite 4	50	V. high	V. good	Low	10
Carbides	70	V. high	V. good	Low	10
Alumina	80	V. high	V. good	Low	12

Note: the hardness of the steels is of the material after tempering at the recommended temperature.

Tool coatings

It is the surface of a tool which is subject to the most arduous environment in use and thus changing the surface properties can markedly affect tool performance. One way of doing this is to deposit a layer of different material on to it. Surface-hardening methods such as carburising, nitriding and induction hardening is one way by which the surface properties of steels can be changed by modifying the surface structure of the steels. Chemical and physical vapour deposition are two methods that can be used to deposit thin hard layers, e.g. carbides or nitrides.

With *chemical vapour deposition (CVD)*, gases pass into a heated chamber where they react and produce a chemical deposit on the components such as a cemented carbide tool. For example, titanium chloride, nitrogen and hydrogen gases might react to coat a component with titanium nitride, a very hard-wearing material. Another common coating is titanium carbide. High temperatures are involved, typically about 900 to 1000°C. The deposited layers are typically about 0.005 mm thick. Typically, a titanium carbide-coated carbide has a tool life two to three times longer than the uncoated carbide. Multiple layers, e.g. nitride and carbide coatings, might be used to improve the performance even more.

Physical vapour deposition (PVD) (Figure 5.37) coats the tool material by evaporating or sputtering the coating material. With the evaporation method only the evaporant is heated and the component being coated remains cold. With the sputtering method both the evaporant and the component remain cold with a high voltage being connected between evaporant and component and

Activity

Make a list of the qualities needed for the following common applications and then suggest appropriate tool material:

(a) drilling through 10 mm thick steel plate,
(b) drilling through a brick wall,
(c) cutting through a wooden kitchen work top,
(d) cutting through a concrete section.

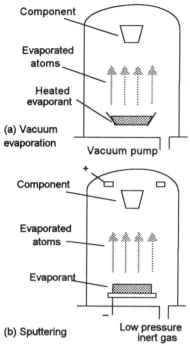

(a) Vacuum evaporation

(b) Sputtering

Figure 5.37 *PVD*

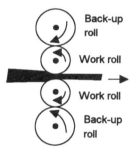

Figure 5.38 *Four-high rolling mill*

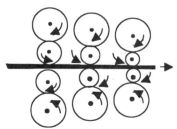

Figure 5.39 *Three-stand tandem mill*

the resulting electric field being responsible for dragging the evaporant onto the component.

CVD is mostly used for coating simply shaped cemented carbide tools while PVD, because high temperatures are not involved, is used for coating complex tools which would distort as a result of the high temperatures involved and also the heat treatment that has to follow.

5.1.6 Case study

Consider the processes involved in the production of the walls of an aluminium alloy beverage can; see the Section in Chapter 3 on aluminium alloys for a discussion of materials. The sequence is:

1 Aluminium from the smelters with aluminium obtained from recycled cans is melted in a furnace and the composition adjusted to give the required aluminium alloy. A grain refiner is added; this provides particles which act as nuclei for crystallisation. The liquid metal is then degassed; this introduces fine bubbles of a nitrogen–chlorine–argon mixture which reacts with dissolved hydrogen and removes it to the atmosphere. Ingots are then cast.

2 In preparing ingots for hot rolling the ingots are homogenised; this involves heating to 570°C and slowly cooling to 510°C in order that diffusion processes can occur and formability is improved as a result of a homogeneous structure being produced. The ingot is then cooled to the hot rolling temperature.

3 The hot rolling involves a number of stages. A four-high breakdown mill (Figure 5.38) is used to reduce the ingot at 450°C to a plate about 40 mm thick and emerges from the rollers at about 300°C. The tandem rolling mill at about 275°C is then used to reduce the plate to sheet about 2.5 mm by steadily reducing the thickness through a sequence of six sets of four-high rollers.

4 During the hot rolling a deformed grain structure is produced and this is removed by an annealing process involving heating to 350°C for about 2 hours at 350°C for recrystallisation.

5 Cold rolling is then used to produce sheet about 0.3 mm thick This takes place in a number of stages, e.g. a tandem arrangement involving three successive set of rollers (Figure 5.39) or perhaps multiple passes through a single set of rollers. The aluminium alloy has then, typically, a 0.2% proof stress of 290 MPa, a tensile strength of 310 MPa and a percentage elongation of 10%.

6 Circular discs of about 140 mm diameter are sheared from the sheet, i.e. blanking. The discs are then deep drawn to give a cup shape. The cup is redrawn and ironed three times to form the can shape. Redrawing is necessary for the very deep cup shape that is required. Ironing is like deep drawing but the

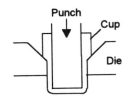

Figure 5.40   *Ironing*

internal diameter remains unchanged with the wall thickness being reduced (Figure 5.40).

7   The top of the cup walls are trimmed by a rotary shear and the can then degreased ready for its painted decoration.

Problems 5.1

1 Explain how, in casting, the nature of the mould material, whether sand or metal, affects the micro-structure of the casting.

2 What factors determine the fluidity of alloys?

3 Explain how chill, columnar and equiaxed crystals grow during casting.

4 Describe the process of sand casting and the characteristic structure of the castings produced.

5 Describe how centrifugal casting can be used to produce large diameter pipes.

6 Suggest a casting method to be used to produce a casting from a high melting point alloy and for which high dimensional accuracy is required.

7 Distinguish between cold- and hot-working processes.

8 Describe the effect on the mechanical properties of a metal of cold working.

9 Explain what is meant by the recrystallisation temperature for cold-worked materials.

10 Describe how the mechanical properties of a cold-worked material change as its temperature is raised from room temperature to about 0.6 times the melting point temperature on the kelvin scale.

11 What factors affect the recrystallisation temperature of a metal?

12 Zinc has a melting point of 419°C. Estimate the recrystallisation temperature.

13 Magnesium has a melting point of 651°C. What order of temperatures would be required to (a) stress relieve, (b) anneal a cold-worked piece of magnesium?

14 How does the temperature at which hot working is carried out determine the grain size and so the mechanical properties?

15 A brass, 65% copper and 35% zinc, has a recrystallisation temperature of 300°C after having been cold worked so that the cross-sectional area has been reduced by 40%.
   (a) How will further cold working change the structure and the properties of the brass?
   (b) To what temperature should the brass be heated to give stress relief.
   (c) To what temperature should the brass be heated to anneal it and give a relatively small grain size?
   (d) How would the grain size and the mechanical properties change if the annealing temperature used for (c) was exceeded by 100°C.

16 Why are the mechanical properties of a rolled metal different in the direction of rolling from those at right angles to this direction.

17 How does a cold-rolled product differ from a hot-rolled product?

18 Explain how extrusion can be used to produce long continuous lengths of uniform cross-section.

19 What shaping process might be used to produce a can, perhaps a Coca Cola can?

20 Explain how powder processing can be used to manufacture products.

21 Explain the term machinability and the difference between materials with good and poor machinability.

22 Explain how electrical discharge machining is used to manufacture products.

23 Suggest the types of tool steels that might be suitable for (a) a punch- press die, (b) a high-speed cutting tool, (c) a hot extrusion die for brass, (d) a blanking die.

5.2 Forming processes with polymers

This section presents an overview of the main forming processes used with polymers, the limitations they put on the form of the products and the types of polymer that can be used.

Many polymer-forming processes are essentially two stage, the first stage being the production of the polymer in a powder, granule or sheet form and the second stage being the shaping of this material into the required shape. The first stage can involve the mixing with the polymer of suitable additives and other polymers in order that the finished material should have the required properties. Second-stage processes for thermoplastics generally involve heating the powder, granule or sheet material until it softens, shaping the softened material to the required shape and then cooling it. For thermosets the second-stage processes involve forming the thermosetting materials to the required shape and then heating them so that they undergo a chemical change to cross-link polymer chains into a highly linked polymer. The main second-stage processes used for forming polymers can be grouped under the general headings of:

- *Moulding*
  This includes injection moulding, reaction injection moulding, compression moulding and transfer moulding.

- *Forming*
  This includes such processes as extrusion, vacuum forming, blow moulding and calendering.

- *Cutting*

The choice of process will depend on factors such as:

- The quantity of items required.

- The size of the items.

- The rate at which the items are to be produced, i.e. cycle time.

- The requirements for holes, inserts, enclosed volumes, threads.

- The type of material being used, i.e. whether thermoplastic or thermoset.

Viscosity of polymers

For liquid for which the flow can be described as *Newtonian flow*, we can imagine the flow as taking place in parallel layers (Figure 5.35), such flow being termed streamline or laminar. The velocity gradient dv/dx produced depends on the shearing force per unit area F/A, i.e. shear stress, the constant of proportionality being called the coefficient of viscosity:

$$\frac{F}{A} = \eta \frac{dv}{dx} \qquad\qquad [2]$$

or:

$$\text{shear stress} = \eta \times \text{shear strain rate} \qquad\qquad [3]$$

where dv/dx is shear strain rate.

For Newtonian liquids, the coefficient of viscosity is independent of the shear stress, the shear strain rate being proportional to the shear stress (Figure 5.36). However, for many real liquids and in particular polymer melts, this is not the case. For most polymer melts, the coefficient of viscosity decreases as the shear strain rate increases, i.e. they become less viscous. Shearing tends to drag out normally entangled polymer chains and so make it easier for them to slide past each other. Thus polymer processing machines which employ high shear strain rates, e.g. screws in injection moulding or extrusion, will make the polymer flow more easily than those which just force polymer melts through a nozzle.

5.2.1 Moulding

Moulding uses a hollow mould to form the product. The main processes are injection moulding, reaction injection moulding, compression moulding and transfer moulding.

Injection moulding

A widely used process for thermoplastics, though it is also used for rubbers, thermosets and composites, is *injection moulding*. With this process, the polymer raw material is pushed into a cylinder by a screw or plunger, heated and then pushed, i.e. injected, into the cold metal mould (Figure 5.37). The pressure on the material in the mould is maintained while it cools and sets. The mould is then opened and the component extracted, and then the entire process repeats itself. The sequence is thus:

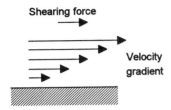

Figure 5.35 *Newtonian flow when flow takes place as layers sliding over each other*

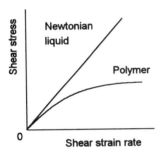

Figure 5.36 *Newtonian liquid and polymer melt*

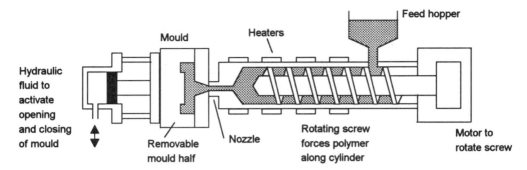

Figure 5.37 *Injection moulding*

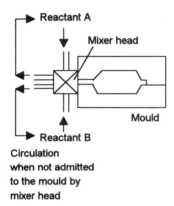

Figure 5.38 *Reaction injection moulding*

1   Mould closed.

2   Nozzle forward.

3   Screw forward.

4   Hold injection pressure.

5   Screw rotates and retracts.

6   Nozzle retracts.

7   Mould opens and moulding ejected.

8   Procedure then repeats

High production rates can be achieved and complex shapes with inserts, threads, holes, etc. can be produced. The process is particularly useful for small components. Typical products are beer or milk bottle crates, toys, control knobs for electronic equipment, tool handles, pipe fittings.

Foam plastic components can be produced by this method. Inert gases are dissolved in the molten polymer. When the hot polymer cools, the gases come out of solution and expand to form a cellular structure. A solid skin is produced where the molten plastic comes in contact with the cold mould surface.

Reaction injection moulding

Reaction injection moulding involves the reactants being combined in the mould to react and produce the polymer (Figure 5.38). The choice of materials that are processed in this way is determined by the reaction time, this must be short, e.g. 30 s, so that cycle times are short. It is mainly used for polyurethanes (mainly elastomeric), polyamides and polypropylene oxide and composites incorporating glass fibres or flakes. The preheated reactants are injected at high speed into a closed mould where they fill the mould and combine to produce the finished product. This method is used for large automotive parts such as spoilers, bumpers and front and rear fascia.

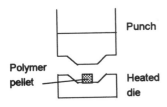

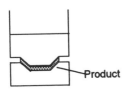

(a) Punch brought down to compress pellet and initiate polymerisation reaction

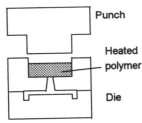

(b) Product then ejected and cycle repeated

Figure 5.38   *Compression moulding*

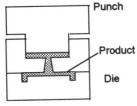

(a) Punch brought down to force polymer into die

(b) Product then ejected and cycle repeated

Figure 5.39   *Transfer moulding*

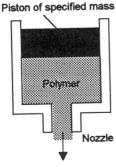

Figure 5.40   *Melt flow indexer*

Compression moulding

Compression moulding (Figure 5.38) is widely used for thermosets. The powdered polymer is compressed between the two parts of the mould and heated under pressure to initiate the polymerisation reaction. The process is limited to relatively simple shapes.

Transfer moulding differs from compression moulding in that the powdered polymer is heated in a chamber before being transferred by a plunger into the heated mould (Figure 5.39).

5.2.2 Forming

Forming processes involve the flow of a polymer through a die to form the required shape, e.g. in extrusion when long lengths of constant section are extruded. Polymer viscosity is thus an important property, determining, among other things, the speed with which the flow can take place and hence the output from the machine. A high production speed would be facilitated by a low viscosity. However, a high viscosity is required if the hot material extruded through the die is not to stretch too much under its own weight while cooling. A balance has thus to be struck between these two demands.

A more convenient measure of viscosity than the coefficient of viscosity for use with polymers is the *melt flow index*. This is the mass of polymer, in grams, which is extruded in a given time through a standard-size nozzle (Figure 5.40) under defined conditions of temperature and pressure. The higher the viscosity of a polymer the longer the time it will take to flow through the nozzle and so the smaller the amount of material extruded in the time, hence a high viscosity means a low melt flow index. The viscosity, and hence melt flow index, of a polymer depends on the molecular chain length. The longer the chain the more chance there is of chains becoming tangled and hence the higher the viscosity and so the lower the melt flow index. Since the molecular weight of a polymer is related to its chain length, the higher the molecular weight the higher the viscosity and the lower the melt flow index. Because the melt flow index is defined with different conditions for different polymers, there is no simple way of comparing melt flow index values for different polymeric materials. However, as a rough guide, for the blow moulding process the value of the melt flow index should not be more than about seven.

Extrusion

A very wide variety of plastic products are made from extruded sections, e.g. curtain rails, household guttering, window frames, polythene bags and film. *Extrusion* (Figure 5.41) is the forcing of the molten polymer through a die. The polymer is fed into a screw mechanism which takes the polymer through the heated zone and forces it out through the die. In the case of an extruded product such as curtain rail, the extruded material is just cooled.

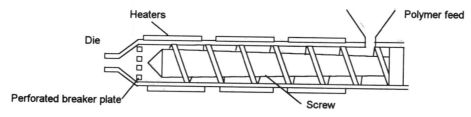

Figure 5.41 *Extrusion*

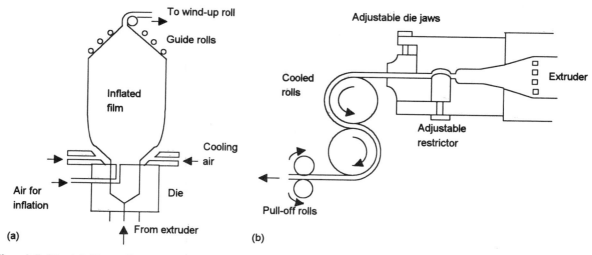

Figure 5.42 *(a) Blown film extrusion, (b) flat film extrusion*

If thin film or sheet is required, a die may be used which gives an extruded cylinder of material. This cylindrical extruded material is inflated by compressed air while still hot to give a tubular sleeve of thin film (Figure 5.42(a)). The expansion of the material is accompanied by a reduction in thickness. Such film can readily be converted into bags. Polyethylene is readily processed to give tubular sleeves by this method but polypropylene presents a problem in that the rate of cooling is inadequate to prevent crystallisation and so the film is opaque and rather brittle. Flat film extrusion (Figure 5.42(b)) can be produced using a slit-die. The rate of cooling, by the use of rollers, can be made fast enough to prevent crystallisation occurring with polypropylene.

Blow moulding

Blow moulding is a process used widely for the production of hollow articles such as plastic bottles. Containers from as small as 10^{-6} m^3 to as large as 2 m^3 can be produced. With *extrusion blow* moulding the process involves the extrusion of a hollow thick-walled tube which is then clamped in a mould (Figure 5.43). Pressure is applied to the inside of the tube to inflate it so that it fills the mould.

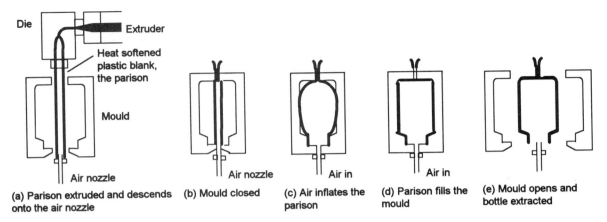

(a) Parison extruded and descends onto the air nozzle

(b) Mould closed

(c) Air inflates the parison

(d) Parison fills the mould

(e) Mould opens and bottle extracted

Figure 5.43 *Blow moulding*

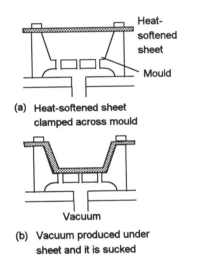

(a) Heat-softened sheet clamped across mould

(b) Vacuum produced under sheet and it is sucked against the mould

Figure 5.43 *Vacuum forming*

Activity

Cut a 1.5 litre lemonade bottle into two identical halves by cutting from its mouth down to its base. Now measure the thickness of the section along the whole length. What can you say about the effect of blow moulding on shape?

Vacuum forming

Vacuum forming is a common method of *thermoforming*. It uses a vacuum on one side of a sheet of heat-softened thermoplastic to force it against a cooled mould and hence produce the required shape (Figure 5.43). Sheets, such as 6 mm thick acrylic, are likely to be preheated in ovens before being clamped but thinner sheets of, say, 2 mm thick are likely to be heated by radiant heaters positioned over the mould. Vacuum forming can have a high output rate, but dimensional accuracy is not too good and such items as holes, threads and enclosed shapes cannot be produced. The method is used for the production of large shaped objects but not very small items.

Calendering

Calendering is a process used to form thermoplastic films, sheets and coated fabrics. The most common use has been for plasticised PVC. Calendering consists of feeding a heated paste-like mass of the plastic into the gap between two rolls, termed nip rolls. It is squeezed into a film which then passes over cooling rolls before being wound round a wind-up roll (Figure 5.44). As shown in the figure, this process can also be used to coat a fabric with a polymer.

5.2.3 Cutting

The processes used to shape a polymer generally produce the finished article with no further, or little, need for machining or any other process. With injection moulding, compression moulding and blow moulding there is a need to cut off gates and flashing; with extrusion lengths have to be cut off. As with metals, single-point and multi-point cutting tools can be used with polymers.

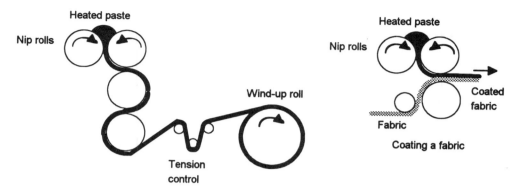

Figure 5.44 *Calendering*

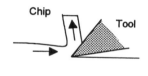

Figure 5.45   *Continuous flow chips*

Several different types of chip can be formed. With slow-speed cutting of a polymer which shows rubber-like elasticity and large elongations before fracture, e.g. polyethylene, continuous flow chips (Figure 5.45) are produced with the polymer just flowing up the surface of the tool. With a less elastic material such as polystyrene, continuous shear types of chips are produced like those occurring with the cutting of metals. Discontinuous shear types of chips are produced when the shear strain at the shear plane becomes larger than the limiting rupture strain. The machined surface is then excessively rough as a result of chips being sheared off. Cracking can also occur round the point of the tool. It is thus desirable to select cutting conditions which result in the formation of continuous chips.

Polymers tend to have low melting points and thus machining conditions, which do not result in high temperatures being produced, are vital if the material is not to soften and deform.

5.2.4 Processing polymers

Issues that can occur when processing polymers are:

- *Consequences of thermal expansion*
  Polymers have high thermal expansivities which result in an amorphous polymer contracting in volume by about 6% and a crystalline polymer by more than twice this when cooling from the processing temperature to room temperature. A volume change of 6% means a change in linear dimensions of 2% if the material is free to contract in all dimensions. Such changes have to be allowed for when the dimensions of moulds and dies are considered. However, the problem is rarely as simple as this, mainly because the ways by which a product can shrink are often restricted by the mould or die.

- *Residual stresses*
  Problems can occur as a result of different parts of a polymer product cooling at different rates during processing. Thus, for example, if a rectangular block of polymer is cooling, the

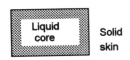

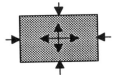

(a) Skin forms as outer
layers cool faster than core

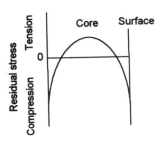

(b) Stresses develop as
core solidifies

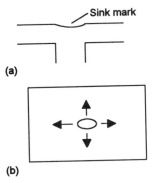

(c) Residual stresses produced

Figure 5.46 *Residual stresses
as a result of cooling*

Sink mark

(a)

(b)

Figure 5.47 *(a) Sink marks,
(b) voids*

outer layers of the material will cool more rapidly than the inner core. This leads to a skin developing and, as the skin is stiffer than the core, it restrains the contraction of the core. This leads to the skin being compressed by the endeavours of the core to shrink and so compressive stresses develop in the skin. The core, since it is restrained by the skin from contracting, is in a state of tension. The net result is a system of *residual stresses* (Figure 5.46). The stresses can have the consequence of *sink marks* developing on the surfaces and might become high enough to cause *voids* to develop in the core (Figure 5.47). Another consequence of the residual stresses is that if the product is machined and the surface layers removed, the equilibrium between the surface and core stresses is disturbed and the product will distort.

• *Crystallisation*
  If a polymer is capable of crystallising then the rate of cooling occurring during the processing can be significant. Too slow a rate of cooling can result in crystallisation and a decrease in transparency. The injection blow moulding of polyethylene terephthalate bottles, the polymer being capable of crystallising, has to be cooled at a sufficiently high rate for insignificant crystallisation to occur to impair the transparency of the product.

• *Orientation*
  Another effect that can occur with polymers during their cooling from the liquid state is *orientation*. Polymer molecules can adopt an alignment in a particular direction if they are subject to a unidirectional stress during cooling from the liquid state. Orientation is greatest in products where the polymer solidifies while the stresses that shape the melt are still acting. If there are glass fibres present in a polymer melt, these can become aligned, like the polymer molecules, during the processing and give a directionality of properties. Thus in the extrusion process, when the material is forced to flow through the die to form the product, e.g. a rod, tube or sheet, orientation can occur as a result of the flow direction and also the slight tension applied by perhaps rollers to pull the product away. The orientated product has a higher tensile modulus and strength in the direction of the pull but a much lower modulus and strength in the transverse direction. With injection moulding, as the melt flows into the cold-mould cavity, unidirectional shear forces develop and result in orientation, the maximum shear forces and hence orientation being at the surfaces of the moulding where the polymer melt is in contact with the cold surface.

  In *stretch-blow moulding*, a combination of blowing and axial stretching can be used to produce a biaxially oriented polymer and so bottles with improved clarity, rigidity and toughness. The hot parison is stretched longitudinally before being blown. The result is a two-way stretching, longitudinal and radial, and so biaxial orientation. This is particularly used in the production of polyethylene terephthalate bottles, widely used for 'fizzy' drinks. Biaxially orientated film is produced by subjecting hot extruded film to a transverse stretching. The result is film with improved clarity and toughness. This is particularly used with films of polymers such as polypropylene and polyethylene terephthalate.

• **Weld lines**

A line of weakness, known as a *weld line*, can form in processing when two flowing streams of polymer melt meet and join together. This can occur if, within a mould, the polymer melt flows round an obstacle to join up on the other side, e.g. round a pillar which is there to produce a hole in the resulting moulding. Another possibility for a weld line occurring is in extrusion blow moulding when the still hot, thick-walled tube has its ends pressed together to form the base of the container. The weld line is a weakness because with a single melt flow the polymer molecules are orientated along the direction of flow at the flow boundary, so that when two flows meet and there is some flow at the flow boundary at right angles to the original direction (Figure 5.47) the direction of orientation changes and we are left with just the lower transverse strength of the polymer at that point. Also there can be inadequate fusion at the junction due to insufficient molecules diffusing across the junction and so bridging it. A weld line is thus a line of weakness.

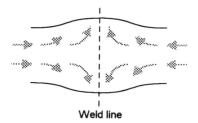

Figure 5.47 *Production of a weld line*

Comparison of polymer processes

Injection moulding and extrusion are the most widely used processes, injection moulding for the mass production of small items, often with intricate shapes, and extrusion for products which are required in continuous lengths or which are fabricated from materials of constant cross-section. The following are some of the factors involved in choosing a process:

• **Rate of production**

Cycle times are typically: injection moulding and blow 10–60 s, compression moulding (cycle time 20–600 s), rotational moulding 70– 1200 s, thermoforming (cycle time 10–60 s).

• **Capital investment required**

Injection moulding requires the highest capital investment, with extrusion and blow moulding requiring less capital. Rotational moulding, compression moulding, transfer moulding, thermoforming and casting require the least capital investment.

• **Most economic production run**

Injection moulding, extrusion and blow moulding are economic only with large production runs. Thermoforming, rotational moulding and machining are used with small production runs. Table 5.8 indicates the minimum output that is likely to be required to make processes economic.

Table 5.8 *Minimum output*

Process	Economic output number
Machining	From 1 to 100 items
Rotational moulding	From 100 to 1000 items
Sheet forming	From 100 to 1000 items
Extrusion	Length 300 to 3000 m
Blow moulding	From 1000 to 10 000 items
Injection moulding	From 10 000 to 100 000 items

Activity

Look at common household and personal items such as:
(a) the cover of a mobile phone,
(b) the body of a plastic kettle,
(c) a plastic bucket,
(d) a plastic ruler.
In each case make notes on the likely manufacturing process, other manufacturing processes that could be used, and why the particular process has been chosen.

- *Surface finish*
  Injection moulding, blow moulding, rotational moulding, thermoforming, transfer and compression moulding, and casting all give very good surface finishes. Extrusion gives only a fairly good surface finish.

- *Metal inserts during the process*
  These are possible with injection moulding, rotational moulding, transfer moulding and casting.

- *Dimensional accuracy*
  Injection moulding and transfer moulding are very good. Compression moulding and casting are good; extrusion is fairly poor.

- *Item size*
  Injection moulding and machining are the best for very small items. Section thicknesses of the order of 1 mm can be obtained with injection moulding, forming and extrusion.

- *Enclosed hollow shapes*
  Blow moulding and rotational moulding can be used.

- *Intricate, complex shapes*
  Injection moulding, blow moulding, transfer moulding and casting can be used.

- *Threads*
  Threads can be produced with injection moulding, blow moulding, casting and machining.

- *Large formed sheets*
  Thermoforming can be used.

5.2.6 Case studies

The following case studies are to illustrate how products are produced from polymers.

Plastic gears

Plastic gears can be made in a number of ways. They might be made by machine cutting from rod or tube, these having been manufactured from the raw polymer by extrusion. Nylon, acetal and polyethylene can be used to make gears in this way.

Another process that can be used is injection moulding in which the required shape is directly obtained as a result of the process. The polymers used with this method are thermoplastics with the most commonly used crystalline polymers being nylon 6, 6.6, 11 or 12, and homopolymer and copolymer acetals. Commonly used amorphous polymers are polycarbonate and styrene modified polyphenylene oxide. Crystalline polymers have better frictional, wear and fatigue properties but amorphous polymers give lower mould shrinkage. Thus amorphous polymers might be preferred if dimensional accuracy is essential. All the polymers used are likely to be reinforced with short-length glass fibres in order to improve the stiffness and strength, increase the maximum service temperature and reduce the coefficient of expansion. The frictional wear properties can be improved by the addition to the polymer of internal lubricants, e.g. PTFE, silicone fluids, molybdenum sulphide. For example, acetal rubbing against acetal has a static coefficient of friction of 0.19 and a dynamic coefficient of 0.15; when both contain 20% PTFE the static coefficient becomes 0.10 and the dynamic coefficient 0.09.

Shrink-wrapped goods

In a supermarket, fresh meat is generally found in trays covered in shrink wrapped or stretch wrapped PVC film. In order that the meat with maintain the required colour associated with fresh meat it is necessary for the wrapping to allow oxygen to permeate through it to the meat. It also has to be tough, able to withstand low temperatures, be clear and be shrinkable. Thin films of PVC with plasticisation gives these properties. The film can be produced by calendering or extrusion.

In situations where shrink wrapping is required but the material has not to be permeable to oxygen or water vapour, orientated polypropylene film can be used. This has the advantage over low density polyethylene film of having greater clarity and sparkle and thus is widely used for the wrapping or toiletry items. Low density polyethylene film is likely to be used for the shrink wrapping of whole pallet loads where clarity and sparkly is not a requisite.

Problems 5.2

1 State a process that could be used for the production of:
   (a) plastic guttering for house roofs,
   (b) plastic bags,
   (c) the plastic tubing for ball-point pens,
   (d) plastic rulers,
   (e) a small lightweight, plastic toy train,
   (f) a plastic tea cup and saucer,
   (g) the plastic body for a camera,
   (h) a hollow plastic container for liquids,
   (i) a plastic milk bottle,
   (j) a car bumper,
   (k) the rubber sealing strip used for double glazing.
2 Describe the types of products produced by the following polymer processes: (a) extrusion, (b) injection moulding, (c) calendering, (d) thermoforming.
3 What are the main types of processes used with thermosetting polymers?
4 Explain what is meant by sink marks and weld lines in polymer products.
5 Explain how orientation can occur with a polymeric material and its effect on the properties.
6 It is proposed that the product design shown in Figure 5.48 should be produced by injection moulding. What problem could you envisage occurring?

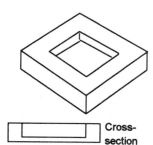

Cross-section

Figure 5.48 *Problem 6*

5.3 Forming processes with ceramics

This section presents an overview of the main forming processes used with ceramics and semiconductor electronics. In general, most ceramic products are made by moulding a powdered mass, with a binder if necessary, into the required shape and then heating it to a high temperature to develop the bonding between the particles and form the product. As most ceramics are both hard and brittle, the shaping process has generally to be the final shape as machining and further working cannot be used.

The methods used for forming ceramics can be grouped as:

- *Wet shaping*
  A wet-mixed mass is formed to the required shape by such methods as slip casting, tape casting and extrusion and then fired.

- *Dry shaping*
  This involves powders being compressed before being fired and includes die pressing, reaction bonding and injection moulding.

- *Manufacture of glass and glass ceramics*
  Glass melts are produced by heating to melting point the mixed raw materials before shaping them by some moulding machinery.

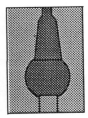
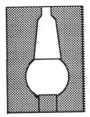

(a) Assemble mould   (b) Fill with slip

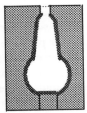

(c) Drain liquid   (d) Trim

(e) Remove mould
and fire

Figure 5.49   *Slip casting*

5.3.1 Wet shaping

The traditional method of *wet shaping* used in the pottery trade involves the shaping of a wet clay-like mass into the required shape, e.g. on a potter's wheel, and then firing it. Such methods are used for largish shapes with axial symmetry, e.g. pots, or hand-shaped pieces.

Slip casting

With *slip casting*, a suspension of clay in water is poured into a porous mould (Figure 5.49). Water is absorbed by the walls of the mould and so the suspension immediately adjacent to the mould walls turns into a soft solid. When a sufficient layer has built up, the remaining suspension is poured out to leave a hollow clay form which is then removed from the mould and fired. This method is used for products such wash basins and other sanitary way, and thin-walled components.

Tape casting

Tape casting is used with alumina ceramics to produce thin, flat, uniform strip which is cut into wafer-like pieces for use as electronic substrates. The wet ceramic is fed onto a continuous moving belt to the required thickness, dried and cut up before being fired.

Activity

Look at some item of ceramic sanitary ware, such as a bathroom basin or cover for a toilet cistern. What can you say about the finish on the outer and inner surfaces?

Extrusion

Extrusion through a suitable die can be used if the product has a constant cross-section. Typical products are long thin rods and tubes.

5.3.2 Dry shaping

Die pressing is commonly used with oxide, carbide and nitride engineering ceramics for the formation of small shapes such as electronic ceramic components. The powdered raw material is packed into a die and then compressed (the method is the same as that used for sintering metals). After compaction the green compact is ejected from the die and then fired to sinter the particles.

Isostatic pressing can be used to overcome the problem of uneven packing of the powder by virtue of die pressing involving pressure being applied uniaxially. This involves a rubber mould bag being filled with the powder and then placed in a chamber of pressure-transmitting fluid. When this is pressurised, the mould is squashed from all directions and so leads to a more uniform packing which is independent of direction.

Many non-oxide ceramics do not sinter as well as a green compact and have to be sintered under pressure and in an inert or non-oxidising atmosphere. *Hot pressing* or *hot isostatic pressing* is thus used. Hot pressing is usually carried out with a graphite die, possibly lined with boron nitride, and is generally limited to simple shapes (Figure 5.50). Hot isostatic pressing involves a squashable die with pressure and heating (Figure 5.51)

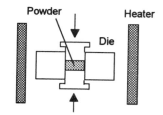

Figure 5.50   *Hot pressing*

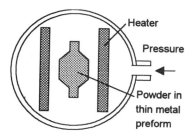

Figure 5.51   *Hot isostatic pressing*

Reaction bonding

A commonly used method for making a non-oxide ceramic is *reaction bonding*. In this process the ceramic is formed in situ by a chemical reaction between the components occurring at the same time as sintering. For example, silicon nitride is formed by heating silicon powder in nitrogen gas; silicon carbide by heating a mixture of silicon and carbon powders. A silicon carbide product is made by mixing silicon and carbon powders with a polymer, the polymer being to give a plastic form which will hold its shape. The product is then formed using the methods normally used for polymers, e.g. extrusion, pressing. The resulting object is then heated to burn out the polymer and then the temperature raised further for the reaction between the silicon and carbon to take place and sintering to occur.

Chemical vapour deposition can be used when the ceramic is formed by the reaction between two vapours, e.g. gases containing silicon and carbon, to produce a ceramic coating on an object.

Injection moulding

Injection moulding (the method is essentially the same as that used for polymers) is used for complex irregular shapes with holes, channels, etc.

5.3.3 Manufacture of glass and glass ceramic products

Glasses are produced by heating the appropriate raw ingredients to give a viscous melt. Figure 5.51 shows how the viscosity of typical glasses varies with temperature and the working range of viscosity required for different processes.

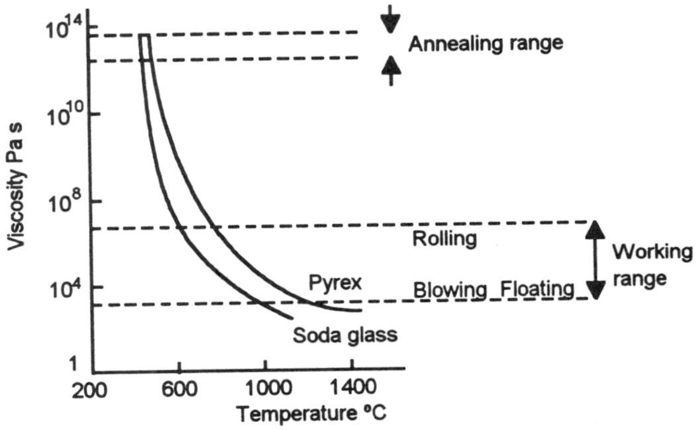

Figure 5.51 *The variation of the viscosity of glasses with temperature*

Activity

Referring to Figure 5.51, estimate the temperatures at which Pyrex glass (a) will begin to flow freely, (b) can be annealed to remove residual stresses.

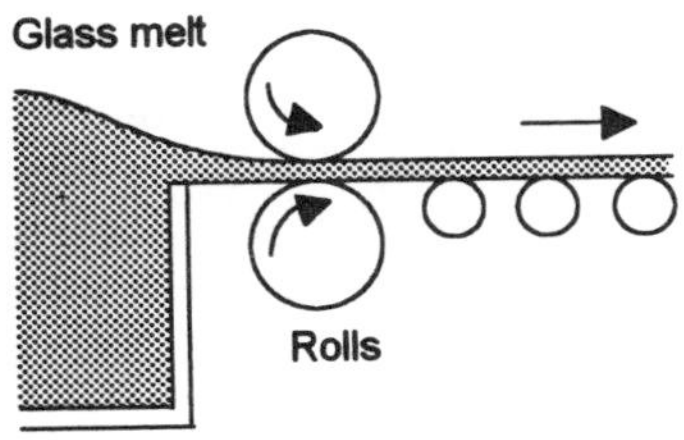

Figure 5.52 *Rolling glass*

Sheet glass can then be made by drawing glass from the melt and flattening it to the sheet shape by passing it between rolls (Figure 5.52). The product is not perfectly flat and parallel and does contain some imperfections. However, most sheet glass is now made by the *float process*. This gives flat, parallel, distortion-free glass. The float process involves the molten glass flowing in a continuous strip on the surface of molten tin (Figure 5.53). For window glass, when it starts flowing onto the molten tin it is at about 1050°C with a viscosity of about 10^3 Pa s and when it leaves the tin as a formed solid sheet it is about 600°C.

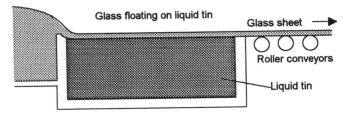

Figure 5.53 *Float glass process*

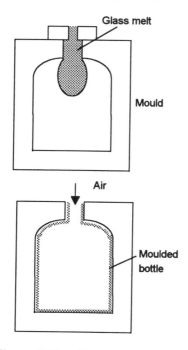

Figure 5.54   *Blow moulding*

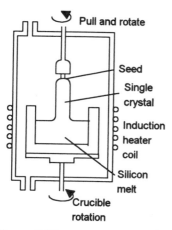

Figure 5.55   *Czochralski crystal puller*

Blow moulding can be used for the production of glass bottles and other glass containers. The glass melt, at a temperature at which its viscosity is just right, flows through an orifice and has a large drop of glass sheared off in a bottle-shaped mould and then air blown into it to cause the glass to expand out to form the bottle (Figure 5.54).

Pressing and extruding can also be used to form glass products. The manufacture of glass ceramics is discussed in Chapter 4.

The cooling of glass from its melt temperature to room temperature is fairly rapid and results in stresses developing in the glass. To eliminate these an *annealing* process is used, the glass being heated to an appropriate temperature and then slowly cooled. For soda glass this temperature is about 500°C. At this temperature sufficient movement of the molecules is possible for the stresses to be relieved without the product losing its shape.

5.3.4 Manufacture of integrated circuits

Integrated circuits involving large numbers of electronic components all on a single, minute, silicon chip are an integral feature of modern life, whether it be as a microprocessor in a computer or the control chip in a domestic washing machine. The following is an outline of the types of processes used to produce such chips.

Purification of silica

Silica can be extracted from silica sand to give a product with about 98% purity. It does, however, require to be much purer than this for use in electronics since the dopants introduced to control the properties of silicon are generally only in concentrations of about 1 part in 10^7. The required purity can be achieved by the use of *zone refining*. This is based on the fact that the solubility of impurities in the material is greater when the material is liquid than when solid. Thus by moving the zone of melting along the length of a silicon rod, so the impurities can be swept up in the molten region.

Crystal growth

In addition to high purity, a single crystal is required, rather than a polycrystalline structure, since the dislocations at grain boundaries and the change in grain orientation can affect the electrical conductivity. A single crystal of silicon can be produced by the *Czochralski technique* (Figure 5.55). Polycrystalline silicon is heated to melting and an existing seed crystal with the required orientation is mounted above the melt surface and brought into contact with it. It is then rotated as it is slowly drawn upwards. As the melt freezes on the crystal it does so in the same orientation and a single crystal rod is gradually created.

Slice preparation

The single rod crystal is cut up into slices of about 500 to 1000 μm thickness by means of a diamond saw. The cutting action leaves some surface damage on the slices and this is removed by lapping. This involves rubbing slices between two plane parallel rotating steel discs while using a fine abrasive, usually alumina. Residual damage is then removed by etching the surfaces with a suitable chemical. The final slice thickness may be about 250 to 500 μm.

Epitaxial growth

The surface of the slices still, however, contains defects and it is the surface layers of a silicon chip in which the circuits are 'fabricated'. *Chemical vapour deposition (CVD)* (Figure 5.56) can be used to improve the surface by depositing silicon vapour onto the surface. Silicon tetrachloride $SiCl_4$ is reduced by hydrogen to deposit silicon atoms on a hot silicon surface.

$$SiCl_4 + 2H_2 \rightarrow Si + 4HCl$$

If the temperature is above about 1100°C, the deposited layer has the same orientation as the substrate. This process is termed *epitaxy* and the deposited layer the *epitaxial layer*.

The epitaxial layer can readily be doped by the inclusion of dopant atoms in the silicon vapour. In this way we can obtain a silicon chip with an n-type or p-type epitaxial layer in which we can then construct transistors and other circuit elements.

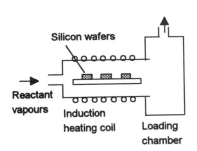

Figure 5.56   *Chemical vapour deposition*

Silicon dioxide

Silicon dioxide SiO_2 has a resistivity of about 10^{15} Ω m, compared with silicon with a resistivity of about 10^3 Ω m. Thus silicon dioxide is a very good insulator. It can be grown on the surface of silicon, bonding extremely well to it, and providing an electrical insulating layer. It also has a much lower diffusion coefficient than silicon for dopants such as boron and phosphorus. This means that silicon dioxide can be used as a mask for the selective diffusion of dopants into silicon.

Silicon dioxide layers can be produced in a number of ways. At temperatures of the order of 1000°C, silicon reacts with oxygen to give silicon dioxide. Silicon can also react with water vapour at about 1000°C to give silicon dioxide and hydrogen (Figure 5.57).

$$Si + O_2 \rightarrow SiO_2$$

$$Si + 2H_2O \rightarrow SiO_2 + 2H_2$$

Thus by passing oxygen or water vapour over silicon slices at about 1000°C, silicon dioxide layers can be produced. Such oxides are said to be *thermally grown* and *dry oxides* if oxygen is used and *wet oxides* if water vapour is used. A problem with such methods is that they require high temperatures, also it involves

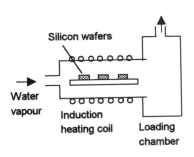

Figure 5.57   *Wet oxide vapour deposition*

converting the surface layers of the silicon into oxide: it 'consumes' silicon. While this may present no problem at an early stage in the manufacture of an integrated circuit, it can disturb existing layers if carried out at a later stage when some layers of the circuit have already been produced.

A process which avoids this problem is chemical vapour deposition using a reaction such as that between silane SiH_4 and oxygen.

$$SiH_4 + 2O_2 \rightarrow SiO_2 + 2H_2O$$

Fairly thick layers of silicon dioxide can be produced with the temperature as low as 400°C. This method also deposits the silicon dioxide on the silicon and does not 'consume' the silicon surface layer.

Doping

The introduction of dopant atoms into a silicon slice can be by *thermal diffusion*. For atoms to diffuse into the surface layers of a silicon slice there must be a concentration gradient, i.e. we need a large concentration of dopant atoms at the surface of the silicon slice. This can be done by exposing the surface to a vapour containing the dopant atoms. An alternative is to deposit on the surface a solid layer, silicon dioxide, containing the dopant and then heating so that diffusion can occur from this layer into the silicon.

A problem with thermal diffusion is that more than one diffusion cycle is likely to be necessary to build up an integrated circuit and the high temperature required for each diffusion causes a redistribution of dopant layers already laid down.

An alternative to thermal diffusion is *ion implantation*. This method involves firing the dopant atoms at high speeds into the silicon slice (Figure 5.58). The point at which the beam hits the slice and the area which it covers can be controlled by applying potential differences to the deflection plates, the depth to which the beam penetrates by controlling the velocity of the ions and the concentration of dopant implanted by controlling the ion current. This method thus allows fine control of the amount of dopant, its depth of penetration and the area in which the implantation occurs. As it is a comparatively low-temperature process, it does not excessively disturb previous processing stages.

A problem with ion implantation is that when surfaces are bombarded with high-velocity ions, in penetrating the surface the beam leaves a trail of damage to the crystal lattice. Also the ions may not end up lodged in sites in the crystal lattice. They need to be so lodged if they are to be active in affecting the electrical conductivity of the silicon. A low-temperature annealing process is thus used to recover the single-crystal structure and enable the ions to be nudged into lattice sites.

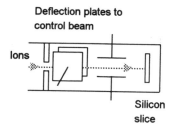

Deflection plates to control beam

Ions

Silicon slice

Figure 5.58   *Ion implanter*

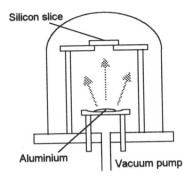

Figure 5.59 *Vacuum deposition*

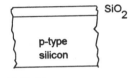

(a) Initially

(b) Coated with photoresist

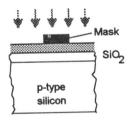

(c) Cover with mask and
expose to UV light

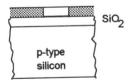

(d) Dissolve unexposed
photoresist

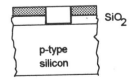

(e) Etch the uncovered
silicon dioxide

Figure 5.60 *Stages in
opening a window*

Electrical connections

Electrical connections need to be made between components in a circuit and also between the circuit and the outside world. This can be achieved by depositing metal on the surface. Aluminium is widely used. The standard method that tends to be used is *vacuum deposition* (Figure 5.59). Small pieces of the metal placed in a tungsten or molybdenum foil boat, or in contact with a filament of such material, which is heated by an electrical current being passed through it. The aluminium melts and vaporises and, because it is in a vacuum chamber, coats any exposed area of silicon slice.

Photolithography and masking

To produce an integrated circuit we need to be able to develop patterns of n- and p-type silicon and aluminium tracks. The procedure for allowing selected parts of the surface of a silicon slice to be ion implanted or diffused with n- or p-type dopants and have metal deposited involves photo-lithography and masking. Thus if we have a silicon dioxide layer on silicon and wish to open a window in the silicon dioxide to enable a dopant to be introduced, the procedure is as follows (Figure 5.60):

1. The initial slice may be p-type silicon which has a surface layer of silicon dioxide (Figure 5.60(a)).

2. Coat the silicon dioxide layer with a negative photoresist. This is initially a monomer or short-chain polymer which is soluble in a solvent (Figure 5.60(b)).

3. Cover the photoresist with a mask which contains the pattern to be transferred to the photoresist. Then expose the photoresist to ultraviolet light (Figure 5.60(c)). Exposure of the negative photoresist to ultraviolet light results in polymerisation or cross-linking of the short chains and a considerable reduction in solubility in the solvent.

4. Develop the pattern in the photoresist by dissolving the non-exposed resist in a solvent (Figure 5.60(d)).

5. Apply an etch to remove the silicon dioxide layer which is not covered by the photoresist (Figure 5.60(e)). Hydrofluoric acid is the standard etch used for silicon dioxide.

6. The result is a window in the silicon dioxide through which we might introduce, for example, n-type dopant.

7. The above procedure can be repeated many times in order to build up the required circuit and components. As an illustration, Figure 5.61 shows the layers that can be used for a resistor. The aluminium provides the connections to the ends of the resistor which are n^+-type silicon. This has been heavily doped and has resulted in such a large number of negative charge carriers that it acts as a good conductor and also provides a region of good contact between the aluminium

and the n-type material. The resistance element is the n-type strip. This might be a channel about 100 µm long and 10 µm wide.

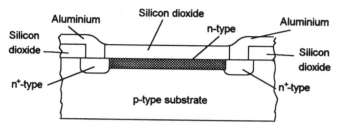

Figure 17.61 *An integrated resistor*

5.3.5 Case studies

See earlier in this chapter for a discussion of the use of chemical and physical vapour deposition to coat tools with ceramic layers and give a much harder tool. Also see Chapter 4 for a discussion of the production of a ceramic glass cooker top.

Domestic ceramics

Domestic ceramics, e.g. plates, are based on clays which are mixed with water to become plastic and formable. They can then be injection moulded, extruded, machined or moulded on a potter's wheel. Items such as plates can be produced by pressing in a mould. Firing is then used to drive off the water, sinter the powder particles and form a silicate glass surrounding the crystalline components of the clay. Glazing is then used to improve the surface appearance. This involves coating the surface with a slurry of powdered silicate glass and then refiring. The glass then flows over the surface and gives it a smooth surface.

Problems 5.3

1 Describe how slip casting can be used to produce a component such as a ceramic tube.
2 Suggest a method that could be used to produce a small, simple, silicon nitride ceramic former for an electronic component.
3 Describe the principle of the float process used for producing sheet glass.
4 Describe how glass bottles can be produced.
5 What are the properties of silicon dioxide which make it so useful in the production of integrated circuits?
6 What are the disadvantages of thermally grown silicon dioxide layers in integrated circuit production compared with those produced by chemical vapour deposition?
7 Compare the methods of doping silicon by diffusion and ion implantation.
8 Suggest the process steps that might be used to produce the part of an integrated circuit structure shown in Figure 5.62.

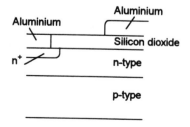

Figure 5.62 *Problem 8*

5.4 Forming processes with composites

This chapter aims to present an overview of the main processes used to form composites with polymer, metal and ceramic matrixes. Fibre-reinforced composites can be produced in a number of ways, depending on whether the fibres are to be continuous or discontinuous throughout the matrix and whether there is to be any specific orientation of the fibres in the product.

5.4.1 Polymer matrix

The following are some of the methods commonly used for the manufacture of fibre-reinforced polymer matrix composites:

- *Hand lay-up*

  The fibres might be in the form of chopped-strand mats or woven fabrics, the fibres being long or short and any required orientation. The mat or fabric is placed in a mould or on a former shaped to the form required of the finished product (Figure 5.63). A liquid thermosetting resin is then mixed with a curing agent and applied with a brush or roller to the fabric. Layers of fabric impregnated with the resin are used to build up the required thickness. Curing, i.e. waiting for the thermosetting polymer bonds to form a network, is usually at room temperature. Such a method is labour intensive but particularly suited to one-offs or small production runs of such items as the hulls of boats.

- *Spray-up*

  This method involves chopped fibres, resin and hardener being sprayed on to a mould. To remove trapped air, the sprayed composite has to be rolled before the resin cures. This method only involves short, randomly orientated fibres and thus gives a lower strength composite than hand lay-up, though the production rate is higher. Such a method is used for items such as sinks and baths.

- *Sheet moulding*

  Layers of fibres are pre-impregnated with resin and partially cured, such a material being referred to as sheet moulding compound (SMC). The fibres can be long or short and orientated. These sheets have a shelf-life of some three to six months at room temperature. The sheets are stacked on the open mould surface and then forced into the mould and the required shape before being fully cured. Figure 5.64 illustrates this when a pair of matched dies are used. This forcing can also be done by vacuum forming, the air between a sheet and a mould surface being removed and so the atmospheric pressure on the other side of the sheet forces it against the mould surface. This method is used for long production runs of such items as doors and panels.

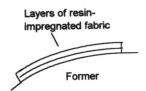

Layers of resin-impregnated fabric

Former

Figure 5.63 *Hand lay-up*

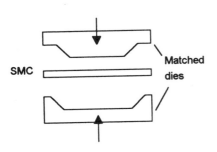

SMC

Matched dies

Figure 5.64 *Sheet moulding*

- **Dough moulding**

  This is similar to sheet moulding but involves using dough moulding compound (DMC). This is a blend of short fibres and resin that has the consistency of bread dough or putty. The DMC is pressed into an open mould and then cured.

- **Prepeg moulding**

  Prepeg materials are long fibres that have already been coated with a partially cured resin. They are formed into the required shape and then either a vacuum bag or light pressure (an autoclave) used to consolidate the layers during curing.

- **Resin transfer moulding**

  This involves fibres and resin being mixed and then injected under pressure into a closed mould before being cured (Figure 5.65). This method is used for such products as fan blades, water tanks, seating, bus shelters and machine cabinets.

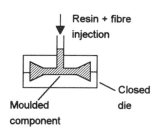

Figure 5.65   *Resin transfer moulding*

- **Pultrusion**

  This method is used for the production of long lengths of uniform cross-section rods, tubes or I-sections. Continuous lengths of fibre reinforcement are passed through a bath of resin and then pulled through a heated die to give the required shape product (Figure 5.66). The reinforcement may be woven fibres, a number of strands collected into a bundle with little or no twist (termed a roving), biaxial materials or random mat. Where unidirectional fibres are used, the longitudinal properties will be significantly different from the transverse properties, e.g. for 75% glass fibres in polyester, a longitudinal tensile strength of 1000 MPa and a transverse tensile strength of 40 MPa. If biaxially balanced fibres are used the strength may be of the order of 200 MPa in both longitudinal and transverse directions. The process is used to produce long lengths of constant-profile sections such as rods, tubes, angle sections, channels, etc. Composites with up to 65% of volume as fibres give composites with a high stiffness, high strength and high strength to weight ratio.

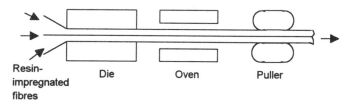

Figure 5.66   *Pultrusion*

- **Filament winding**

  Continuous lengths of fibres are passed through resin and laid out in the required directions on a mandrel (Figure 5.67), this being termed *wet winding*, or pre-impregnated fibres used, this being termed *dry winding*. This can be done using a computer-controlled system so that the fibres are laid down in

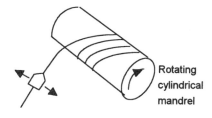

Figure 5.67   *Filament winding*

a predetermined manner to give the required orientations. After curing, the mandrel is removed. Products made using this method are pressure vessels, helicopter blades and storage tanks.

5.4.2 Metal matrix composites

The methods used for manufacturing metal matrix composites can be grouped as:

- **Solid state methods**

  There are two basic methods, *diffusion bonding* and *powder processing*. With diffusion bonding, a fibre mat containing the fibres held in place by a polymer binder is sandwiched between two sheets of foil (Figure 5.68). A number of such sandwiches might be used and then the stack is hot pressed in a die to form the product. It is an expensive process and is generally limited to simple shapes such as plates and tubes. With powder processing, discontinuous fibres or particles are mixed with metal powder and the mixture then hot pressed.

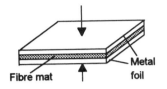
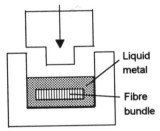

Figure 5.68   *Diffusion bonding*

- **Liquid state methods**

  A simple form of liquid state method is just to mix the fibres or particles with the liquid metal and then cast the metal in one of the usual ways. Uniform mixing is, however, difficult to achieve. Another method is known as *squeeze casting*. This involves the die cavity having a bundle of fibres or fibre mat inserted and then molten metal poured in (Figure 5.69). Pressure is then applied and forces the metal into the fibres.

- **Deposition methods**

  With these methods, the matrix material is vapour deposited or electroplated onto the fibres, the coated fibres often then being hot pressed. An example of a vapour-deposited composite is silicon carbide reinforcement of aluminium. An example of an electroplated composite is tungsten reinforcement of nickel.

(a) Fibre bundle in liquid metal

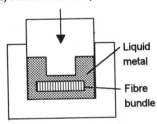

(b) Metal impregnates fibres

Figure 5.69   *Squeeze casting*

5.4.3 Ceramic matrix composites

The methods used for producing ceramic matrix composites include:

- **Wet shaping**

  By incorporating particles or chopped fibres with a suspension of the ceramic in water, the wet shaping methods of slip casting, tape casting and extrusion can be used.

- **Continuous fibres and hot pressing**

  A method (Figure 5.70) that is widely used for glass matrix composites involves drawing fibres through a mixture of powdered glass in water and a water soluble resin binder. The resulting impregnated fibres are then wound onto a mandrel in the required alignment to give a monolayer tape, cut into slices and then stacked to give laminates which, after the resin is burnt off are hot pressed. The method is also used for glass fibre-reinforced ceramics, the glass fibres being impregnated with ceramic.

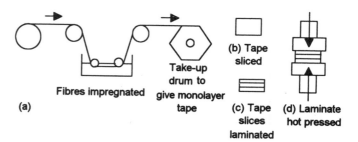

Figure 5.70 *Production of continuous fibre-reinforced glass*

• *Conventional dry pressing*

Die pressing can be used with the matrix constituent, particle reinforcement and a binder and then pressed and fired or hot pressed. There is difficulty in obtaining a homogeneous distribution of reinforcement throughout the product.

• *Vapour deposition*

Chemical vapour deposition involves coating the fibres by the matrix material. It is particularly used for carbon fibres in a carbon matrix, the vapour being produced by using methane CH_4 as the vapour which then breaks down when in contact with the heated preform of fibres to give deposits of carbon on the fibres.

$$CH_4 \rightarrow C + 2H_2$$

The method has also been used for other ceramic matrices, e.g. silicon carbide:

$$SiCl_4 + CH_4 \rightarrow SiC + 4HCl$$

5.4.4 Case studies

The following case studies show how the above manufacturing processes can be used to produce particular products.

Golf club shafts

Consider the use of fibre-reinforced plastic for a golf club shaft. The loading experienced when a club is used is both flexural and torsional and, under such conditions, the stiffness of the shaft has to be comparable to that of a steel shaft. To provide both flexural and torsional stiffness, a shaft can be produced by providing fibres laid down in the way shown in Figure 5.71. Such a composite can

---

Activity

Compare the forming methods used for polymer matrix, metal matrix and ceramic matrix composites. Make a list of similarities and of differences.

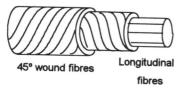

45° wound fibres   Longitudinal fibres

Figure 5.71 *Golf club shaft*

be produced by taking three sheets of fibre reinforcement which has been pre impregnated with catalysed resin, forming them into the required shape and then consolidating them by utilising a vacuum in vacuum bag moulding or pressure in autoclave moulding and curing. Accurate alignment of the fibres is possible with such a method since there is minimal flow during moulding.

Tennis rackets

The function of a tennis racket is to transmit power from the arm of the player to a tennis ball. The requirements for the frame and handle of a racket are a high strength, high stiffness, low weight, tough and able to withstand impact loading, durable and does not creep or warp as a result of exposure to temperature or humidity changes, and can be processed into the required shape. Another requirement is the ability to damp out vibrations; when the ball hits the strings, the impact leads to vibrations of the racket and these are transmitted through the frame of the racket to the arm of the player. If these vibrations are not reduced in amplitude in this transmission, the elbow of the player can suffer some damage, known as tennis elbow. Cost will be a factor when considering tennis rackets for the general population but less a requirement for rackets for professionals.

The requirement for high strength and low weight can be translated into a requirement for a high value of strength/density, i.e. specific strength. Similarly the requirement for high stiffness and low weight into a requirement for a high value of modulus/density, i.e. specific modulus. Possibilities would seem to be wood, metals and composites. Table 5.9 shows typical values for some possible materials.

Table 5.9 *Materials for tennis rackets*

Material	Specific strength MPa/Mg m^{-3}	Specific stiffness GPa/Mg m^{-3}	Relative toughness	Relative vibration damping	Relative cost
Woods					
Ash	107	20	Good	Good	Low
Hickory	105	21	Good	Good	Low
Aluminium alloys					
Al–Cu alloy, precipitation hardened	15	25	Good	Poor	Medium
Al–Mg alloy, annealed	54	25	Good	Poor	Medium
Steels					
Mn steel, quenched and tempered	90	27	Good	Poor	Medium
Ni–Cr–Mo steel, quenched and tempered	115	27	Good	Poor	Medium
Composites					
Epoxy + 60% carbon	890	90	Medium	Medium	High
Epoxy + 70% glass	750	25	Medium	Medium	High

Wood has the advantages that it is tough, has good specific strength, good damping properties for vibrations and is cheap. The specific stiffness could be better. Warping could be a problem. However, this can be overcome by using laminated wood, i.e. several pieces of wood with their fibres in different directions bonded together to give a laminate. This combining together of pieces of wood also gives a method by which the shape of the racket can be obtained.

Aluminium alloys have the advantages of toughness and good specific stiffness. They are, however, more expensive than wood. Another problem is that they have very poor vibration damping. Aluminium can be protected against corrosion attack by damp environments by anodising. An aluminium racket could be made by bending extruded hollow sections into the required shape.

Steels can give high specific strengths and high specific stiffness. The steels with these high strengths are likely to be comparable in price with the aluminium alloys. Problems are, however, the very poor vibration damping and the poor corrosion resistance in a damp environment. A steel racket could be made by bending extruded hollow sections into the required shape.

Composite materials can be made which have the advantages of very high specific strengths, very high specific stiffnesses, reasonable vibration damping and tolerable toughness. A composite racket can be made by injecting a melt of a polymer containing carbon fibres (nylon 66 with 30% by volume short-length carbon fibres) into a racket-shaped mould. This would give a racket with a solid composite for the frame and handle. The procedure that can then be adopted to improve the properties is, while the racket is still in the mould and only the outer skin of the composite has solidified, to pour out the liquid core so that when the racket solidifies there is a hollow tube. An alternative method of obtaining a hollow core is to position a bismuth–tin core into the moulding tool, inject the melt of the polymer with carbon fibres round it and then melt out the metal by immersion in an oil bath. The core can then be filled with a polyurethane foam to improve the vibration damping of the racket.

In comparing the above, the composite material racket gives the best properties but is considerably more expensive than the others. For cheapness and properties, wood is probably the next best material, followed by aluminium alloys with steel being the worst.

Problems 5.4

1 Suggest methods that can be used to make the following products: (a) a one-off boat hull with glass fibre-reinforced polymer, (b) mass production of door panels for lorries in glass fibre-reinforced polymer, (c) mass production of fan blades in glass fibre-reinforced polymer, (d) nose cones for missiles in continuous carbon fibre-reinforced carbon.

2 Explain the method of squeeze casting used for metal matrix composites.

5.5 Joining materials

This section is an overview of the main methods that are used for joining components. We can classify the methods as:

- *Chemical methods*
  Sticking a stamp on an envelope is an example of a chemical joining method, i.e. an adhesive, to join a stamp to an envelope. A chemical reaction is used to effect the joint.

- *Physical methods*
  Soldering a wire to an electrical contact is an example of such a method. In this group of methods we can include soldering, brazing and welding. These methods depend on changes of state from liquid to solid to make the joint.

- *Mechanical methods*
  Nailing or screwing two pieces of wood together is an example of a mechanical method being used to join two components. The term *fastening systems* is often used for these joining methods. They depend on stresses set up by the fastener.

The factors that determine the joining process to be chosen for a particular application are:

- *The materials involved*
  For example, are the two components being joined the same material or dissimilar materials?

- *The shape of the components being joined*
  For example, are the two components perhaps sheets or complex three-dimensional shapes?

- *Permanent or temporary joint*
  Do the components have to be separated at some time? For example, a welded joint can be considered to be reasonably permanent but two materials held together by a bolt can be considered to be easily separated and then reconnected.

- *Limitations imposed by the environment*
  For example, a joint between dissimilar materials can result in thermal expansion mismatch when the temperature changes and result in stresses being developed which might cause the joint to fail.

- *Cost*

5.5.1 Adhesives

An *adhesive* can be defined as any substance that is placed between two surfaces in order to hold the two surfaces together by means of a chemical reaction. For an adhesive to work it must wet the surfaces being joined. This means that it should spread over the surface and not remain in globules (Figure 5.72).

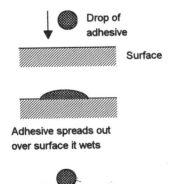

Drop of adhesive

Surface

Adhesive spreads out over surface it wets

Adhesive remains as a globule on surface it does not wet

Figure 5.72 *Wetting a surface*

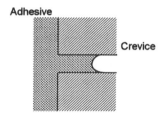

Adhesive

Crevice

Figure 5.73 *Capillary flow into a crevice*

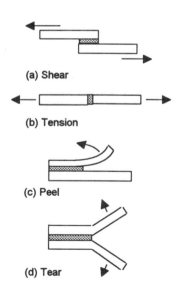

(a) Shear

(b) Tension

(c) Peel

(d) Tear

Figure 5.74 *Loading modes*

When two surfaces are bonded together by an adhesive between them, then the surfaces can be considered as being held together by intermolecular forces between molecules in the adhesive and in the surface and also by mechanical bonding resulting from adhesive having flowed into crevices in the surface and, following solidification of the adhesive, giving the interlocking adhesive protrusions. For this reason, surfaces are generally roughened before the adhesive is applied.

For mechanical interlocking to occur, the adhesive has to penetrate into surface crevices. When a capillary tube is dipped into water, the water rises up the tube. The same situation arises with the movement of liquid adhesive into crevices (Figure 5.73).

Joint strength

The joint strength of an adhesive joint depends on the way the joint is loaded (Figure 5.74). The maximum strength tends to be when the joint is loaded in shear, with tension, peel and tear modes being generally much weaker.

Typically, a room temperature-cured epoxy resin adhesive has a shear strength at room temperature of about 17 MPa, the strengths for other failure modes being lower. Thus, if a joint is required to withstand a load of, say, 500 N then with the joint loading in shear the bonded area must at least be $500/(17 \times 10^6)$ m^2, i.e. 29 mm^2.

Failure of an adhesive bonded joint can occur due to either the bonds between the solid and the adhesive failing or the bonds within the adhesive layer failing. For many adhesives, the bonds between it and the solid are stronger than those within the adhesive itself. For this reason, thin layers of adhesive are preferable to thick layers.

The strength of an adhesive bonded joint thus depends on the type of adhesive used, the mode of loading, the area that is bonded and the thickness of the adhesive layer. In addition, the maximum strength of a joint depends on the curing time and temperature, and in some instances the pressure applied during the curing.

Types of adhesives

Adhesives can be classified according to the type of chemical involved:

- ### Natural adhesives
  Vegetable glues made from plant starches are typical examples of natural adhesives. These types are used on postage stamps and envelopes. However, such adhesives give bonds with poor strength which are susceptible to fungal attack and are also weakened by moisture. They set as a result of solvent evaporation.

- ### Elastomers
  Elastomeric adhesives, e.g. natural rubber, butyl, nitrile, neoprene and polyurethane, are based on synthetic rubbers and set as a result of solvent evaporation. Strong joints are not

produced as they have low shear strength. The adhesive is inclined to creep. These adhesives are mainly used for unstressed joints and flexible bonds with plastics and rubbers.

- ### *Thermoplastics*

  These include a number of different setting types. An important group are polyamides, these are applied hot, solidify and bond on cooling. They are widely used with metals, plastics, wood, etc. and have a wide application in rapid assembly work such as furniture assembly and the production of plastic film laminates.

  Another group are the acrylic acid diesters which set when air is excluded, the reaction being one of the build-up of molecular chain length. Cyanoacrylates, the 'super-glues', set in the presence of moisture with the reaction taking place in seconds. This makes them very useful for rapid assembly of small components.

  Other forms of thermoplastic adhesives, e.g. polyvinyl acetate, set by solvent evaporation.

  In general, thermoplastic adhesives have low shear strength and under high loads are subject to creep, so that they are generally used in assemblies subject to low stresses. They have poor to good resistance to water but good resistance to oil. They tend to be used for bonding mainly non-metals such as wood and paper.

- ### *Thermosets*

  These set as a result of the build-up of molecular chains to give a rigid cross-linked bond. Epoxy resins, such as Araldite, are one of the most widely used thermoset adhesives. They are two-part adhesives, in that setting only starts to occur when the two components of the adhesive are brought together. They will bond almost anything and give strong bonds which are resistant to water, oil and solvents.

  Phenolic resins are another example of thermoset adhesives. Heat and pressure are necessary for setting. They have good strength and resistance to water, oil and solvents and are widely used for bonding plywood.

- ### *Two-polymer types*

  Thermosets by themselves give brittle joints, but combined with a thermoplastic or elastomer give a more flexible joint. Phenolic resins with nitrile or neoprene rubbers have high shear strength, excellent peel strength, good resistance to water, oils and solvents and good creep properties. Phenolic resins with polyvinyl acetate, a thermoplastic, give similar bond strengths but with even better resistance to water, oils and solvents. These adhesives are used for bonding laminates and metals. Joints using them can be subjected to high stresses and can often operate satisfactorily up to temperatures of about 200°C.

Activity

Take a look at several commercially available glues, such as 'super glue' and wood glue, and compare the guidelines given on the packaging as to use. How is each glue to be used and what type of bonds can be achieved by each glue?

Advantages and disadvantages of adhesives

The use of adhesives to bond materials can have the advantages over other joining methods of:

* Dissimilar materials can be joined together, e.g. metals to polymers. The adhesive bond can act as an insulator between different metals and reduce galvanic corrosion.

* Jointing can take place over large areas.

* A uniform distribution of stress over the entire bonded area is produced with a minimum of stress concentration.

* The bond is generally permanent.

* Joining can be carried out at room temperature or temperatures close to it. This prevents problems occurring as a result of a high temperature producing structural changes in the bonded materials.

* A smooth finish is obtained.

Disadvantages are:

* The optimum bond strength is not produced immediately, a curing time being necessary.

* The bond can be affected by environmental factors such as heat, cold and humidity. Many adhesives cannot be used above about 200°C.

* Adhesive bonds are susceptible to peeling.

5.5.2 Soldering and brazing

Soldering and *brazing* are bonding processes which involve the use of a bonding alloy which has a lower melting point than that of the materials being joined and alloys with them when melted. Solders have lower melting points than brazes, the term solder being used if it melts below 450°C and braze if it melts above.

The soldering or brazing process requires the solder or braze to wet the surfaces being joined and alloy with them. Cleaning of the surfaces is thus necessary. *Fluxes* are then used to dissolve surface contaminant films and prepare the surfaces for soldering/brazing. The flux also serves the function of protecting the molten solder/braze from oxidation during the process. The procedure for soldering or brazing (Figure 5.75) is thus to raise the temperature so that the flux melts, dissolves any surface contamination, and wets the components being joined. As the temperature is further increased, the solder or braze melts and displaces the flux from the surfaces being bonded. After cooling, the contaminated flux is then removed to avoid the possibility of it corroding the metal.

Soldered and brazed joints tend to be weak in tension and thus joints need to be designed to reduce tensile stresses to a minimum.

Key point

Soldering and *brazing* are bonding processes which involve the use of a bonding alloy which has a lower melting point than that of the materials being joined and alloys with them when melted.

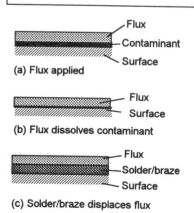

(a) Flux applied

(b) Flux dissolves contaminant

(c) Solder/braze displaces flux

Figure 5.75 *Flux*

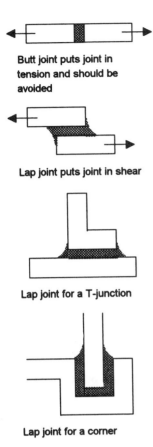

Butt joint puts joint in
tension and should be
avoided

Lap joint puts joint in shear

Lap joint for a T-junction

Lap joint for a corner

Figure 5.76 *Joint design*

The use of lap-joints (Figure 5.76) enables the joints to be put in shear rather than tension and are thus to be preferred to butt joints.

Soldering

The joining agent, the solder, is different from the two materials being joined but alloys locally with them. The joining agent, the solder, is heated together with the materials being joined until it melts and alloys with their surfaces. On cooling, the alloy solidifies and forms a bond between the two materials. By definition, the highest temperature which is necessary for soldering is 450°C, though temperatures less than 300°C are often used. The traditional solders are alloys of tin and lead. The composition used depends on the metals being joined and the type of joint concerned. (See Chapter 3 for the phase diagram and discussion for lead–tin alloys). The eutectic composition, 61.9% tin–38.1% lead, changes from liquid to solid without any change in temperature. Alloys with different compositions go through a 'pasty' phase when changing from liquid to solid, there being a mixture of liquid and solid existing over a range of temperatures. A solder with the eutectic composition is commonly used for electrical work where automation is employed, no 'pasty' phase being desirable in case the soldered components move while the solder is setting. Plumber's solder has a 50–50 composition since some degree of movement is desirable while the solder is setting and also it allows for the joint to be wiped smooth during the setting. Above 100°C, a 95% tin–5% antimony solder may be used.

Brazing

Brazing is similar to soldering but involves temperatures above 425°C. Brazing can be used with aluminium and its alloys, nickel and copper alloys, cast iron, steels and other metals such as the high-temperature superalloys. Dissimilar metals can be joined.

The procedure for brazing is similar to that for soldering, but the end result is a stronger joint. Brazing may be achieved by techniques which just involve heating a band of metal on either side of the joint by means of, perhaps, a torch or by heating the entire component in a furnace. The temperatures used for brazing can affect the mechanical properties of hardened and tempered steels or precipitation-hardened alloys. This can mean, in the case of a torch-brazing operation, that a band of metal on either side of the joint is effectively re-tempered by the brazing operation and thus will have different properties to the rest of the metal. With furnace brazing, the entire component will be effectively re-tempered. Likewise, similar changes in properties can occur with precipitation-hardened alloys.

The term 'braze' comes from the original use of brass as the substance to make the joint. Such alloys are still used for joining steels, but now a wide variety of brazes exists with the choice being made according to the metals being joined and whether torch (Table 5.10) or furnace brazing is used.

Table 5.10 *Recommended brazes for torch brazing*

Metals being joined	Recommended braze
Aluminium and some aluminium alloys	Aluminium–silicon, aluminium–silicon–copper
Copper	Silver, bronze, copper–phosphorus, silver–copper, copper–silver–zinc, copper–silver–phosphorus, copper–silver–zinc–cadmium–nickel, copper–silver–zinc–cadmium, silver–copper–palladium, gold–copper
Copper-based alloys	Silver, copper–phosphorus, silver–copper, copper–silver–zinc, copper–silver–phosphorus, copper–silver–zinc–cadmium– nickel, copper–silver–zinc–cadmium, silver–copper–palladium, gold–copper
Mild steel, carbon steels and alloy steels	Copper–zinc, bronze, copper–silver–zinc, copper–silver–cadmium–nickel, copper–silver–zinc–cadmium, nickel–chromium–boron, nickel–chromium–silicon–boron, silver–palladium–manganese, nickel–palladium–manganese, copper–palladium, silver–copper–palladium
Nickel-based alloys	Copper–zinc, bronze, copper–silver–zinc, copper–silver–zinc–cadmium–nickel, copper–silver–zinc–cadmium, nickel–chromium–silicon–boron, nickel–silicon–boron, silver–palladium–manganese, nickel–palladium–manganese, copper–palladium, silver–copper–palladium
Cobalt-based alloys	Bronze

For aluminium and its alloys, the aluminium–silicon eutectic composition (12.6% silicon) is the basis of the brazes generally used. This has a melting point of 577°C. The strength of joints with this braze usually exceeds that of the metals joined.

For copper and its alloys, the traditional braze is the eutectic composition silver–copper alloy. This has a composition of 72% silver and a melting point of 780°C. The strength of joints with this braze usually exceeds that of the metals joined. The addition of zinc to give a copper–silver–zinc alloy with a eutectic composition of 20% copper, 55% silver and 25% zinc and a melting point of 677°C enables components to be joined at a lower temperature and thus tends to be used where the microstructures of the joined alloys are too unstable when heated to the higher temperature of the silver–copper braze. The addition of cadmium reduces the melting temperature even further. With silver–copper or copper–silver–zinc brazes and ferrous materials, the strength of the joints tends to be about 235 to 380 MPa.

For the nickel-based superalloys, brazes based on the nickel–palladium eutectic composition alloy (40% palladium) are often used. This has a eutectic melting temperature of 1235°C and excellent oxidation resistance, a requirement for a superalloy component such as a turbine blade to be used at high temperatures.

For the brazing of carbide tool inserts to carbon steel tool holders, the brazing alloy used is the copper–phosphorus eutectic (7.5% phosphorus).

The standard flux which has been used for centuries, and still is used for metals such as mild steel, copper and brass, is borax. Alkaline-halide-type fluxes are used with aluminium and aluminium-base alloys.

The strength of a brazed joint is markedly affected by the joint clearance, i.e. the separation of the two joint material faces. The maximum strength tends to occur with joint clearances of the order of 0.1 mm, greater clearances resulting in a considerable reduction in strength. The reason for this is that a degree of alloying occurs between the brazing material and the metal or metals being joined. If the clearance is too large, not all the material in the joint changes to the new alloy and so the strength is not increased. With joint clearances less than 0.1 mm it is difficult to obtain sound joints and so the strength is not so high.

5.5.3 Welding

With brazing and soldering, the joint is effected by inserting a metal between the two metal surfaces being joined, the inserted metal having a lower melting point than that of the materials being joined. With *fusion welding*, the joint is effected directly between the parts being joined by the application of heat to melt the interfaces and so cause the materials to fuse together. Fusion welding thus requires the melting point of the components to be exceeded. With *solid-state welding*, the joint is effected by applying pressure to bring the interfaces of the joint materials into intimate contact and so fuse the two together. Ductile materials such as copper or aluminium can be joined by this method at room temperature. Most materials, however, require heat to render them sufficiently ductile.

Welding processes are capable of producing high-strength joints. The temperatures involved in making the welds may, however, cause detrimental changes in the materials being joined. These may be local distortions due to uneven thermal expansion, residual stresses, or micro- structural changes.

Welding processes

There are four main types of process used for fusion welding:

- *Electric arcs*

  With electric arc welding, an arc is produced between the workpiece and an electrode (Figure 5.77). When the arc is struck, the tip of the electrode melts and globules of it are projected across onto the joint. Temperatures of the order of 20 000 K are produced with currents between the electrode and workpiece of the order of 200 A. Most electrodes are coated with a flux which vaporises and provides a protective shield to reduce oxidation of the molten metal. An alternative method of providing such shielding is to use a gas-shielded arc in which an inert gas is directed over the weld pool. Arc welding gives high quality welds, is a flexible method, and is low in cost. It is used for joints on bridges, piping, ships, etc.

- *Electrical resistance*

  With electrical resistance heating, the high temperatures necessary for welding are produced by passing an electric

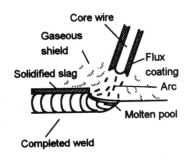

Figure 5.77   *Basics of arc welding*

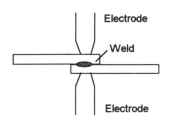

Figure 5.78 *Basics of resistance welding*

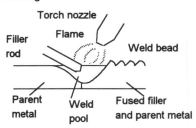

Figure 5.79 *Oxy-acetylene welding*

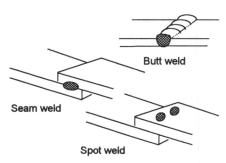

Figure 5.78 *Forms of weld joint*

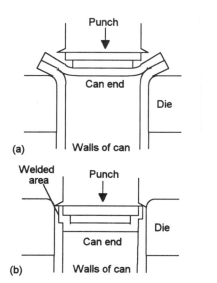

Figure 5.79 *Sealing cans by cold pressure welding*

current across the interface of the joint and producing resistance heating in the electrical resistance of the joint (Figure 5.78). Electrical resistance welding can be used to give butt welds between two surfaces which butt up to each other, seam welds (a line of welded material between two sheets) or spot welds in which the weld occurs on just small regions, the spots. Spot welding is used in the car industry for bodywork; seam welding in sheet metal fabrication.

- **Radiation**
  With radiation heating, the high temperatures occur as the result of focusing a beam of electrons, in a vacuum or low pressure, onto the joint area. This method can make deep narrow welds. An alternative is to use a laser to focus a beam of radiant energy onto the joint. Unlike the electron beam method, this does not require a vacuum chamber.

- **Thermochemical**
  Thermochemical welding uses a chemical reaction to generate the heat. One form of this uses a thermit reaction to generate heat, another has oxygen and some fuel gas such as acetylene combining in a flame (Figure 5.79). A filler rod is used to provide extra metal for the joint. Thermit welding is used for the repair of iron and steel castings, railway lines, shafts, etc.

Because of the high temperatures involved with fusion welding there is generally the need to protect the hot metal from attack by the atmosphere, this being termed *shielding*. The shielding may be provided by the welding being carried out in a vacuum, by the weld area being swept clear of air by shielding gases such as argon, helium or carbon dioxide or by the weld area being coated with a mixture of metal oxides and silicates which produce a glass-like flux. Figure 5.78 shows some forms that joints might take.

There are a number of forms of solid-state welding:

- **Pressure welding**
  This involves a ductile material being pressed against a similar or dissimilar metal; aluminium and copper can be welded at room temperature by this method. This method is used for cladding sheets with a thin layer of some other metal, e.g. aluminium alloy sheets are often clad with aluminium to improve corrosion resistance. It is also a common process for sealing cans (Figure 5.79).

- **Friction welding**
  This involves sliding one material surface, under pressure, over the other. The friction breaks up any surface films and softens the surfaces by virtue of the rise in temperature produced by the friction.

- **Explosive welding**
  With explosive welding, the two surfaces are impacted together by an explosive charge.

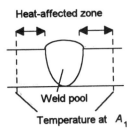

Figure 5.80 *The heat-affected zone*

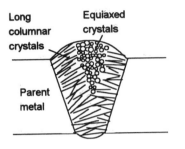

Figure 5.81 *Weld micro-structure*

Welds in steel

Figure 5.80 shows a cross-section of a weld between two steel plates. Molten steel is produced between the two plates in the welding process and during this melting and then solidification, heat is conducted into the plates on either side of the weld. The term *heat-affected zone* is used to describe those parts of the plates which have their temperatures raised above the critical point A_1.

In the weld pool the metal shows the typical cast-metal structure. Thus for a single-run weld, where we have the weld pool cooling with its sides in contact with the colder parent metal, there is a central zone of equiaxed crystals surrounded by long columnar crystals (Figure 5.81).

In the heat-affected zone the microstructure depends on the temperatures reached and steel concerned. Figure 5.82 shows the types of temperatures that may be realised in the weld pool and heat-affected zone and the effect on the microstructure for a 0.3% plain carbon steel.

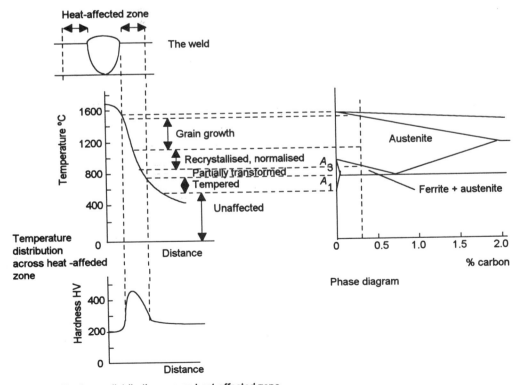

Figure 5.82 *The effect of welding heat on the hardness of a 0.3% plain carbon steel*

When a 0.3% carbon steel is being welded, we have the following situations occurring:

- Close to the weld pool the temperatures rise to well above the A_3 temperature and the metal is heated into the austenitic state. With the low-carbon steel, when the temperature rises to above A_3, grain growth occurs. The closer to the weld pool the higher the temperature and so the larger the grain size produced. Thus near the weld pool there is a coarse grain structure. With increasing distance from the weld pool, the peak temperature decreases and so the grain structure is less coarse. The rate of cooling decreases with increasing distance from the weld pool. Martensite can form where the rate of cooling is fast enough and thus gives a much harder material.

- For the region just above the A_3 temperature, the microstructure is fully transformed into austenite and recrystallisation occurs. Cooling may result in the metal becoming normalised.

- At the edge of the heat-affected zone the temperature is A_1. Thus for the region between the A_3 and A_1 temperatures, i.e. near the edge of the heat-affected zone, only partial transformation to austenite takes place.

- Where the temperature does not exceed 723°C, the eutectoid temperature, only tempering can occur. Such tempering results in some loss of hardness.

The weld itself is likely to be less hard than the parent metal because of the loss of some alloying elements during the melting process. For low-alloy steels and other than low-carbon plain carbon steels, the rate of cooling is likely to be such that martensite forms in that part of the heat-affected zone that reaches the highest temperature. The result of such changes is a marked increase in hardness and brittleness. This can lead to *embrittlement* and a risk of cracking.

The hardness produced depends on the hardenability of the steel, the cooling rate and to some extent on the prior grain size. The hardenability of a steel depends on the elements present in the alloy, different elements having different effects. An overall measure is the *carbon equivalent*, the greater this is the greater will be the hardenability and hence the greater the hardness in the heat-affected zone. A number of formulae exist for calculating the carbon equivalent (see Section 3.1.8), a common one being:

$$\text{carbon equivalent \%} = \%C + \frac{\%Mn}{6} + \frac{\%Cr + \%Mo + \%V}{5} + \frac{\%Ni + \%Cu}{15} \qquad [4]$$

Steels with a very low carbon equivalent, less that 0.2%, can be welded easily because the austenite transformation on cooling occurs at a fairly high temperature and martensite is not produced. Between 0.2 and 0.4% carbon equivalent, components have to be

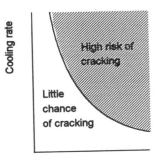

Figure 5.83 *Effect of carbon equivalent on risk of cracking*

welded with due consideration being given to the rate of cooling in order to minimise the martensite produced. Above 0.4% carbon equivalent, welding is virtually impossible because of the difficulty in making welds which do not crack. Figure 5.83 illustrates the way the risk of cracking depends on the carbon equivalent. The hardness that can be tolerated before cracking occurs depends on the hydrogen content of the weld. With low hydrogen content higher hardness can be tolerated.

A factor influencing the cooling rate is the mass of material being welded; the cooling rate depends on the temperature difference being the weld zone and the rest of the material and so the thicker the joint the faster the cooling. The likelihood of martensite forming can be reduced by reducing the cooling rate. This can be done by preheating the whole structure in the vicinity of the joint before welding, thus reducing the rate of cooling into the parent metal.

In describing the above properties of welds it has been assumed that there was just a single-pass weld. In practice most welds are multi-pass, each succeeding pass forming a fresh heat-affected zone in previously deposited weld metal and previously formed heat-affected zones.

Example

Figure 5.84 shows the TTT diagram for a 0.4% carbon–3.5% nickel–0.2% molybdenum steel and the cooling curve that describes the cooling of the material close to the weld pool. What would you expect the outcome of the cooling to be and how can the situation be improved?

The steel has a carbon equivalent of 0.4 + 3.5/15 + 0.2/5 = 0.67% and so is considered not weldable without special treatment. The cooling curve indicates that martensite will be produced and so the risk of cracking is high. If the component is preheated, the rate of cooling into the parent material can be reduced and no martensite produced. The result could be the modified rate of cooling shown in Figure 5.85.

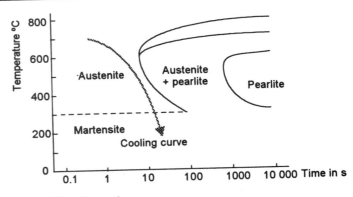

Figure 5.84 *Example*

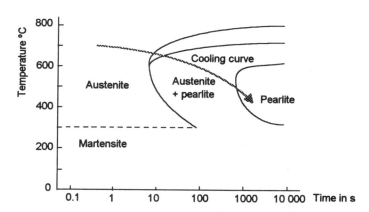

Figure 5.85 *Example*

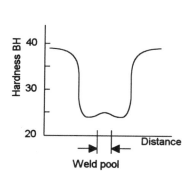

Figure 5.86 *Hardness plot for non-heat-treatable aluminium alloy*

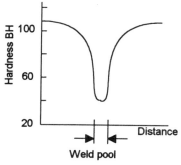

Figure 5.87 *Hardness plot for a heat-treatable aluminium alloy*

Welds in aluminium and copper alloys

The alloys of aluminium and copper are difficult to weld because of the high thermal conductivity, high thermal expansion and a tendency to give porous welds. The high thermal conductivity means that there is a high rate of cooling and so there is difficulty in heating up the parent metal round the weld zone to a high enough temperature to give complete fusion with the weld pool. The high coefficient of expansion means that significant residual stresses are produced as a result of the weld area expansion being constrained by the colder surrounding parent metal. In the case of aluminium the porosity arises because the metal in the molten state absorbs hydrogen from such sources as the welding flame, fluxes and atmospheric moisture and when the weld cools this hydrogen is released. In the case of copper the porosity arises from the evolution of steam during cooling, this resulting from the dissolved oxygen in the copper reacting with hydrogen absorbed from the welding flame, fluxes and atmospheric moisture.

Figure 5.86 shows the results of a hardness traverse of a weld between two plates of a non-heat-treatable work-hardened aluminium alloy. Within the heat-affected zone, the alloy is fully or partially annealed by the temperatures produced during the welding. The result is that the work-hardened material is much softer in the weld region than in the unaffected material. Since the tensile strength is related to the hardness, there is a drop in strength. This effect is in the main irreversible, although the strength of the weld pool area can be improved by rolling or hammering.

Figure 5.87 shows the result of a hardness traverse of a weld between two plates of a heat-treatable aluminium alloy. Softening and a reduction in tensile strength occurs within the heat-affected zone. However, the strength may be almost completely recoverable by solution treatment and ageing of the component.

Fusion welding of aluminium is mainly used with pure aluminium, the non-heat-treatable aluminium–manganese and aluminium–magnesium alloys, and the aluminium–magnesium–

silicon and aluminium–zinc–magnesium heat-treatable alloys. With the heat-treatable alloys there is an uncontrolled solution of the constituents during the welding and this is then followed by uncontrolled precipitation during cooling. The higher strength aluminium–copper–magnesium and duraluminium alloys cannot be effectively fusion welded and such alloys are thus normally joined by riveting.

Copper oxide reacts with hydrogen at temperatures above about 500°C to give copper and steam. A consequence of this is that the steam gives rise to porosity in the weld pool, as well as causing fissures to develop and embrittlement in the heat-affected zone close to the weld pool. To reduce this effect the copper has to be deoxidised prior to welding. This can be achieved by the presence of small amounts of phosphorus in the copper. Hence, welds of acceptable strength are possible in phosphorus-deoxidised copper. Brasses are difficult to weld because of the volatilisation of zinc, this leading to porosity. Tin bronzes, aluminium bronzes and silicon bronzes can be welded without the porosity problem occurring; aluminium bronzes, however, are susceptible to cracking during welding. Cupronickels can be welded without porosity if a deoxidised alloy, or filler rods containing deoxidant, are used. The strength of work-hardened or age-hardened copper alloys is reduced by welding. After welding, the strength of a fusion welded joint in a hard temper copper is probably about that of the annealed metal.

Welding defects

The defects that can be introduced during welding include:

- **Cracks**
  Several possible types of cracks are possible (Figure 5.88). These include *hot cracking* which can arise from when the weld solidifies, such cracks propagating inwards along the centre line of the weld, and *hydrogen cracking* in the heat-affected zone where the propensity has arisen because of martensite being formed.

- **Incomplete penetration**
  This has the fused zone not extending through the plate (Figure 5.89) and occurs when there has been too little heat during the welding process.

- **Undercut**
  Figure 5.90 shows undercutting. It can occur when the welding causes the burning away of the side walls of the parent material at the weld as a result of perhaps too high a current or too large an electrode or it is inclined at too shallow an angle or insufficient time has been allowed for the depositing of metal. Undercutting weakens the joint and creates a slag trap.

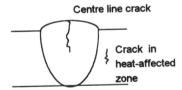

Figure 5.88   *Examples of cracks*

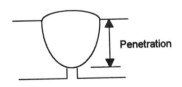

Figure 5.89   *Incomplete penetration*

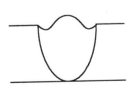

Figure 5.90   *Undercut*

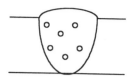

Figure 5.91   *Porosity*

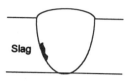

Figure 5.92   *Entrapped slag*

Figure 5.93   *Points where lack of fusion can occur*

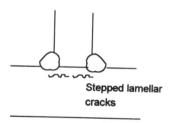

Figure 5.94   *Lamellar cracking*

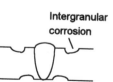

Figure 5.95   *Weld decay*

- *Porosity*

  This consists of small cavities in the weld (Figure 5.91) and arises from gas entrapped in it. The gas evolution may arise from such causes as foreign matter such as grease on the parent material, moisture absorbed by the flux-coated electrodes or gas absorbed by the molten weld pool from the surrounding atmosphere and released during solidification.

- *Slag entrapment*

  Slag is non-metallic particles trapped in the weld material (Figure 5.92). Slag can originate from the slag formed by the electrode coating, from dirty joint surfaces or from mill scale on the surface of the parent material. Entrapment may arise with multi-pass welding from inadequate removal of the slag after a previous run.

- *Incomplete fusion*

  This is the failure to fuse together adjacent layers of a weld or adjacent weld metal and the parent metal (Figure 5.93). It arises from the failure to raise the metal to its melting point and can be caused by a small electrode being used on thick cold plate, too low a current, the welding speed too high or scale or dirt on the joint surface.

- *Lamellar tearing*

  These are cracks associated with restrained corner or tee joints where the fusion boundary is more or less parallel to the plate surface and occur in the parent metal close to the heat-affected zone (Figure 5.94).

Corrosion of welds

Because of differences in the microstructure and composition of welded areas compared with the parent metal, selective corrosion of welds is likely in many corrosive environments. Thus welded carbon steels exposed to a marine environment may show corrosion more markedly in the weld material or heat-affected zone than in the parent metal.

Heating a stainless steel to about 500 to 700°C can lead to the precipitation of carbides at grain boundaries. This results in the removal of chromium from grains to the boundaries and hence a reduction in corrosion resistance and so intergranular corrosion. This effect is known as *weld decay* since such effects occur during the welding of stainless steels (Figure 5.95). The defect can be overcome by heat treating the steel or using a stabilised stainless steel, i.e. one which includes niobium or titanium.

5.5.4 Fastening systems

Fasteners provide a clamping force between two pieces of material. A wide variety of types of fastener exist and the materials used to make them. The types of fastener available can be classified as threaded, non-threaded and special purpose. Steel is probably the

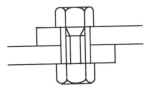

Figure 5.96   *Example of a threaded fastener*

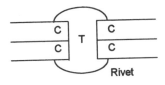

T = tension, C = compression

Figure 5.97   *A riveted joint*

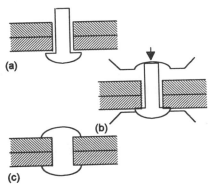

Figure 5.98   *Stages in riveting*

most common material used, although aluminium alloys, brass and nickel are amongst other metals used. Aluminium alloy fasteners have the advantage over steel of being much lighter, non-magnetic and more corrosion resistant. Nickel has the particular advantage of strength at high temperatures.

The choice of fastener will depend on a number of factors:

- *Environmental*
  For example: high, normal and low temperature; corrosive, abrasive and oxidising conditions.

- *Nature of the external loading on the fastener*
  For example: tension, compression, shear, cyclic, impact.

- *Life and service requirements*
  For example: the need for perhaps frequent assembly and disassembly.

- *Design of the components being joined and types of materials*
  For example: stress concentrations; the elastic, plastic and thermal properties of the materials.

- *Quantity of fasteners required and cost*

Threaded fasteners

With a threaded fastener, the clamping force holding the two pieces of material together is produced by a torque being applied to the fastener to stretch it and this is maintained during the service life of the fastener. The fastener is in tension and the two pieces of material in compression (Figure 5.96). Bolts mated with nuts and screws with threads in the material are examples of this type of fastener. Low- or medium-carbon steel bolts will typically have a minimum proof strength of about 225 to 400 Pa, the proof strength being the maximum stress the bolt can withstand without acquiring a permanent set. For quenched and tempered medium-carbon bolts the minimum proof stress is 600 to 650 MPa and quenched and tempered alloy steel about 970 MPa. Threaded fasteners are particularly useful for joining components that are likely to need to be dismounted during the life of the product.

Non-threaded fasteners

Rivets, eyelets, nails and pins are examples of non-threaded fasteners. Figure 5.97 shows a riveted joint as an example, the shank of the rivet having been put in tension and the two pieces of joined material in compression. Rivets can be used for joining dissimilar or similar materials, both metallic and non-metallic, to give permanent joints.

Figure 5.98 shows the basic stages that are typical of a riveting process. When the force applied to the rivet is sufficiently high, plastic deformation occurs and the shank of the rivet increases in diameter as its length decreases. That part of the shank within the

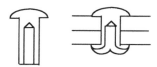

Figure 5.99   *Tubular rivet*

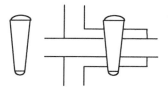

Figure 5.100   *Taper pin*

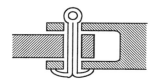

Figure 5.101   *Split cotter pin*

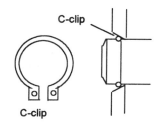

Figure 5.102   *C-clip*

hole increases in diameter until it fills the hole, and the unsupported part of the shank outside the hole continues to deform until a head is formed. A ductile material has to be used for the rivet material, e.g. mild steel, copper, brass, aluminium, aluminium alloy. For some materials the required ductility is obtained in the cold state, in other cases the riveting takes place hot. Where the riveting force might damage or distort the materials being joined, tubular or semi-tubular rivets (Figure 5.99) might be used instead of solid rivets.

Pins, either in the solid or tubular forms, are widely used for fastening. For example, *taper pins* (Figure 5.100) are used to join wheels onto the ends of shafts, the pin just being driven through holes in the two parts until it is fully home and giving a tight fit. Taper pins are usually supplied in mild steel but can be obtained in alloy steel, stainless steel or brass. Another form of pin is the split *cotter pin* (Figure 5.101). This is used where freedom of movement in the joint is required or as a locking device for slotted nuts on bolts.

There is a wide variety of forms of *spring-retaining clips*. A simple form is a C-clip which is used to lock and retain components on shafts, the clip generally fitting into a groove on the shaft (Figure 5.102). Most C-clips are made from hardened and tempered carbon steel.

Fatigue properties of fastened joints

The use, for example, of bolts or rivets as fasteners for joints can introduce fretting damage. *Fretting* is the wear process that occurs at the areas of contact of two metals undergoing small amplitude cyclic slip. On steels this damage may be visible as the red oxide or iron, on aluminium as black oxides. The damage produced is referred to as *galling* or *scuffing*. Such damage can lower the fatigue strength by factors as high as three for aluminium alloys. One form of anti-fret treatment is to separate the metal surfaces by using PTFE shims or a coating of a paint containing the solid lubricant molybdenum disulphide.

5.5.5 Joining methods for plastics

The joining methods that can be used with plastics can be considered to fall into four main groups:

- *Welding*
  Thermoplastics can be welded by a number of methods, all involving the melting of the interface between the plastic surfaces being joined.

- *Adhesive bonding*
  The adhesive may be elastomers, thermoplastics, thermosets or two-polymer types. Some plastics, e.g. nylons, polystyrene and PVC, may be bonded by the use of a solvent to soften the

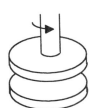

Figure 5.102   *Spin welding*

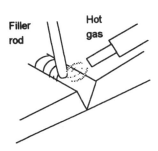

Figure 5.103   *Hot-gas welding*

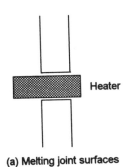

(a) Melting joint surfaces

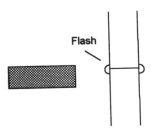

(b) Heater removed and the surfaces pressed together

Figure 5.104   *Hot-plate welding*

interfaces of the joint, then light pressure is applied to bring the surfaces into close contact.

- **Riveting**
  Metal or thermoplastic rivets can be used for joints between plastics and metals.

- **Press and snap fits**
  This can be used for both permanent and recoverable assemblies and depends on the ability of plastics to be subject to quite severe elastic distortion and return to their original shape when the load is removed.

- **Thread systems**
  Screw threads and self-tapping screws are widely used.

Welding plastics

There are a number of methods that can be used to weld plastics, the following being some of the more commonly used ones:

- **Spin welding and vibration welding**
  These are friction welding methods in which the weld is produced by the frictional heat developed at the interface between two thermoplastic materials. Spin welding is only suitable for circular components and involves holding the lower part in a jig while the upper part is brought into contact with it while rotating at high speed (Figure 5.102). The friction at the interface causes the plastic to melt and the surfaces to fuse together. This is a reliable method of producing pressure and vacuum-tight joints having a bond which is almost as strong as the parent material. Vibration welding can be applied to non-circular shapes, the method involving an oscillatory motion rather than a rotation to produce the melting and consequential fusing.

- **Hot-gas welding, hot-plate welding, hot-wire welding**
  These all involve the melting of the interface of the joint by direct heating. Hot-gas welding involves heat being applied using a welding torch to blow hot gas on to the joint and uses a filler rod to supply molten plastic to fill the joint (Figure 5.103). It is similar in principle to oxy-acetylene welding with metals. The process requires a skilled operator if high-strength joints are to be produced: too little heating of the joint area and the weld will be weak; too much heating and the plastic will degrade and a poor weld will be produced. This process is used for the fabrication of large containers.
  In hot-plate welding, the faces of the parts to be joined are pressed against a heated plate, the plate being coated with PTFE to prevent sticking, to become melted and then the two surfaces being joined by being pressed together (Figure 5.104). By using specially designed heaters, three-dimensional

shapes can be welded by this method. Nylons cannot be welded this way since they oxidise when the melted resin is exposed to air. Hot-plate welding is used for the on-site joining of thermoplastic pipes.

With hot-wire welding, constant pressure is applied to the joint while an electrically heated wire passes through the joint, melting the plastic at the joint surfaces and so forming a weld.

- *Ultrasonic welding*

  This involves an input to the joint area of high-frequency vibrations, of the order of 20 kHz. These cause the two surfaces of the joint to vibrate against each other and, as a result of friction, melt. The process is fast, some machines being capable of bonding some 30 parts per minute by this method, and can be automated. Weld strength is consistently high and the process is very versatile. Some thermoplastics cannot, however, be welded by this method.

- *Dielectric welding*

  This method is used for the welding together of thin sheets of plastic. The materials are placed between the plates of a capacitor and act as the dielectric. A high-frequency alternating voltage applied to the capacitor results in a high-frequency electric field in the plastic. This is the principle of the microwave oven. This heats the plastic and gives bonding. The process has been used for the production of upholstery, imitation leather, luggage and inflatables.

Adhesive bonding of plastics

The main types of adhesives used with plastics are:

- *Epoxy resins*

  Epoxy resins are thermosetting adhesives and usually involve two components, the resin and the hardener, which have to be combined for the bond to be made. Generally they are used with thermosetting materials, producing good bonds with those materials, but are not so useful with thermoplastics in that poorer bonds are produced.

- *Acrylic acid diesters*

  These set when air is excluded and are termed *anaerobic*. The surfaces to be bonded are coated and then brought together under light pressure, this excluding air. They form good bonds with thermosets but are not suitable for use with the common engineering thermoplastics.

- *Two-part acrylics*

  These involve a hardener being used with the acrylic resin, one surface of the joint being coated with the hardener and the other with the resin. The two surfaces are then brought together for the bond to be made. These adhesives will bond to almost anything and are used with thermosets and many thermoplastics.

Activity

Look at the welding methods used with metals and with plastics. List the similarities and the differences between the methods and suggest reasons for the differences.

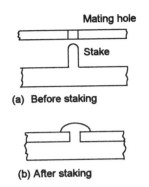

Figure 5.105 *Ultrasonic staking*

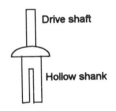

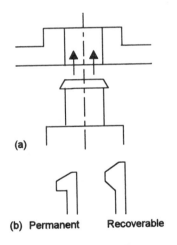

Figure 5.106 *Plastic rivet*

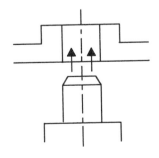

Figure 5.107 *Snap fit*

Figure 5.108 *Press fit*

- *Cyanocrylates*

The cyanocrylates, the 'super-glues' will form good bonds with thermosets and virtually all thermoplastics.

Riveting plastics

Riveting using metal rivets can be used to make joints between plastics and between plastics and metals. Tubular or bifurcated rivets are generally used. The problem with riveting is that some plastics show a pronounced delayed response to deformation; they are said to exhibit a 'memory'. This means that though the rivet was initially fitted tightly, a less than tight fit might develop with time following the release of the load on the rivet. Only those plastics which have good ductility and are not brittle, have high strength and good resistance to creep, are suitable for riveting.

One form of riveting is known as *ultrasonic staking*. One of the parts is made with integral protecting rivets or 'stakes' (Figure 5.105). These, when mated with the other part, project through holes in it. The protecting part of the stake is deformed by the application of ultrasonic energy into a mushroom-like head and so clamps the two parts together. This type of joining is frequently used when metal parts have to be attached to a plastic, the plastic with the stakes and the metal part containing the holes.

Figure 5.106 shows another form of plastic rivet. This is a one-piece injection moulding having a hollow rivet shank which terminates in flared split prongs. A solid drive pin protrudes from the head of the rivet. After insertion in the hole, the drive pin is hit and driven through the rivet head and into the hollow shank, forcing the prongs out against the hole walls.

Press and snap fits

Plastics can be subject to quite severe elastic distortion and still return to their original shape when the load is removed. Press and snap fits rely on this characteristic.

A common form of snap fit is the hook joint. Figure 5.107(a) shows an example of such a joint. When the component is pushed into the hole, the end is deformed so that it can slide through the hole until it emerges from the other end. Then it expands and locks the component in position. The type of hook end shown in the figure gives a permanent joint in that it is not possible to disengage it. Figure 5.107(b) shows how the hook end varies when it is designed for use as a permanent joint and a recoverable joint.

Figure 5.108 shows a press-fit joint; the component is a tight fit in the hole, there being no mechanical interlocking.

Snap fits are stronger and more dependable than press fits since they rely on a mechanical interlocking of two components as well as friction, whereas press fits rely only on friction.

Thread systems with plastics

Screw threads are one means of joining plastic components. The main problem that can arise is the system coming loose due to the

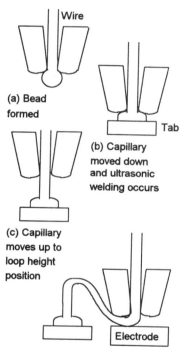

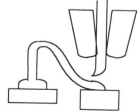

Figure 5.109 *Connections with microcircuits*

creep of the plastic when under load. Self-tapping screws, in which the screw cuts its own thread when screwed into the plastic, are widely used.

5.5.6 Case studies

The following are case studies to illustrate some of the issues involved in joining materials.

Railway tracks

To give the continuous forms of railway track which is now the case, lengths of rail are joined together to give lengths of about 90 m before being taken to the track where they are to be laid. The steel used for the rail has about 0.7% carbon, 1.0% manganese and 0.3% silicon. This gives a carbon equivalent of about 0.9%. Steels with above 0.4% carbon equivalent are difficult to weld since the rate of cooling that occurs leads to martensite production and potentially cracks. However, by preheating, the rate of cooling in air can be kept low enough for complete transformation to pearlite. In the case of the production of long lengths of rail, the procedure adopted is to preheat, carry out the welding and then within minutes, while still hot, the deformed metal from around the welded area is removed. Spray cooling by a jet of cold air is then used followed by a water quench; this procedure gives a cooling rate which is sufficiently slow for the rail to become completely pearlitic. The rail is then shaped to the required profile by a press and grinding then used to get it to the correct dimensions.

Connections with microcircuits

With microcircuits we have the problem of making connections using wire which is only about 20 μm in diameter between tabs on chips which are even thinner to electrode pins which might be a few millimetres in diameter. The method used is ultrasonic welding. The wire is fed through an alumina capillary and an electric discharge used to form a bead at the end of the wire (Figure 5.109). This is then moved into contact with a tab and then ultrasonic energy is transmitted down the capillary from a piezoelectric transducer to form the weld. The capillary tube is then moved over the electrode and a second ultrasonic weld made to it. A spark discharge through the pin is then used to melt the wire and leave a solid bead on the free wire which is ready for the next tab weld.

The Challenger disaster

In January 1986 the space shuttle Challenger exploded shortly after lift off with the seven people on board all loosing their lives. The reason for the explosion was found to be the failure of the seals of a joint (Figure 5.110) in the tubular steel case of one of the solid fuel booster rockets. This had allowed hot gases to escape and weaken adjacent structural parts. The firing of the booster

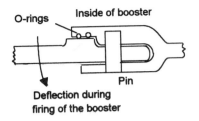

O-rings

Inside of booster

Deflection during
firing of the booster

Pin

Figure 5.110 *Joint with Challenger booster*

produced an increase in pressure inside the rocket casing which had the effect of deflecting the case and open a gap at the joint. The O-rings should have expanded to fill the gap but, at the low temperatures occurring on the day of the disaster, the rubber had lost its elasticity. The joint thus leaked.

Problems 5.5

1 Discuss the merits of joining by means of adhesives and the limitations of the method.
2 List examples of the types of applications for which the following adhesives might be used: (a) natural adhesives, (b) elastomers, (c) polyamides, (d) cyanacrylates, (e) phenolic resins, (e) two-polymer such as phenolic resin with neoprene.
3 Describe the ways in which adhesive joints can be loaded and which is the one most likely to give the strongest joint.
4 An adhesive has a shear strength of 10 MPa. What bonded area is required if a joint made with this adhesive has to withstand a shear load of 1 kN?
5 How does soldering differ from brazing?
6 What are the functions of the flux in soldering and brazing?
7 Why, for soldered or brazed joints, are lap joints to be preferred to butt joints?
8 The steel used for railway lines has a carbon equivalent of about 0.9%. What procedure is necessary to enable good welds to be made?
9 Why is 'shielding' necessary with welding operations?
10 Explain how pressure welding can be used to fit the end cap on a tin.
11 Describe and explain the hardness and structural changes that can occur with the welding of steel plates.
12 Explain how porosity arises in the welding of aluminium sheets.
13 Why are there problems in welding high-carbon steels?
14 Explain the reasons for preheating joints before welding.
15 Explain undercutting with welds and how it occurs.
16 State the factors that have to be considered in choosing a fastener.
17 Explain what fretting is and its significance for fastened joints.
18 When might a tubular rivet be used in preference to a solid rivet?
19 Explain how hot-plate welding is used to join two plastic components.
20 Suggest joining methods that might be used (a) to fix a metal clip on the face of a sheet of plastic, (b) to join two thermoplastic pipes, (c) to bond two sheets of thermoset materials together face to face.
21 How do snap-fit fasteners differ from press-fit fasteners?

6 Failure of materials

Summary

In dealing with material, engineers are invariably involved with considering their deterioration and failure. For example, there is the damage to the bodywork of a car when it is involved in a collision, the classic case of the failure of the Comet aircraft in the 1950s as a result of repeated flexing of its fuselage, the deformation of a material with time which limits which materials can be used at the temperatures occurring with turbine blades, and the rusting of cars. This chapter is a consideration of the mechanisms of failure by direct fracture, fatigue, and creep; also there is a consideration of the environmental stability and wear of materials.

Objectives

By the end of this part, the reader should be able to:

*   distinguish between brittle and ductile fracture and describe the failure of metals, polymers, ceramics and composites;
*   recognise the role in fractures of stress concentrations and temperature;
*   explain the principle of fracture toughness;
*   explain how fatigue failure occurs and use S-N graphs to discuss the factors affecting fatigue properties;
*   explain how creep arises and the factors affecting creep;
*   discuss the environmental factors that can result in the degradation of metals, explaining how they arise and prevention methods;
*   discuss the environmental stability of polymers and ceramics;
*   Explain how wear occurs and compare the properties of the different materials used for bearing surfaces.

6.1 Fracture

In July 1962, just fifteen months after it had opened, the Kings Bridge in Melbourne suffered a partial collapse as a loaded vehicle of about 45 000 kg was crossing. One of the spans had collapsed. Examination showed that four girders had fractures starting from cracks in the parent metal at fillet welds. In December 1965, the Sea Gem, an offshore drilling rig that was operating in the North Sea collapsed and nineteen people died. Investigation showed that

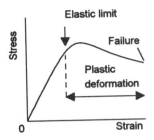

Figure 6.1 *Stress–strain graph for a ductile material*

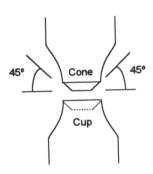

Figure 6.2 *Ductile failure*

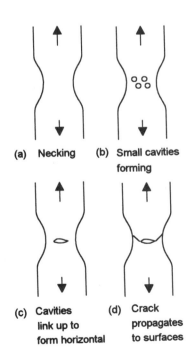

(a) Necking (b) Small cavities forming

(c) Cavities link up to form horizontal crack (d) Crack propagates to surfaces

Figure 6.3 *Stages in ductile fracture*

tie bars had fractured. These are examples of fractures occurring on large structures; there are, however, many examples of fractures on small structures and components. For example, there is the damage to the bodywork of a car after a collision, the drinking glass that has broken after being dropped, the wooden shaft of a hammer that breaks while the hammer is being used, etc.

6.1.1 Types of fracture

With metals the basic modes of a failure can be considered to be ductile failure when a material yields as a result of excessive deformation, brittle fracture, fatigue and creep. Other ductile materials such as some polymers can also fail as a result of yielding and excessive deformation but most other non-metallic materials tend to fail by brittle fracture rather than excessive yielding.

Ductile fracture

When a ductile material has a gradually increasing tensile stress applied, it behaves elastically up to a limiting stress and then beyond that stress plastic deformation occurs (Figure 6.1). As the stress is increased the cross-sectional area of the material is reduced and a necked region is produced. With a ductile material there is a considerable amount of plastic deformation before failure occurs in the necked region as a result of excessive yielding. When it occurs, the fracture (Figure 6.2) shows a typical cone and cup formation with the surfaces of the fractured material dull or fibrous. This is because, under the action of the increasing stress, small internal cracks form which gradually grow in size until there is an internal, almost horizontal, crack. The final fracture occurs when the material shears at an angle of 45° to the axis of the direct stress. This type of failure is known as *ductile failure*.

With ductile failure the sequence of events is considered to be:

1 Following elastic strain the material becomes plastically deformed and a neck forms (Figure 6.3(a)).

2 Within the neck, small cavities or voids are formed (Figure 6.3(b)). These develop as a result of the stress causing small particles of impurities or other discontinuities in the material to either fracture or separate from the metal matrix. The more such nuclei there are available to trigger the development of these cavities, the less the material will extend before fracture and so the less ductile the material. Thus increasing the purity of a material increases its ductility.

3 These cavities then link up to form an internal crack which spreads across the material in a direction at right angles to applied tensile stress (Figure 6.3(c)).

4 The crack finally propagates to the material surface by shearing in a direction which is approximately at 45° to the

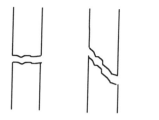

Figure 6.4   *Brittle tensile fracture*

Key points

With ductile fracture there is a considerable amount of plastic deformation before failure occurs in a necked region. With brittle fracture, there is little plastic deformation before fracture and metals have a fracture surface which appears bright and granular

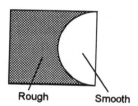

Rough   Smooth

Figure 6.5   *Fracture surface for a brittle polymer*

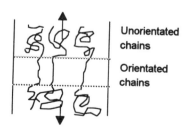

Unorientated chains

Orientated chains

Figure 6.6   *Chain orientation*

applied stress to give a fracture in the typical form of a cup and cone (Figure 6.3(d)).

Brittle fracture

If you drop a china cup and it breaks, it is possible to pick up the pieces and stick them back together again and have something which still looks like a cup. The china cup has failed by what is termed *brittle fracture*. Brittle fracture is the main mode of failure for glass and ceramics. With a brittle fracture the material fractures before plastic deformation has occurred.

Figure 6.4 shows possible forms of brittle tensile failure for metals. The surfaces of the fractured material appear bright and granular due to the reflection of light from individual crystal surfaces. This is because the fracture has grains within the material cleaving along planes of atoms.

We can consider the sequence of events leading to brittle fracture to be:

1   When stress is applied the bonds between atoms and between grains in the material are elastically strained.

2   At some critical stress the bonds break, remember the material is brittle and there is no plastic deformation and hence small-scale slip, and a crack propagates through the material to give fracture.

Polymers

Brittle failure with polymeric materials is a common form of failure with materials below their glass transition temperature, i.e. amorphous polymers. The resulting fracture surfaces show a mirror-like region, where the crack has grown slowly, surrounded by a region which is rough and coarse where the crack has propagated at speed (Figure 6.5).

In an amorphous polymer the chains are arranged randomly with no orientation. When stress is applied it can cause localised chain slippage and an orientation of molecule chains (Figure 6.6) with the result that the applied stress causes small voids to form between the aligned molecules and fine cracks, termed *crazing*, are formed. This is what constitutes the mirror-like region. Because of the inherent weakness of the material in the crazed region it serves as a place for cracks to propagate from and cause the material to fracture. Initially the crack grows by the growth of the voids along the midpoint of the craze. These then coalesce to produce a crack which then travels through the material by the growth of voids ahead of the advancing crack tip. This part of the fracture surface shows as the rougher region.

With crystalline polymers, the application of stress results in the folded molecular chains becoming unfolded and aligned. The result is then considerable, permanent deformation and necking. Prior to the material yielding and necking starting, the material is quite likely to begin to show a cloudy appearance. This is due to small voids being produced within the material. Further stress

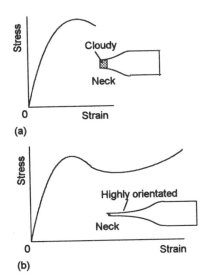

(a)

(b)

Figure 6.7 *Fracture with ductile materials: (a) unplasticised PVC, (b) polypropylene*

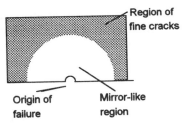

Figure 6.8 *Typical fracture surface for a ceramic*

Activity

Look at examples of material failures such as:
(a) a broken piece of chalk,
(b) a broken polystyrene ruler,
(c) a torn piece of polythene film,
(d) a broken wooden ruler.
Try to describe the failures in terms described in Section 6.1.1.

causes these voids to coalesce to produce a crack which then travels through the material by the growth of voids ahead of the advancing crack tip. Figure 6.7 shows typical forms of stress–strain graphs for polymers showing this form of failure.

Activity

Get hold of a plastic rule or discarded CD case. Holding onto opposite ends, gently flex the component back-and-forth. You may hear some crackling noises, followed by the appearance of fine lines running at right angles to the flexing axis. These lines are not cracks, but crazes.

Ceramics

Ceramics are brittle materials, whether glassy or crystalline. Typically a fractured ceramic shows around the origin of the crack a mirror-like region bordered by a misty region containing numerous micro cracks (Figure 6.8). In some cases the mirror-like region may extend over the entire surface.

Composites

The fracture surface appearances and mechanisms for composites depend on the fracture characteristics of the matrix and reinforcement materials and on the effectiveness of the bonding between the two. Thus, for example, for a glass-fibre reinforced polymer, depending on the strength of the bonds between fibres and polymer, the fibres may break first and then a crack propagate in shear along the fibre–matrix interface. Eventually the load which had been mainly carried by the fibres is transferred to the matrix which then fails. Alternatively the matrix may fracture first and the entire load is then transferred to the fibres which carry the increasing load until they break. The result is a fractured surface with lengths of fibre sticking out from it, rather like bristles out from a brush.

6.1.2 Factors affecting failure

This section is a discussion of the factors affecting failure.

Stress concentration

If you want to break a small piece of material, one way is to make a small notch in the surface of the material and then apply a force. The presence of a notch, or any sudden change in section of a piece of material, can very significantly change the stress at which fracture occurs. The notch or sudden change in section produces what are called *stress concentrations*. They disturb the normal stress distribution and produce local concentrations of stress.

The amount by which the stress is raised depends on the depth of the notch, or change in section, and the radius of the tip of the notch. The greater the depth of the notch the greater the amount by which the stress is raised. The smaller the radius of the tip of the notch the greater the amount by which the stress is raised. An approximate relationship that has been derived for the stress at the tip of a notch is:

$$\text{stress at end of notch} = \sigma_a \left(1 + 2 \sqrt{\frac{L}{r}} \right) \quad [1]$$

where σ_a is the stress applied to the material as a whole, L the length of the notch and r its radius. The increase in stress due to the notch is thus:

$$\text{increase in stress} = 2 \sqrt{\frac{L}{r}} \quad [2]$$

This increase in stress is termed the *stress concentration factor*.

A crack in a brittle material will have quite a pointed tip and hence a small radius. Such a crack thus produces a large increase in stress at its tip. One way of arresting the progress of such a crack is to drill a hole at the end of the crack to increase its radius and so reduce the stress concentration. A crack in a ductile material is less likely to lead to failure than in a brittle material because a high stress concentration at the end of a notch leads to plastic flow and so an increase in the radius of the tip of the notch. The result is then a decrease in the stress concentration.

Speed of loading

Another factor which can affect the behaviour of a material is the speed of loading. A sharp blow to the material may lead to fracture where the same stress applied more slowly would not. With a very high rate of application of stress there may be insufficient time for plastic deformation of a material to occur and so what was, under normal conditions, a ductile material behaves as though it were brittle.

The Charpy and Izod tests give a measure of the behaviour of a notched sample of material when subject to a sudden impact load. The results are expressed in terms of the energy needed to break a standard size test piece; the smaller the energy needed the easier it is for failure to occur with shock loads in service. The smaller energies are associated with materials which are termed brittle; ductile materials needing higher energies for fracture to occur.

Temperature

The temperature of a material can affect its behaviour when subject to stress. Many metals which are ductile at high temperatures are brittle at low temperatures. For example, a steel may behave as a ductile material above, say, 0°C but below that temperature it becomes brittle. Figure 6.9 shows how the impact test results might vary with the temperature at which a test piece was tested for such a material. The *ductile–brittle transition temperature* is thus of importance in determining how a material

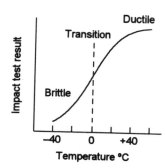

Figure 6.9 *Ductile–brittle transition*

will behave in service. Table 6.1 gives examples of these transition temperatures for two steels.

Table 6.1 *Ductile–brittle transition temperatures*

BS, AISI ref.	Composition %	Heat treatment	Transition temp. °C
817M40, 4340	0.4 C, 0.8 Mn, 1.7 Ni, 0.9 Cr, 0.2 Mo	Quenched + tempered 540°C	−90
		Quenched + tempered 650°C	−120
805M20, 8620	0.2 C, 0.9 Mn, 0.6 Ni, 0.7 Cr, 0.2 Mo	Quenched + tempered 425°C	−30
		Quenched + tempered 650°C	−125

The transition temperature with a steel is affected by the alloying elements in the steel. Manganese and nickel reduce the transition temperature. Thus for low-temperature work, a steel with these alloying elements is to be preferred. Carbon, nitrogen and phosphorus increase the transition temperature.

Thermal shocks

Pouring hot water into a cold glass can cause the glass to crack. This is a case of thermal shock loading. The layer of glass in contact with the hot water endeavours to expand but is restrained by the colder outer layers of the glass, these layers not heating up quickly because of the poor thermal conductivity of glass. The result is the setting up of stresses which can be sufficiently high to cause failure of the brittle glass.

6.1.3 Fracture toughness

One measure of the fracture toughness of a material is in terms of the *stress intensity factor* at the tip of a crack that is required for it to propagate. This indicates how much bigger the stress is in the vicinity of a notch, crack or change in section compared to an area remote from such a discontinuity. The critical stress intensity factor K_c, often termed the *fracture toughness*, is when crack propagation can occur. The smaller the value of K_c the less tough the material.

The critical stress intensity factor K_c is a function of the material and plate thickness concerned. The thickness factor is because the form of crack propagation is influenced by the thickness of the plate. In thin plates, failure is by shear on planes at 45° to the tensile forces across the crack (Figure 6.10). Thicker plate shows a central flat fracture with 45° shear fractures at the sides; the thicker the plate the greater the amount of central flat fracture. The effect of this on the value of the critical stress intensity factor is shown by the graph in Figure 6.11. High values of K_c occur with

Key point

Fracture toughness can be defined as being a measure of the resistance of a material to fracture, i.e. a measure of the ability of a material to resist crack propagation.

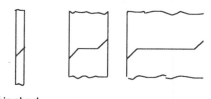

Thin sheet with just 45° shear fracture

Thick sheets have a central flat fracture surface

Figure 6.10 *Effect of thickness on crack propagation*

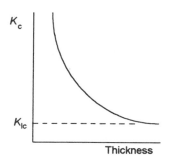

Figure 6.11 *Effect of plate thickness on the critical stress intensity factor*

thin sheets and decrease as the sheet thickness is increased to become almost constant at large thicknesses. At such large thicknesses, the portion of the fracture area which has sheared is very small, most of the fracture being flat and at right angles to the tensile forces. This lower limiting value of the critical stress intensity factor is called the *plane strain fracture toughness* and is denoted by K_{Ic}. This factor is solely a property of the material. It is the value commonly used in design for all but the very thin sheets; it being the lowest value of the critical stress intensity factor and hence the safest value to use. The lower the value of K_{Ic} the less tough the material is assumed to be. Table 6.2 gives some typical values.

Table 6.2 K_{Ic} *values*

Material	$K_{IC} \text{ MN m}^{-3/2}$
Metals	
Aluminium alloys	20 to 160
Copper alloys	50 to 110
Magnesium alloys	10 to 11
Nickel alloys	60 to 110
Steels	25 to 150
Titanium alloys	40 to 150
Polymers	
Cast acrylic sheet	2
General purpose polystyrene	1
Polypropylene	3
Ceramics	
Alumina	5
Silicon carbide	4
Soda-lime glass	0.7

Within a given type of metal alloy there is an inverse relationship between yield stress and toughness, the higher the yield stress the lower the toughness. Thus if, for instance, the yield strength of low alloy, quenched and tempered steels is pushed up by metallurgical means then the toughness declines. Steels become less tough with increasing carbon content and larger grain size.

The toughness of plastics is improved by incorporating rubber or another tougher polymer, copolymerisation, or incorporating tough fibres. For example, styrene–acrylonitrile (SAN) is brittle and far from tough. It can, however, be toughened with the rubber polybutadiene to give the much tougher polymer acrylonitrile–butadiene–styrene (ABS).

6.1.4 Non-destructive testing

Defects such as cracks, inclusions and porosity may be introduced during the manufacture of components or as a result of degradation during service. *Non-destructive testing (NDT)* is the name given to the various techniques which allow inspection of

Key point

Factors affecting fracture toughness:

Composition of the material
Different alloy systems have different fracture toughness. Thus, for example, many aluminium alloys have lower values of plane strain toughness than steels. Within each alloy system there are, however, some alloying elements which markedly reduce toughness, e.g. phosphorus and sulphur in steels.

Heat treatment
Heat treatment can markedly affect the fracture toughness of a material. Thus, for example, the toughness of a steel is markedly affected by changes in tempering temperature.

Material thickness
See Figure 6.11.

Service conditions
Service conditions such as temperature, corrosive environment and fluctuating loads can all affect fracture toughness.

the material to detect the presence, location and size of such defects without impairing the ability of the tested component to function. Destructive tests are those such as tensile testing or impact tests in which samples of the material are cut from the component, so destroying its ability to function.

Non-destructive tests include:

- ### *Visual inspection*

  Examination of a component by the naked eye can reveal large surface defects. The use of a magnifying lens or microscope enables smaller surface defects to be identified. Optical inspection probes which can be inserted into cavities can be used to enable the surfaces of cavities to be examined. Such probes consist of a viewing eyepiece lens at one end of an optical fibre and an objective lens at the other. Light to illuminate the surface being viewed is also conveyed from the viewing end to the viewed end by another optical fibre, generally concentrically arranged round the viewing optical fibre. Visual inspection can only find defects which break the surface.

- ### *Dye penetrant*

  This method is used to make more easily visible defects that break the surface of a component such as cracks and zones of surface porosity. The procedure is first to clean the surface so that it is completely free from contaminants such as oil, water and grease. The dye penetrant is then applied by brush, spray or immersion so that a film of the penetrant is formed over the component surface. The penetrant is drawn into the defects by capillary attraction. After allowing time for this to happen, the excess penetrant is removed from the surface by the use of water or special solvents. With a coloured dye penetrant, the next stage of the process is development to clearly reveal the presence of the dye in cracks. This consists of spraying the surface with chalk dust, generally suspended in a volatile carrier fluid. The penetrant liquid in the cracks is drawn into the pores of the chalk and results in some spread of the liquid (Figure 6.12). This magnifies the apparent width of cracks and makes them more clearly visible. With a fluorescent dye penetrant, the surface is viewed by ultraviolet light to make the dye-penetrated cracks visible. This method can be used for surface defects with metals, many plastics, glass and glass ceramics.

- ### *Magnetic particle inspection*

  This method can be used for the detection of defects which break the surface or are close to the surface of ferromagnetic materials. When a ferromagnetic component is magnetised, any discontinuity that is approximately at right angles to the magnetising field direction will distort the magnetic field lines and if at the surface or close to the surface will result in the formation of a 'leakage field' (Figure 6.13). The magnetic field in a material is generally produced by passing a heavy

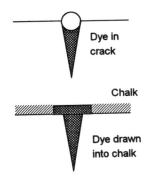

Figure 6.12 *Development of dye in cracks*

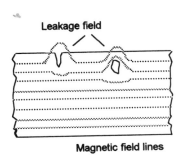

Figure 6.13 *Leakage field*

current through the component, by placing it in a coil through which a current passes or making it part of a magnetic circuit. Magnetic particles, a fine powder of metals or metal oxides, is then sprayed over the surface, either dry in air or a gas or wet on some liquid suspension. The particles 'stick' to the regions of leakage flux and thus render them clearly visible.

- ### Eddy current systems

If an alternating current is passed through a coil an alternating magnetic field is produced. This will induce eddy currents within any conducting material in the vicinity. These eddy currents will also produce a magnetic field. This produces a back e.m.f. in the magnetising coil and so alters its impedance. If the conducting material contains a crack, the flow pattern of the eddy currents is altered and this causes a change in the magnetic field they produce and consequently a change in the coil impedance. This method can detect surface defects and some subsurface defects with metals.

- ### Ultrasonics

Ultrasonics is widely used for the detection of internal defects in materials. Audible sound waves have frequencies ranging from about 20 Hz to 20 kHz. The waves used in ultrasonics are way beyond the audible region, having frequencies ranging from about 0.5 MHz to 20 MHz. Figure 6.14 shows the basic principles of the ultrasonic method for detecting internal defects. The probe contains a piezo-electric crystal which, when an alternating potential difference is applied across it, transmits ultrasonic waves into the material. Some of these pass through to the rear side of the material before being reflected back to the probe, others are reflected back from internal defects. In some probes two piezoelectric crystals are used, one to transmit the waves and one to detect the reflections. However, a single crystal can be used for both purposes.

The simplest method, termed the A-scan, involves a pulse of ultrasonic waves being transmitted into the material and the reflections detected from the backwall and the defect. The initial pulse and the reflected pulses are then displaced on a cathode ray tube (Figure 6.15). The linear distance across the screen from the initial pulse of the reflections is proportional to the distance of those reflecting surfaces from the probe. With the B-scan, the information from a series of A-scans is obtained as the probe is moved in a straight line across the surface of the material and stored in a memory before being displayed on a cathode ray tube screen as a two-dimensional view of the defects in the material (Figure 6.16). With the C-scan, the probe is moved over the top surface plane of the material. The result is a display showing a map of the defects at a particular depth in the material.

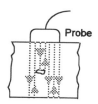

Figure 6.14 *Basic principle of ultrasonic method*

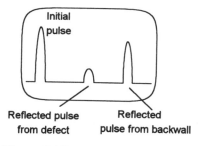

Figure 6.15 *CRT display*

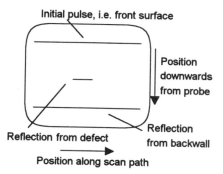

Figure 6.16 *B-scan*

Key points

In considering the failure of a material there are a number of key questions that need to be addressed:

- Was the material to specification?
- Was the right material chosen for the task?
- Was the material correctly heat treated?
- Was the situation in which the material was used, and hence the properties thought necessary for the material, wrongly diagnosed?
- Were factors such as stress, stress concentrations, potential flaw sizes considered in the design?
- Did an abnormal situation occur, perhaps as a result of human error, and was the material therefore subject to unforeseen conditions?
- Has the assembly of the structure, e.g. welding, been correctly carried out?

The evidence on which answers to the above questions can be produced is derived from a consideration of the type of failure, tests on the material, and a consideration of the situation occurring when the failure happened and preceding the event.

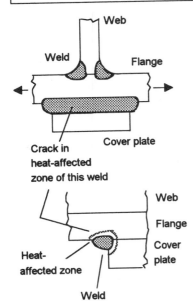

Figure 6.17   *The initial crack zone*

- *Radiography*

This is based on the use of X-rays or gamma radiation and is the same technique as that used for obtaining X-rays of parts of the human body in, say, looking for the fracture of a bone. Radiography can be used with most materials for the detection of both internal and surface defects.

Activity

Make up a chart listing where each of the NDT methods described in Section 6.1.4 may be used and what the limitations are.

6.1.5 Case studies

The Kings bridge, Melbourne

In July 1962, just fifteen months after it had been opened, the Kings bridge in Melbourne, a 700 m long elevated four-lane motorway bridge, suffered a partial collapse as a loaded vehicle of about 45 000 kg was crossing. One of the spans had collapsed. Examination of the bridge showed that four girders had typical brittle fractures. It was established that the ultimate collapse was just the last stage in an ever-increasing pattern of fractures which had been occurring over many months. Every crack had started in the heat-affected zone of a weld (Figure 6.17).

The material used was a high tensile, fusion welding quality, structural steel to British Standard 968 (1941). The composition specified for the steel was 0.23% max. carbon, 1.8% max. manganese, 1.0% max. chromium and the sum of the manganese and chromium to be 2.0% max. Samples of the steel used to produce the girders were analysed and indicated a composition at the limits indicated by the maxima (Table 6.3). Samples from the failed girders indicated compositions outside the limits.

Table 6.3 *Composition of samples*

	% C	% Mn	% Cr	%(Mn + Cr)
According to specification	0.23 max.	1.8 max.	1.0 max.	2.0 max.
Material sample, batch 55	0.21	1.70	0.23	1.93
Material sample, batch 56	0.23	1.58	0.24	1.82
Girder from batch 55	0.25	1.75	0.25	2.00
Girder from batch 56	0.26	1.70	0.25	1.95

Thus while samples taken from the material in these batches had indicated a composition near the upper limit of the specification, what had not been taken into account was that the test pieces were only samples and there would be scatter about those values, the result being that the composition of the actual material used for the girders was outside the specification limits. The effect of the high carbon and high manganese contents was to make it more difficult to produce crack-free welds. The carbon equivalent of the samples of the material was about 0.5, while that taken from the girders was about 0.6. Both these values are high. In addition, the welding techniques used were found to be unsatisfactory and could also lead to cracking.

The collapse of the bridge occurred after a cold night when the temperature was about 2°C. Izod impact tests on samples from the girders indicated that at 0°C the material had impact strengths in the range 3 to 20 J. It is likely that when the fractures occurred the bridge was still close to 0°C and thus the low impact values coupled with the initial cracks in the welds led to a brittle fracture.

The Sea Gem oil rig

The Sea Gem was an offshore drilling rig operating in the North Sea. It consisted of a rectangular pontoon supported by ten steel tubular legs on each side. Each of the legs had a pneumatic jacking system by which it could be lowered to the sea bed and the pontoon raised clear of the water. The arrangement allowed the rig to be moved from one location to another. In December 1965 during procedures preliminary to jacking down the pontoon so that the rig could be moved, the rig collapsed. Nineteen people died.

Investigation showed that the disaster was initiated by brittle fracture of tie bars. The tie bars formed the suspension links transferring the weight of the pontoon to the ten legs. The tie bars had been flame cut from 76 mm steel plates to the shape shown in Figure 6.18. The upper ends of the tie bars had relatively small fillet radii of 4.8 mm. Charpy tests on the steel gave low impact strengths at 0°C of 10 to 30 J. On the day concerned, the temperature was about 3°C. Prior to being used in the North Sea, the rig had been used in the Gulf of Mexico and the Middle East where temperatures are higher. Under such conditions, the impact strengths of the material was much higher. Figure 6.19 shows how the impact strength of the tie bar material varied with temperature. In the North Sea the tie bar material was brittle.

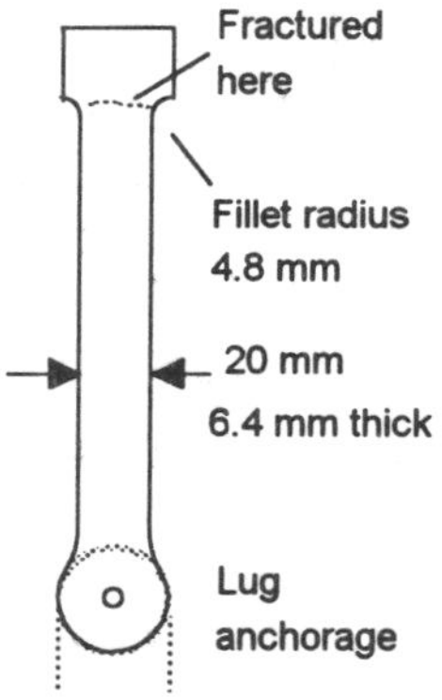

Investigations of the upper end of the tie bars showed that the small fillet radii gave a stress concentration factor of 7. This factor, together with the brittle state of the material, was responsible for the material failing during the stresses imposed by the jacking-down operation. After one or more tie bars had failed, the resulting distribution of forces led to the rapid collapse of the entire structure.

Ammonia plant pressure vessel

In December 1965 a large thick-walled cylindrical pressure vessel was being given a hydraulic test at the manufacturer's works, prior to installation in an ammonia plant at Immingham, when it fractured catastrophically with one segment weighing some 2000 kg being thrown a distance of about 46 m. The pressure vessel was about 16 m long with an internal diameter of 1.7 m and was made

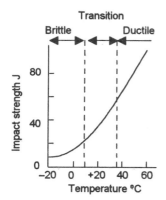

Figure 6.18   *Tie bar*

Figure 6.19   *Impact strength of tie bar material*

of 149 mm thick plates of Mn–Cr–Mo–V steel welded to a forged-end fitting. The material had been furnace stress relieved and the welds locally stress relieved. The design pressure was 35.2 MPa for the vessel with a hydrostatic test pressure of 50 MPa. The ambient temperature at the time of failure was 7°C.

Failure was found to have started in the weld region joining the forged-end fitting to the plate vessel. The origin of the failure was micro cracks in the heat-affected zone. Two of the three micro cracks had propagated and led to the brittle fracture. Analysis of the end forging showed that considerable segregation of carbon and manganese had occurred and thus the forging was heavily banded. Table 6.4 shows the composition of a band and the cast analysis of the forging in relation to the material specification.

Table 6.4 *Composition of forging*

	% Mn	% Cr	% Mo	% V
As specification	1.5	0.70	0.28	0.17
Cast analysis	1.48	0.83	0.29	0.20
In the band	1.88–2.00	0.80–0.82	0.32–0.39	0.25
Outside the band	1.53–1.59	0.69–0.71	0.25–0.21	0.20

As a result of the high carbon and manganese within a band, where the heat-affected zone was in a band the hardness went as high as 400 to 470 HV. The welding technique was based on the assumption that the carbon content was less than 0.20% and so the preheat was discontinued immediately on completion of the weld. Since the carbon content in a band was higher than this, the rate of cooling coupled with the hydrogen not being able to disperse sufficiently resulted in cracking. Thus micro cracks occurred in the banded regions of the heat-affected zone. While this accounted for the micro cracks, it could not explain why the cracks propagated.

The weld metal was found to have very low impact values. Charpy tests indicated values of the order of 10 J instead of the anticipated 40 J or more at the failed temperature of 8°C. Heating samples of the weld metal from the vessel to 650°C for six hours did, however, result in the higher impact values being obtained. It was considered that this could only be explained if the stress-relieving treatment of the welds had not been properly carried out. Thus the failure of the pressure vessel can be attributed to segregation in the forging and incorrectly carried out stress relief of the welds.

Problems 6.1

1 Given a fractured metal specimen, what would you look for in order to determine whether the fracture was ductile or brittle?

2 Explain how the cup and cone form of fractured surface occurs.

3 Describe the appearance of the fractured surface for brittle fracture with a polymer.

4   How does the presence of a notch or an abrupt change in section have an effect on the failure behaviour?

5   In 1944 in Cleveland, Ohio, a steel tank holding liquefied natural gas fractured and caused 128 deaths and considerable damage. The temperature to which the steel was exposed was very low, of the order of –160°C. What effect might this low temperature have had on the steel?

6   Drilling a hole at the end of a small crack is used to prevent the crack propagating. Explain.

7   Explain what is meant by fracture toughness.

8   Suggest non-destructive testing methods that might be used to identify (a) surface defects with a metal, (b) surface defects with a plastic, (c) internal defects in a metal.

6.2 Fatigue

> ## Key point
>
> Fatigue failure of a material occurs at a stress much less than the normal failure stress of the material and is the result of repeated or cyclic stressing.

If you take a paper clip and repeatedly flex it back and forth, it does not last very long before it breaks, even though you have not applied a very high stress. In service many components undergo thousands, often millions, of changes of stress. Some are repeatedly stressed and unstressed, while some undergo alternating stresses of compression and tension. For others the stress may fluctuate about some value (see Figure 6.23). Many materials subject to such conditions fail, even though the maximum stress in any one stress change is less than the fracture stress determined by a simple tensile test. Such a failure, as a result of repeated stressing, is called a *fatigue failure*.

The source of the alternating stresses can be due to the conditions of use of a component. Thus, in the case of an aircraft, the changes of pressure between the cabin and the outside of the aircraft every time it flies subject the cabin skin to repeated stressing. Components such as a crown and wheel pinion are subject to repeated stressing by the very way in which they are used. Other components may receive their stressing 'accidentally'. For example, vibration of a component, and hence alternating stresses, may occur as a result of the transmission of vibration from some machine nearby. It has been said that fatigue causes at least 80% of the failures in modern engineering components.

6.2.1 Fatigue failure

A macroscopic examination of the surfaces of components that have failed by cyclic loading shows distinct surface markings which are characteristic of fatigue failure. Figure 6.20 shows the various stages involved in fatigue failure.

1   A fatigue crack often starts at some point of stress concentration (Figure 6.20(a)). This point of origin of the failure can be seen on the failed material as a smooth, flat, semicircular or elliptical region, and is often referred to as the nucleus.

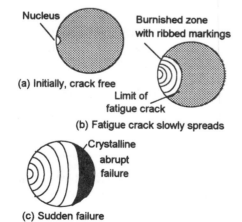

Nucleus

Burnished zone with ribbed markings

(a) Initially, crack free

Limit of fatigue crack

(b) Fatigue crack slowly spreads

Crystalline abrupt failure

(c) Sudden failure

Figure 6.20   *Stages in fatigue failure*

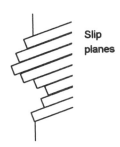

Slip planes

Figure 6.21 *Slip planes providing a nucleus*

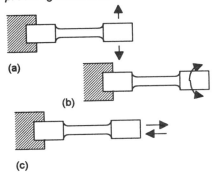

(a)

(b)

(c)

Figure 6.22 *Test modes*

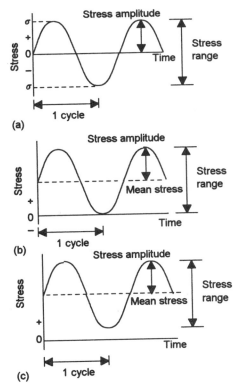

(a)

(b)

(c)

Figure 6.23 *Alternating stress (a) about zero stress, (b) from zero to some maximum, (c) about some stress value*

2 Surrounding the nucleus is a burnished zone with ribbed markings, resembling seashell markings or the marks left on a beach by the tide (Figure 6.20(b)). These markings are produced by the crack propagating relatively slowly through the material and the resulting fractured surfaces rubbing together during the alternating stressing of the component.

3 When the crack is long enough, it spontaneously propagates to give a sudden abrupt fracture of the remaining material (Figure 6.20(c)).

We can explain fatigue failure as being the result of slip, the direction of which reverses with the stress cycle. Thus slip may occur in one direction on one slip plane and the reverse way on an adjacent slip plane during the reverse stress cycle. The above stages in the growth of fatigue cracks are thus:

1 *The nucleus and initial crack*
This is the initiation zone of the fatigue crack. It can be some initial surface defect of the material or arise from the slip causing surface upheavals at the free surface termination of bands of slip (Figure 6.21). Further application of stress results in the nucleus turning into a crack. The repeated slip that occurs results in an increase in dislocation density as a result of the plastic deformation and causes the material to work harden.

2 *Crack propagation*
The fatigue crack slowly propagates with the metal extruded from the slip bands forming the ridges in the burnished zone.

3 *Failure*
The stage 2 crack continues growing until, when the critical crack length is reached and there is insufficient area remaining to support the applied load, there is a complete brittle or ductile failure of the material.

6.2.2 Fatigue tests

Fatigue tests can be carried out in a number of ways, the way used being the one needed to simulate the type of stress changes that will occur to the material of the component when in service. For example, there are bending-stress machines which bend a test piece of the material alternately one way and then the other (Figure 6.22(a)), torsional-fatigue machines which twist the test piece alternately one way and then the other (Figure 6.22(b)) and another type which produces alternating tension and compression by direct stressing (Figure 6.22(c)).

The tests can be carried out with stresses which alternate about zero stress (Figure 6.23(a)), apply a repeated stress which varies from zero to some maximum stress (Figure 6.23(b)) or apply a stress which varies about some stress value and does not necessarily reach zero at all (Figure 6.23(c)). With (a), the stress varies between $+\sigma$ and $-\sigma$, tensile stress being denoted by a positive sign and compressive by a negative sign. The stress range is thus 2σ and the mean stress zero. With (b), the mean stress is half the stress range. With (c), the mean stress is more than half the range. The following are standard definitions used to describe the variables:

stress range = maximum stress – minimum stress [1]

stress amplitude S = ½(maximum stress – minimum stress) [2]

mean stress = ½(maximum stress + minimum stress) [3]

load ratio = $\dfrac{\text{maximum stress}}{\text{minimum stress}}$ [4]

S-N graphs

During the fatigue tests, the machine is kept running with a particular stress range, alternating the stress until the test piece fails. The number of cycles of stressing up to failure is recorded. The test is repeated for different stress ranges. Such tests enable S-N graphs to be plotted. The vertical axis is the stress amplitude S, i.e. half the stress range, and the horizontal axis is the number of cycles N to failure. Figure 6.24 shows examples of S-N graphs.

The number of cycles of stress reversal that a specimen can sustain before failure occurs depends on the stress amplitude, the bigger the stress amplitude the smaller the number of cycles that can be sustained. The *fatigue limit* is the stress amplitude at a particular number of cycles which will result in failure. For Figure 6.24(a), there is a stress amplitude S_D, called the *endurance limit*, for which the material will endure an infinite number of stress cycles with smaller stress amplitudes. For any stress amplitude greater than the endurance limit, failure will occur if the material undergoes a sufficient number of stress cycles. For Figure 6.24(b), there is no stress amplitude at which failure cannot occur. For such materials a fatigue limit S_N may be quoted for a particular number of cycles N.

It must be recognised that there is a relatively large scatter of results in any fatigue test. Thus an S-N graph is drawn through data points which are the mean value of the number of cycles for a number of tests at each stress value. This variation must be considered in interpreting such graphs.

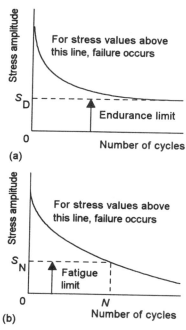

Figure 6.24 *Typical S-N graphs for (a) a steel, (b) an aluminium alloy*

Key points

The S/N graph is a plot of the stress amplitude against the number of cycles to failure.

The *fatigue limit* is the stress amplitude at a particular number of cycles which will result in failure.

The *endurance limit* is the limiting stress amplitude for which a material will endure an infinite number of stress cycles with smaller stress amplitudes.

Example

The following are fatigue test results for an aluminium alloy test piece. Plot the S-N graph and determine the fatigue limit for 10^8 cycles.

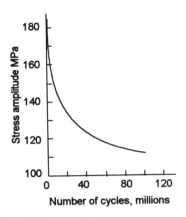

Figure 6.25 *Example*

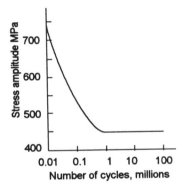

Figure 6.26 *Example*

Stress amplitude MPa	185	155	145	120	115
Number of cycles before failure × 10⁶	1	5	10	50	100

Figure 6.25 shows the resulting *S-N* graph. The fatigue limit for 10^8 cycles is 115 MPa. Extrapolation of the graph appears to indicate that there is no stress amplitude below which there would be no failure.

Example

Plot the *S-N* graph and determine the endurance limit for the steel test piece giving the following fatigue data:

Stress amplitude MPa	750	550	450	450	450
Number of cycles before failure × 10⁶	0.01	0.1	1	10	100

Figure 6.26 shows the *S-N* graph, note the use of a logarithmic scale for the number of cycles. For one million, ten million and one hundred million cycles, the stress amplitude for failure is the same, namely 450 MPa. For stress amplitudes less than this value there appears to be no number of cycles which will result in failure. Thus the endurance limit is 450 MPa.

Effect of mean stress

For any particular value of the mean stress it is possible to determine an *S-N* graph. When the mean stress is zero we have an alternating stress like that shown in Figure 6.23(a) and can obtain a value for the fatigue limit at a particular stress amplitude. However, when the mean stress is equal to the tensile strength of the material, then the fatigue limit is zero since the material will fail without any stress cycles being undertaken. Between these limits of mean stress, increasing the mean stress decreases the fatigue limit.

Factors affecting fatigue properties

The main factors affecting the fatigue properties of a component are:

* **Stress concentrations**
  Stress concentrations are caused by such design features as sudden changes in cross-section, keyways, holes and sharp corners. Figure 6.27 shows the effect of the stress

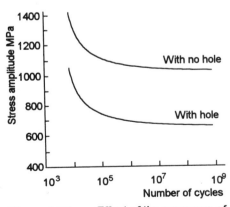

Figure 6.27 *Effect of the presence of a small hole; at every stress amplitude, fewer cycles are needed to reach failure and so there is a lower endurance limit*

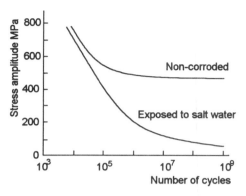

Figure 6.28   *Effect of exposure to salt water for a steel. At every stress amplitude the number of cycles to reach faulure has been reduced. The non-corroded steel has an endurance limit of about 450 MPa, the corroded steel has no such limit*

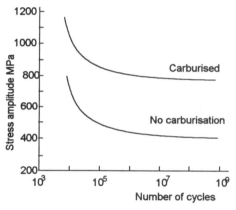

Figure 6.29   *Effect of carburisation on the S-N graph for a steel*

concentration produced by a small hole on the *S-N* graph for a steel. Stress concentrations are also caused by surface scratches, dents, machining marks and corrosion which have left the surface in a roughened form. Figure 6.28 shows the effect of corrosion on the *S-N* graph for a steel. Stress concentrations reduce the fatigue lifetime for a component.

• *Residual stresses*

Residual stresses can be produced by many fabrication and finishing processes. If the stresses produced are such that the surfaces have compressive residual stresses then the fatigue properties are improved, but if tensile surface residual stresses are produced then poorer fatigue properties result. The case hardening of steels by carburising results in compressive surface residual stresses and so improves the fatigue resistance (Figure 6.29). Many machining processes result in the production of tensile surface residual stresses and so result in poorer fatigue resistance.

• *Temperature*

An increase in temperature can lead to a reduction in fatigue properties as a consequence of oxidation or corrosion of the metal surface increasing. For example, the nickel–chromium alloy Nimonic 90 undergoes surface degradation at temperatures of about 700 to 800°C and, as a consequence, there is poorer fatigue performance at these temperatures.

• *Microstructure of alloy*

The microstructure of an alloy is a factor in determining the fatigue properties. This is because the origins of fatigue failure are extremely localised, involving slip at crystal planes. Because of this, the composition of an alloy and its grain size can affect its fatigue properties. Inclusions, such as lead in steel, can act as nuclei for fatigue failure and so impair fatigue properties.

6.2.2 Metals and fatigue resistance

Steels typically have an endurance limit which is generally about 0.4 to 0.5 times the tensile strength of the material. Inclusions in steels can impair the fatigue properties, thus steels with lead or sulphur present to enhance machinability are to be avoided if good fatigue properties are required. The optimum structure for good fatigue properties for steels is tempered martensite. Cast steels and cast irons tend to have relatively low endurance limits.

With steels there is generally a fatigue limit below which fatigue failure will not occur, regardless of how many load cycles occur. However, with non-ferrous alloys this is generally not the case and a fatigue limit is thus quoted for a specific number of cycles, usually 10^7 or 10^8 cycles. This fatigue limit for aluminium alloys is generally about 0.3 to 0.4 times the tensile strength of the material and for copper alloys about 0.4 to 0.5 times the tensile strength of the material. Table 6.5 gives some typical values of tensile strength and endurance/fatigue limits for a range of alloys.

6.2.3 The fatigue properties of polymers

Fatigue tests can be carried out on polymers in the same way as on metals. A factor not present with metals is that when a polymer is subject to an alternating stress it becomes significantly warmer. The faster the stress is alternated, i.e. the higher the frequency of the alternating stress, the greater the temperature rise. This causes the elastic modulus to decrease and at high enough frequencies this may be to such an extent that failure occurs. Thus fatigue in polymers is very much frequency dependent. Under very high-frequency alternating stresses the temperature rise may be large enough to melt the polymer. Thus with polymers, fatigue failure may be either as a consequence of fatigue crack initiation and propagation or by polymer softening which occurs to such an extent that the polymer component can no longer support the load.

A very good example of a fatigue resistant polymer is polypropylene. This material is therefore used where cyclic loading is expected, e.g. the one-piece pencil case where a thin section acts as a hinge between the lid and base or the flip-top bottle top on a detergent bottle.

Key point

The fatigue behaviour of polymers is very sensitive to the frequency of the stress variations since at high frequencies the energy introduced leads to heating of the polymer.

Activity

Flex a strip of a plastic rapidly back-and-forth until it breaks. Then feel the broken edges. Are they hotter than the surroundings?

Table 6.5 *Tensile strength and endurance/fatigue limits for alloys*

Material	Condition	Tensile strength MPa	Endurance/ fatigue limit MPa
Steels, endurance limit			
0.1% carbon steel	Normalised	360	190
0.6% carbon steel	Normalised	740	320
Alloy steels, 3.5% Ni, 1.55% Cr	Oil quenched + tempered 350°C	850	470
Stainless steel, 18% Cr, 8% Ni	Annealed	560	240
Aluminium alloys, fatigue limit with 5×10^8 cycles			
Wrought alloy, 1.0% Mg, 0.27% Cu,	Annealed	130	60
0.60% Si, 0.20% Cr	Sol. treated + aged	315	100
Casting alloy, 5% Mg	As cast	170	50
Copper alloys, fatigue limit with 10^8 cycles			
Brass, 70% Cu, 30% Zn	Hard	76	21
Brass, 90% Cu, 10% Zn	Hard	60	23
Cupro-nickel, 30% Ni	Hard	510	230

6.2.3 The fatigue properties of composites

The fatigue properties of composite materials depend on such factors as the interaction between the mechanical properties of the matrix and the reinforcement, the strength of the bond between the two, the volume fractions of the two, the direction and type of loading, the loading frequency and the temperature. For a random discontinuous glass fibre-reinforced polymer composite, the various stages of failure might be:

1 Development of micro cracks as a result of debonding of reinforcement from matrix.

2 The micro cracks propagate in the matrix and result in the matrix resin cracking.

3 Finally the cracks may have propagated sufficiently for separation of the reinforcement from the matrix.

6.2 Case study

In 1953 and 1954 crashes occurred of the relatively new aircraft the Comet. The Comet was one of the earliest aircraft to have a pressurised fuselage, constructed from aluminium, so that the passengers could travel in comfort. The 1953 crash occurred in a severe tropical storm near Calcutta, the 1954 crash in the Mediterranean sea near Naples. It was not possible to recover parts of the 1953 crash because it had gone down in water 1000 m deep; however, it was possible to recover parts of the 1954 crash by dredging them from a depth of about 180 m in the Mediterranean Sea. The result was a vast mosaic of pieces which had to be fitted together like a jigsaw. The problem was then to identify what was the original failure and what was damage as a consequence of the aircraft crashing. At a small countersunk hole near the corner of a window, drilled to take a fastener, evidence was found of fatigue damage and emanating from this were what were considered to be the cracks responsible for the initial failure (Figure 6.30).

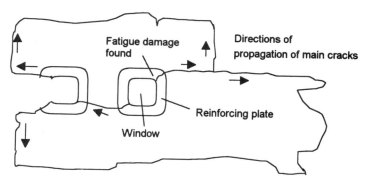

Figure 6.30 *Source of failure*

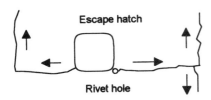

Escape hatch

Rivet hole

Figure 6.31   *Sources of failure*

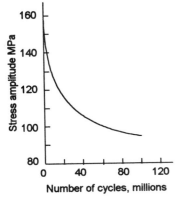

Figure 6.32   *Problem 4*

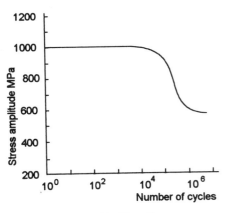

Figure 6.33   *Problem 7*

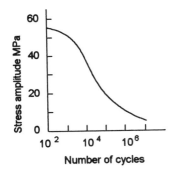

Figure 6.34   *Problem 8*

As part of the investigation, a pressure cabin section was subjected to pressure tests to simulate the effects of flights. The fuselage of a pressurised aircraft is rather like a cylindrical vessel which is pressurised every time the aircraft climbs and pressure reduced every time it descends. In the simulation, the fuselage was immersed in a water tank and the pressure cycles obtained by pumping water into and out of the fuselage. The fuselage used for the test had already made 1230 pressurised flights before the test. It failed after the equivalent of a further 1830 such flights. Failure occurred in a similar place to the crashed aircraft and started from a small area of fatigue damage (Figure 6.31).

The findings of the investigation were that the fundamental cause of the aircraft crash was fatigue failure at a small hole near a window, the small hole acting as a stress raiser. A crack then slowly spread until it reached the critical length and then self-propagated.

Problems 6.2

1 Explain what is meant by fatigue failure.
2 Describe the various stages in the failure of a component by fatigue.
3 Explain the terms 'fatigue limit' and 'endurance limit'.
4 Figure 6.32 shows the *S-N* graph for an aluminium alloy:
  (a) For how many stress cycles could a stress amplitude of 120 MPa be sustained before failure?
  (b) What would be the maximum stress amplitude that could be applied if the component made of this material is to withstand 50 million stress cycles?
  (c) What should the limiting stress be if the number of stress cycles is not likely to exceed 10 million?
5 What is the fatigue limit for the uncarburised steel giving the *S-N* graph in Figure 6.29.
6 The following data are for the nickel–chromium alloy Nimonic 90. Determine the endurance limit.

Stress amplitude MPa	750	480	350	320	320
Number of cycles before failure $\times 10^6$	0.1	1	10	100	1000

7 Figure 6.33 shows the *S-N* graph for a nickel-based alloy Inconel 718.
  (a) What is the endurance limit?
  (b) What is the significance of the constant stress amplitude part of the graph from 10^0 to 10^4 cycles?
  (c) The graph is for the material at 600°C. The tensile strength at that temperature is 1000 MPa. Discuss the effect of fatigue on the limiting stress that can be applied to this material.
8 What is the fatigue limit of the unplasticised PVC at 10^6 cycles which gave the *S-N* graph in Figure 6.34.

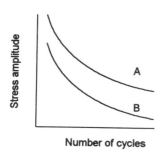

Figure 6.35   *Problem 12*

9 The following data were obtained for cast acrylic plastic when tested with a square waveform at 0.5 Hz. Why specify the frequency? What is the fatigue limit at 10^6 cycles?

Stress amplitude MPa	70	62	58	55	41	31
Number of cycles before failure	10^2	10^3	10^4	10^5	10^6	10^7

10 What is the effect on the *S-N* graph for a component of different mean stresses about which the alternating stresses occur?

11 What factors are likely to contribute to the onset of fatigue failure with a component?

12 Figure 6.35 shows two *S-N* graphs, one for a material in an unnotched state and the other for the same material with a notch. Which of the graphs would you expect to represent each condition?

6.3 Creep

Key point

Creep is the deformation of a material with time when subject to constant stress.

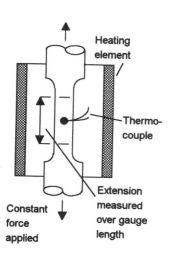

Figure 6.36   *A creep test*

There are many situations where a piece of material is exposed to a stress for a protracted period of time. The stress–strain data obtained from the conventional tensile test generally refers to a situation where the stresses are applied for quite short intervals of time and so the strain results refer only to immediate values resulting from the stresses. Suppose stress was applied to a piece of material and the stress remained acting on the material for a long time, what would be the result? If you tried such an experiment with a strip of lead or perhaps a thermoplastic, you would find that the strain would increase with time. The material increases in length with time even though the stress remains constant. This phenomenon is called *creep*. Creep can be defined as the deformation of a material with the passage of time when the material is subject to a constant stress. For metals, other than the very soft metals like lead, creep effects are negligible at ordinary temperatures but become significant at high temperatures. For polymers, creep is often quite significant at ordinary temperatures and even more noticeable at higher temperatures.

6.3.1 Creep data

Figure 6.36 shows the essential features of a creep test. A constant stress is applied to the test piece, sometimes by the simple method of suspending loads from it. Because creep tests with metals are usually performed at high temperatures a thermostatically controlled heater surrounds the test piece; the temperature of the test piece is generally measured by a thermocouple. Figure 6.37 shows the general form of results from a creep test. Following an 'instantaneous' elastic strain region, the curve generally has three parts. During the *primary creep* period the strain is changing but the rate at which it is changing with time decreases. During the

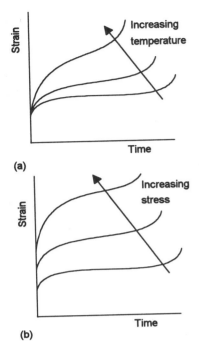

(a)

(b)

Figure 6.38 *(a) Different temperatures but constant stress, (d) different stresses but constant temperature*

secondary creep period the strain increases steadily with time at a constant rate. During the *tertiary creep* period the rate at which the strain is changing increases and eventually causes failure.

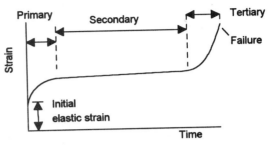

Figure 6.37 *Typical creep graph for a metal*

A family of graphs can be produced that show the creep for different initial stresses at a particular temperature (Figure 6.38(a)) and for different temperatures for a particular initial stress (Figure 6.38(b)).

Stress to rupture

Because of creep an initial stress, which did not produce early failure, can result in failure after some period of time. Such an initial stress is referred to as the *stress to rupture* in some particular time. Thus an acrylic plastic may have a rupture stress of 50 MPa, if loaded at room temperature, in one week. The value of the stress to rupture depending, for a particular material, on the temperature and the time.

Table 6.6 shows the stress to rupture data that might be quoted for a 0.2% plain carbon steel. The data means that at 400°C the carbon steel will rupture after 1000 hours if the stress is 295 MPa. If the steel at 400°C is required to last for 10 000 hours then the stress must be below 147 MPa. If, however, the temperature is 500°C, then to last 100 000 hours the stress must be below 30 MPa. The stress to rupture a material in a particular time depends on the temperature.

Table 6.6 *Stress to rupture data for a 0.2% plain carbon steel*

Time in hours	Rupture stress MPa	
	400°C	500°C
1 000	295	118
10 000	225	59
100 000	147	30

Stress relaxation

So far in this chapter, in the discussion of creep we have assumed that the stress remained constant and the strain varied with time. However, there are situations where the strain is kept constant, e.g. a bolt clamping two plates together. If a bolt is tightened up then a

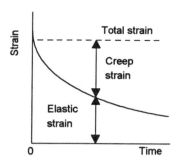

Figure 6.39   *Replacement of elastic strain by creep strain with time*

consequence of creep is that the bolt will gradually become less tight with time. Creep causes the stress applied by the bolt to relax with time.

Consider a bolt which is tightened onto a rigid component. The total strain for a strained bolt will be constant and not vary with time since the length of the stretched shank remains constant. The total strain at any time is the sum of the elastic strain ε_{el} and the creep strain ε_{cr}.

$$\text{total strain} = \varepsilon_{el} + \varepsilon_{cr} \qquad [5]$$

As the creep strain increases with time so the elastic strain decreases. Figure 6.39 shows the general picture. The elastic strain is σ/E, where σ is the stress applied to the bolt and E its modulus of elasticity. Thus as the elastic strain decreases so the stress decreases.

6.3.2 Creep with metals

Following the initial instantaneous strain, primary creep occurs as a result of the movements of dislocations. Initially the dislocations can move easily and quickly, hence the rapid increase in strain with time. However, dislocation pile-ups start to occur at grain boundaries, i.e. *work hardening* occurs, and this slows down the rate of creep. Work hardening occurs more rapidly than recovery processes. The term *recovery* is used to describe the effects of diffusion resulting in the 'unlocking' of dislocations tied up with obstructions and the mutual annihilation of dislocations on meeting each other. The amount of recovery that occurs depends on the temperature; the higher the temperature the greater is the amount of recovery. Therefore, at low temperatures when there is little recovery the work hardening quite rapidly reduces the rate of creep. At higher temperatures the work-hardening effect is reduced by recovery and so the creep rate is less reduced.

During the secondary creep stage, the rate of recovery is sufficiently fast to balance the rate of work hardening. As a result, the material creeps at a steady rate. In addition to dislocation movements there is also, during this stage of creep, some sliding of grains past each. This is referred to as *grain boundary slide*. This mechanism becomes more significant the higher the temperature. Fine-grained materials contain more grain boundaries per unit volume of a material than coarse-grained materials and thus it is more difficult for significant grain boundary slide to occur with fine-grained materials. Thus fine-grained materials tend to be more creep resistant than coarse-grained ones.

With tertiary creep, the rate of strain increases rapidly with time and finally results in fracture. This accelerating creep rate occurs when voids or micro cracks occur at grain boundaries. These are the result of vacancies migrating to such areas and grain boundary slide. The voids grow and link up so that finally the material fails at the grain boundaries.

Table 6.7 *Effect of alloy changes on creep resistance*

Material composition %	Condition	Rupture stress MPa				Oxidation limit °C
		at 700°C		at 800°C		
		1000 hours	10 000 hours	1000 hours	10 000 hours	
80 Ni, 20 Cr	½ h at 1050°C, air cool	100	59	31	18	900
80 Ni, 20 Cr, + Co, Al, Ti	8 h at 1080°C, air cool + 16 h at 700°C, air cool	370	245	140	70	900

Key points

Factors affecting creep with metals

Temperature
The higher the temperature the greater the creep at a particular stress (Figure 6.38(a)).

Stress
The higher the stress at a particular temperature, the greater the creep (Figure 6.38(b)).

Metal
Figure 6.40 shows how the stress to rupture different materials in 1000 hours varies with temperature. Aluminium alloys fail at quite low stresses when the temperature rises above 200°C. Titanium alloys can be used at higher temperatures before the stress to rupture drops to very low values, while stainless steel and nickel–chromium alloys offer better resistance to creep.

Alloy composition
The creep behaviour of an alloy can be affected by the addition of quite small amounts of precipitation-forming elements.

Grain size
Creep resistance increases with decreasing grain size.

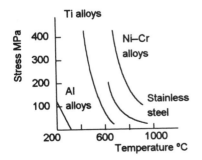

Figure 6.40 *Effect of metal on stress to rupture values*

Creep-resistant metals

The movement of dislocations is an essential feature of creep, hence creep can be reduced by reducing such movement. One mechanism is the use of alloying elements which give rise to dispersion hardening. Finely dispersed precipitates are barriers to the movement of dislocations. The Nimonic series of alloys has excellent creep resistance and is based on an 80% nickel–20% chromium alloy with the addition of small amounts of titanium, aluminium, carbon or other elements to form fine precipitates. Table 6.7 shows the effects of these alloying additions on the creep properties. The Nimonic alloys also have excellent resistance to corrosion at high temperatures and so this, combined with their high creep resistance, makes them useful for high-temperature applications such as gas turbine blades.

6.3.3 Creep with polymers

While creep is only generally significant for metals at high temperatures, it can be significant with polymers at normal temperatures. The creep behaviour of a polymer depends on the temperature and stress level, just like metals. It also depends on the type of plastic involved; flexible plastics show more creep than stiff ones. Generally thermosets have higher temperature resistance than thermoplastics; however, the addition of suitable fillers and fibres can improve the temperature properties of thermoplastics.

Figure 6.41 shows how the strain on a sample of polyacetal at 20°C varies with time for different stresses. The higher the stress the greater the creep. As can be seen from the graph, the polymer creeps quite significantly in a period of just over a week, even at relatively low stresses. Often such graphs have a log scale for both the strain and the time.

Data presentation for polymers

In addition to the form of graph shown in Figure 6.41, there are three additional ways by which creep data can be presented for polymers:

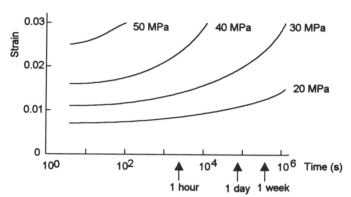

Figure 6.41 *Creep behaviour of polyacetal at different stresses, the temperature being constant*

- *Isochronous stress–strain graphs*

    This is the stress–strain graph for a constant time. Based on the Figure 6.41 graph of strain against time at different stresses, a vertical line is drawn at some particular time and the stresses needed for the different strains read from the graph. Then a graph of stress against strain can be produced for that time (Figure 6.42).

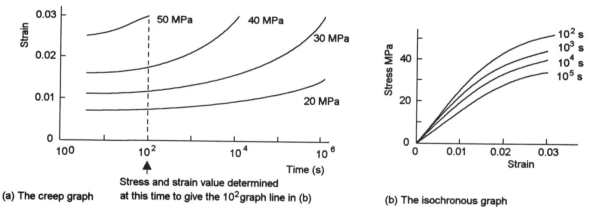

Figure 6.42 *Obtaining the isochronous stress–strain graph*

- *Creep modulus–time graphs*

    For a specific time, the quantity obtained by dividing the stress by the strain for the isochronous stress–strain graph can be calculated and is called the *creep modulus*.

    $$\text{creep modulus } E_c = \frac{\sigma}{\varepsilon} \qquad [6]$$

    where σ is the constant applied stress and ε is the strain which is a function of time. The strain is usually specified and the corresponding stress is read off the graph. The creep modulus

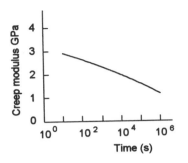

Figure 6.43 *Creep modulus*

is not the same as the elastic (Young's) modulus though it can be used to compare the stiffness of plastics. The creep modulus varies both with time and strain and Figure 6.43 shows how, at 0.5% strain and 20°C, it varies with time for the polyacetal described in Figures 6.41 and 6.42.

- *Isometric stress–time graphs*

The isometric stress–time graph is based on data taken from the strain–time graph at constant strain. Figure 6.44 shows how such a graph can be obtained by drawing horizontal lines on the strain–time graph.

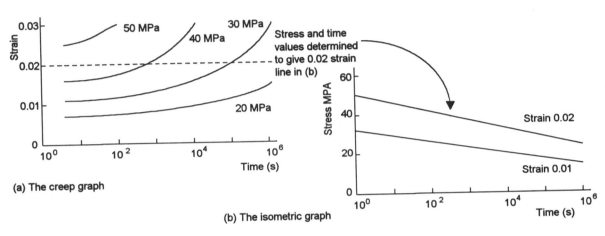

Figure 6.44 *Obtaining the isometric stress–time graph*

Example

Using Figures 6.41, 6.42, 6.43 and 6.44, determine the maximum load a rod of polyacetal with a cross-sectional area of 3×10^{-5} m² can carry for a year if the maximum strain is to be 0.02 after one week. One week is 6.0×10^5 seconds.

Using Figure 6.41, an allowable strain of 0.02 at 3.15×10^7 s corresponds to a stress of about 28 MPa. Using Figure 6.42, an allowable strain of 3.15×10^7 s also corresponds to a stress of 28 MPa. Figure 6.43 gives a creep modulus of about 1.5 GPa at a time of 1 week. This then also indicates a stress of $1.4 \times 10^9 \times 0.02 = 28$ MPa. Using Figure 6.44, an allowable strain of 0.02 at 3.15×10^7 s corresponds to a stress of about 28 MPa. Thus the maximum load is $28 \times 10^6 \times 3 \times 10^{-5} = 840$ N.

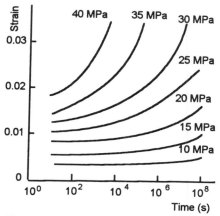

Figure 6.45 *Example*

Recovery from creep

On removing the load from a polymer, the material can recover most, or even all, of the strain given sufficient time. This is different from metals where the strain produced by creep is not recoverable. A consequence of creep and recovery with plastics is that where a component is subject to an intermittent load, under load the component creeps but when the load is removed the component recovers.

The time taken to recover depends on the initial strain and the time for which the material was creeping under the load. For this reason such data is often given in the form of a graph of fractional recovery of strain with reduced time, where:

fractional recovery of strain =
$$\frac{\text{strain recovered}}{\text{creep strain at time load removed}} \qquad [7]$$

A fractional recovery strain value of 1 means that the material has completely recovered and is back to its original size.

reduced time =
$$\frac{\text{recovery time}}{\text{time under load before recovery starts}} \qquad [8]$$

A reduced time value of 1 means that the recovery time is the same as the time it was under load and during which creep occurred.

As an illustration, Figure 6.46 shows a graph for the recovery from creep of nylon 66.

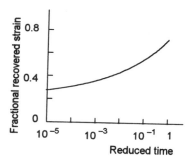

Figure 6.46 *Recovery from creep of nylon 66*

6.3.4 Case studies

Table 6.8 indicates typical temperature limitations for a range of materials and the following case studies illustrate the effects of creep on the choice of materials for products.

Table 6.8 *Temperature limitations of materials*

Temperature limit (°C)	Materials
Room temp. to 150	Few thermoplastics are recommended for prolonged use above about 100°C. Glass-filled nylon can, however, be used up to 150°C. The only engineering metal which has limits within this range is lead.
150 to 400	Magnesium and aluminium alloys can in general only be used up to about 200°C, though some specific alloys can be used to higher temperatures. For example, the aluminium alloy LM13 (AA336.0) is used for pistons in engines and experiences temperatures of the order of 200 to 250°C while some cast aluminium bronzes can be used up to about 400°C with wrought aluminium bronzes up to about 300°C. Plain carbon and manganese–carbon steels are widely used for temperatures in this range.
400 to 600	Plain carbon and manganese-carbon steels cannot be used above about 400–450°C. For such temperatures low-alloy steels are used. For temperatures up to about 500°C a carbon–0.5% Mo steel might be used, up to about 525°C a 1% Cr–0.5% Mo steel, up to about 550°C a 0.5% Cr–Mo–V steel, and up to about 600°C a steel with 5 to 12% Cr. Titanium alloys are also widely used in this temperature range. The alpha–beta alloy 6% Al–4% V (IMI318) is used up to about 450°C. Near alpha alloys can be used to higher temperatures, e.g. the alloy IMI 829 is used up to about 600°C.
600 to 1000	Metals most widely used in this temperature range are the austenitic stainless steels, Ni–Cr and Ni–Cr–Fe alloys, and cobalt base alloys. Austenitic stainless steels with 18% Cr–8% Ni can be used up to about 750°C. A range of high-temperature alloys based on the nickel–chromium base are able to maintain their strength, resistance to creep and oxidation resistance at high temperatures, e.g. Nimonic series alloys such as Nimonic 90 which can be used up to about 900°C, Nimonic 901 to about 1000°C. Another series of high-temperature alloys are the Ni–Cr–Fe alloys, such as the Inconel and Incoloy series. For example, Inconel 600 can be used up to virtually 1000°C and Incoloy 800H to 700°C.
Above 1000	The materials which can be used at temperatures in excess of 1000°C are the refractory metals, i.e. molybdenum, niobium, tantalum and tungsten, and ceramics. The refractory metals, and their alloys, can be used at temperatures in excess of 1500°C. Surface protection is one of the main problems facing the use of these alloys at high temperatures. Ceramics can also be used at such high temperatures but tend to suffer from the problems of being hard, brittle and vulnerable to thermal shock. Alumina is used in furnaces up to about 1600°C, silicon nitride to about 1200°C and silicon carbide to about 1500°C.

Turbine blade

Turbine blades rotate at speeds of the order of 10 000 rev/min for long periods at high temperatures. Centripetal forces act on the blades and result in considerable loading along the axis of a blade. This combination of loading and temperature means that creep is a significant problem. The specification for a blade material is that it should withstand a stress of 250 MPa, the type of stress likely to be experienced on take-off, at 850°C for 30 hours with no more than 0.1% irreversible creep strain.

Nickel alloys can be designed to meet this specification. The basic feature is the use of alloying to obstruct the movement of dislocations; a typical alloy might have 10% cobalt, 10% tungsten, 9% chromium, 5.5% aluminium, 2.5% tantalum, 1.5% titanium,

1.5% hafnium, 0.25% iron, 1.25% molybdenum, 0.15% carbon, 0.1% silicon, 0.1% manganese, 0.05% copper, 0.05% zirconium, 0.015% boron and small traces of sulphur and lead. Nickel crystallises as a face centred cubic close-packed structure and the aim of the alloying to give high creep resistance is to hinder the movement of dislocations. Cobalt, tungsten and chromium are soluble in nickel but have ions which are different in size from nickel ions and so give solid solution strengthening; strains set up round the substitutional atoms hinder the movement of dislocations. Aluminium and titanium form stable compounds with nickel and give extremely hard nodules lodged within the nickel structure. Molybdenum, tantalum, tungsten and titanium form carbides with the carbon and these give hard insoluble precipitates which obstruct the movement of dislocations. Boron, zirconium and hafnium accumulate at grain boundaries and make it more difficult to shear grains apart. Chromium also forms an oxide layer on the blade surface and this improves corrosion resistance.

Soldered joints

The solder used for making electrical connections is a tin–lead alloy with a yield stress of about 30 MPa. Typically, creep failure occurs at room temperature after 1000 hours at 4.5 MPa and when at 80°C the failure after 1000 hours is at 1.4 MPa. This is because, even at room temperature, the alloy is not far below its melting point.

The creep does, however, have a benefit with soldered joints. As a result of frequent current changes through soldered joints, they are subject to thermal fatigue. The diffusion process that occurs with creep heals the damage which occurs with the fatigue and so increases the fatigue endurance limit.

Plastic container

Consider the design of a plastic, e.g. low density polythene, container to hold liquid with the container having a domed base (Figure 6.47). For a container to hold a large volume of liquid, there is significant loading on the base as a result of the weight of the liquid and so creep could be expected to gradually flatten it with time. To minimise the effect of creep, the base thickness can be increased and as small a dome radius as possible used.

Figure 6.47   *The domed base*

6.3 Problems

1 Explain what is meant by 'creep'.
2 Describe the form of a typical strain–time graph resulting from a creep test and the various stages involved in the creep.
3 Describe the effects of (a) increased stress and (b) increased temperature on the creep behaviour of materials.

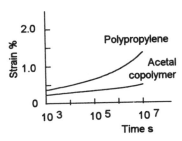

Figure 6.48   *Problem 4*

4 Figure 6.48 shows how the strain changes with time for two different polymers when subject to a constant stress. Which material creeps most?

5 Figure 6.49 shows a graph of the stress to rupture and the stress to give 1% creep at 10 000 h for a nickel–chromium alloy. Estimate the stress that will result in a (a) rupture, (b) 1% creep at a temperature of 900°C after 10 000 h.

6 Figure 6.45 shows the creep graph for PVC at 20°C. If a PVC circular rod is to be subject to a constant load of 800 N, what will its diameter have to be if the maximum strain after one year is to be 0.01?

7 If PVC is to be used for rain water guttering, on the basis of the data given in Figure 6.45, discuss what might happen to the guttering during a storm, particularly if the down pipe becomes blocked.

8 Estimate the creep modulus after 100 s for the PVC giving Figure 6.45 when subject to a stress of (a) 25 MPa, (b) 40 MPa.

6.4 Environmental stability of materials

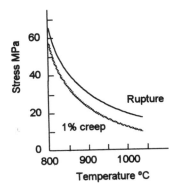

Figure 6.49   *Problem 5*

The car owner can rightly be concerned about rust patches appearing on the bodywork of the car, as the rust not only makes the bodywork look shoddy but also indicates a mechanical weakening of the material. If it continues, eventually a hole might develop. The steel used for the car bodywork has thus changed with time due to an interaction between it and the environment. *Rust* is the term commonly used for the corrosion of iron. The term *corrosion* is used to describe an unintentional chemical reaction between a metal and its environment which results in degradation as a result of the removal of the metal or its conversion into an oxide or some other compound. The environments to which metals can be commonly exposed are the air, water and marine; in addition there are more specialised environments such as steam, high temperatures and chemicals. The possibility of such corrosion is therefore a factor that has to be taken into account when a material is selected for a particular purpose. But it is not only metals that can be affected by their environment, so also can polymers and ceramics. This section is primarily about corrosion with metals with a brief discussion of the environmental stability of polymers and ceramics.

6.4.1 Dry corrosion

The term *dry corrosion* is used to describe that corrosion which arises as a result of a reaction between a metal and a gas, e.g. the oxygen in the air. Most metals react, at moderate temperatures, only slowly with the oxygen in the air to build up of a layer of oxide on the surface of the material. At elevated temperatures, however, the rate of oxidation increases. The initial layer of oxide produced on a metal surface can sometimes form a layer which prevents further oxygen coming into contact with the metal and so

Activity

Look at examples of rusted objects, such as (a) steel chains by the seaside, (b) parts of a car, (c) objects in the house or garage. Make a note of the conditions to which each item has been exposed.

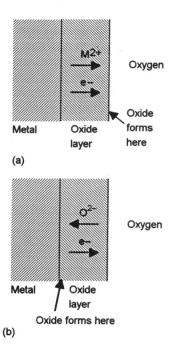

(a)

(b)

Figure 6.50 *Oxide formation*

stops any further oxidation from taking place. Aluminium is an example of a metal that builds up an oxide layer at room temperature which is a very effective barrier against further oxidation.

Oxide layer growth

The reaction between a metal and oxygen can be summarised as:

$$\text{metal} + \text{oxygen} \rightarrow \text{metal oxide}$$

Metal oxides are generally ionic compounds and so in the above reaction we have metal and oxygen ions combining to give the metal oxide. The action is better described as taking place in stages, the following illustrating the case where divalent metal M ions are involved:

1   The metal atoms form ions and free electrons

$$M \rightarrow M^{2+} + 2e^-$$

2   The oxygen molecules pick up electrons to become oxygen ions

$$\tfrac{1}{2}O_2 + 2e \rightarrow O^{2-}$$

3   Metal ions and oxygen ions combine to form an ionic oxide

$$M^{2+} + O^{2-} \rightarrow MO$$

Initially, with a clean metal surface exposed to oxygen, the oxygen molecules dissociate into ions which then bond with metal ions in the surface of the metal to form an oxide layer. If this layer is porous, the oxygen is able to pass through the oxide layer to reach the underlying metal surface and the reaction continues. However, usually the film is not porous and for oxidation to continue it is necessary for metal and oxygen ions to come into contact as a result of oxygen ions diffusing through the surface oxide layers to come into contact with the metal or metal ions diffusing through the oxide layer to the outer surface to come into contact with oxygen. In one case the oxide forms at the metal–oxide interface, while in the other it forms at the oxide–oxygen interface (Figure 6.50).

The rate of continued oxidation depends on:

*   The rate of supply of oxygen to the outer surface.

*   The rate at which the reactants forming the oxide diffuse through surface oxide layers.

*   The permeability or otherwise of the surface oxide layers.

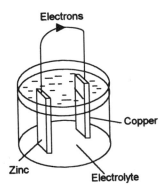

Electrons

Copper

Zinc

Electrolyte

Figure 6.51 *A simple cell*

6.4.2 Aqueous corrosion

A simple electrical cell could be just a plate of copper and one of zinc dipping into an electrolyte (Figure 6.51). Such a cell gives a potential difference between the two metals; it can be measured with a voltmeter or used to light a lamp. Different pairs of metals give different potential differences. Thus a zinc–copper cell gives a potential difference of about 1.1 V and an iron–copper cell about 0.8 V. A zinc–iron cell gives a potential difference of about 0.3 V; this value is, however, the differences between the potential differences of the zinc–copper and iron–copper cells. It is as though we had a cell made up with zinc–copper–iron.

By tabulating values of the potential differences between the various metals and a standard, hydrogen, a table can be produced from which the potential differences between any pair of metals can be forecast. Table 23.1 gives the potential differences relative to hydrogen for some of the more common elements. The table lists what is called the *standard e.m.f. series* and shows the e.m.f. developed when a metal atom changes to an ion.

Table 6.9 *Standard e.m.f. series*

Metal	Potential difference V
Gold → +3 ion	+1.5
Silver → +1 ion	+0.80
Copper → + 2 ion	+0.34
Hydrogen → +1 ion	0.00
Lead → +2 ion	−0.13
Tin → +2 ion	−0.14
Nickel → +2 ion	−0.25
Iron → +2 ion	−0.44
Zinc → +2 ion	−0.76
Aluminium → +3 ion	−1.66
Magnesium → +2 ion	−2.36
Sodium → +1 ion	−2.71

Consider a silver–aluminium cell. If we think of this as being a silver–hydrogen–aluminium cell then we have a silver–hydrogen cell in series with a hydrogen–aluminium cell, i.e. an aluminium–hydrogen cell connected the other way round. The potential difference of the resulting cell is then $+0.80 - (-1.67) = 2.47$ V.

Anode and cathode reactions

When metals ionise we have a reaction involving the loss of electrons from electrically neutral metal atoms to form positive metal ions:

$$M \rightarrow M^{n+} + ne^-$$

where n is determined by the valence of the metal and is usually 1, 2 or 3. Thus with a metal immersed in an electrolyte, its ions can go into solution, leaving behind electrons which freely move in the solid (they are in its conduction band). This separation of charge causes a potential difference to be produced between the metal and the surrounding ions. The metal ions will attract polar water molecules from an aqueous electrolyte and so become partially neutralised, thus more metal ions can leave the metal before equilibrium is reached when the potential difference is large enough to prevent further metal ions leaving the metal. If the electrolyte is an aqueous solution of a salt of the metal, then fewer metal ions can leave the metal before the equilibrium potential difference is attained. The term *contact potential* is used for this equilibrium potential difference between the metal and the electrolyte. Some metals ionise more readily than others and produce larger contact potentials.

When we have two dissimilar metals immersed in an electrolyte with external electrical connection between the two, one of the metals ionises more readily than the other and so produces a larger potential difference between it and the electrolyte. This results in electrons moving from the anode to the cathode and reacting with the metal ions there to give neutral metal atoms, so preventing the ions of the other metal from moving into the electrolyte. Thus at the anode we have:

$$M \rightarrow M^{n+} + ne^-$$

and at the cathode:

$$M^{n+} + ne^- \rightarrow M$$

The net result, for the metals, is as shown in Figure 6.52. At the anode, the more easily ionisable metal, metal ions move into solution and electrons move through the external circuit to the other metal. The anode thus loses metal, i.e. corrodes. The cathode does not lose any metal ions into solution, i.e. it does not corrode. Other reactions involving the electrolyte will also occur, e.g. the evolution of hydrogen gas.

The term *galvanic series* is used for the listing of metals in order of corrosion tendency in a particular environment, e.g. sea water. It is similar to the standard e.m.f. series given in Table 6.9, though it is not precisely the same because the type of electrolyte affects the ability of the metal ions to go into solution.

Dissimilar metal corrosion

If a copper pipe is connected to a galvanised steel tank, perhaps the cold water storage tank in your home, a cell is created (Figure 6.53) and corrosion follows. Galvanised steel is zinc coated steel; there is thus a copper–zinc cell. With such a cell the copper is the cathode and the zinc the anode; the zinc thus corrodes and so exposes the iron. The result is corrosion of the iron and hence the

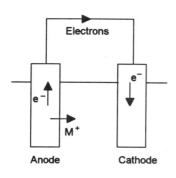

Figure 6.52 *Cell action*

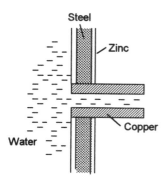

Figure 6.53 *Copper pipe connected to galvanised steel tank*

overall result of connecting the copper pipe to the tank is likely to end up with a leaking tank.

Stainless steel and mild steel form a cell, the mild steel being the anode. Thus if a stainless steel trim, on, say, a car, is in electrical contact with mild steel bodywork, then the bodywork will corrode more rapidly than if no stainless steel trim was used. The electrical connection between the stainless steel may be through water gathering at the junction between the two. With oxygen-free water the cell potential difference may be only 0.15 V but if it was sea water containing sodium and chlorine ions and oxygen, the potential difference could become as high as 0.7 V. So the use of salt in Britain to melt ice on the roads can lead to greater corrosion.

It is not only with two separate metals that cells can be produced and corrosion occur. An alloy or a metal containing impurities can give rise to galvanic cells within itself. For example, brass is an alloy of copper and zinc and the copper and zinc can form a cell with the zinc as the anode. Thus the zinc corrodes. The effect is called *dezincification* and is one example of *demetallification*. After such corrosion, the remaining metal is likely to be porous and lacking in mechanical strength. A similar type of corrosion takes place for carbon steels in the pearlitic condition, the cementite in the steel acting as the cathode and the ferrite as the anode. The ferrite is thus corroded. Cast iron is a mixture of iron and graphite; the graphite acts as the cathode and the iron as the anode of a cell. The iron is thus corroded, the effect being called *graphitisation*.

Concentration cells

Variations in the concentration of the electrolyte in contact with a metal can lead to corrosion. That part of the metal in contact with the more concentrated electrolyte acts as a cathode while the part in contact with the more dilute electrolyte acts as an anode and so corrodes the most. Such a cell is called a *concentration cell*.

Another type of concentration cell is produced if there are variations in the amount of oxygen dissolved in the water in contact with a metal. That part of the metal in contact with the water having the greatest concentration of oxygen acts as a cathode while the part of the metal in contact with the water having the least concentration acts as an anode and so corrodes the most. A drop of water on a level surface is likely to have a higher concentration of dissolved oxygen near its edges where it is in contact with the air than in the centre of the drop (Figure 6.54). The metal at the centre of the drop acts as an anode and so corrodes the most.

A similar situation arises with a piece of iron dipping into water. The water nearer the surface can have its oxygen replenished more readily than deeper levels of water. The result is that the iron near to the water surface becomes a cathode and that lower down an anode (Figure 6.55). The corrosion is thus most pronounced towards the lower end of the piece of iron.

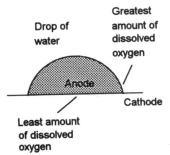

Figure 6.54 *Water drop on a metal surface*

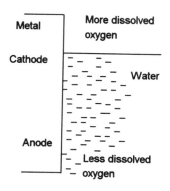

Figure 6.55 *Metal dipping into water*

Key point

Iron rusts when the environment contains both oxygen and moisture but not with oxygen alone or moisture alone.

Rusting

Iron rusts when the environment contains both oxygen and moisture but not with oxygen alone or moisture alone. Iron nails kept in a container with dry air or in oxygen-free water, e.g. boiled water, do not rust. The reaction, an example of a concentration cell, can be considered to occur in a number of stages:

1 Atoms in the surface of the metal lose electrons and become positive ions:

$$Fe \rightarrow Fe^{2+} + 2e^-$$

The area of the metal where this occurs becomes an anode.

2 The liberated electrons move through the metal to where they can combine with water and oxygen molecules to give negative hydroxyl ions. This area becomes the cathode.

$$\tfrac{1}{2}O_2 + H_2O + 2e^- \rightarrow 2OH^-$$

3 The negative hydroxyl ions combine with the metal ions to give hydrated ion oxide.

$$Fe^{2+} + 2OH^- \rightarrow Fe(OH)_2$$

This oxide, termed *rust*, is friable and powdery and so does not protect the metal surface from further attack. Indeed it can hold a layer of water close to the metal surface and accentuate the reaction in air.

Factors affecting aqueous corrosion

Corrosion of a metal in contact with water depends on a number of factors:

* *The metal*
  Gold does not corrode in water, iron does. The position of a metal in the standard e.m.f. series is an indicator of the likelihood of that metal corroding in water. The more negative its potential the more likely it is to behave as an anode and thus the greater the tendency to corrode.

* *Metallurgical factors*
  Aqueous corrosion requires an anode and a cathode to occur. These can result from such structural differences in metals as chemical segregation occurring within the metal (as with dezincification with brass), inclusions, local differences in grain size.

* *Surface defects*
  Surface defects can allow water to come into contact with a metal as a consequence of removing a protective coating.

- *Stress*

  Variations in stress within a metal of a component can lead to the production of cells and hence corrosion. A component which has part of it heavily cold-worked and part less worked will contain internal stresses which can result in the heavily worked part acting as an anode and the less-worked part as a cathode. Therefore the heavily worked part corrodes most.

- *The environment*

  Rusting of iron requires the presence of both water and oxygen, hence iron in oxygen-free water does not rust. Chemically active pollutants, especially those that are soluble in water, in the environment can have a marked effect on corrosion. Man-made pollutants such as oxides of carbon and sulphur, produced in the combustion of fuels, dissolve in water to give acids which readily attack metals and many other materials. Marine environments also are particularly corrosive, due to the high concentration of salt from the sea. The sodium chloride dissociates into its ions in water and the chloride ions combine with the metal ions to produce chlorides of the metals which are soluble in water and thus cannot act as a protective layer on the surface of the metal as a non-soluble oxide might do. The salt may also destroy any protective oxide layer that has been acquired by a metal.

- *Temperature*

  Increasing the temperature increases the rate at which dissolved gases diffuse through water and so increases the reaction rate.

- *Aeration*

  Aeration of water can affect the corrosion rate, increasing the supply of oxygen for, say, the rusting reaction.

- *Anodic reactions*

  Reactions may occur between the anodic region of a metal and the electrolyte which result in the development of surface films of oxides or absorbed gases which then form a barrier to the movement of metal ions into the electrolyte. Further corrosion is thus prevented. This action is said to lead to *passivity*. Many metals form passive films and thus do not give the electrode potentials that would be expected of them from the galvanic series. Thus, to take account of this, galvanic series are available for specific corrosive environments, e.g. in flowing sea water.

6.4.3 Types of corrosion

There are many types of corrosion attack; the following are some of the more common types classified according to the appearance of the corroded material or the type of environment involved:

Activity

You need three small containers, such as the plastic canisters that photographic film comes in, and some steel nails. Then set up the following conditions:

1. Fill one canister with boiled water that has been cooled, put a nail in it and then close the canister with its lid.
2. Fill a second canister with tap water, put a nail in it and leave the canister open.
3. Fill a third container with tap water, add some salt, put a nail in it and leave the canister open.

You may try some other conditions. Leave the samples for a period of two weeks. Then remove the nails and examine their condition. What can you conclude, bearing in mind the terms used and the discussion in this section of the book.

- *Uniform corrosion*

  This is where the entire surface of a metal exposed to the corrosive environment is corroded to the same extent.

- *Two-metal corrosion*

  This is often termed *galvanic corrosion* and occurs when two different metals are in electrical contact to give an electric cell.

- *Pit and crevice corrosion*

  Pit corrosion can start as a surface discontinuity. Within the pit the ingress of oxygen is restricted and so a localised concentration cell can develop. A crevice is just another form of pit. Aluminium alloys, magnesium alloys and steels are susceptible to pitting, nickel alloys, titanium alloys and stainless steels less so; copper alloys are highly resistant.

- *Selective leaching*

  This is where only certain parts of an alloy are attacked. Dezincification of brass is an example of this, a cell forming between the copper and the zinc in the alloy.

- *Intergranular corrosion*

  This is one form of selective leaching where the grain boundaries are more reactive to chemical attack than other parts of a metal. For example, an 18% chromium–8% nickel austenitic stainless steel will have chromium carbides precipitated at the grain boundaries if heated for long enough between about 500 and 800°C. When the chromium carbides form at the grain boundaries they deplete the regions adjacent of chromium. These depleted regions become anodic to the rest of the grain which is not so depleted. The result is a cell and corrosion of the depleted regions.

- *Stress corrosion*

  This form of corrosion takes place when both stress and a corrosive environment are present, taking the form of a network of fine cracks that propagate through the material. Fairly specific conditions are necessary for this to occur. Examples of this are brass exposed to ammonia, mild steel exposed to nitrates and some aluminium and titanium alloys in salt water.

  There is no single explanation of stress corrosion. One explanation which fits some circumstances is that the cracks are initiated at some pit or other discontinuity on the metal surface. When the crack area is subject to tensile stresses tending to open the crack, stress concentration can lead to high stresses at the tip of the crack. This can result in slip and the exposure of 'fresh' metal which is anodic to the metal which has been exposed for some time to the environment.

- *Corrosion fatigue*

  Fatigue failure can be accelerated by corrosive environments, the rate at which cracks propagate being increased.

- *Fretting corrosion*

  This is a particular form of corrosion fatigue that can occur when closely fitting metal surfaces are subject to slight oscillatory slip in a corrosive environment. Due to the surfaces not being perfect, contacts between the surfaces occur at a few high points. The result is high stresses at these points. This leads to localised plastic flow and cold welding. The motion, however, ruptures these welds and loose metal particles are produced. This would occur regardless of the environment. However, a corrosive environment such as air (oxygen) leads to these particles oxidising, making them even more abrasive and increasing the damage to the surfaces.

- *Erosion corrosion*

  The term *erosion* is used to describe the wear produced as a result of the relative motion past a surface of particles or bubbles in a liquid. The result of such abrasive action, essentially being a mechanical process, can be to break down protective coatings on metals and lead to corrosion by concentration cells.

- *Cavitation corrosion*

  This is a form of erosion corrosion that is produced by the formation and collapse of bubbles in a liquid near a metal surface. Rapidly collapsing bubbles can cause very high localised pressures which can damage metal surfaces.

- *Microbiological corrosion*

  This is corrosion produced by the activity of micro-organisms. For example, bacterial activity in sea water can lead to an increase in the level of hydrogen sulphide dissolved in the water. This can have a serious, corrosive effect on steel.

6.4.4 Corrosion prevention

Methods of preventing corrosion, or reducing it, can be summarised as:

- *Selection of appropriate materials*

  Care needs to be taken not to use two different metals in close proximity that can form a cell, particularly if they are far apart in the galvanic series. The metal which is the anode will become corroded. However, there are situations where the introduction of a dissimilar metal can be used for protection. Pieces of magnesium or zinc placed close to buried iron pipes (Figure 6.56) can protect the pipes in that the magnesium or zinc becomes the anode and the iron the cathode and so the magnesium or zinc corrodes and the pipe is protected. Similarly, the steel hull of a ship can be protected b   the water line by fixing pieces of magnesium or zinc to i

  Selection of an appropriate metal for a specific en can do much to reduce corrosion. For example, co

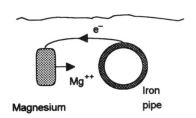

Figure 6.56   *Sacrificial protection of a pipe*

develops a protective layer which then insulates it from further attack and hence the use of copper for domestic water pipes as it offers high resistance to corrosion in such situations. Dezincification can occur with some copper alloys and such corrosion might mean changing to an alloy with a lower zinc content or to a cupronickel. Aluminium readily develops a durable protective layer in air and this protects it from further corrosion. Iron and steel can corrode in many environments and thus there may be a need to change to a steel which has had chromium added to it to improve its corrosion resistance. The addition of nickel to an iron–aluminium alloy can result in an alloy with very high resistance to corrosion in sea water.

• *Selection of appropriate design*

Potential crevices (Figure 6.57) which can hold water or some other electrolyte and so permit a cell to function, perhaps by bringing two dissimilar metals in electrical contact or by producing a concentration cell, should be avoided in the design. Suitable design can also do much to reduce the incidence of stress corrosion.

• *Modification of the environment*

Corrosion can be prevented or reduced by modification of the environment in which a metal is situated. Thus in the case of a packaged item, an impervious packaging can be used so that water vapour cannot come into contact with the metal. Residual water vapour within the package can be removed by including a desiccant such as silica gel within the package.

Where the environment adjacent to the metal is a liquid it is possible to add certain compounds to the liquid so that corrosion is inhibited, such additives being called *inhibitors*. In the case of water in steel radiators or boilers, compounds which provide chromate or phosphate ions may be used as inhibitors.

• *Use of protective coatings*

One way of isolating a metal surface from the environment is to cover the metal with a coating which is impervious to oxygen, water or pollutants. Coatings of grease, with perhaps the inclusion of specific corrosion inhibitors, can be used to give a temporary protective coating. Plasticised bitumens or resins can be used to give a harder but still temporary coating; organic polymers or rubber latex can be applied to give coatings which can be stripped off when required. One of the most common coatings is paint, different types having different resistances to corrosive environments. Thus some paints have a good resistance to acids while others are good for water.

Metallic coatings can be applied to metals to protect them by acting as a sacrificial anode. Thus galvanised steel is zinc-coated steel. The zinc acts as an anode with the steel being the cathode. If the surface layer is broken (Figure 6.58),

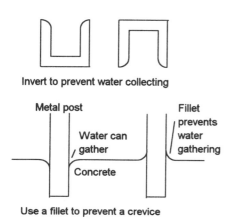

Invert to prevent water collecting

Metal post

Water can gather

Concrete

Fillet prevents water gathering

Use a fillet to prevent a crevice

Figure 6.57   *Ways of reducing corrosion*

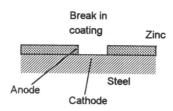

Break in coating

Zinc

Anode

Steel

Cathode

Figure 6.58   *Galvanised steel*

the zinc corrodes rather than the steel. In some cases the metal coating applied to a metal is not sacrificial, e.g. chromium plating. Chromium is often applied to a steel by electroplating, often over an electroplated base coating of nickel. Both the chromium and nickel are, however, more cathodic than the steel and the corrosion protection is based on the better corrosion properties of chromium and the electroplated layers not being breached.

Another form of coating used with metals is a *chemical conversion coating*. Such a coating involves chemically producing changes in the surface layers of the metal. The application of solutions of chromates to some metals, e.g. aluminium, magnesium and copper, can result in the formation of a corrosion-resistant surface layer. The layer is composed of a mixture of oxides of chromium and the metal. Steel surfaces are often treated with phosphoric acid or solutions containing phosphate ions, the process being called *phosphating*. The treatment results in the removal of surface rust and the formation of a protective phosphate coating. The coating, though offering some corrosion protection, is generally used as a precursor for other coatings such as paint. Aluminium in the atmosphere generally has an oxide surface layer which offers some corrosion resistance. This can be improved by thickening the layer by an electrolytic process, the treatment being called *anodising*.

Coatings applied to the surfaces of metals can have one or more functions: to protect against corrosion, to protect against abrasion and wear, to provide electrical or thermal insulation or conductivity, to improve the appearance of the surface.

6.4.5 The environmental stability of polymers

While some polymers are highly resistant to chemical attack, others are liable to stain, craze, soften, swell or dissolve completely. Thus, for instance, nylon shows little degradation with weak acids but is attacked by strong acids; it is resistant to alkalis and organic solvents. Polystyrene, however, though showing similar properties with acids and alkalis, is attacked by organic solvents. Polymers are generally resistant to water, hence their wide use for containers and pipes. However, there is generally a small amount of water absorption.

Polymers are generally affected by exposure to the atmosphere and sunlight. Ultraviolet light, present in sunlight, can cause a breakdown of the bonds in the polymer molecular chains and result in surface cracking. For this reason, plastics often have a UV inhibitor mixed with the polymer when the material is compounded. Deterioration in colour and transparency can also occur. The term *environmental stress cracking* is used for the crack growth in polymers caused by a combination of tensile stress and an environmental liquid or gas, many polymers crazing. The term stress corrosion is used for the similar corrosion with metals.

Activity

Look at real examples of corrosion protection, such as (a) the use of painting on car bodies, (b) the use of grease on bicycle chains, (c) galvanising of steel gates, (d) anodising of decorative aluminium objects. For each case, list reasons for the use of the given method using the information presented in this section of the book.

Natural rubber is resistant to most acids and alkalis. However, its resistance to petroleum products is poor. It also deteriorates rapidly in sunlight. Some synthetic rubbers, such as neoprene, are more resistant to petroleum products and sunlight. Ozone can cause cracking of natural rubber and many synthetic rubbers. Neoprene is one of the rubbers having a high resistance to ozone.

Polymeric materials can provide fire hazards in that they can help spread a fire and produce a dense, toxic smoke. Many polymers burn very easily and produce the toxic gas carbon monoxide; the nitrogen containing polymers, e.g. polyurethane, produces hydrogen cyanide. Polyurethane foams burn very readily because of the high surface to volume ratio.

6.4.6 The environmental stability of ceramics

Ceramics are relatively stable when exposed to the atmosphere, though the presence of sulphur dioxide in the atmosphere and its subsequent change to sulphuric acid can result in deterioration of ceramics. Thus building materials such as stone and brick can be severely damaged by exposure to an industrial atmosphere in which sulphur dioxide is present.

Damage to ceramics may also result from the freezing, and consequential expansion, of water which has become absorbed into pores of the material. Also, the low thermal conductivity can result in large thermal gradients being set up and hence considerable stresses when the hotter material is restricted from expanding by the colder material. This can lead to flaking of the surface.

6.4.7 Case studies

The following is a consideration of the corrosion resistance of particular materials and case studies illustrate the effects that can arise from corrosion. You can, no doubt, easily add examples of everyday occurrences where corrosion has played a part, e.g. the rusting of a car or bicycle, the corrosion at the base of a steel fence post, the galvanised steel domestic cold water tank which starts to leak, the plastic sheeting used for garden cloches which discolours and becomes brittle, the rubber gasket that perishes.

Corrosion resistance of materials

Carbon steels and low-alloy steels are not particularly corrosion resistant, rust being the evidence of such corrosion. In an industrial atmosphere, in fresh and sea water, plain carbon steels and low-alloy steels have poor resistance. Painting, by providing a protective coating of the surface, can reduce such corrosion. The addition of chromium to steel can markedly improve its corrosion resistance. Steels with 4–6% chromium have good resistance in an industrial atmosphere, in fresh and sea water, while stainless steels have an excellent resistance in an industrial atmosphere and fresh water but can suffer some corrosion in sea water. The corrosion resistance of grey cast iron is good in an industrial atmosphere but not so good in fresh or sea water, though still better than that of plain carbon steels.

Aluminium when exposed to air develops an oxide layer on its surface which then protects the substrate from further attack. Wrought alloys are often clad with thin sheets of pure aluminium or an aluminium alloy to enhance the corrosion resistance of such alloys. Thus in air, aluminium and its alloys have good corrosion resistance. When immersed in fresh or sea water, most aluminium alloys offer good corrosion resistance, though there are some exceptions which must be clad in order to have good corrosion resistance. Copper in air forms a protective green layer which protects it from further attack and thus gives good corrosion resistance. Copper has also good corrosion resistance in fresh and sea water, hence the widespread use of copper piping for water distribution systems and central heating systems. Copper alloys likewise have good corrosion resistance in industrial atmospheres, fresh and sea water through demetallification can occur with some alloys, e.g. dezincification of brass with more than 15% zinc. Nickel and its alloys have excellent resistance to corrosion in industrial air, fresh and sea water. Titanium and its alloys have excellent resistance, probably the best resistance of all metals, in industrial air, fresh and sea water and are thus widely used where corrosion could be a problem.

Plastics do not corrode in the same way as metals and thus, in general, have excellent corrosion resistance. Hence, for example, the increasing use of plastic pipes for the transmission of water and other chemicals. Polymers can deteriorate as a result of exposure to ultraviolet radiation, e.g. that in the rays from the sun, heat and mechanical stress. To reduce such effects, specific additives are used as fillers in the formulation of a plastic.

Most ceramic materials show excellent corrosion resistance. Glasses are exceedingly stable and resistant to attack, hence the widespread use of glass containers. Enamels, made of silicate and borosilicate glasses, are widely used as coatings to protect steels and cast irons from corrosive attack.

Table 6.10 gives a rough indication of the corrosion resistance of materials to different environments.

Falling lamp-posts in London

In December 1979, it was reported that there was a problem with falling lamp-posts in central London. Severe corrosion was occurring at the base of lamp-posts. The reason for this was considered to be the daily depositing of urine at the base of the posts by dogs when taken for walks by their owners. The urine had considerably increased the rate of corrosion.

The Kohlbrand bridge

The Kohlbrand bridge is a suspension bridge of length 520 m and is part of a link road between the port of Hamburg and a major motorway. It was built in 1974.

Table 6.10 *Corrosion resistance in various environments*

Corrosion resistance	Material
Aerated water	
High resistance	All ceramics, glasses, lead alloys, alloy steels, titanium alloys, nickel alloys, copper alloys, PTFE, polypropylene, nylon, epoxies, polystyrene, PVC
Medium resistance	Aluminium alloys, polythene, polyesters
Low resistance	Carbon steels
Salt water	
High resistance	All ceramics, glasses, lead alloys, stainless steels, titanium alloys, nickel alloys, copper alloys, PTFE, polypropylene, nylon, epoxies, polystyrene, PVC, polythene
Medium resistance	Aluminium alloys, polyesters
Low resistance	Low-alloy steels, carbon steels
UV radiation	
High resistance	All ceramics, glasses, all alloys
Medium resistance	Epoxies, polyesters, polypropylene, polystyrene, HD polyethylene, polymers with UV inhibitor
Low resistance	Nylon, PVC, many elastomers
Strong acids	
High resistance	Glasses, alumina, silicon carbide, silica, PTFE, PVC, polythene, epoxies, elastomers, lead alloys, titanium alloys, nickel alloys, stainless steels
Medium resistance	Magnesium oxide, aluminium alloys
Low resistance	Carbon steels, polystyrene, polyurethane, nylon, polyesters
Strong alkalis	
High resistance	Alumina, nickel alloys, steels, titanium alloys, nylon, polythene, polystyrene, PTFE, PVC, polypropylene, epoxies
Medium resistance	Silicon carbide, copper alloys, zinc alloys, elastomers, polyesters
Low resistance	Glasses, aluminium alloys
Organic solvents	
High resistance	All ceramics, glasses, all alloys, PTFE, polypropylene
Medium resistance	Polythene, nylon, epoxies
Low resistance	Polystyrene, PVC, polyesters, ABS, most elastomers

In 1976 the first signs of corrosion were detected in the suspension cables and attempts were made to halt the process by coating the strands in the cable with a plastic sealant. But within a year of this remedial action the cables near the bridge road deck and the anchor points began to expand as the rate of corrosion accelerated. In 1978 a survey showed that rain had completely permeated the cable strands from top to bottom and that the salt applied to the road surface in winter had also worked its way into the stays. In addition it is thought that vibration from heavy vehicles crossing the bridge hastened the corrosion. As a consequence of the rapidly deteriorating situation, all the cables were replaced, with the cable stays being heavily galvanised to give better protection against the salt. In addition, vibration dampers are being used in the cable anchor points to damp down vibration.

Failure of Mount Pleasant bridge

In December 1967 the Mount Pleasant bridge in West Virginia in the United States failed without warning and 46 people died. The investigation found that the failure could be attributed to the brittle fracture of an eyebar in its head end. Examination of the fractured surface showed the fracture originating from a heavily corroded region. Sulphur was found to be present on the fracture surface and it was considered that hydrogen sulphide in the air had been responsible for a stress corrosion crack, which then resulted in the brittle fracture that followed. The sensitivity of the bridge steel to hydrogen sulphide stress corrosion cracking was verified by tests on notched specimens of the steel.

Problems 6.4

1 It has been observed that cars in a dry desert part of a country remain remarkably free of rust when compared with cars in a damp climate such as England. Offer an explanation for this.

2 Why does the de-aeration of water in a boiler reduce corrosion?

3 What criteria should be used if corrosion is to be minimised when two dissimilar metal pipes are joined together?

4 Aluminium pipes are to be used to carry water into a water tank. Possible materials for the tank are copper or galvanised steel. Which material would you advocate if corrosion is to be minimised?

5 It is found that for a junction between mild steel and copper in a sea water environment that the mild steel corrodes rather than the copper. With a mild steel–aluminium junction in the same environment it is found that the aluminium corrodes more than the mild steel. Explain the above observations.

6 Explain the process of dezincification that occurs with brasses.

7 Explain why a drop of water on a steel surface is found to give a small rust patch in the centre of the drop area and not round the circumference of the drop where air, water and metal are together.

8 Explain how pitting occurs.

9 Explain how pieces of magnesium placed close to buried pipes are used to reduce the corrosion of the iron.

10 Compare the use of zinc and tin as protective coatings for steel.

11 In an article on underwater equipment design the author states that good design required the avoidance of crevices and high stress concentration and also the number of different metals should be kept to a minimum. Give explanations for these criteria.

12 1 kg of a sacrificial magnesium anode attached to the steel hull of a ship is found to completely corrode in 50 days. What is the average current produced between the magnesium and the hull in that time? Magnesium is divalent, the Avogadro constant is 6.0×10^{23}, the molar mass of magnesium is 24.3 g and the charge on an electron is 1.6×10^{-19} C.

13 A sacrificial magnesium anode corrodes with an average current of 1.0 A for 100 days. What will be the loss of magnesium from the anode in that time? Magnesium is divalent, the Avogadro constant is 6.0×10^{23}, the molar mass of magnesium is 24.3 g and the charge on an electron is 1.6×10^{-19} C.

6.5 Wear

Key point

Wear is the progressive loss of material from surfaces as a result of contact, such as sliding or rolling contact, with other surfaces.

Wear is the progressive loss of material from surfaces as a result of contact with other surfaces. It can occur as a result of sliding or rolling contact between surfaces or from the movement of fluids containing particles over surfaces. Because wear is a surface effect, surface treatments and coatings play an important role in improving wear resistance. Lubrication can be considered to be a way of keeping surfaces apart and so reducing wear.

Mild steels have poor wear resistance. However, increasing the carbon content increases the wear resistance. Surface hardenable carbon or low-alloy steels enable wear resistance to be improved as a result of surface treatments such as carburising, cyaniding or carbonitriding. Even better wear resistance is provided by nitriding medium-carbon chromium or chromium–aluminium steels, or by surface hardening high-carbon high-chromium steels. Grey cast iron has good wear resistance for many applications. Better wear resistance is, however, provided by white irons. Among non-ferrous alloys, beryllium coppers and cobalt-base alloys, such as Stellite, offer particularly good wear resistance.

6.5.1 Bearing materials

Metallic materials for use as bearing surfaces need to be hard and wear resistant, with a low coefficient of friction, but at the same time sufficiently tough. When one surface slides over another, the frictional force is proportional to the normal force and is independent of the apparent area of contact between the sliding surfaces. The term 'apparent area' has been used because no matter how smooth a surface, on an atomic scale it is irregular, and contact between two sliding surfaces only occurs at a limited number of discrete points. It is these small, real, contact areas that have to carry the load between surfaces. Because the real areas are so small, the pressure at the contact points will be very high, even under light loading. With metals, the pressure will generally be high enough to cause appreciable plastic deformation and adhesion between the two surfaces at these points. This is termed cold welding for metals. When surfaces slide over each other, these junctions have to be sheared. The frictional force thus arises from the force to shear junctions and the force required to plough the asperities of one surface through those of the other surface.

Bearing materials can be classified, in the main, into four categories:

- *Whitemetals*

  These are tin-base or lead-base alloys with the addition of mainly antimony or copper. They have a microstructure of hard intermetallic compounds of tin and antimony embedded in a soft matrix. The hard particles support the load, since the asperities penetrate the softer material, but the greater area of contact between the surfaces is with the soft material. Thus sliding takes place within a thin smeared film of the softer material. Whitemetals have relatively low fatigue strength and this can limit their use to low-load conditions. Reducing the thickness of the bearing material can improve the fatigue properties but does require care because of the size of the hard intermetallic particles. Tin-base alloys resist corrosion better, have higher thermal conductivity, have higher modulus of elasticity and higher yield stress but are significantly more expensive than lead-base alloys. Both forms of alloy are relatively soft.

- *Copper-base alloys*

  These offer a wider range of strength and hardness than whitemetals. They include tin bronzes with between 10 and 18% tin, leaded tin-bronzes containing 1 or 2% lead, phosphor bronzes and copper–lead alloys containing about 25 to 30% lead. The properties of the copper–lead alloys depend on the lead content, the higher the amount of lead the lower the fatigue strength but the better the sliding properties. They have poorer corrosion resistance than whitemetals but better wear resistance, a higher modulus of elasticity and better fatigue resistance. Bearings are manufactured by casting onto steel strip or sintering copper and lead on the strip. The bronzes have higher strengths, hardness, modulus of elasticity and better fatigue resistance than the copper–lead alloys and the whitemetals, so being used for high load-bearing loads.

- *Aluminium-base alloys*

  Aluminium–tin alloys, with about 5 to 7% tin, 1% copper, 1% nickel and small amounts of other elements, give bearing materials with a high fatigue strength, hardness and strength which makes them suitable for high-load bearings. However, they have the disadvantage of a high thermal expansivity which can lead to loose bushes or even seizure against other surfaces.

- *Non-metallic bearing materials*

  Polymers suitable for bearing materials include phenolics, nylon, acetal and PTFE. For some applications the polymers have fillers, e.g. graphite-filled nylon, PTFE with a silicon lubricant, acetal with PTFE filler. In addition to the fillers used to decrease the coefficient of friction, other fillers such as glass fibres are added to increase strength and dimensional stability. Polymers have the advantage of a very low coefficient of friction but the disadvantage of a low thermal conductivity. Polymer rubbing against polymer can lead to high rates of wear, but polymer against steel gives a very low wear rate. Polymers have thermal expansivities much greater than metals and so can present problems, e.g. a higher running clearance between surfaces is needed. They tend to be used under low load conditions where they have the advantage of being cheap. Typical applications are: PTFE bearings in car steering linkages and food processing equipment; phenolics for marine propeller shafts; acetals for electrical appliances.

- *Metal–non-metallic bearing materials*

  Graphite-impregnated metals and PTFE-impregnated metals are widely used bearing materials. Such materials are able to utilise the load-bearing and temperature advantages of metals with the low coefficient of friction and soft properties of the non-metals. Thus graphite-impregnated metals rubbing against a steel mating surface can be used with load pressures up to about 40 MPa and operating temperatures up to 500°C while PTFE impregnated metals can be used with loads up to 100 MPa and temperatures of 250°C.

Table 6.12 gives simple comparisons of the various types of bearing materials. The ability of a bearing material to dissipate heat resulting from friction can be important in preventing a rise in temperature to a level which could result in seizure of a bearing, degradation of a lubricant or even melting of the bearing material.

A bearing material is inevitably a compromise between the opposing requirements of softness and high strength. One way of achieving strength with a relatively soft bearing material is to use the soft material as a lining on a steel backing, e.g. whitemetals, aluminium or copper-base alloys as a thin layer on a steel backing. Plastics when bonded to a steel backing can be used at higher speeds than otherwise would be possible, because the steel is able to dissipate heat better than the plastic alone and also the thinner the layer of plastic the smaller the amount by which it will expand.

Table 6.12 *Comparison of bearing materials (also continued on next page)*

Material	Hardness BH	Yield stress MPa	Strength MPa	Fatigue strength MPa	Elastic modulus GPa	Density Mg/m^3
Tin-base whitemetal	17–25	30–65	70–120	25–35	51–53	7.3–7.7
Lead-base whitemetal	15–20	20–60	40–110	22–30	29	9.6–10
Copper–lead	20–40	40–60	50–90	40–50	75	9.3–9.5
Phosphor bronze	70–150	130–230	280–420	90–120	80–95	8.8
Leaded tin bronze	50–80	80–150	160–300	80	95	8.8
Aluminium-base	70–75	50–90	140–210	130–170	73	2.9
Polymers	5–20		20–80	5–40	1–10	1.0–1.3

Table 6.12 *Continued*

Material	Thermal conductivity $W\,m^{-1}\,K^{-1}$	Relative corrosion resistance	Relative wear resistance	Relative cost
Tin-base whitemetal	50	5	2	7
Lead-base whitemetal	24	4	3	1
Copper–lead	42	3	5	1.5
Phosphor bronze	42	2	5	2
Leaded tin bronze	42	2	3	2
Aluminium-base	160	3	2	1.5
Polymers	0.1	5	5	0.3

Note: the larger the number for relative corrosion resistance and for relative wear resistance the better the resistance.

Activity

With reference to the discussion of bearing materials in Section 6.5, list the features that characterise a very good bearing material.

Case study

Consider the requirements for a light-load bearing when the load is steady and the speed is high. The high speed means high temperatures can be produced and so a high thermal conductivity is desirable. This rules out polymeric materials and lead-base whitemetals. The light load means that a high-strength material is not necessary. This means that we need not consider the highly priced copper-base alloys since there is no point paying for strength if it is not required. Because the load is steady, fatigue strength is not a particular requirement. The choice thus appears to be between aluminium-base alloys and tin-base whitemetals. The aluminium-base alloys have the advantages of a higher modulus of elasticity and much lower costs and so seem likely to be the optimum material.

Problems 6.5

1 Explain how frictional forces arise.
2 Explain how the microstructure of white metals is used to improve wear resistance.
3 Why is the low thermal conductivity of polymers a problem when they are used as bearing materials?

7 Selection of materials

Summary

In considering the behaviour of products and designing products, engineers are concerned with the selection of materials and processes. For example, in considering the materials and processes that might be used for the mass production of small toy cars for very young children, what materials and manufacturing processes could be used and how, in a particular situation, can we decide which ones to use? This chapter considers case studies illustrating the choices made of both materials and processes and the criteria that can be used to determine the optimum material and process.

Objectives

By the end of this part, the reader should be able to:

- produce reasoned arguments for the choice of materials and processes for particular products;
- use property values, material property indices, cost per unit property factors to determine the optimum material for a particular product.

7.1 Selection

Design, materials and manufacturing interact. The designed shape and the materials dictate the choice of manufacturing process. Slender metal shapes can be made by drawing or rolling but not by casting. Many processes require ductile materials, e.g. rolling, forging and drawing. For some shapes, polymers may require stiffening by the incorporation of fibres. By alignment of fibres in a composite material, the properties in such materials can be enhanced in a particular direction.

7.1.1 Which material?

A number of questions need to be answered before a decision can be made as to the specification required of a material and hence a decision as to the optimum material for a particular task. The questions can be grouped under four general headings:

* *What properties are required?*

* *What are the processing requirements and their implications for the choice of material?*

* *What is the availability of materials?*

* *What is the cost?*

The Key points indicates the type of issues that are likely to be considered in trying to arrive at answers to the above general questions.

7.1.2 Which processing method?

In making a decision about the manufacturing process to be used for a product there are a number of questions that have to be answered:

* *What is the material?*
  The type of material to be used affects the choice of processing method. For example, if casting is to be used and the material has a high melting point then the process must be either sand casting or investment casting.

* *What is the shape?*
  The shape of the product is generally a vital factor in determining which type of process can be used. For example, a product in the form of a tube could be produced by centrifugal casting, drawing or extrusion but not generally by other methods.

* *What type of detail is involved?*
  Is the product to have holes, threads, inserts, hollow sections, fine detail, etc.? For example, forging could not be used if there was a requirement for hollow sections.

* *What dimensional accuracy and tolerances are required?*
  High accuracy would rule out sand casting, though investment casting might well be suitable.

* *Are any finishing processes to be used?*
  Is the process used to give the product its final finished state or will there have to be an extra finishing process. For example, planing will not produce as smooth a surface as grinding.

* *What quantities are involved?*
  Is the product a one-off, a small batch, a large batch or continuous production? While some processes are economic for small quantities, others do not become economic until large quantities are involved. For example, open die forging

---

Key points

Properties
What mechanical properties are required?
At what temperatures?
What chemical properties are required?
Consider the environment and the possibility of corrosion.
What thermal properties are required?
Could expansion be a problem?
What electrical properties are required?
A good conductor of electricity or perhaps an insulator?
What magnetic properties are required?
Soft or hard magnetic properties or perhaps non-magnetic?
What dimensional conditions are required?
A good surface finish, dimensional stability, a particular size or shape?

Processing parameters
Are there any special processing requirements which will limit the choice of material?
For example, cast or perhaps extruded?
Are there any material treatment requirements?
For example, annealed or perhaps solution hardened?
Are there any special tooling requirements?
Does the hardness required of a material mean special cutting tools are required?

Availability
Is the material readily available?
Already in store, or perhaps quickly obtainable from normal suppliers?
Are there any ordering problems?
Only available from special suppliers? Is there a minimum order quantity?
What form, e.g. bars or sheet, is the material usually supplied in?

Cost
What is the cost of the raw material?
Could a cheaper material be used?
What quantity is required?
Per week, per month, per year? What stocking policy should be adopted for the material?
What are the cost implications of the process requirements?
High initial expenditure? Running costs high or low? Skilled labour required?
What are the cost penalties for overspecification?
If the material is stronger than is required, will this significantly increase the cost? If manufactured to a higher quality than is required, what will be the cost implications?
What will be the cost of the processed material?

could be economic for small numbers but closed die forging would not be economic unless large numbers were produced.

7.2 Case studies

The following case studies are to illustrate the interaction between the consideration of materials and manufacturing processes in the design of products.

7.2.1 A crane hook

Consider the design of a hook (Figure 7.1) for use with a crane. The evolution of the product design specification leads to a load capacity of 150 Mg (1.5 MN) being specified with the conditions of service being in temperatures which can fall as low as –20°C and up to +40°C. The hook will be connected to the crane by a 200 mm pin through a hole in the hook and so needs to be wear resistant. Only small numbers of such hooks are required.

In arriving at a design we need to consider the stresses involved in the hook and so arrive at views on the size and materials. The load will subject the hook to both direct tensile loading and a bending moment. The shank section of the hook will be subject to predominantly a direct tensile load. At the eye of the hook there will be a direct tensile load and a stress concentration factor to be taken into account. In addition we have to consider that the load carried by the hook may not be always directly downwards but some sideways movement can occur and so the hook must be able to resist bending in the sideways direction. If we consider the use of steel for the hook then we might want a design with the stresses limited to a maximum of 80 MPa, this allowing a reasonable safety factor. By carrying out stress analysis for the hook, a proposal for the size of the hook can be evolved.

The steel for the hook will need to have a yield strength, if we assume a safety factor of 4, of about 320 MPa. This is easily achieved with a plain medium carbon steel, e.g. 0.3 to 0.4%, without the need to go for a more expensive alloy steel. A higher carbon steel would have impact properties which would be too low. BS 080M36 (AISI 1035) has, when quenched and tempered at 550 to 660°C, a yield stress of 400 MPa, a hardness of about 200 BH and an Izod impact value of 25 J. However, if welding is required in the manufacture, then a high-strength low-alloy structural steel, e.g. 0.2% carbon with about 1% manganese, could be used. This would probably increase the cost by a factor of about 1.2 times that of the plain carbon steel.

The hook could be manufactured by casting, by forging, or by building up the thickness from several layers of plate which are riveted together to give the required thickness. The last method is likely to be the cheapest, particularly for small numbers. The inner surface of the hook might be made more wear resistant by a weld deposited hard-facing alloy; the eye of the hook might be made more wear resistant by press-fitting a high-carbon steel bushing which is welded round its circumference to lock it into place.

Figure 7.1   *Crane hook*

7.2.2 A kettle

Consider the design brief for a product for boiling water in small volumes, up to say 2 litres, for making coffee or tea in the home. It needs to be easily used, be safe in use, not contaminate or discolour the water in any way, have a pleasing appearance, be capable of mass production and comparable in price to existing methods.

As part of the conceptual design, we can consider whether the heat source for the hot water should be generated internally in the product, as with an electric kettle, or externally, as with a kettle which is to stand on the hot ring of a cooker. For ease of use, but not necessarily cost, an internal source of heat seems more desirable. Thus we are led to the idea of an internal, insulated, electric heater. The material for the kettle will have to withstand temperatures up to 100°C and not contaminate the water. The kettle will also have to have walls which are stiff enough, and show no significant creep at such temperatures, without being so thick that the appearance is not pleasing. We might use a metal, e.g. an aluminium alloy or stainless steel, or a polymer. Both can give products which are light and with a pleasing appearance. The choice of polymer is restricted by the need for it to not contaminate the water and have a high enough modulus of elasticity and good creep resistance. Nylons are not feasible because they absorb water, polycarbonates and polyesters deteriorate when in contact with water and most thermosetting polymers will contaminate water with time. To meet the stiffness criterion we are led to the idea of using a polymer which is not amorphous at the temperatures reached but partially crystalline. A possible material is polypropylene which has been stiffened by the incorporation of about 10% talc. Polymers have the advantage over metal of being able to be coloured, though cannot be made as shiny as a metal.

A number of kettle shapes might be considered, e.g. jug shaped or perhaps dome-shaped (Figure 7.2), with the additional features of an external indicator of the level of the water in the kettle, a lid which is latched, a filter to remove hard water deposits, and a method of reducing the build-up of scale on the hot electric element by having the element in a sealed base.

What process might be feasible for the manufacture of the kettle body? For an aluminium alloy kettle body we might consider a casting process, e.g. gravity die, pressure die or investment casting, or forming it by drawing. Pressure die casting has a faster throughput than the other casting methods, gives a surface which has a good texture, and seems the most likely casting method. The stages in the process would be: cast, clean and polish. To form the kettle body by drawing will require a number of drawing and ironing stages to get the cup depth and required wall thickness. The material needs to be ductile and the shape produced has to be relatively simple. The stages in the process would be: blank shapes from a sheet, draw, anneal, iron, anneal, clean and polish, form the spout by a separate process, cut a hole for the spout and attach it. The spout might be made by a one-stage casting process or by a

Figure 7.2   *Kettle shapes*

number of stages involving forming the shape from sheet and then joining. Soldering with aluminium can present problems, a particular problem being that aluminium forms a protective surface oxide layer and this has to be broken down before soldering can be used. Stainless steel does not present the soldering problem that aluminium does. Casting has thus the advantage of requiring less operations than forming and so is likely to be the optimum method; however, forming is used to make some jug kettles where the simple shape lends itself to a drawing/ironing process.

For a polymer kettle body we might consider injection moulding or rotational moulding. Cost-wise, short production runs would favour rotational moulding while long runs favour injection moulding. The spout might be moulded with the body as a single entity.

7.2.3 A car head lamp reflector

The design requirements for a car head lamp reflector are that it reflects light, is a shape such that main beam and dipped beam directional light beams can be produced, is a shape that can fit with a head lamp which is styled to conform with the aerodynamic shape required for the front end of a modern car, can be made in quantity at an economic price, and is of a material that maintains its reflecting ability and shape at the temperatures likely to be realised in a sealed head lamp unit.

One way that the main beam and dipped beams can be produced is to use a two filament bulb with a parabolic reflector (Figure 7.3), switching between the filaments to give the required beam. The filament for the main beam is located at the focus of the parabola and the filament for the dipped beam is ahead of the focus, the lower part of the light beam being blocked off by a cup since it would be directed in an upwards direction and dazzle oncoming motorists. An alternative reflector shape which avoids the loss of the lower part of the dipped beam, and so is more efficient and gives a brighter beam, is to use a reflector which does not have a constant focal length. Thus the reflector shape might be of the form shown in Figure 7.4, the inner part of the reflector having a longer focal length than the outer part. Such a design enables head lamps to be shallower and fit chassis holes which are other than circular. The design of such lamp reflectors requires very close design and control over the curvatures of the reflector.

The traditional material used for head lamp reflectors was steel which was pressed to the required shape. This presents problems with the more complex shapes required of modern head lamps with, say, rectangular shape and varying focal length. Die casting of metal offers a possible solution but is expensive. An alternative, which is comparatively cheap, is to use a polymer and manufacture the shape by moulding. The polymer will need to be resistant to the temperatures generated in a sealed head lamp and not deteriorate or significantly creep; the temperatures can be as high as 200°C. The high temperature requirement suggests that a

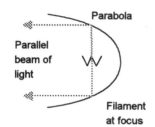

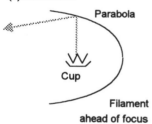

Figure 7.3 *Parabolic reflector*

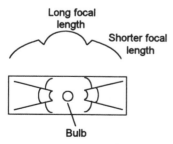

Figure 7.4 *Variable focal length reflector in a rectangular case*

thermoset, rather than thermoplastic, is required. A dough moulding compound (DMC) of a polyester resin containing talc and short-length glass fibres has been used with injection moulding. The moulded reflector is coated with a lacquer and then given its reflective surface by the vapour deposition of aluminium in a vacuum; the lacquer gives a smooth surface and also prevents outgassing from the plastic during the vapour deposition of the aluminium.

7.2.4 A transformer core

Basically a transformer consists of two coils wrapped around a soft iron yoke so that when an alternating current is passed through one coil, a changing magnetic flux is produced which links the other coil and induces in it an alternating e.m.f. The aim is to get the maximum amount of flux linking the secondary coil and for losses to be minimised. Figure 7.5 shows the basic constructional details of transformers.

Thus, the core needs to have high permeability and low coercivity with a high value of flux needed to saturate it. The low coercivity is needed if hysteresis losses are to be kept low, this loss being proportional to the area enclosed by the $B–H$ curve. The core should be made with a a large cross-section and higher relative permeability material in order to reduce the reluctance of the magnetic circuit, reluctance = (length of core)/(relative permeability × permeability of free space × cross-sectional area), and so reduce the magnetisation current needed to produce the flux density. However, the hysteresis loss is proportional to the mass of the core, and so is the cost, and thus a balance has to be achieved.

Losses can also occur as a result of eddy currents induced in the core and so these currents need to be kept as low as possible. The currents through the core will depend on its electrical resistance, the higher the resistance the smaller the current. Since resistance = (resistivity × length)/(cross-sectional area), a small cross-sectional area increases the resistance. For this reason the core is made of laminations. The power loss due to the eddy current is proportional to the square of the lamination thickness and so as thin a lamination as possible should be used. The laminations need to be electrically insulated from each other and so each needs coating with insulation. Typically, laminations are about 0.3 mm thick and produced by rolling a suitable alloy. A material with the required magnetic properties but also with as high a resistivity as possible is thus required.

The material that is generally used is iron with 3 to 4% silicon. The effect of the silicon is to increase the electrical resistivity of the iron. A resistivity of about 1×10^{-7} Ω m for pure iron becoming about 4×10^{-7} Ω m with 3% silicon. Above 3% silicon gives an alloy which is too brittle to cold roll, the ductile–brittle transition occurring at room temperature when the percentage silicon reaches about 3.5%. Orientating the grains in the alloy so that they are aligned with the magnetising field reduces the coercivity and hence the hysteresis loss. Orientation can be

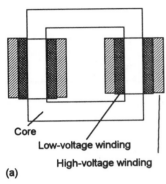

Core

Low-voltage winding

High-voltage winding

(a)

Core

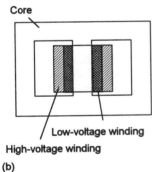

Low-voltage winding

High-voltage winding

(b)

Figure 7.5 *Transformer construction (a) core, (b) shell*

achieved by cold rolling to build up dislocations and then annealing; the iron then recrystallises with dislocation free grains and a high degree of grain orientation. Grain orientated 3% silicon iron has an initial relative permeability of 1500, a maximum relative permeability of about 40 000, a coercivity of 8 A/m and a saturation magnetic induction of 2 T.

The processing steps for the production of the laminates are hot rolling to a thickness of about 2 mm, cold rolling to produce gain alignment and reduction in thickness with two stages likely to be used with an intermediate annealing as work hardening occurs, annealing to promote recrystallisation, decarburisation by heating in wet hydrogen (the carbon content has to be kept low, below 0.01%, to avoid a phase transition occurring during cooling), and then coating with magnesium oxide and annealing in an atmosphere of dry hydrogen to give a surface insulating coating.

7.2.5 A car wheel

The road wheel for a car has to be able to withstand the loads experienced during use (these are not only vertical loads but loading as a result of bumping along roads and cornering), kerb impacts, have nut torque retention, have an acceptable appearance, be light and cheap when mass produced. Assuming no redesign of cars, the wheel has to fit within the existing space and be attached in the same way as currently occurs and conform with British Standards (BS AU 50).

Mild steel or an aluminium alloy could be used with the method of manufacture in two pieces, the rim and the hub, being pressing or casting. Another possibility is to use a fibre reinforced polymer with the wheel being manufactured in a single entity (Figure 7.6) by hot press moulding. To give the required strength and toughness a mixture of continuous and random glass fibres has been used with vinyl ester. The continuous fibres were aligned in the direction of the principal hoop and radial stresses and assembled with the ester and the random fibres in a former before being moulded. The bolt and valve holes were then drilled. Such a wheel is found to offer a 10 to 15% reduction in weight over an aluminium wheel and 30 to 45% reduction over a steel wheel.

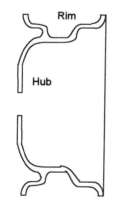

Figure 7.6   *Road wheel*

7.2.6 Fizzy drink bottle

Consider the properties required of a bottle to be used to contain, say, a fizzy drink:

- High impact strength, not brittle, tough.

- Good barrier properties, i.e. the drink must not seep out through the container wall or lose its fizz due to a loss of carbon dioxide pressure.

- A material which will not taint the drink.

- Relatively stiff so that the bottle retains its shape.

- Able to withstand the pressure due to the carbon dioxide without deforming.

- Transparent and clear.

- Capable of being blow moulded.

- Cheap since the bottle is not intended to be reused (unlike milk bottles which are intended to be reused a number of times).

- Light weight.

The blow moulding requirement means that the material must be a thermoplastic. With regard to impact properties the choice of material is restricted to those polymers which are tough or, at least, tough when not notched. Such materials that are clear and transparent limits the choice to low-density polyethylene, PVC or polyethylene terephthalate.

The bottles are to contain fizzy drinks, i.e. liquids containing carbon dioxide under pressure. Table 7.1 gives data concerning the permeability to water and carbon dioxide of the three materials. Despite being the tougher material, low-density polyethylene would not be a suitable material for a fizzy drink container because of its high permeability to carbon dioxide. The drink would not retain its fizz. Polyethylene terephthalate appears to be the best choice. This material is widely used for carbonated drink containers; PVC is used for non-carbonated drinks, e.g. wine.

Table 7.1 *Permeabilities to water and carbon dioxide*

Polymer	Permeability 10^{-8} mol mN^{-1} s^{-1}	
	To water	To carbon dioxide
Low-density polyethylene	30	5700
PVC	40	98
Polyethylene terephthalate	60	30

7.3 Selection criteria

In considering the materials, from the whole range of materials that exist, that can be used for a product, a number of methods have evolved to attempt to determine the optimum material and processing method for a particular product. The following is a brief consideration of such methods, illustrated by case studies.

7.3.1 Material properties

Often in considering the properties required of a material we have to deal with requirements requiring a combination of properties. For example, in considering a structural member to be subject to an axial load, we might want the material which will allow the maximum load per unit mass of member. The mass m is the product of the density ρ and the volume of the member and thus for a length L of cross-sectional area A is $LA\rho$. The area that can be used depends on the yield stress σ_y, being F/σ_y. Thus we can write:

$$m = LA\rho = L\rho\frac{F}{\sigma_y}$$

and so:

$$\frac{F}{m} = \frac{1}{L} \times \frac{\sigma_y}{\rho} \tag{1}$$

Thus, if we take L to be constant, the parameter determining the load per unit mass is σ_y/ρ. This is termed a *performance index*. A performance index is a grouping of properties which, when maximised, give some maximum performance of a material.

As another example, the material to be selected for a cantilever of length L might need to be chosen to give, for the minimum mass, as stiff a cantilever as possible. This means maximising the force F per unit deflection y at the free end. For a cantilever:

$$\frac{F}{y} = \frac{F}{FL^3/3EI}$$

If we assume a square cross-section of side b then $I = b^4/12$ and so:

$$\frac{F}{y} = \frac{3Eb^4/12}{L^3} = \frac{Eb^4}{4L^3} \tag{2}$$

The mass m of the cantilever is $b^2 L\rho$, where ρ is the density. Hence, using equation [2], we can write the mass as:

$$m = L\rho\left(\frac{4L^3}{E} \times \frac{F}{y}\right)^{1/2} = 2L^{5/2}\left(\frac{F}{y}\right)^{1/2}\frac{\rho}{E^{1/2}}$$

and so:

$$\frac{(F/y)^{1/2}}{m} = \frac{1}{2L^{5/2}} \times \frac{E^{1/2}}{\rho} \tag{3}$$

Taking the length to be constant, the mass is minimised and the stiffness maximised when $E^{1/2}/\rho$ is maximised. This is then the performance index.

Strength or yield stress to density and elastic modulus to density ratios are both commonly used as general performance indices for materials, the intention being to maximise such ratios in order to obtain the best performance for the least mass. These may not give the best performance in all loading situations. Table 7.2 give the indices for maximising the strength to mass and stiffness to mass ratios for a range of different loading conditions where failure is considered to be due to excessive deflection.

Table 7.2 *Performance indices*

Component	Maximise stiffness	Maximise strength
Tie, i.e. tensile strut	$\dfrac{E}{\rho}$	$\dfrac{\sigma_y}{\rho}$
Beam	$\dfrac{E^{1/2}}{\rho}$	$\dfrac{\sigma_y^{2/3}}{\rho}$
Column, i.e. compressive strut	$\dfrac{E^{1/2}}{\rho}$	$\dfrac{\sigma_y}{\rho}$
Plate, loaded externally or by self weight in bending	$\dfrac{E^{1/3}}{\rho}$	$\dfrac{\sigma_y}{\rho}$
Cylinder with internal pressure	$\dfrac{E}{\rho}$	$\dfrac{\sigma_y}{\rho}$
Spherical shell with internal pressure	$\dfrac{E}{(1-v)\rho}$	$\dfrac{\sigma_y}{\rho}$

E is the modulus of elasticity, ρ the density, σ_y the yield stress (sometimes the tensile strength is used) and v is Poisson's ratio.

Activity

The components listed in Table 7.2 may not seem like components encountered in everyday objects. However, the steps of a ladder are beams subject to bending when under load. Make a list of other examples of beams that are commonly found.

Other performance indices might be derived for a spring to have the maximum energy storage per unit mass, a diaphragm to have the maximum deflection per unit pressure, a beam to have the maximum flexural vibration frequency, a component to have the maximum energy storage for a given temperature and time, a component to have the minimum thermal distortion. Table 7.3 shows examples of such indices.

Table 7.3 *Performance indices*

Component and attribute	Performance index to be maximised
Spring: energy per unit mass to be maximised	$\dfrac{\sigma_y^2}{E\rho}$
Diaphragm: deflection per unit pressure to be maximised	$\dfrac{\sigma_y^{3/2}}{E}$
Beam: flexural vibration frequency to be maximised	$\dfrac{E^{1/2}}{\rho}$
Plate: flexural vibration frequency to be maximised	$\dfrac{E^{1/3}}{\rho}$
Energy storage for given temperature and time to be maximised	$\dfrac{\lambda}{a^{1/2}}$
Thermal distortion to be minimised	$\dfrac{\lambda}{a}$
Thermal shock resistance to be maximised	$\dfrac{\sigma_y}{Ea}$

E is the modulus of elasticity, ρ the density, σ_y the yield stress, λ the thermal conductivity, a the thermal diffusivity and a the coefficient of thermal expansion.

Example

Determine the type of material which will give the best energy-storage spring when the mass is to be minimised.

The following lists some possible materials for springs and typical values of the performance index $\sigma_y^2/E\rho$.

Material	Yield stress MPa	Tensile modulus GPa	Density Mg/m³	Index kJ/kg
Spring steel	600	200	7.8	2
Glass	50	70	2	18
Wood	100	10	0.5	2
Nylon	70	2	1	2
Rubber	20	0.004	1	100

For the criteria used, rubber is the clear favourite with glass being the next best. Wood may seem a surprising option but remember bows and arrows.

Property charts

One way of considering materials for a particular property or property index is to use charts with the materials indicated against their properties or index. Thus there might be a chart of the form shown in Figure 7.7 where the materials are indicated against both their values of elastic modulus and their density.

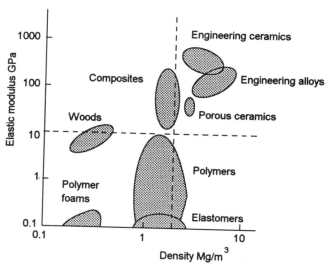

Figure 7.7 *Property chart*

To select materials which have at least a tensile modulus of, say, 10 GPa a line is drawn across the chart at that value and all the materials above that line form the subset from which a material can be selected. If we also have the requirement that the density must be less than 2 Mg/m^3 then we draw a line on the chart at that value and all the materials to the left of that line form the subset with that criterion. The subset of materials with both criteria are thus those in the upper left quadrant.

If we wanted a cantilever with the stiffness maximised and the mass minimised, then equation [3] indicates that we want the performance index $C = E^{1/2}/\rho$ maximised. Because we have plotted a log–log graph in Figure 7.7:

$$\lg C = \tfrac{1}{2}\lg E - \lg \rho$$

$$\lg E = 2\lg \rho + 2\lg C \tag{4}$$

Thus for a particular performance index we have a straight line on the graph of gradient 2. There will be a family of straight lines with this gradient. Figure 7.8 shows one such line; the other lines in the family will be parallel to it.

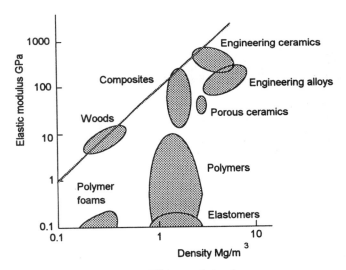

Figure 7.8 *Guide line for $E^{1/2}/\rho$ maximised*

Such a chart can also be used if a material is required in the form of a beam with the stiffness maximised. This requires the determination of the materials with values of E/ρ greater than some critical value. Suppose, for example, the subset of materials is required for which E/ρ is greater than 1000.

$$\lg E - \log \rho = \lg 1000$$

$$\lg E = \lg \rho + 3 \tag{5}$$

This will be a line on the chart with a constant slope of 1 and an intercept of 3 with the $\lg E$ axis. Figure 7.9 shows such a line and the subset thus indicated.

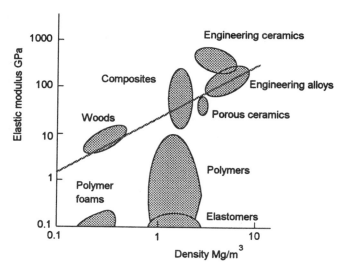

Figure 7.9 *Line for E/ρ = 1000*

Within the areas for each type of material, points can be located for individual alloys, polymers, woods, etc. and so the chart used to indicate which material would fit the imposed criteria. A range of such charts can be found in *Materials Selection in Mechanical Design* (Pergamon Press 1992) by M. F. Ashby.

Case study: aircraft skins

Aircraft fuselages are constructed using H, T and L section members to which a metal skin is attached. One way of considering a fuselage is as a pressurised cylinder. Mass is an important consideration for materials used for aircraft, the aim being to have the minimum mass possible. Thus, as indicated by Table 7.2 for a pressurised cylinder, a performance index that can be used in determining the materials is σ_y/ρ, i.e. the specific strength. There are other factors involved in determining possible materials and thus the list of properties required includes:

* *Lightness*
 The lighter an aircraft the greater the range, speed or payload.

* *Reliability*
 Failure of an aircraft material puts the occupants of the aircraft at risk. However, reliability cannot be ensured by using a large safety factor in the design since this would lead to an increase in weight. Thus reliability has to be a tight quality control of materials and good design.

* *High specific proof or yield stress*
 The 0.2% proof stress per unit mass or yield stress per unit mass should be as high as possible for the temperatures experienced by the aircraft skin in flight.

* *High fatigue strength*
 In flight and during take-off and landing, aircraft are subject to buffeting and vibratory stresses.

- *High creep resistance*
  There must be high creep resistance at the temperatures that can occur at the skin during flight.

- *Good corrosion resistance*
  Failure of the material must not occur as a result of corrosion, particularly taking into account that aircraft do fly in marine environments.

- *Ease of forming sheets*
  The sheets used for the skin will have to be formed to the shapes required of the various parts of the surface. This means reasonable ductility, though hardening can occur after forming.

- *Ease of joining sheets*
  Sheets will need to be welded or riveted to form the large surfaces required.

- *Cost*
  The lowest cost material, after processing, is required which is consistent with the required properties.

Aircraft speeds of the order of 500 m/s result in a skin temperature of about 120°C, while supersonic speeds of the order of 800 m/s give temperatures of about 300°C. Military aircraft can attain speeds as high as 2000 m/s and give skin temperatures of the order of 500°C. Thus the speed planned for the aircraft determines the temperature range over which the material must have the required properties.

Possible materials might be aluminium alloys, carbon steels, stainless steels, nickel alloys and titanium alloys. Figure 7.10 shows these materials on a properties chart of strength plotted against density and Table 7.4 shows the properties possible with such materials at room temperature. For normal passenger aircraft with subsonic speeds, all the materials are feasible on the basis of the maximum temperature at which they can be used. However, on the basis of the specific yield stress, aluminium alloys and titanium alloys are the best. On the basis of cost of material, aluminium alloys are preferable to titanium. Processing costs for titanium are also higher than for aluminium alloys. A point worth noting is that carbon steels are not suitable, despite their low cost, as their high density results in a low specific yield stress.

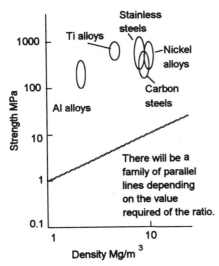

Figure 7.10 *Properties chart and guidline for σ_y/ρ*

Table 7.4 *Typical properties*

Material	Density Mg/m³	Yield stress MPa	Specific yield strength MPa/Mg m⁻³	Max. use temperature °C	Relative cost per unit sheet area
Aluminium alloys	2.8	200–400	71–140	200	2
Carbon steels	7.8	200–500	26–64	350	1
Stainless steels	7.7	300–1000	39–130	700	6
Nickel alloys	8.9	300–900	34–100	1000	6
Titanium alloys	4.5	700–1100	156–244	600	10

Table 7.5 *Properties of titanium alloys*

Alloy composition %	Condition	Yield stress MPa			
		20°C	300°C	400°C	500°C
90 Ti, 8 Al, 1 Mo, 1 V	Annealed	970	630	570	520
	Solution treated + aged	1200	780	710	650
90 Ti, 6 Al, 4 V	Annealed	940	660	580	430
	Solution treated + aged	1100	710	630	490

To obtain a high yield stress the aluminium alloy chosen would be a heat-treatable wrought alloy to allow the material to be formed in the soft condition before being hardened by precipitation hardening. A possible alloy would be one having 4.0% copper, 0.8% magnesium, 0.5% silicon and 0.7% manganese. When soft this has a 0.2% proof stress of 90 MPa and when hard 400 MPa. Aluminium alloys present one major problem, they do not make good welds and thus rivets have to be used to joint sheets.

For a higher speed aircraft, aluminium alloy is not suitable since it does not retain good mechanical properties at the higher temperatures. For such aircraft titanium has to be considered, despite its higher costs. Unlike aluminium, titanium does make good welds. Table 7.5 shows possible alloys and their properties.

On the basis of the mechanical properties, the first of the alloys in Table 7.5 would appear to be the best choice. However, this alloy is susceptible to stress corrosion in salt water environments. Aircraft flying over the sea do encounter moisture containing salt and so the second alloy in the table, which does not have this problem, is to be preferred.

For very high speeds the choice has to narrow down to nickel alloys since they possess good mechanical properties at high temperatures.

7.3.2 Cost factors

Since low cost is often a requirement, one way of comparing the properties of a subset of materials is on the basis of cost per unit property or group of properties. This is particularly useful where there is one property or group of properties which is the main requirement.

Case study: car bodywork

The properties required of the material used for the bodywork of cars includes:

- Can be formed to the shapes required.

- Has a smooth and shiny surface.

- Corrosion is not too significant.

- Is not brittle, being sufficiently tough to withstand small knocks, and relatively stiff.

• Is cheap, taking into account the costs of raw materials, processing and finishing.

Processing will be a key factor in determining the material to be used. For metals, forming from sheet is the obvious method. Hot forming does present the problem of an unacceptable surface finish and so a material has to be chosen which allows for cold forming, i.e. a highly ductile material such as low-carbon steels or aluminium alloys (Table 7.6).

Table 7.6 *Ductilities of carbon steel and aluminium alloys*

Material	Percentage elongation
0.1% carbon steel	42
0.2% carbon steel	37
0.3% carbon steel	32
1.25% Mn, aluminium alloy	30
2.24% Mn, aluminium alloy	22

From the data in Table 7.6 it can be established that carbon steels and aluminium alloys could be used, both having enough ductility to enable sheet to be formed. In addition, both are reasonably tough. Aluminium alloys have the advantage of lower densities and so could lead to lower weight cars. The carbon steels do, however, have the advantages of work hardening more rapidly than the aluminium alloys. A material that work hardens rapidly is less likely to form a neck. In the case of cold forming of sheet, this would show as a thin region in the formed sheet which would clearly not be desirable. The great advantages of carbon steel, outweighing all other considerations, is that they are much cheaper. Thus the optimum material is a low-carbon steel; in practice a steel with less than 1% carbon is used.

Polymeric materials could be used for car bodywork. The problem with such materials is obtaining enough stiffness. One way of overcoming this is to form a composite material with glass fibre mat or cloth in a matrix of a thermoset. Unfortunately such a process of building up bodywork is essentially a manual rather than a machine process and so is slow. While it can be used for one-off bodies it is not suitable for mass production. Another possibility is to form glass-reinforced panels by hot pressing from sheet-moulding compound and then fitting the panels to a steel frame. This does enable a mass production method to be used. Such a method has been used for lorry cabs. Another possibility is to produce a sandwich type of composite for panels. This could be a foamed plastic between plastic or metal sheets. While such composite materials can give an appropriate stiffness, the costs tend to be higher than using steel.

Activity

Using data tables make up a property chart for tensile modulus against yield stress.

Table 7.7 *Cost per unit property comparisons*

Material	Relative cost/ m³ for sheet	Tensile modulus GPa	Tensile strength MPa	Cost per unit stiffness	Cost per unit strength
Low-carbon steel	1.0	220	1000	0.005	0.001
Aluminium alloy (Mn)	2.2	70	200	0.03	0.01
Polypropylene	0.2	1–2	30–40	0.1–0.2	0.005–0.007
ABS	0.8	1–3	17–58	0.3–0.8	0.01–0.05
Polyester	2.0	2–4	20–70	0.5–1.0	0.03–0.1
Polyester–glass cloth	3.0	20	300	0.15	0.01

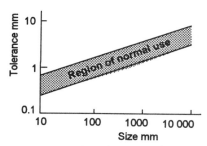

Figure 7.11 *Capabilities with sand casting*

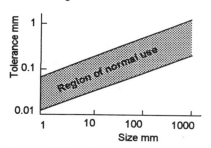

Figure 7.12 *Capabilities with pressure die casting*

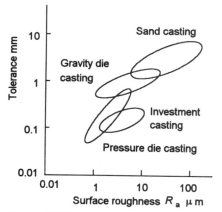

Figure 7.13 *Capabilities of casting processes with metals*

Table 7.7 shows a comparison of a number of materials on the basis of cost per unit property. If cost per unit stiffness is taken as the prime factor then low-carbon steel is the obvious choice. If, however, cost per unit strength was the prime factor then the choice of low-carbon steel is still the choice but not so clear cut.

7.4 Selection of processes

In considering the selection of process for a product, the choice is made on the basis of considering the answers to the questions posed in Section 7.1.2.

Charts are often used to show the capability, for a particular material, of a process. Thus, to show the tolerances possible for sand casting with aluminium and magnesium alloys we might have a chart of the form shown in Figure 7.11. With pressure die casting for aluminium and magnesium alloys we might have a chart of the form shown in Figure 7.12 As the two charts indicate, we can obtain much better tolerances with pressure die casting aluminium and magnesium alloys than sand casting them. K.G. Swift and J.B. Booker in their book *Process Selection* (Arnold 1997) give charts for each type of process.

Charts showing the capabilities of a number of processes may be superimposed on top of each other, the normal use region for each being within a box For example, Figure 7.13 shows the capabilities for tolerances and surface roughness with metals. Examples of such charts are given in *Materials Selection in Mechanical Design* by M.F. Ashby (Pergamon 1992).

Case study: car bumper

Consider the problem of determining the optimum moulding process for a car bumper to be manufactured from a plastic such as polypropylene with incorporated rubber particles or polyurethane incorporating glass flakes. The rubber particles decrease the ductile–brittle transition to below the –10°C it otherwise would be; the glass flakes increase the inherently low stiffness of the polyurethane. Possible processes are injection moulding, reaction injection moulding, compression moulding or perhaps the spray-up method used for composites.

All the processes use permanent moulds though the mould cost would be highest, by a long way, for the injection moulding, less for the reaction moulding, even less for the compression moulding and cheapest for the spray-up method. However, the injection moulding would give the highest production rate, the spray-up method the lowest. The labour costs would be the lowest with the moulding processes which are largely automatic but highest with the spray-up method. As a consequence, the spray-up method turns out only to be useful when prototypes are being made or very small production runs involved. Because of the large numbers likely to be required. Injection moulding could well be the most economic and is indeed the method used for the mass car production market. Reaction injection moulding is used where smaller runs of a model are anticipated.

Case study: ringbolt

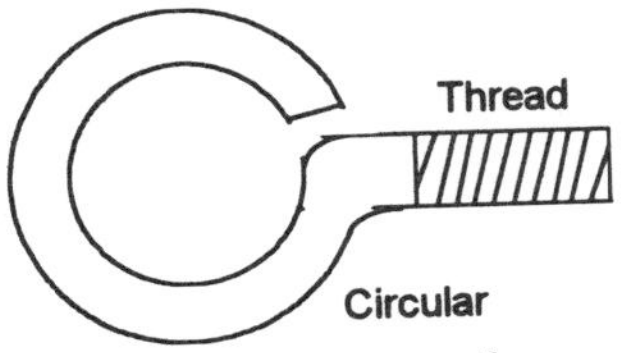

Consider the processes that could be used to manufacture a ringbolt (Figure 7.14) from a reasonably ductile metal. One possibility is to use bar stock, produce the thread by cutting or rolling, and then form the ring by bending round a die. Another possibility is to use sand casting, die casting or shell moulding to form the shape and then cut the thread. Another possibility is to use upset forging from bar stock and then thread cutting; upset forging involves the heated metal stock being gripped by dies and end pressed into the required shape. Sand casting might be ruled out on the basis of poor surface finish or the numbers required. The most likely methods to be used are the ones involving bending round a die or forging. Die costs are likely to be lowest and production rates for the production of the ring shape of a few hundred per hour are feasible.

Figure 7.14   *Ringbolt*

Problems

1 Propose materials and manufacturing methods that might be used for:
   (a) the windows in an aircraft,
   (b) the seat-back fold-away table in an aircraft,
   (c) a tin opener,
   (d) a pan for domestic cooking,
   (e) the red rear light lens for a car,
   (f) sliding patio door frames.

2 Determine the type of material which when used as a diaphragm will give the maximum diaphragm deflection for a pressure difference across it. Consider glasses, spring steel, nylon and rubber.

3 Determine the type of material that will give the minimum thermal distortion when the temperature changes. Consider aluminium, copper, diamond and silicon carbide.

Solutions to problems

Chapter 1

1.1

1 (a) Y, (b) Y
2 (a) Strong and brittle, (b) strong and tough
3 20 MPa
4 0.67%
5 50 kN
6 12%
7 50 kN
8 The bronze is stronger and more ductile
9 Stronger in compression, brittle
10 (a) C, (b) B, (c) A, (d) A
11 (d)
12 0.0125 Ω
13 X
14 22%
15 X paramagnetic, Y diamagnetic, Z ferromagnetic
16 Y
17 Low coercivity, small area
18 These might include (a) ease of forming in one piece, easily cleaned, stain resistance, waterproof, (b) stiff, strong, cheap, (c) leak proof, suitable for hot liquids, cheap, not easily broken, (d) good conductor, flexible, (e) cheap to make, wear resistant during handling, stiff, (f) withstands changing forces, stiff, strong, withstands impact forces, (g) attractive appearance, cheap to form
19 (a) Stainless steel, (b) wood, (c) china (ceramic), (d) copper, (e) alloys of copper (cupro-nickel or bronze depending on the colour of the coins), (f) steel, (g) plastics, e.g. ABS
20 (a) Modulus of elasticity, (b) ductility, percentage elongation, (c) fracture toughness, (d) strength, (e) electrical resistivity (conductivity), (f) thermal conductivity, (g) corrosive properties
21 Brittle, must not be subject to sudden forces or changes in temperature
22 128 MPa/Mg m^{-3}, 33 MPa/Mg m^{-3}, 0.78 £/MPa, 6 £/MPa
23 Polymer lower density than the others, steel ferromagnetic and aluminium not ferromagnetic

1.2

1 (a) 420 MPa, (b) 62%, (c) 18–40%, (d) 355 MPa, (e) 510 MPa, (f) 1020-1070 kg/m³, (g) 3.0–4.5 MPa m$^{-1/2}$, (h) 20 MPa, (i) 11–13 × 10^{-2} K^{-1}, (j) 1.4–3.1 GPa
2 Cast iron 0.014 GPa/kg m^{-3}, Al alloy 0.027 GPa/kg m^{-3}, PVC 0.002 GPa/kg m^{-3}
3 Steel 220 GPa, Al alloy 71 GPa, polypropylene 1–2 GPa, composite 20 GPa
4 120 MPa/Mg m^{-3}
5 1.4–3.1 GPa, in the high range of modulus values for plastics
6 470–570 MPa, 170–280 MPa, 18–35%
7 150M36: 620–770 MPa, 400 MPa, 18%; 530M40: 700–850 MPa, 525 MPa, 17%
8 LM6
9 Polyacetal

1.3

1 (a) 61 GPa, (b) 380 MPa
2 (a) 10.8 MPa, (b) 1.1 GPa
3 (a) 660 MPa, (b) 425 MPa, (c) 200 GPa
4 31.1%
5 (a) 480 MPa, (b) 167 GPa
6 300 MPa, 280 MPa
7 2.5 GPa, 80 MPa
8 18.55%
9 Stronger and less ductile
10 (a) Titanium alloy, (b) nickel alloy
11 Cellulose acetate
12 Becoming more ductile
13 Becoming more brittle
14 As the temperature drops becoming more brittle
15 Becoming more ductile
16 HV 198
17 HV 275
18 HV 71
19 HB 217
20 HB 57
21 Ni–Cr alloy most corrosion resistant
22 Increasing carbon reduces oxidation, increasing chromium reduces oxidation
23 Industrial pollutants more damaging than marine conditions, with rural surroundings being least corrosive
24 (a) Hardness, (b) impact, (c) tensile test for the modulus of elasticity, (d) impact, or tensile or hardness, test, (e) bend test
25 0.52 T, 40 kA/m, 4, 24, 0.6 T
26 0.6 T, 1 A/m, 80 000, 0.7 T

Chapter 2

2.1

1 (a) 32, (b) 42, (c) 32

2 (a) $1s^22s^22p^63s^23p^64s^2$, (b) $1s^22s^22p^63s^23p^63d^6$, (c) $1s^2$, (d) $1s^22s^22p^63s^23p^6$

3 See Section 2.1.2

4 (a) Y, (b) X

5 Crystalline: long-range orderly arrangement of particles, amorphous: no long-range order

6 See Section 2.1.4

7 7901 kg/m³

8 0.2851 nm

9 9040 kg/m³

10 2708 kg/m³

11 0.3148 nm

12 0.0486 nm³

13 −3.3%

14 Simple, 0.387

15 As caesium chloride, 0.725

16 0.039 nm

17 0.028 nm

18 1.78 h

19 32.4 h

20 7.4 h

21 1.3% 10^{-11} m²/s

22 10.7 h

2.2

1 Orderly regions of particles in a polycrystalline material

2 Between the planes of closest packed atoms, see Section 2.2.1

3 1.2 MPa

4 3.5 MPa

5 See Figure 1.61 and associated text

6 See Section 2.2.2

7 Block slip indicates metals stronger than actually occur because entire planes have to slip. Dislocations give lower yield stresses because slip only has to occur atom by atom

8 5.1×10^{25} /m³

9 6.7×10^{23} /m³

10 Body–centred

11 See Figure 2.72 and associated text

12 See Figure 2.72 and associated text

13 See Figure 2.75 and associated text

2.3

1 See Section 2.3.1

2 (a) Yes, (b) No, (c) No

3 About 1320°C, 1280°C

4 (a) Liquid + alpha, (b) liquid only

5 (a) About 2325°C, 2150°C, (b) one, the liquid phase, (c) between 60 and 80% MgO

6 See Section 2.3.3

7 Solid ¼, liquid ¾

8   Solid 0.62, liquid 0.38
9   115°C, 990°C
10 (a) 76% of 48% tin in liquid phase, 24% of 15% tin in alpha phase, (b) 61% alpha, 39% beta
11 5%
12 (a) 100% liquid, (b) 68% alpha, 32% liquid, (c) 71% alpha, 29% beta
13 (a) 1200% liquid, (b) 22.8% beta, 77.2% liquid, (c) 31% alpha, 69% beta
14 52% tin–48% lead
15 (a) Liquid, (b) alpha, (c) alpha + beta, (d) liquid, (e) liquid + alpha, (f) alpha + beta, (g) liquid, (h) liquid + beta, (i) alpha + beta
16 77% alpha 1.2% silicon, 23% liquid 9.0% silicon
17 30.5% alpha 1.2% silicon, 61.5% liquid 9.0% silicon
18 577°C
19 Eutectic composition, i.e. 12.6% silicon
20 (a) Liquid, (b) alpha + liquid, (c) solid alpha, (d) alpha + beta
21 2.2% alpha, 97.8% beta
22 97.8% alpha, 2.2% beta
23 71.9% silver, 28.1% copper
24 alpha + beta, 79.5% alpha 7.9% silver, 92.1% copper, 20.5% beta 91.2% silver, 8.8% copper
25 See Section 2.3.5
26 (a) Saturated solid solution with coarse precipitate, (b) supersaturated solid solution

2.4

1   5714
2   Linear regular chains, bendy and smooth, not easily bonded to other molecules and chains
3   Stiffer, less flexible
4   See Figure 2.132 and associated text
5   Below: stiff; above: flexible
6   Becomes stiff and brittle
7   Glass transition temperature
8   930 kg/m^3
9   Necking occurs when molecular chains align and this requires less force than stretching the bonds in the aligned chains
10 The drawn polymer has orientated chains
11 Makes chains slide past each other more easily, reduces tensile strength but increases ductility
12 (a) Longer chains more tangled and so stiffer, (b) more crystallinity, more van der Waals bonds and so stiffer, (c) more cross-links and hence stiffer
13 0.127×10^{-6} m
14 No, randomness might reduce tendency to pack in an orderly manner

2.5

1 This gives the ratio Al to O of 2 to 3 which satisfies the bonding requirement in Al_2O_3
2 See Figure 2.144, (a) Figure 2.144(a), (b) Figure 2.146, (c) Figure 2.147
3 See Figure 2.148 and associated text
4 Small crystals have formed and scatter light
5 63%, eutectic composition and so lowest temperature at which it is liquid
6 See Section 2.5.4

2.6

1 Good conductor; it is silver
2 7.4×10^{-5} m/s
3 4.0×10^{-3} m^2 V^{-1} s^{-1}
4 0.0041 m/s
5 2.2 S/m
6 624 S/m
7 (a) p, (b) n, (c) n, (d) p
8 9.26×10^{21} /m^3
9 Like barium titanate, see Figure 2.183 and associated text
10 Polar groups, because of larger mass, are slower to respond to alternating fields and so the relative permittivity drops as the frequency increases
11 See Section 2.6.5
12 See Section 2.6.5
13 Below ferromagnetic, above only weakly magnetic
14 0.645 T
15 0.32 T
16 $8\mu_B$
17 $25\mu_B$

Chapter 3

3.1

1 Ferrite in pearlite, cementite in pearlite (see Figures 3.7–9)
2 (a) 89% alpha, 11% cementite, (b) 71% alpha, 29% cementite
3 11.4%
4 Pearlite with laminar structure of ferrite and cementite
5 See Figure 3.10
6 See Section 3.1.2
7 See Section 3.1.2, reduce hardness
8 See Section 3.1.3
9 (a) mild/high carbon, about 0.6 C, (b) mild steel, about 0.1 C, (c) high carbon, about 0.8 C, (d) mild steel, about 0.1 C, (e) high carbon, about 0.95 C
10 See Section 3.1.4
11 See Table 3.2
12 Sulphur and lead
13 Mo, Ni, V

14 See Table 3.2, in particular Ni solution harden and reduce grain size, Cr formation of carbide, Mo reduce chance of brittlensss after tempering
15 High Cr steels, see Section 3.1.5
16 See Section 3.1.5
17 See Section 3.1.5
18 See Section 3.1.5
19 (a) Grey or white, (b) how close to eutectic and hence more chance of grey than white, (c) Si, P: carbon equivalent, Mn, S: improve chance of grey by stabilising cementite
20 See Section 3.1.8
21 Determines rate of cooling and hence grey or white
22 Better ductility and strength
23 Graphite clusters rather than flakes
24 2.4%
25 Makes more ductile but weaker
26 (a) SG ferritic, (b) SG pearlitic, (c) white, (d) grey, (e) grey

3.2

1 See Section 3.2.1
2 See Section 3.2.2
3 See (a) Figure 3.33, (b) Figure 3.34, (c) Figure 3.36
4 Start at about 10 s and finish about 100 s
5 Push 'nose' to the right, see Figure 3.37 and associated text
6 See Section 3.22
7 See Figure 3.39 and associated text
8 About (a) 30°C/s, (b) 10 to 20°C/s
9 See Section 3.2.3
10 35 HRC, 30 HRC
11 About (a) 400 HV, (b) 300 HV
12 See Section 3.2.3
13 About 56 HRC
14 See Sections (a) 3.2.5, (b) 3.2.5, (c) 3.2.5, (d) 3.2.5
15 See Figure 3.64
16 (a) Martensite, (b) ferrite and pearlite, (c) tempered martensite, (d) bainite, (e) coarse pearlite and ferrite, (f) martensite
17 (a) Martensite, (b) tempered martensite, (c) bainite, (d) coarse pearlite, (e) coarse pearlite and ferrite
18 See Table 3.17
19 Selective heating

3.3

1 (a) Increase strength, (b) working fragments the grains and increases strength but decreases ductility
2 Increases strength due to solid solution strengthening
3 See Section 3.3.1 and Figure 3.73
4 Different compositions, 3004 more ductile and 5182 harder. When melted, the melt will be a mixture and so neither alloy.

Possible solutions involve separation prior to melting or during melting as a result of different melting temperatures

5 See Figure 3.76 and associated text
6 See Figure 3.76 and associated text
7 (a) About 600°C, 577°C, (b) about 550°C, 546°C
8 Where low mass is important, e.g. aircraft parts
9 Low density, high strength maintained to high temperatures
10 See Section 11.3.3
11 Intercrystalline corrosion
12 An atmosphere containing hydrogen can result in steam and cause cracking
13 See Section 3.3.5 and Figures 3.81 and 3.82
14 See Figure 3.82 and associated text: (a) alpha, (b) alpha, (c) alpha + beta'
15 About (a) 46% zinc, (b) 35% zinc, (c) 30% zinc
16 (a) Alpha solid solution; duplex: intermetallic compounds present in solid solution, (c) alpha: ductile; duplex: less ductile and higher strength
17 See Figure 3.83 and associated text
18 (a) Complete solid solubility, (b) all compositions
19 Excellent corrosion resistance, strength at high temperature
20 Main properties are likely to be: (a) low melting point, lack of corrosion of die, (b) high strengh to weight ratio, (c) very ductile, not corroded, (d) strength at high temperatures, low density, (e) easily bent, good corrosion resistance, (f) ductile for deep drawing, (g) high strength to weight ratio, (h) strength at high temperatures, (i) decorative, not corroded, easily machined, (j) good thermal conductivity, (k) good electrical conductivity, (l) good corrosion/erosion resistance

Chapter 4

4.1

1 LDPE: flexible, tough; HDPE: stronger, stiffer but less tough
2 General purpose: brittle, transparent; toughened: improved impact resistance, decreased stiffness, strength and transparency
3 Side groups giving more stiffness and strength
4 PVC
5 Soften
6 Resistance to chemical attack, low coefficient of friction
7 Less brittle
8 Do not soften when heated, stronger, stiffer
9 Phenol formaldehyde, stiff, strong, fairly brittle
10 Stiff, fairly brittle
11 See Figure 4.19 and associated text
12 (a) Phenol formaldehyde, (b) plasticised PVC, (c) butyl rubber, (d) polyester, (e) acrylic, (f) plasticised PVC, (g) CAP, (h) ABS, (i) EVA, (j) melamine formaldehyde, (k) phenol formaldehyde, (l) natural rubber, (m) NBR, (n) polyurethane
13 (a) ABS, (b) nylon 6.6, (c) PET, (d) NBR

4.2

1 See Section 4.2.1
2 Clay + feldspar + filler, see Section 4.2.1
3 Some components melt to a viscous liquid which solidifies as a glass
4 Hard wearing, can be used at high temperature
5 Laminar structure for boron nitride crystals
6 See Section 4.2.3
7 (a) 334 K, (b) 601 W/m, (c) 245×10^{-6} m^2 K s^{-1}
8 (a) Glasses have poor shock resistance, soften at too low a temperature, (b) steels have high thermal conductivity which would result in the entire top becoming hot
9 73 MPa

4.3

1 For example: strengthens a weak material, roughens a brittle material, modifies stiffness, improves strength to weight and strength to stiffness ratios
2 36.4 GPa, 6.97 GPa
3 182.2 GPa, 7.2 GPa
4 194 GPa, 103 GPa
5 About 1250 MPa
6 About 1800 MPa
7 95.3%
8 12.8
9 The minimum length at which the tensile stress in the fibre reaches the maximum value
10 0.27 mm, 1863 MPa
11 2.84 Mg/m^3
12 131 GPa
13 See equation [27]
14 240 kg/m^3, 0.10 GPa
15 7.6 GPa
16 2.2 GPa
17 5.1 GPa
18 As given in the problem
19 As given in the problem

Chapter 5

5.1

1 Determines rate of cooling and hence grain size
2 Temperature above liquidus, freezing range
3 See Figure 5.5 and associated text
4 See text associated with Figure 5.6
5 See text associated with Figure 5.9
6 Investment casting
7 See Section 5.1.2
8 See Table 5.4 and associated text
9 Crystals begin to grow from nuclei in the most heavily deformed parts of the metal

10 See Figure 5.15
11 See text associated with Figure 5.15
12 About 231 K
13 (a) Less than 300°C, (b) 300°C to 682°C, (c) More than 682°C
14 Smaller the greater the amount of cold work
15 (a) Distorted grains and harder, (b) up to 300°C, (c) 573K to 955 K, (d) bigger grains and hence lower strength and hardness but higher percentage elongation
16 Orientation of grains
17 Harder, better surface finish
18 See Figure 5.19 and asociated text
19 Deep drawing
20 See Section 5.1.3
21 The ease of machining
22 See Figure 5.33 and associated text
23 (a) BD, (b) BT or BM, (c) BH, (d) BO

5.2

1 (a) Extrusion, (b) blown film extrusion, (c) extrusion or injection moulding, (d) extrusion, (e) injection moulding, (f) injection moulding, (g) injection moulding, (h) blow moulding, (i) blow moulding, (j) reaction injection moulding, (k) extrusion
2 (a) Long constant section lengths, (b) components with inserts, threads, holes, etc., (c) sheet, (f) contoured sheet
3 Moulding
4 See text associated with Figures 5.47 and 5.48
5 See Section 5.2.4
6 The thicker rim would take longer to cool than the inner part and so distortion could occur

5.3

1 See Section 5.3.1
2 Die pressing or reaction bonding
3 See Section 5.3.3
4 Blow moulding, see Figure 5.54 and associated text
5 Insulation, low diffusion coefficient for many dopants
6 See Section 5.3.4
7 See Section 5.3.4
8 Epitaxial growth of n-type layer on p-type slice, growth of silicon dioxide layer, window produced in silicon dioxide layer, diffusion of n^+ layer through window, deposit aluminium over entire exposed surface, use photolithograph to selectively etch away part of aluminium

5.4

1 (a) Hand lay-up, (b) sheet moulding, (c) resin transfer moulding, (d) chemical vapour deposition

2 See Section 5.4.2

5.5

1 See Section 5.5.1
2 See Section 5.5.1
3 See Figure 5.74, shear
4 0.0001 m^2
5 Soldering below 425°C, brazing above
6 Dissolve surface contaminant films and protect molten metal from oxidation
7 Soldered/brazed joints weaker in tension but stronger in shear
8 Preheating
9 To reduce attack of hot metal by atmosphere
10 See Figure 5.79 and associated text
11 See Figure 5.82 and associated text
12 Absorption of hydrogen
13 Martensite and cracking
14 Reduce rate of cooling and hence martensite formation
15 See Figure 5.90
16 See Section 5.5.4
17 See Section 5.5.4
18 Where riveting forces might damage or distort material
19 See Section 5.5.5
20 (a) Ultrasonic staking, (b) hot plate welding, (c) adhesive
21 Snap fit has mechanical interlocking as well as friction, press fit only friction

Chapter 6

6.1

1 See Section 6.1.1
2 See Figure 6.2 and associated text
3 See Figure 6.5
4 Stress concentration
5 Made it brittle
6 Increase tip radius and reduce stress concentration
8 See Section 6.1.3

6.2

1 Failure as a result of repeated stress
2 See Figure 6.20
3 See Section Figure 6.24 and associated text
4 About (a) 10^7, (b) 10^8, (c) 120 MPa
5 400 MPa
6 320 MPa
7 (a) 600 MPa, (b) no crack propagation, (c) more than 10^4 cycles limit use of material
8 10 MPa
9 Results in heating, 41 MPa
10 See Section 6.2.2
11 (a) and (b) not fail, but (b) near limit

12 A = unnotched, B = notched

6.3

1 Deformation of a material with time when subject to constant strain
2 See Figure 6.37 and associated text
3 See Figure 6.38
4 Polypropylene
5 About (a) 30 MPa, (b) 20 MPa
7 Creep which increases if blocked for some time and also increases with the stress
8 (a) 2.5 GPa, (b) 2.0 GPa

6.4

1 See Section 6.4.2 and section on rusting
2 Oxygen is needed
3 Use metals close together in galvanic series, prevent electrical contact
4 Galvanised steel, closer to aluminium in galvanic series
5 The galvanic series with sea water has mild steel anodic with respect to copper and aluminium anodic with respect to mild steel
6 See Section 6.4.3
7 See text associated with Figure 6.54
8 See Section 6.4.3
9 Magnesium acts as anode
10 Zinc is sacrificial, tin is just for better corrosion-resistant layer
11 Localised cells, stress corrosion, galvanic cells
12 1.8 A
13 0.91 kg

Chapter 7

7.1

1 (a) A flat circular shape window mounted with a rubber gasket in a metal frame, the gasket distributing more uniformly the stresses at the edges and so reducing the chances of fracturing the window. With a flat circular shape the difference in pressure between the two sides will bend it and put one side in tension; to avoid this a domed shape can be used and so, like the arch of a bridge, will only be in compression and so a material, such as glass, can be used which is much stronger in compression than in tension.
(b) You might consider a basic design involving the table and two brackets, one at each end, to which the table is hinged at one end and which at the other are hinged to the back of the seat. Loads likely to be on the table are typically up to about 200 N. The table could be a moulded using a polymer as a single entity with its rim and the linkages for connections to the brackets. The arms need to be able to resist a significant bending moment and so might be made from aluminium

which is in the form of a channel section to give a reasonable second moment of area and be light.

(c) The functions required are a point to create a hole and then a cutting edge to cut round the edge of the lid. One possible design would be the old-fashioned can opener involving a sharp pointed blade which is levered round the edge of the can. Such an arrangement could be made by stamping out the blade and the handle from sheet steel, bending the handle to the required shape and then riveting the blade to the handle. The more modern form involves a cutting wheel which is used to pierce and cut round the can edge, handles being pressed together to give the piercing action and then the wheel being rotated to give the can edge cutting. Such an arrangement is likely to involve the two handles, a wing handle for the rotation, the cutting wheel, a bush and a pivot rivet. The handles and wheel could be produced by stamping from sheet steel.

(d) A number of materials could be considered for the pan, e.g. an aluminium alloy, stainless steel, cast iron. An aluminium alloy or stainless steel could be selected to be ductile enough to be drawn and ironed from sheet. Cast iron would involve casting. The handle needs to be made of material which will not get as hot as the pan itself and will remain stiff and strong enough when effectively acting as a cantilver with an end load. The handle might be a polymer, a thermoset such as phenol formaldehyde (Bakelite) with a wood flour filler, it has a high enough maximum service temperature of 150°C, a tensile modulus of about 5 to 8 GPa and a tensile strength of about 50 MPa. It can be compression moulded to the required shape and attached by means of a screw to the pan body. British Standards (BS 6743) gives a standard specification for the performance of handles attached to cookware.

(e) The lens needs to be transparent, red, and have reasonable impact strength. This would suggest that a polymer is better than glass. A material that is used is an acrylic (PMMA) with added red dye to give a moulding compound which can be formed by injection moulding. The process and material also lends itself to being suitably shaped and have integral fixing holes.

(f) The frames need to be light, rigid, with good corrosion resistance and have an attractive appearance. An aluminium alloy can be used, with the required section being manufactured in long lengths by extrusion and anodised to improve the corrosion resistance and appearance.

2 Rubber
3 Diamond

Index